PROGRESS IN

Nucleic Acid Research and Molecular Biology

Volume 61

PROGRESS IN
Nucleic Acid Research and Molecular Biology

edited by

KIVIE MOLDAVE

Department of Molecular Biology and Biochemistry
University of California, Irvine
Irvine, California

Volume 61

ACADEMIC PRESS
San Diego   London   Boston   New York
Sydney   Tokyo   Toronto

This book is printed on acid-free paper. ∞

Copyright © 1998 by ACADEMIC PRESS

All Rights Reserved.
   No part of this publication may be reproduced or transmitted in any form or by any means, electronic or mechanical, including photocopy, recording, or any information storage and retrieval system, without permission in writing from the Publisher.
   The appearance of the code at the bottom of the first page of a chapter in this book indicates the Publisher's consent that copies of the chapter may be made for personal or internal use of specific clients. This consent is given on the condition, however, that the copier pay the stated per copy fee through the Copyright Clearance Center, Inc. (222 Rosewood Drive, Danvers, Massachusetts 01923), for copying beyond that permitted by Sections 107 or 108 of the U.S. Copyright Law. This consent does not extend to other kinds of copying, such as copying for general distribution, for advertising or promotional purposes, for creating new collective works, or for resale. Copy fees for pre-1998 chapters are as shown on the title pages. If no fee code appears on the title page, the copy fee is the same as for current chapters.
0079-6603/98 $25.00

Academic Press
a division of Harcourt Brace & Company
525 B Street, Suite 1900, San Diego, California 92101-4495, USA
http://www.apnet.com

Academic Press Limited
24-28 Oval Road, London NW1 7DX, UK
http://www.hbuk.co.uk/ap/

International Standard Book Number: 0-12-540061-6

PRINTED IN THE UNITED STATES OF AMERICA
98 99 00 01 02 03 BB 9 8 7 6 5 4 3 2 1

Contents

Some Articles Planned for Future Volumes ix

Neurofilaments in Health and Disease 1
Jean-Pierre Julien and Walter E. Mushynski

I. Neurofilament Structure 2
II. Neurofilament Phosphorylation 4
III. Neurofilament Functions 7
IV. Neurofilaments in Diseases 10
V. Mechanisms of Neurofilament-Induced Pathology 14
VI. Future Directions 19
    References 20

Regulation of Cytochrome P450 Gene Transcription by Phenobarbital 25
Byron Kemper

I. Introduction 26
II. Phenobarbital Induction of Drug Metabolism 27
III. Induction of P450s in *Bacillus megaterium* 30
IV. Proximal Promoter 35
V. Distal Phenobarbital-Responsive Enhancer 41
VI. Perspectives and Models of Phenobarbital Induction of P450 Genes .. 55
    References 62

RNA and Protein Interactions Modulated by Protein Arginine Methylation 65
Jonathan D. Gary and Steven Clarke

I. Introduction 66
II. Background 68
III. Identification of Protein Arginine N-Methyltransferase Genes 78
IV. Methyl-Accepting Substrates for Protein Arginine
    N-Methyltransferases 88
V. Determination of Arginine Methyltransferase Substrates 100

VI. Functional Roles for the Protein Arginine N-Methyltransferase
   and Arginine Methylation 111
VII. Future Directions 123
   References 124

Genetic Regulation of Phospholipid Metabolism: Yeast as a Model Eukaryote ... 133

Susan A. Henry and Jana L. Patton-Vogt

I. Introduction 134
II. Regulation of Phospholipid Metabolism in Yeast 141
III. Model for the Regulation of Genes Containing UAS_{INO} 165
IV. Summary and Future Directions 172
   References 173

Inosine-5'-Monophosphate Dehydrogenase: Regulation of Expression and Role in Cellular Proliferation and T Lymphocyte Activation ... 181

Albert G. Zimmermann, Jing-Jin Gu, Josée Laliberté, and Beverly S. Mitchell

I. Introduction 182
II. IMPDH Activity and Cellular Proliferation 183
III. *De Novo* Purine Biosynthesis and Salvage 184
IV. Cellular IMPDH Activity Comprises Activities of Two Isozymes ... 190
V. IMPDH Activity and Cellular Signaling 200
VI. *In Vitro* Effects of IMPDH Inhibition on the Immune System ... 201
VII. Clinical Applications for IMPDH Inhibitors 202
VIII. Summary 203
   References 204

Structure and Function Analysis of *Pseudomonas* Plant Cell Wall Hydrolases ... 211

Geoffrey P. Hazlewood and Harry J. Gilbert

I. Introduction 212
II. Composition and Structure of Plant Cell Walls 213
III. Plant Cell Wall Hydrolases 215
IV. *Pseudomonas fluorescens* subsp. *cellulosa* 217
V. Architecture of *Pseudomonas* Plant Cell Wall Hydrolases 218

VI. Function of Noncatalytic Domains	222
VII. Structures and Catalytic Mechanisms of *Pseudomonas* Plant Cell Wall Hydrolases	231
References	239

Regulation of the Spatiotemporal Pattern of Expression of the Glutamine Synthetase Gene 243

Heleen Lie-Venema, Theodorus B. M. Hakvoort, Formijn J. van Hemert, Antoon F. M. Moorman, and Wouter H. Lamers

I. Introduction	244
II. Spatiotemporal Aspects of Glutamine Synthetase Expression	246
III. Levels of Regulation of Glutamine Synthetase Expression	266
IV. Glutamine Synthetase Gene: Evolution, Structure, and Regulatory Regions	273
V. Concluding Remarks	293
References	295

Structural Organization and Transcription Regulation of Nuclear Genes Encoding the Mammalian Cytochrome *c* Oxidase Complex 309

Nibedita Lenka, C. Vijayasarathy, Jayati Mullick, and Narayan G. Avadhani

I. Introduction	311
II. Sequence Properties of Nuclear-Encoded Subunits	313
III. Structure of Nuclear-Encoded Genes	320
IV. Mechanisms of Coordinate Regulation	332
V. Summary and Future Direction	337
References	338

Control of Meiotic Recombination in *Schizosaccharomyces pombe* 345

Mary E. Fox and Gerald R. Smith

I. Control of Entry into Meiosis	348
II. Nuclear Cytology during Meiosis	352
III. Meiotic Recombination Hotspots	355

IV. Recombination-Deficient Mutants and Genes	361
V. Region-Specific Control of Meiotic Recombination	369
VI. Conclusion and Perspective	373
Note Added in Proof	375
References	375

The Nucleosome: A Powerful Regulator of Transcription 379

Alan P. Wolffe and Hitoshi Kurumizaka

I. The Nucleosome Core	381
II. The Nucleosome	390
III. Transcription Factor Access to Nucleosomal DNA	400
IV. Concluding Remarks	418
References	418

INDEX 423

Some Articles Planned for Future Volumes

Mechanisms of Growth Hormone-Regulated Transcription
    Nils Billestrup

The Molecular Biology of Cyclic Nucleotide Phosphodiesterases
    Marco Conti and Catherine Jin

Tissue Transglutaminase—Retinoid Regulation and Gene Expression
    Peter J. A. Davies and Shakid Mian

Genetic Approaches to Structural Analysis of Membrane Transport Systems
    Wolfgang Epstein

Intron-Encoded snRNAs
    Maurille J. Fournier and E. Stuart Maxwell

Regulation of Mammalian Ribosomal Gene Transcription by RNA Polymerase I
    Ingrid Grummt

The Nature of DNA Replication Origins in Higher Eukaryotic Organisms
    Joel A. Huberman and William C. Burhans

A Kaleidoscopic View of the Transcriptional Machinery in the Nucleolus
    Samson T. Jacob

Function and Regulatory Properties of the MEK Kinase Family
    Gary L. Johnson, P. Gerwins, C. A. Lange-Carter, A. Gardner, M. Russell, and R. R. Villiancourt

Structure and Function Characteristic of Dyrk, a Novel Subfamily of Protein Kinases with Dual Specificity
    Hans-Georg Joost and Walter Becker

DNA Repair and Chromatin Structure in Genetic Diseases
    Muriel W. Lambert and W. Clark Lambert

Mammalian DNA Polymerase Delta: Structure and Function
    Marietta Y. W. T. Lee

The Initiation of DNA Base Excision Repair of Dipyrimidine Photoproducts
    R. Stephen Lloyd

Translation Initiation Factors in Eukaryotic Protein Biosynthesis
    Umadas Maitra

DNA Helicases: Roles in DNA Metabolism
  STEVEN W. MATSON AND DANIEL W. BEAM

Specificity of Eukaryotic Type II Topoisomerase: Influence of Drugs, DNA Structure, and Local Sequence
  MARK T. MULLER AND JEFFREY SPITZNER

Immunoanalysis of DNA Damage and Repair Using Monoclonal Antibodies
  MANFRED F. RAJEWSKY

DNA Methyltransferases
  NORBERT O. REICH, BARRETT ALLAN, AND JAMES FLYNN

Positive and Negative Transcriptional Regulation by the Retinoblastoma Tumor Suppressor Protein
  PAUL D. ROBBINS AND JOHN HOROWITZ

Organization and Expression of the Chicken α-Globin Genes
  KLAUS SCHERRER AND FELIX R. TARGA

Mechanism of Regulatory GTPase in Protein Biosynthesis
  MATHIAS SPRINZL AND ROLF HILGENFELD

Regulation of *Bacillus subtilis* Pyrimidine Biosynthetic Operon by Transcriptional Attenuation: Control of Gene Expression by an mRNA-Binding Protein
  ROBERT L. SWITZER, ROBERT J. TURNER, AND YANG LU

The Role of the TATA Box-Binding Protein (TBP)-Associated Factors, TAFs, in the Regulation of Eukaryotic Gene Expression
  P. ANTHONY WEIL AND ALLYSON M. CAMPBELL

Neurofilaments in Health and Disease

JEAN-PIERRE JULIEN[*,1] AND
WALTER E. MUSHYNSKI[†]

*Centre for Research in Neuroscience
McGill University
Montreal, Canada H3G 1A4
†Biochemistry Department
McGill University
Montreal, Canada H3G 1Y6

I. Neurofilament Structure	2
II. Neurofilament Phosphorylation	4
III. Neurofilament Functions	7
A. Axon Caliber	7
B. NF-H as a Modulator of Axonal Transport	7
C. Neurofilament Requirement for Efficient Axonal Regeneration	9
IV. Neurofilaments in Diseases	10
A. Neurofilament Accumulations in Human Diseases	10
B. Transgenic Mouse Models with Neurofilament Abnormalities	11
C. Codon Deletions in the NF-H Genes of ALS Patients	13
V. Mechanisms of Neurofilament-Induced Pathology	14
A. Disruption of Axonal Transport	14
B. A Link between SOD1 and Neurofilaments?	16
VI. Future Directions	19
References	20

This article reviews current knowledge of neurofilament structure, phosphorylation, and function and neurofilament involvement in disease. Neurofilaments are obligate heteropolymers requiring the NF-L subunit together with either the NF-M or the NF-H subunit for polymer formation. Neurofilaments are very dynamic structures; they contain phosphorylation sites for a large number of protein kinases, including protein kinase A (PKA), protein kinase C (PKC), cyclin-dependent kinase 5 (Cdk5), extracellular signal regulated kinase (ERK), glycogen synthase kinase-3 (GSK-3), and stress-activated protein kinase γ (SAPKγ). Most of the neurofilament phosphorylation sites, located in tail regions of NF-M and NF-H, consist of the repeat sequence motif, Lys-Ser-Pro (KSP). In addition to the well-established role of neurofilaments in the control of axon caliber, there is growing evidence based on transgenic mouse studies that neurofilaments can affect the dynamics and perhaps the function of other cytoskeletal elements, such as microtubules and actin filaments. Perturbations in phosphorylation or in metabo-

---

[1] To whom correspondence should be addressed.

lism of neurofilaments are frequently observed in neurodegenerative diseases. A down-regulation of mRNA encoding neurofilament proteins and the presence of neurofilament deposits are common features of human neurodegenerative diseases, including amyotrophic lateral sclerosis (ALS), Parkinson's disease, and Alzheimer's disease. Although the extent to which neurofilament abnormalities contribute to pathogenesis in these human diseases remains unknown, emerging evidence, based primarily on transgenic mouse studies and on the discovery of deletion mutations in the NF-H gene of some ALS cases, suggests that disorganized neurofilaments can provoke selective degeneration and death of neurons. An interference of axonal transport by disorganized neurofilaments has been proposed as one possible mechanism of neurofilament-induced pathology. Other factors that can potentially lead to the accumulation of neurofilaments will be discussed as well as the emerging evidence for neurofilaments as being possible targets of oxidative damage by mutations in the superoxide dismutase enzyme (SOD1); such mutations are responsible for ~20% of familial ALS cases
© 1998 Academic Press

Neurofilaments are major elements of the neuronal cytoskeleton and are particularly abundant in large myelinated axons of motor and sensory neurons. Like other types of intermediate filaments, neurofilaments are very dynamic structures capable of subunit exchange. In addition to their well-established role in the control of axon caliber, there is growing evidence that neurofilaments can affect the dynamics and perhaps the function of other cytoskeletal elements, such as microtubules and actin filaments. Perturbations in the normal metabolism of neurofilaments are also associated with various diseases. A down-regulation of mRNA encoding neurofilament proteins and the presence of neurofilament deposits are common features of human neurodegenerative diseases, including amyotrophic lateral sclerosis (ALS), Parkinson's disease, and Alzheimer's disease. Although the extent to which neurofilament abnormalities contribute to pathogenesis in these human diseases remains unknown, emerging evidence, based primarily on transgenic mouse studies, suggests that disorganized neurofilaments can cause the selective degeneration and death of motor neurons.

Current knowledge of neurofilament structure, phosphorylation, and function is reviewed in Sections I–III. In Sections IV and V we discuss the evidence based on genetic and transgenic mouse studies indicating that neurofilaments may play a key role in motor neuron disease.

I. Neurofilament Structure

Neurofilaments are assembled by the copolymerization of three intermediate filament proteins, NF-L (61 kDa), NF-M (90 kDa), and NF-H (110 kDa)

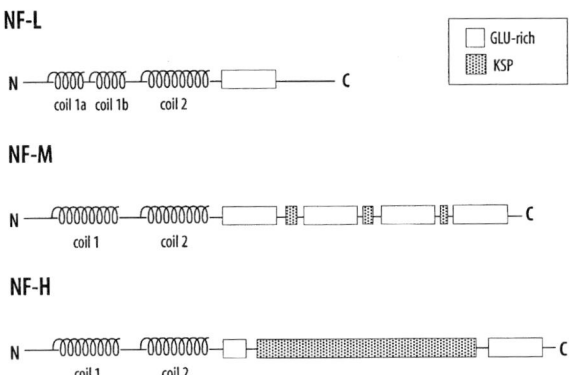

FIG. 1. Structure of neurofilament proteins. The three neurofilament subunits have a central domain that consists of approximately 310 amino acids and that is involved in the formation of coiled-coil structures. The current model of an intermediate filament is that two coiled-coil dimers of protein subunit line up in a staggered fashion to form an antiparallel tetramer. Eight tetramers packed together are required to make a 10-nm filament. Neurofilaments are obligate heteropolymers and there is evidence for the existence of two types of heterotetramers in neurofilaments, one containing NF-L and NF-M and the other containing NF-L and NF-H.

(1). The three neurofilament subunits share with other members of the intermediate protein filament family a central domain, of approximately 310 amino acids, which is involved in the formation of coiled-coil dimers (Fig. 1). Two coiled-coil dimers then line up in a staggered fashion to form an antiparallel tetramer (2). The subsequent chronology of linear and lateral associations between tetramers is difficult to discern, although protofilaments consisting of tetramers linked end to end, and protofibrils consisting of two laterally associated protofilaments, have been proposed. Neurofilaments are obligate heteropolymers requiring NF-L together with either NF-M or NF-H for polymer formation (3, 4). There is evidence for the existence of two types of heterotetrameric units in neurofilaments, one containing NF-L and NF-M and the other NF-L and NF-H (5).

A striking feature of neurofilament proteins is their extensive carboxy-terminal tail domains, characterized by a high content of charged amino acids, such as Glu and Lys, as well as phosphoserine in the case of NF-M and NF-H (6). Studies of the *in vitro* assembly of neurofilaments indicate that purified NF-M or NF-H alone will not form intermediate filaments and that NF-M and NF-H tail domains form side-arm projections of about 55 and 63 nm in length, respectively, extending from the neurofilament axis (7). Most of the phosphorylation sites in NF-M and NF-H are located in tail region subdomains containing multiple copies of the sequence motif Lys-Ser-Pro (KSP)

(*10–13*). There are 5 to 12 copies of the KSP motif in the tail domain of NF-M from various mammalian species (*12, 13*), whereas the number in NF-H ranges from 43 to 44 in humans (*10, 14*) to over 50 in the mouse (*11*), rat (*16*), and rabbit (*16*). NF-H and NF-M also contain one or a few SP motifs without an N-terminally adjacent lysine.

The propensity for this repetitive region of the NF-H gene to undergo unequal crossing-over (*16*) may explain interspecies size variations in NF-H (*17*) as well as the existence of deletion mutations within this subdomain (*14*). The KSP motif can be highly conserved, also occurring in multiple copies in the large subunit of the squid (*18*). However, most of the SP-containing motifs in chicken NF-M lack the adjacent lysine residue (*19*).

II. Neurofilament Phosphorylation

NF-H and NF-M are among the most highly phosphorylated proteins in the nervous system (*20, 21*). The notion that phosphorylation is involved in various aspects of neurofilament metabolism is supported by several lines of evidence. These include the presence of different classes of phosphorylation sites in the two end domains of neurofilament subunits (*22*) as well as the association with neurofilaments of several protein kinases (*23*) and protein phosphatase-2A (*24*). Furthermore, the subcellular phosphorylation pattern for neurofilaments is altered in several neurodegenerative diseases, suggesting that aberrant neurofilament phosphorylation may play a role in the underlying pathogenic mechanism (*25–27*).

Head domain phosphorylation, particularly that of NF-L, by second-messenger-dependent kinases has been implicated in the regulation of neurofilament assembly (*22*). There is compelling evidence that protein kinase A in particular is involved in modulating filament assembly by phosphorylating Ser-2 and Ser-55 in the head domain of NF-L (*28*). Protein phosphatase-2A has been implicated in the dephosphorylation of these protein kinase A sites (*28, 29*), suggesting that neurofilament dynamics are modulated through the antagonistic effects of the two enzymes.

The disrupted neurofilament network in okadaic acid-treated dorsal root ganglion (DRG) neurons recovers rapidly following removal of the phosphatase inhibitor from the culture medium (*30*). This, as well as other evidence (*22*), indicates that phosphate moieties in the head domain turn over rapidly. The comparatively slow turnover of phosphate groups in the tail domain (*31, 32*) suggests a different role for phosphorylation in this segment of neurofilament proteins. Indeed, phosphorylation sites in the tail domain appear to be functionally heterogeneous, having been implicated in neurofilament–microtubule interactions (33, 34), in neurofilament gelation (36), and in the axonal expansion that accompanies myelination (36–40).

The availability of monoclonal antibodies that could distinguish between phosphorylated and unphosphorylated KSP repeats in NF-H and NF-M (41, 42) led to the discovery that these two subunits are more highly phosphorylated in the axon than in the perikaryon. This normal phosphorylation pattern is altered in several neurodegenerative diseases (25–27), wherein perikaryal neurofilaments are characteristically hyperphosphorylated and often form aggregates.

In view of the prominence of KSP motifs in NF-M and NF-H, it is not surprising that several neuronal enzymes belonging to the superfamily of proline-directed kinases have been implicated in the phosphorylation of NF-M and NF-H. The identified kinases in this category include cyclin-dependent kinase 5 (Cdk6) (43, 44), extracellular signal regulated kinase (ERK) (45, 46), glycogen synthase kinase-3 (GSK-3) (47), and stress-activated protein kinase γ (SAPKγ) (48). Table I lists the relevant protein kinases, their consensus phosphorylation site sequences, and their effects on NF-H mobility and reactivity with phosphorylation-dependent antibodies. Because KSP (and XSP) motifs in a variety of sequence contexts are phosphorylated *in vivo* (49, 50), it appears that several kinases may be involved in tail domain phosphorylation.

TABLE I
PROLINE-DIRECTED PROTEIN KINASES IMPLICATED IN NEUROFILAMENT PHOSPHORYLATION

Kinase	Phosphorylation sequences		Effect of phosphorylation on NF-H	
	Consensus	Known sites in NF-H	Gel mobility	SMI 31 reactivity
Cdk5/p35	X(S/T)PXK	KSPAK[a]	Decrease	Increase[b]
ERK	PX(S/T)PX	Not determined	Partial decrease	Increase[c]
GSK-3	S-X$_3$-S(OPO$_3$) and SP[d]	KS°PPVKS°PEAK AKS°PVSK,KAES°PVK[e]	Slight decrease	Increase[f]
SAPKγ	SP(D/E)[g]	KSPAEA[h]	Decrease	Increase

[a]From Ref. 44.
[b]From Ref. 115.
[c]From Ref. 46.
[d]From Ref. 116.
[e]From Ref. 117.
[f]From Ref. 118.
[g]From Ref. 119.
[h]From Ref. 48.

The implication of SAPKγ in the hyperphosphorylation of perikaryal NF-H (*48*) provides a new approach for determining whether aberrant phosphorylation of perikaryal neurofilaments has an adverse effect on neuronal integrity. The SAPK pathway plays a pivotal role in cellular responses to inflammatory cytokines and to various stress stimuli. Responses such as growth arrest, apoptosis, or activation of immune and reticuloendothelial cells are apparently mediated through phosphorylation of transcription factors, c-JUN, ATF-2, and Elk-1, by SAPKs (*51*). The activation of SAPK is generally initiated at the plasma membrane and can involve second messengers such as ceramide (*52*) as well as Rho family GTPases (*51*). Conceivably, the aberrant phosphorylation of neurofilaments may occur incidentally during transfer of activated SAPK through the cytoplasm to the nucleus. Neurons in particular would be prone to this type of incidental phosphorylation due to the presence in the perikaryon of proteins, such as NF-M and NF-H, containing multiple SP motifs.

Low levels of aberrant tail domain phosphorylation might normally be reversed by protein phosphatases. Acute stressing of DRG neurons increases phosphorylation of NF-H to levels normally seen in axons, as judged by retarded mobility on sodium dodecyl sulfate–polyacrylamide gel electrophoresis (SDS-PAGE) (*48*). Because it takes about 1–2 days for the mobility of perikaryal NF-H to return to its normal level following removal of stressing agent (B. I. Giasson and W. E. Mushynski, unpublished results), protein phosphatase activity in the neuronal perikaryon may be limited. Protein phosphatases-2A (*53*) and -2B (calcineurin) (*54*) have been implicated in the dephosphorylation of KSP sites in NF-H. It is interesting to note that SOD1 has a protective effect on calcineurin and may thus play a role in signal transduction (*55*).

Subjecting neurons to some chronic form of stress might result in prolonged hyperphosphorylation of perikaryal NF-H, with possible adverse effects. Phosphorylation of the NF-H tail domain by cdc-2-related protein kinases inhibits neurofilament–microtubule interactions (*33, 34*). It is not known whether such interactions are involved in driving the axonal transport of neurofilaments. However, increased phosphorylation does slow the rate of neurofilament transport (*40, 56*) and more highly phosphorylated forms NF-M and NF-H associate preferentially with stationary neurofilaments (*26*).

Other evidence that aberrant phosphorylation may cause the perikaryal neurofilaments to accumulate comes from studies on other kinases. Activation of protein kinase C in cultured motor neurons caused the hyperphosphorylation of perikaryal neurofilaments and enlargement of proximal dendritic processes (*57*). Conversely, inhibition of protein kinase C in neuronal cultures was shown to reverse neurofilament aggregation caused by overexpression of NF-L (58). Although protein kinase C does not phosphorylate KSP repeats directly (22), it can activate the SAPK pathway (59).

III. Neurofilament Functions

A. Axon Caliber

It is now well established that neurofilaments act as modulators of the caliber of large myelinated axons. This is an important function because axonal caliber is a determinant of conduction velocity. Unequivocal proof for neurofilament involvement in the control of axonal caliber has been provided by analysis of animals lacking axonal neurofilaments. These include a quail quivering mutant deficient in NF-L protein (60), transgenic mice expressing an NF-H/lacZ fusion construct (61), and NF-L-deficient mice that were generated in J.-P. Julien's laboratory (62). All of these animals showed a dramatic hypotrophy of axons due to the scarcity of neurofilaments. Our analysis of ventral root axons in mice lacking NF-L revealed a reduction of two- to threefold in the caliber of myelinated axons (Fig. 2). This axonal hypotrophy is accompanied by a 50% decrease in conduction velocity.

There is evidence that axonal caliber is modulated not only by the number of neurofilaments but also by local changes in neurofilament phosphorylation. Support for this view comes from studies linking reduced levels of NF-H phosphorylation with a decrease in axon caliber. These include work on the dysmyelinating *Trembler* mouse mutant, on hypomyelinating transgenic mice expressing either a diphtheria toxin A or SV40 large T antigen in Schwann cells, as well as other approaches showing a local effect of myelination on NF-H phosphorylation (36–40).

B. NF-H as a Modulator of Axonal Transport

Following their synthesis in the perikaryon, the three neurofilament proteins are transported down the axon with the slow axonal transport component (1). The classical view that neurofilaments move down the axon as a cohesive network (63, 64) has been challenged by increasing evidence for the dynamic nature of neurofilaments (65–67). Experiments involving photobleaching in regenerating axons suggest that subunit exchange can occur along the axon (68) and another report indicates that neurofilament proteins can be transported in an unpolymerized form along axonal microtubules (69).

There is evidence indicating that the NF-H protein can affect neurofilament transport. First, the appearance of NF-H during postnatal development coincides with a slowing of axonal transport (70). Second, overexpression of

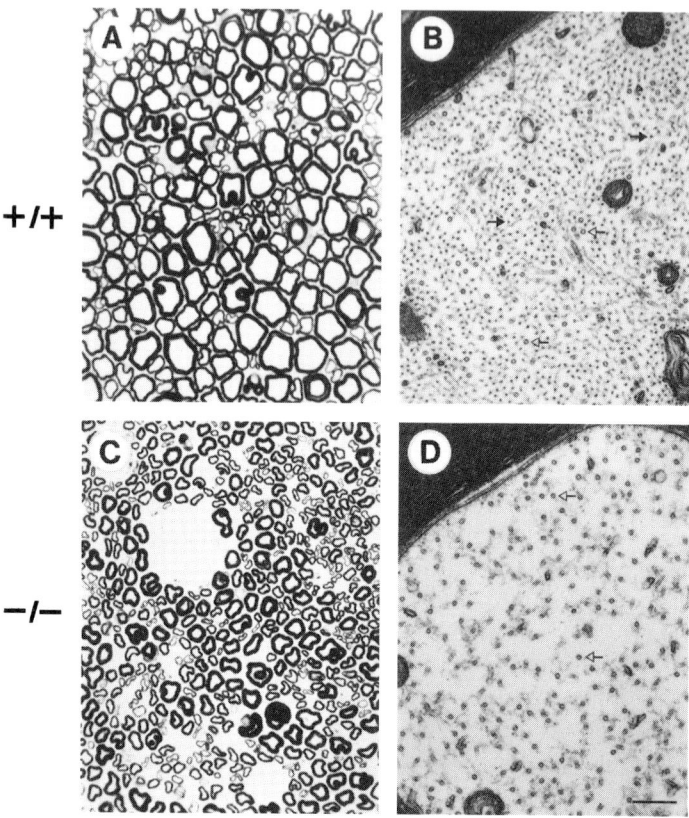

FIG. 2. Reduced calibers of ventral root axons in mice lacking NF-L protein. Light microscopy reveals a two- to threefold reduction in the caliber of large motor axons in NF-L −/− mice (C) as compared to those from normal mice (A). Electron microscopy of transverse sections shows the abundance of neurofilaments in normal axons (B) and the lack of neurofilament structures in axons from the adult NF-L −/− mice (D). Open arrows indicate microtubules and the filled arrows point to neurofilaments. The bar represents 10 μm for light microscopy (A, C) and 0.25 μm for electron microscopy (B, D).

either human or mouse NF-H in transgenic mice provokes a reduction of neurofilament transport (71, 72). Third, as previously mentioned, phosphorylation influences the rate of neurofilament transport and more highly phosphorylated forms of NF-H associate preferentially with stationary neurofilaments (26, 40).

Although neurofilaments and microtubules are highly interdependent structures, the precise mechanism by which NF-H can affect the slow components of axonal transport remains unknown. One possible explanation is that excessive levels of NF-H or hyperphosphorylation of NF-H could alter neurofilament–microtubule interactions with ensuing transport impairment. This is supported by the observation that phosphorylation of NF-H by cdc-2-related protein kinases inhibits NF-H binding to microtubules (33, 34). However, it is not known whether such interactions are involved in driving the axonal transport of neurofilament proteins. Another possible explanation is that NF-H may influence transport by acting as a modulator of axonal microtubule dynamics. Hypophosphorylated NF-H was found to compete with tau, a microtubule-associated protein (MAP) that stabilizes microtubules, for similar binding sites in the carboxy-terminal region of tubulin (34). Accordingly, it is plausible that changes in the level or in the extent of phosphorylation of NF-H could affect the stabilization of microtubules indirectly. This notion is supported by our finding of a twofold increase in the number of microtubules in ventral root axons of mice with a targeted disruption of the NF-H gene (J.-P. Julien and Q. Zhu, unpublished observation). This mouse model will be useful in future studies to determine the effects of downregulation of NF-H levels on axonal transport.

C. Neurofilament Requirement for Efficient Axonal Regeneration

The notion that neurofilaments can affect microtubule dynamics is consistent with our finding of a delay in the regeneration of myelinated axons after crush injury of a peripheral nerve in mice lacking NF-L (62). In the second week after axotomy, the number of newly regenerated axons in the sciatic nerve of NF-L −/− mice corresponded to ~25% of the number of regenerated axons found in normal mice. Although there is widespread agreement that the microtubule and actin filament dynamics are important during axonal outgrowth (73–77), it is conceivable that neurofilaments could provide a scaffold that contributes to the stabilization of newly growing axons. Alternatively, the neurofilament network could support axonal outgrowth indirectly by contributing to microtubule and/or actin filament dynamics and functions. There is evidence for interactions between neurofilaments and microtubules occurring via the NF-H tail domain and the carboxy terminus of tubulin (7, 34). Moreover, neurofilaments have been found to be connected to actin filaments in sensory neurons by a linker protein encoded by a neuronal splice form of the BPAG1 gene (78), which is responsible for the well-known autosomal recessive disease *dystonia musculorum* (79, 80). Future studies will be required to determine how alterations in the level of each neurofilament protein can affect specific functions of the actin and microtubule cytoskeletons in axonal transport and growth cone motility.

IV. Neurofilaments in Diseases

A. Neurofilament Accumulations in Human Diseases

Abnormal accumulations of neurofilaments, often called spheroids or Lewy bodies, have been described in a variety of disorders. Table II lists some pathologies associated with disorganized neurofilaments, including neurodegenerative diseases such as amyotrophic lateral sclerosis, Alzheimer's disease, and Parkinson's disease. The formation of neurofilament deposits has been widely viewed to be a consequence of neuronal dysfunction. The extent to which such disorganized neurofilaments contribute to neurodegeneration in these human diseases is unknown. Interestingly, the neurofilament deposits are often associated with decreases in the levels of mRNA coding for NF-L. For instance, reductions in NF-L mRNA levels of 60 and 70% have been reported in degenerating neurons in ALS (81) and Alzheimer's disease (82), respectively. Whether a decrease in neurofilament content is a factor that could exacerbate the disease process is a possibility that has not been examined systematically. In addition, it is unclear whether

TABLE II
Human Diseases with Abnormal Neurofilament Deposits

Disease	Neurofilament abnormalities	Ref.
ALS (70% of cases)	Neurofilament accumulations in motor neuron	120
	Decline of 60% in NF-L mRNA	81
Parkinson's disease (100% of cases)	Lewy bodies in substantia nigra and locus coeruleus	121
	Declines of 30% NF-L mRNA and 70% NF-H mRNA	122
Alzheimer's disease (20% of cases)	Cortical Lewy bodies; Decline of 70% in NF-L mRNA	82, 123
Lewy body dementia	Cortical Lewy bodies	121
Guam-parkinsonism (100% of cases)	Neurofilament deposits in motor neurons	124
Giant axonal	Neurofilament accumulations in peripheral axons	125
Lhermitte–Duclos disease	Hypertrophy of granule cells of cerebellum with enhanced neurofilament content	126
Peripheral neuropathies	Neurofilament accumulations in peripheral axons that can be induced by various toxic agents, such as IDPN, hexanedione and acrylamide.	127

a decrease in NF-L results in changes in neurofilament composition. Evidence that changes in subunit stoichiometry induced by overexpression of any single neurofilament subunit in transgenic mice can provoke neurofilament accumulations and in some cases cause motor neuron disease will be reviewed in the following section.

B. Transgenic Mouse Models with Neurofilament Abnormalities

In the past few years, several transgenic mouse models with neurofilament accumulations have been generated by different laboratories. As illustrated in Table III, abnormal neurofilament accumulations can be formed by overexpressing any one of the three neurofilament proteins. Several trans-

TABLE III
Transgenic Mice with Abnormal Neurofilament Accumulations

Transgene	Neuronal populations affected	Neurodegenerative disease	Ref.
Human NF-H	Spinal motor neurons and DRG neurons	Yes	71, 83
Mouse NF-H	Spinal motor neurons and DRG neurons	No	72
Mouse NF-H/lacZ	Perikarya of CNS and PNS neurons	No	61
MSV/mouse NF-L	Spinal motor neurons and DRG neurons	Yes	86
MSV/mutant NF-L	Spinal motor neurons and DRG neurons	Yes	87
Human NF-L	Thalamic neurons and cortical neurons	No	90
Human NF-M	Cortical neurons and forebrain neurons	No	91
MSV/mouse NF-M	Spinal motor neurons and DRG neurons	No	89
Other mouse models			
Transgenic bearing SOD1 mutant (HG93A)	Vacuoles in motor neurons and spheroids in proximal motor axons	Yes	99, 106
Wobbler mouse	Motor neurons; increased levels of NF-M mRNA expression by three- to fourfold	Yes	128
BAPG1 knock-out or (dt/dt) mice	DRG sensory neurons; neurofilament aggregates in sensory axons	Yes	78–80

genic mouse lines expressing human NF-H proteins were generated by microinjection of large genomic clones bearing the normal human NF-H gene (83). Two normal NF-H alleles exist in the human population, one bearing 43 KSP phosphorylation site repeats, and the other 44 KSP repeats (14). Both alleles were found to be capable of inducing motor neuron disease when overexpressed in transgenic mice but earlier onset of disease occurred with the allele having 43 KSP repeats. A modest two- to threefold overexpression of this allele in mice provoked a late-onset motor neuron disease, characterized by the presence of aberrant neurofilament accumulations in spinal motor neurons (71). Motor neuron dysfunction in NF-H transgenic mice progresses with age, and is accompanied by atrophy and slow degeneration of distal axons, resulting in secondary muscle degeneration (71). Our interpretation of this effect is that excessive levels of human NF-H protein alter neurofilament properties, resulting in retardation of axonal transport and aberrant aggregation of neurofilaments in the cell body and proximal axon.

Surprisingly, transgenic mice with a four- to fivefold increase in the level of wild-type mouse NF-H protein did not develop a motor neuron disease despite retardation of neurofilament axonal transport, accumulation of neurofilaments in motor neurons, and atrophy of myelinated axons (72). These results indicate that human NF-H (83) is markedly more potent than murine NF-H (72) in causing neurodegeneration. The reason for this discrepancy could well be due to differences in the relative proportions of different sequence motifs in the KSP repeat domain. Human NF-H contains over three times as many cyclin-dependent kinase 5 (Cdk6) consensus sites compared to murine NF-H (84). Conceivably, Cdk5 or related proline-directed kinases in mouse motor neurons phosphorylate human NF-H more extensively than the murine subunit. Furthermore, human NF-H exhibits a high degree of homology with NF-H from the rabbit, an animal species highly susceptible to neurofilamentous pathology induced by aluminum (85). Thus, the enhanced deleterious effect of human NF-H in mice is likely due to the properties of its tail phosphorylation domain, a region that can affect filament interactions and perhaps axonal transport (36).

High-level overexpression of the wild-type mouse NF-L protein in transgenic mice, achieved by use of the strong viral promoter from murine sarcoma virus (MSV), provoked an early-onset motor neuron disease accompanied by massive accumulation of neurofilaments in spinal motor neurons and by muscle atrophy (86). A more severe phenotype was produced by expressing an assembly-disrupting NF-L mutant having a leucine-to-proline substitution near the carboxy-terminal end of the conserved rod domain (87). Although no such NF-L mutations have been reported in human ALS, similar mutations in keratins are the cause of severe forms of genetic skin diseases (88). Expression of this mutant NF-L protein at only 50% of the endogenous NF-L level was sufficient to induce within 4 weeks massive neurofilament accumulations in spinal motor neurons and was accompanied by death of motor neurons, neuronophagia, and severe denervation atrophy of the skeletal muscle.

These transgenic studies with the mouse NF-L and human NF-H genes provided the first demonstration that disorganized neurofilaments can provoke neuronal degeneration and death. The cellular selectivity of this disease is remarkable in that degeneration is restricted to spinal motor neurons even though the NF-L or NF-H transgenes are expressed throughout the nervous system. Nonetheless, these transgenic mouse models should not be considered as absolute replicas of human ALS in that motor neurons from the brain cortex did not develop neurofilamentous pathology. It is likely that this is due to the smaller size and therefore lower neurofilament content of mouse cortical neurons as compared to the corresponding human neurons. Clearly, additional unknown cellular factors may also contribute to the deleterious effects of neurofilament deposits because large neurofilament accumulations were well tolerated within dorsal root ganglion neurons (83, 86, 87)

As shown in Table III, some types of neurofilament deposition appear to be less toxic than others. Not all transgenic mouse lines with neurofilament abnormalities develop motor neuron disease. For instance, no overt phenotypes occur in transgenic mice expressing a mouse NF-H/lacZ fusion gene even though massive neurofilament aggregates are detected in neuronal perikarya throughout the nervous system (61). Also, large neurofilamentous swellings are found in DRG and motor neurons of transgenic mice overexpressing an NF-M transgene, but no axonal degeneration is observed (89). Moreover, transgenic mice with moderate overexpression of the human NF-L (90) or NF-M genes (91) appear normal, although a detailed analysis of the mice reveals the presence of abnormal perikaryal immunoreactivity in some populations of brain neurons. It has not yet been documented whether these neurofilament accumulations cause neurodegeneration with aging.

C. Codon Deletions in the NF-H Genes of ALS Patients

Additional evidence for neurofilament involvement in ALS was provided by the detection of mutations in the KSP phosphorylation domain of NF-H from sporadic cases of ALS (14, 92). The combined results from two laboratories suggest that deletion mutations in the phosphorylation domain of the NF-H gene may be responsible for a small percentage of ALS cases (~1.3%). To date, 7 out of 500 sporadic ALS cases examined were found to have deleted codons in the NF-H gene. These results are summarized in Fig. 3. So far, the search for mutations in the KSP repeat domain of NF-H (93) or in all coding regions of NF-L, NF-M, or NF-H (94) has failed to reveal mutations linked to disease in >100 familial ALS cases. Therefore, it can be concluded that

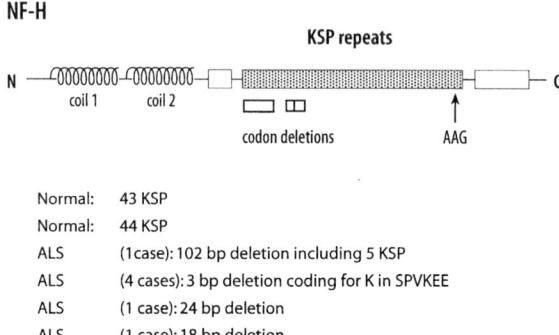

Normal:	43 KSP
Normal:	44 KSP
ALS	(1 case): 102 bp deletion including 5 KSP
ALS	(4 cases): 3 bp deletion coding for K in SPVKEE
ALS	(1 case): 24 bp deletion
ALS	(1 case): 18 bp deletion

FIG. 3. Codon deletions in the NF-H gene of some ALS patients. There are two normal forms of the human NF-H gene, one with 43 KSP and the other with 44 KSP repeats. To date, the combined results from the screening of 562 sporadic ALS cases by two groups revealed codon deletions in the NF-H gene of seven unrelated ALS patients diagnosed as sporadic cases (92). In three ALS patients, the deletions consisted of 102, 24, and 18 bp, resulting in the loss of one or more KSP phosphorylation motifs. Four other ALS cases shared the same 3-bp deletion for the codon AAG (K) in the phosphorylation sequence SPVKEE. No mutations were found in ~500 control samples.

mutations in neurofilament genes as primary causes of disease can account for only a small proportion of sporadic or familial ALS cases. Although the mechanism underlying disease induced by NF-H mutations remains to be elucidated, their occurrence in the KSP-repeat segment again emphasizes the importance of this major NF-H phosphorylation domain. It is noteworthy that, in four ALS cases, the loss of a lysine residue in the sequence SPVKEE converted this Cdk5 consensus phosphorylation sequence (84) to a consensus sequence for a stress-activated protein kinase (SAPK) (48). As previously mentioned, agents that activate SAPKγ were found to cause abnormal hyperphosphorylation of NF-H protein in the perikarya of cultured DRG neurons (48). This finding raises the possibility that aberrant phosphorylation following stress may lead to neurofilament-induced pathology. However, it remains to be established whether chronic activation of SAPKγ can provoke the *in vivo* accumulation of neurofilaments.

V. Mechanisms of Neurofilament-Induced Pathology

A. Disruption of Axonal Transport

Studies with transgenic mice overexpressing human NF-H have provided clues as to how disorganized neurofilaments may contribute to neurodegeneration. Pulse labeling with [^{35}S]methionine of spinal motor neurons in NF-H transgenics revealed axonal transport defects not only for neurofilament proteins but also for other axonal components (71). Microscopic analysis of NF-H transgenic mice further demonstrated a paucity of cytoskeletal elements, mitochondria, and smooth endoplasmic reticulum (SER) in degenerating axons, although mitochondria were detected in motor neuron perikarya (71). These data demonstrated that disorganized neurofilaments can interfere with the delivery of components required for axonal maintenance and that a similar mechanism may contribute to pathogenesis in human ALS (Fig. 4). In agreement with this model is the report of a marked increase in neurofilaments, mitochondria, and lysosomes in the axon hillock of motor neurons in ALS patients, and a decreased content of cytoplasmic organelles in the initial axon segment of chromatolytic neurons (95).

The precise mechanism by which disorganized neurofilaments may interfere with axonal transport is not clearly understood. One simplistic explanation is that abnormal neurofilament accumulations could physically impede axonal transport. However, it is not certain whether the neurofilament deposits are the provoking agent of toxicity. In many transgenic mouse models the presence of large neurofilament aggregates in neuronal perikarya was well tolerated (Table III). In addition, intoxication by β,β′-iminodipropioni-

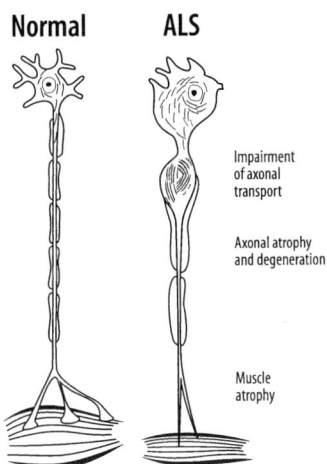

FIG. 4. Model of motor neuron degeneration by neurofilament-induced disruption of axonal transport. The abnormal accumulation of neurofilaments can impede the delivery of components required for axonal maintenance, including cytoskeletal elements and mitochondria. It is currently unclear whether the transport disruption involves a physical block by neurofilament accumulations or alterations of cellular components required for axonal transport, such as changes in microtubule organization, stability, or function.

trile (IDPN) induces neurofilamentous accumulations in proximal motor axons with no effect on fast axonal transport (96). An alternative explanation is that changes in the organization or metabolism of neurofilament proteins could affect the function of other cellular components required for axonal transport. For instance, it is possible that excessive levels of NF-H could affect microtubule stability and function. This would be consistent with *in vitro* binding studies indicating that both NF-H and tau proteins compete for the same binding site in the carboxy-terminal region of tubulin and that this interaction is regulated via phosphorylation of the KSP domain (34).

The disruption of axonal transport by disorganized neurofilaments is a pathological mechanism consistent with several aspects of ALS. It can explain in part the cellular selectivity of ALS pathology. Large motor neurons are particularly vulnerable to neurofilament abnormalities because they synthesize large amounts of neurofilament proteins. Moreover, there is a retardation in the slow axonal transport of cytoskeletal elements during aging (97), a factor that can predispose to the disease. Although ALS is a disease of many etiologies, disorganized neurofilaments may be key intermediates contributing to the neurodegenerative process. Many factors can potentially lead to the disorganization of neurofilaments, including deregulation of neurofilament protein synthesis, defective transport and proteolysis, abnormal phosphorylation, and other protein modifications (Table IV). Neurofilament accumulations can also result from alterations in proteins that interact with neurofilaments, as revealed by the analysis of mice lacking the neuronal isoform of BPAG1, a protein that cross-links actin filaments and neurofilaments (78). BPAG1 belongs to a group of large coiled-coil proteins that includes desmoplakin and plectin. A neuronal splice form, BPAG1n, is expressed specifically in sensory neurons. Disruption of the gene encoding BPAG1 in mice caused the specific degeneration and death of sensory neurons identical to the situation in the autosomal recessive dystonia musculorum (*dt/dt*) mouse mutant. Both the BPAG1 knockout and *dt/dt* mice are characterized by the presence of abnormal neurofilament aggregates in sensory axons (78), but the extent to which neurofilaments contribute to neurodegeneration in these mice is not currently known.

B. A Link between SOD1 and Neurofilaments?

The Cu^{2+}/Zn^{2+} superoxide dismutase (SOD1) is a metalloprotein that converts the superoxide anion (O_2^-) to hydrogen peroxide (H_2O_2), which is then detoxified to water and oxygen by catalase and glutathione peroxidase. Mutations in the SOD1 gene located on chromosome 21 have been found in ~20% of familial ALS cases (~2% total ALS cases) (98). To date, more than 40 different *SOD1* mutations have been identified in familial ALS, the majority being inherited in an autosomal dominant fashion (98). Most of the

TABLE IV
FACTORS POTENTIALLY LEADING TO ACCUMULATION OF NEUROFILAMENTS

Factor	Evidence	Ref.
Neurofilament gene mutations	Codon deletions in the NF-H gene of ALS patients	14
	Severe motor neuron disease in transgenic mice expressing a mutant NF-L assembly-disrupting gene	87
Deregulation of neurofilament gene expression	Transgenic mice overexpressing NF-L, NF-M, or NF-H	83, 86, 90, 91, 72
	Decreased ratio of NF-L to NF-H mRNAs in a dog model HCSMA	110
Altered neurofilament phosphorylation	Aggregation of neurofilaments modulated by PKC inhibitor	58
	Hyperphosphorylation of NF-H protein in neuronal perikarya by stress-activated protein kinase	28
SOD1 mutations	Neurofilament inclusions in FALS and in transgenic mice expressing SOD1 mutants	105, 99, 106
Alterations in neurofilament-associated proteins	Degeneration of sensory neurons in mice lacking a BPAG1 isoform	78
Toxic agents such as IDPN, hexanedione, and acrylamide	Neurofilament accumulations in animal models and cultured cells	129
Aluminum	Neurofibrillary pathology in rabbits	85

SOD1 mutations are missense point mutations and many lines of evidence suggest that SOD1 mutations cause ALS through a mechanism that involves a gain of some adverse function rather than a loss of superoxide dismutase activity. For instance, transgenic mice overexpressing the G93A (*99*), G37R (*100*), or G85R (*101*) mutant *SOD1* developed motor neuron disease even though the SOD activity in these mice was not reduced. Moreover, mice homozygous for the targeted disruption of the *SOD1* gene do not develop motor neuron disease (*102*).

Three mechanisms have been proposed to account for the toxicity of *SOD1* mutations. Beckman *et al.* (*103*) suggested a gain-of-function mechanism in which the mutations would render the copper in the active site of SOD1 more accessible to peroxynitrite, allowing the formation of a nitronium-like intermediate that can nitrate proteins on tyrosine residues. The second proposed mechanism postulates that mutations increase the peroxidase activity of SOD1, resulting in the increased production of hydroxyl radicals from hydrogen peroxide (104). Increases in protein nitration or in the peroxidase activity of SOD1 mutants are not mutually exclusive and both can contribute to the damage of various cellular components. A third mechanism is inferred from the discovery that SOD1 mutations can alter the activity of calcineurin, a phosphatase under the control of Ca^{2+} and calmodulin. Wang and colleagues (55) have reported that normal SOD1 protects calcineurin from inactivation, probably by preventing oxidation of the iron at its catalytic site. These results suggest that SOD1 could play a role in signal-transducing phosphorylation cascades.

How can the involvement of neurofilaments in ALS pathogenesis be reconciled with the discovery of mutations in the *SOD1* gene in one-fifth of familial ALS cases? There are reports suggesting a link between *SOD1* mutations and the formation of neurofilament inclusions. The presence of massive neurofilament accumulations has been described in a case of familial ALS with an *SOD1* mutation at codon 113 (105). In addition, neurofilament inclusions resembling those found in ALS have been observed in transgenic mice expressing mutant SOD1 (106). It also seems plausible that abundant proteins with long half-lives, such as neurofilament proteins, would be favored targets for damage in motor neurons expressing mutant SOD1. This view is supported by the recent report of nitrotyrosine immunoreactivity in NF-L associated with spheroids found in motor neurons from familial ALS cases (107). It has been speculated that nitration of tyrosines in NF-L may impede its assembly and lead to formation of disorganized filaments (107). Alternatively, oxidative modification of neurofilament proteins by altered SOD1 activity could result in the formation of protein cross-links. Cross-linking may involve the formation of dityrosine linkages by reactive nitrogen species, copper-mediated oxidation of sulfhydryl groups, or the production of carbonyls on lysine residues. Carbonyl-related modification of NF-H has been reported in the neurofibrillary pathology of Alzheimer's disease (108), but it remains to be demonstrated in ALS.

The possibility that neurofilaments have a protective role against toxic effects induced by SOD1 mutations or other primary insults cannot be excluded. In this regard, it would be important in future studies to investigate whether the down-regulation of NF-L mRNA detected in neurodegenerative diseases, including ALS (81) and, to a lesser degree, in normal aging (109) is a factor that may contribute to increased vulnerability of neurons to oxidative damage. Finally, the finding that calcineurin activity is modulated in part by SOD1 raises the possibility that *SOD1* mutations could lead to abnormal phosphorylation of neurofilament proteins.

VI. Future Directions

It is expected that recent evidence pointing to a key role for neurofilaments in motor neuron disease will stimulate research on the potential factors and signaling pathways that could lead to their abnormal accumulation in neurons. Factors that can potentially induce neurofilament disorganization are listed in Table IV. These include factors that can affect the regulation of neurofilament protein expression, assembly, transport, proteolysis, and phosphorylation as well as other types of posttranslational modification. Of particular importance are current investigations aiming to clarify the specific types of oxidative modification taking place as a result of SOD1 mutations and their *in vivo* effects on neurofilament proteins. Another aspect that merits further attention is the apparent paradox that levels of NF-L mRNA in ALS are reduced by 60% in motor neurons containing neurofilament accumulations (*81*). A decreased ratio of NF-L to NF-H mRNAs has also been reported in a dog model of motor neuron disease (*110*). These findings raise the possibility that alterations in the stoichiometry of the NF-L, NF-M, and NF-H subunits might lead to a disorganization of neurofilaments. In this regard, the targeted disruption of neurofilament genes in mice provides a powerful approach to study how reduced subunit levels can affect neurofilament metabolism and neuronal function.

Although current evidence suggests that the toxicity of neurofilament deposits is due to a disruption of axonal transport, the reason why certain varieties of disorganized neurofilaments are more toxic than others in specific cell types remains to be determined. The basis for the early and more pronounced vulnerability of motor neurons may be that they contain different amounts and/or types of protein kinases and protein phosphatases compared to sensory neurons. This notion is supported by the finding that motor neurons, but not DRGs, show pathological changes in mice lacking Cdk5 (*111*). Such differences may explain the finding that axonal NF-H in motor neurons is more highly phosphorylated that that in sensory neurons (*112, 113*). This difference may also prevail at the cell body level because microinjection of a monoclonal antibody against hyperphosphorylated NF-H caused the neurofilament network to collapse in motor neurons but not in sensory neurons (*114*). Motor neurons may thus be predisposed to react more adversely in situations promoting aberrant phosphorylation of neurofilaments.

Species differences in the properties of NF-H and the location of mutations found in NF-H from ALS cases have further highlighted the importance of the KSP phosphorylation domain. It is expected that the transgenic mouse approach will be useful in assessing the effects of mutations in the NF-H phosphorylation domain on the organization and metabolism of neurofilaments. Moreover, the cross-breeding of neurofilament gene knock-out mice with mice serving as models of neurodegeneration, such as transgenic mice expressing mutant SOD1, BPAG1 knock-out mice, and wobbler mice (listed in Table III), should provide a means for assessing the contribution of neurofilament proteins to diseases with different etiologies.

ACKNOWLEDGMENTS

This work was supported by the Medical Research Council of Canada (MRC), the ALS Association (United States), the ALS Society of Canada, the American Health Assistance Foundation (United States), and the Canadian NeuroScience Network. J.-P. J. has an MRC senior scholarship.

REFERENCES

1. P. N. Hoffman and R. J. Lasek, *J. Cell Biol.* **66,** 351 (1975).
2. P. M. Steinert and D. R. Roop, *Annu. Rev. Biochem.* **57,** 593 (1988).
3. G. Ching and R. Liem, *J. Cell Biol.* **122,** 1323 (1993).
4. M. K. Lee, Z. Xu., P. C. Wong, and D. W. Cleveland, *J. Cell Biol.* **122,** 1337 (1993).
5. J. A. Cohlberg, H. Hajarian, T. Tran, P. Alipourjeddi, and A. Noveen, *J. Biol. Chem.* **270,** 9334 (1995).
6. J. P. Julien and F. Grosveld, in "The Neuronal Cytoskeleton" (R. D. Burgoyne, ed.), p. 215. Wiley-Liss, New York, 1991.
7. S. Hisanaga and N. Hirokawa, *J. Mol. Biol.* **211,** 871 (1990).
8. J. P. Julien and W. E. Mushynski, *J. Biol. Chem.* **257,** 10467 (1982).
9. J. P. Julien and W. E. Mushynski, *J. Biol. Chem.* **258,** 4919 (1983).
10. J. F. Lees, P. S. Shneidman, S. F. Skuntz, M. J. Carden, and R. A. Lazzarini, *EMBO J.* **7,** 1947 (1988).
11. J. P. Julien, F. Côté, L. Beaudet, M. Sidky, D. Flavell, F. Grosveld, and W. Mushynski, *Gene* **68,** 307 (1988).
12. E. Levy, R. K. H. Liem, P. D'Eustachio, and N. J. Cowan, *Eur. J. Biochem.* **166,** 71 (1987).
13. M. M. Myers, R. A. Lazzarini, V. M.-Y. Lee, W. W. Schlaepfer, and D. L. Nelson, *EMBO J.* **6,** 1617 (1987).
14. D. A. Figlewicz, A. Krizus, M. G. Martinoli, V. Meininger, M. Dib, G. A. Rouleau, and J.-P. Julien, *Hum. Mol. Genet.* **3,** 1757 (1994).
15. S. S. M. Chin and R. K. H. Liem, *J. Neurosci.* **10,** 3714 (1990).
16. D. R. Soppet, L. L. Beasley, and M. B. Willard, *J. Biol. Chem.* **267,** 17354 (1992).
17. C. Kaspi and W. E. Mushynski, *Ann. N.Y. Acad. Sci.* **455,** 794 (1985).
18. J. Way, M. R. Hellmich, H. Jaffe, B. Szaro, H. C. Pant, H. Gainer, and J. Battey, *Proc. Natl. Acad. Sci. U.S.A.* **89,** 6963 (1992).
19. D. Zopf, B. Dineva, H. Betz, and E. D. Gundelfinger. *Nucleic Acid Res.* **18,** 521 (1990).
20. J.-P. Julien and W. E. Mushynski, *J. Biol. Chem.* **257,** 10467 (1982).
21. J.-P. Julien and W. E. Mushynski, *J. Biol. Chem.* **258,** 4919 (1983).
22. R. A. Nixon and H. Sihag, *Trends Neurosci.* **14,** 501 (1991).
23. A. Dosemeci, C. Floyd, and H. C. Pant, *Cell. Mol. Neurobiol.* **10,** 369 (1990).

24. T. Saito, H. Shima, Y. Osawa, M. Nagao, B. A. Hemmings, T. Kishimoto, and S. I. Hisanaga, *Biochemistry* **34,** 7376 (1995).
25. W. W. Schlaepfer, *J. Neuropathol. Exp. Neurol.* **46,** 117 (1987).
26. R. A. Nixon, *Brain Pathol.* **3,** 29 (1993).
27. J. Q. Trojanowski, M. L. Schmidt, R. W. Shin, G. T. Bramblett, D. Rao, and V. M.-Y. Lee, *Brain Pathol.* **3,** 45 (1993).
28. B. I. Giasson, J. A. Cromlish, E. S. Athlan, and W. E. Mushynski, *J. Neurochem.* **66,** 1207 (1996).
29. M. G. Sacher, E. S. Athlan, and W. E. Mushynski, *J. Biol. Chem.* **269,** 14480 (1994).
30. M. G. Sacher, E. S. Athlan, and W. E. Mushynski, *Biochem. Biophys. Res. Commun.* **186,** 524 (1992).
31. R. A. Nixon and S. E. Lewis, *J. Biol. Chem.* **261,** 16298 (1986).
32. R. K. Sihag and R. A. Nixon, *J. Biol. Chem.* **266,** 18861 (1991).
33. S. Hisanaga, M. Kusubata, E. Okumura, and T. Kishimoto, *J. Biol. Chem.* **266,** 21798 (1991).
34. H. Miyasaka, S. Okabe, K. Ishiguro, T. Uchida, and N. Hirokawa, *J. Biol. Chem.* **268,** 22695 (1993).
35. J. Eyer and J.-F. Letterier, *Biochem. J.* **252,** 655 (1988).
36. S. de Waegh, V. M.-Y. Lee, and S. T. Brady, *Cell* **68,** 451 (1992).
37. M. Mata, N. Kupina, and D. J. Fink, *J. Neurocytol.* **21,** 199 (1992).
38. J. S. Cole, A. Messing, J. Q. Trojanowski, and V. M.-Y. Lee, *J. Neurosci.* **14,** 6956 (1994).
39. S.-T. Hsieh, G. J. Kidd, T. O. Crawford, Z. Xu, W.-M. Lin, B. D. Trapp, D. W. Cleveland, and J. W. Griffin, *J. Neurosci.* **14,** 6392 (1994).
40. R. A. Nixon, P. A. Paskevich, R. K. Sihag, and C. Y. Thayer, *J. Cell Biol.* **126,** 1031 (1994).
41. L. A. Sternberger and N. H. Sternberger, *Proc. Natl. Acad. Sci. U.S.A.* **80,** 6126 (1983).
42. V. M.-Y. Lee, L. Otvos, M. J. Carden, M. Hollosi, B. Dietzschold, and R. A. Lazzarini, *Proc. Natl. Acad. Sci. U.S.A.* **85,** 1998 (1988).
43. J. Lew, R. J. Winkfein, H. K. Paudel, and J. H. Wang, *J. Biol. Chem.* **267,** 25922 (1992).
44. K. T. Shetty, W. T. Link, and H. C. Pant, *Proc. Natl. Acad. Sci. U.S.A.* **90,** 6844 (1993).
45. H. M. Roder and V. M. Ingram, *J. Neurosci.* **11,** 3325 (1991).
46. H. M. Roder, F. J. Hoffman, and W. Schröder, *J. Neurochem.* **64,** 2203 (1995).
47. R. J. Guan, B. S. Khatra, and J. A. Cohlberg, *J. Biol. Chem.* **266,** 8262 (1991).
48. B. I. Giasson and W. E. Mushynski, *J. Biol. Chem.* **271,** 30404 (1996).
49. E. Elhanany, H. Jaffe, W. T. Link, D. M. Sheeley, H. Gainer, and H. C. Pant, *J. Neurochem.* **63,** 2324 (1994).
50. G. S. Bennett and R. Quintana, *J. Neurochem.* **68,** 534 (1997).
51. J. M. Kyriakis and J. Avruch, *J. Biol. Chem.* **271,** 24313 (1996).
52. Y. A. Hannun, *J. Biol. Chem.* **269,** 3125 (1994).
53. Veeranna, K. T. Shetty, W. T. Link, H. Jaffe, J. Wang, and H. C. Pant, *J. Neurochem.* **64,** 2681 (1995).
54. M. Mata, P. Honegger, and D. J. Fink, *Cell. Mol. Neurobiol.* **17,** 129 (1997).
55. W. Wang, V. C. Culotta, and C. B. Klee, *Nature (London)* **383,** 434 (1996).
56. D. F. Watson, K. P. Fittro, P. N. Hoffman, and J. W. Griffin, *Brain Res.* **539,** 103 (1991).
57. M. M. Doroudchi and H. D. Durham, *J. Neuropathol. Exp. Neurol.* **55,** 246 (1996).
58. J. E. Carter, J.-M. Gallo, V. E. R. Anderson, B. H. Anderton, and J. Robertson, *J. Neurochem.* **67,** 1997 (1996).
59. D. T. Denhardt, *Biochem. J.* **318,** 729 (1996).
60. H. Yamasaki, G. S. Bennett, C. Itakura, and M. Mizutani, *Lab. Invest.* **66,** 734 (1992).
61. J. Eyer and A. Peterson, *Neuron* **12,** 389 (1994).
62. Q. Zhu, S. Couillard-Després, and J.-P. Julien, *Exp. Neurol.* **148,** 299 (1997).

63. R. J. Lasek, *J. Cell Sci. Suppl.* **5,** 161 (1986).
64. R. J. Lasek, J. A. Garner, and S. T. Brady, *J. Cell Biol.* **99,** 212s (1984).
65. S. S. M. Chin and R. K. H. Liem, *J. Neurosci.* **10,** 3714 (1990).
66. M. J. Monteiro and D. W. Cleveland, *J. Cell Biol.* **108,** 579 (1989).
67. R. A. Nixon, *in* "The Neuronal Cytoskeleton" (R. D. Burgoyne, ed.), p. 283. Wiley-Liss, New York, 1991.
68. S. Okabe, H. Miyasaka, and N. Hirokawa, *J. Cell Biol.* **121,** 375 (1993).
69. S. Terada, T. Nakata, A. C. Peterson, and N. Hirokawa, *Science* **273,** 784 (1996).
70. M. B. Willard and C. Simon, *Cell* **35,** 551 (1983).
71. J. P. Collard, F. Côté, and J.-P. Julien, *Nature (London)* **375,** 61 (1995).
72. J. R. Marszalek, T. L. Williamson, M. K. Lee, Z. Xu, P. N. Hoffman, T. O. Crawford, and D. W. Cleveland, *J. Cell Biol.* **135,** 711 (1996).
73. D. Bamburg, Jr., D. Bray, and K. Chapman, *Nature (London)* **321,** 788 (1986).
74. A. Brown, T. Slaughter, and M. M. Black, *J. Cell Biol.* **119,** 867 (1992).
75. S. Okabe and N. Hirokawa, *J. Cell Biol.* **120,** 1177 (1993).
76. E. M. Tanaka and M. W. Kirschner, *J. Cell Biol.* **115,** 345 (1991).
77. E. Tanaka and J. Sabry, *Cell* **83,** 171 (1995).
78. Y. Yang, J. Dowling, Q.-C. Yu, P. Kouklis, D. W. Cleveland, and E. Fuchs, *Cell* **86,** 655 (1996).
79. A. Brown, G. Bernier, M. Mathieu, J. Rossant, and R Kothary, *Nature Genet.* **10,** 301 (1995).
80. L. Guo, L. Degenstein, J. Dowling, Q.-C. Yu, R. Wollmann, B. Perman, and E. Fuchs, *Cell* **81,** 233 (1995).
81. C. Bergeron, K. Beric-Maskarel, S. Muntasser, L. Weyer, M. J. Somerville, and M. Percy, *J. Neuropathol. Exp. Neurol.* **53,** 221 (1994).
82. D. R. Crapper McLachlan, W. J. Lukiw, L. Wong, C. Bergeron, and N. T. Bech-Hansen, *Mol. Brain Res.* **3,** 255 (1988).
83. F. Côté, J.-F. Collard, and J.-P. Julien, *Cell* **73,** 35 (1993).
84. H. Pant and Veeranna, *Biochem. Cell Biol.* **73,** 575 (1995).
85. M. J. Strong, *J. Neurol. Sci.* **124,** 20 (1994).
86. Z. Xu, L. C. Cork, J. W. Griffin, and D. W. Cleveland, *Cell* **73,** 23 (1993).
87. M. K. Lee, J. R. Marszalek, and D. W. Cleveland, *Neuron* **13,** 975 (1994).
88. E. Fuchs, *J. Cell Biol.* **125,** 511 (1994).
89. P. C. Wong, J. Marszalek, T. O. Crawford, Z. Xu, S. T. Hsieh, J. W. Griffin, and D. W. Cleveland, *J. Cell Biol.* **130,** 1413 (1995).
90. D. Ma, L. Descarries, J.-P. Julien, and G. Doucet, *Neuroscience* **68,** 135 (1995).
91. J. C. Vickers, J. H. Morrison, V. L. Friedrich, G. A. Elder, D. P. Perl, R. N. Katz, and R. A. Lazzarini, *J. Neurosci.* **14,** 5603 (1994).
92. J. P. Julien, *Trends Cell Biol.* **7,** 243 (1997).
93. K. Rooke, D. A. Figlewicz, F. Han, and G. A. Rouleau, *Ann. Neurol.* **46,** 789 (1996).
94. J. D. Vechio, L. I. Bruijn, Z. Xu, R. H. Brown, and D. W. Cleveland, *Ann. Neurol.* **40,** 603 (1996).
95. S. Sasaki and M. Iwata, *Neurology* **47,** 535 (1996).
96. J. W. Griffin, K. E. Fahnestock, D. L. Price, and L. C. Cork, *Ann. Neurol.* **14,** 55 (1983).
97. I. G. McQuarrie, S. T. Brady, and R. J. Lasek, *Neurobiol. Aging* **10,** 359 (1989).
98. R. H. Brown, *Cell* **80,** 687 (1995).
99. M. E. Gurney, H. Pu, A. Y. Chiu, M. C. Dal Canto, C. Y. Polchow, D. D. Alexander, J. Caliendo, A. Hentati, Y. W. Kwon, H.-Xiang Deng, W. Chen, P. Zhai, R. L. Sufit, and T. Siddique, *Science* **264,** 1772 (1994).
100. P. C. Wong, C. A. Pardo, D. R. Borchelt, M. K. Lee, N. G. Copeland, N. A. Jemkins, S. S. Sisodia, D. W. Cleveland, and D. L. Price, *Neuron* **14,** 1105 (1995).

101. M. E. Ripps, G. W. Huntley, P. R. Hof, J. H. Morrison, and J. W. Gordon, *Proc. Natl. Acad. Sci. U.S.A.* **92,** 689 (1995).
102. A. G. Reaume *et al.*, *Nature Genet.* **13,** 43 (1996).
103. J. S. Beckman, M. Carson, C. D. Smith, and W. H. Koppenol, *Nature (London)* **364,** 584 (1993).
104. M. Wideau-pazos *et al.*, *Science* **271,** 515 (1996).
105. G. A. Rouleau *et al.*, *Ann. Neurol.* **39,** 128 (1996).
106. P.-H. Tu, P. Raju, K. A. Robinson, M. E. Gurney, J. Q. Trojanowski, and V. M-Y. Lee, *Proc. Natl. Acad. Sci. U.S.A.* **93,** 3155 (1995).
107. S. M. Chou, H. S. Wang, and K. Komai, *J. Chem. Neuroanat.* **10,** 249 (1996).
108. M. A. Smith *et al.*, *J. Neurochem.* **64,** 2660 (1995).
109. I. M. Parhad, J. N. Scott, L. A. Cellars, J. S. Bains, C. A. Krekoski, and A. W. Clark, *J. Neurosci. Res.* **41,** 355 (1995).
110. N. A. Muma and L. C. Cork, *Lab Invest.* **69,** 436 (1993).
111. T. Oshima, J. M. Ward, C.-G. Huh, G. Longenecker, H. C. Pant, R. O. Brady, L. J. Martin, and A. B. Kullcarni, *Proc. Natl. Acad. Sci. U.S.A.* **93,** 11173 (1996)
112. L. Soussan, A. Admon, A. Aharoni, Y. Cohen, and D. M. Michaelson, *Cell. Mol. Neurobiol.* **16,** 463 (1996).
113. L. Soussan, A. Barzilai, and D. M. Michaelson, *J. Neurochem.* **62,** 770 (1994).
114. H. D. Durham, *J. Neuropathol. Exp. Neurol.* **49,** 582 (1992).
115. S. Hisanaga, M. Kusubata, E. Okumura, and T. Kishimoto, *Cell Motil. Cytoskel.* **31,** 283 (1995).
116. P. J. Kennely and E. G. Krebs, *J. Biol. Chem.* **266,** 15555 (1991).
117. S. D. Yang, J. J. Huang, and T. J. Huang, *J. Neurochem.* **64,** 1848 (1995).
118. S. Guidato, L. H. Tsai, J. Woodgett, and C. C. J. Miller, *J. Neurochem.* **66,** 1698 (1996).
119. B. J. Pulverer, J. M. Kyriakis, J. Avruch, E. Nikolakaki, and J. R. Woodgett, *Nature (London)* **353,** 670 (1991).
120. S. Carpenter, *Neurology* **18,** 841 (1968).
121. M. L. Schmidt, J. M. Murray, V. M. Y. Lee, W. D. Hill, A. Wertkin, and J. Q. Trojanowski, *Am. J. Pathol.* **139,** 53 (1991).
122. W. D. Hill, M. Arai, J. A. Cohen, and J. Q. Trojanowski, *J. Comp. Neurol.* **329,** 328 (1993).
123. G. Perry, M Kawai, M. Tabaton, M. Onorato, P. Mulvihill, P. Richey, A. Morandi, H. A. Connolly, and P. Gambetti, *J. Neurosci.* **11,** 1748 (1991).
124. P. Rodgers-Johnson, R. M. Garruto, R. Yanagihara, K-M. Chen, D. C. Gajdusek, and C. J. Gibbs, *Neurology* **36,** 7 (1986).
125. M. W. Klymkowsky and D. J. Plummer, *J. Cell Biol.* **100,** 245 (1985).
126. A. T. Yachnis, J. Q. Trojanowski, M. Memmo,, and W. W. Schlaepfer, *J. Neuropathol. Exp. Neurol.* **47,** 206 (1988).
127. S. Brimijioin, *in* "Peripheral Neuropathy" (P. J. Dyk, P. K. Thomas, E. H. Lambert, and R. Bunge, eds.), Vol. 1, p. 477. Saunders, Philadelphia, Penn., 1984.
128. R. Pernas-Alonso, A. E. Schaffner, C. Perrone-Capano, A. Orlando, F. Morelli, C. T. Hansen, J. L. Barker, B. Esposito, F. Cacucci, and U. di Porzio, *Mol. Brain. Res.* **38,** 267 (1996).
129. J. W. Griffin, K. E. Fahnestock, D. L. Price, and L. C. Cork., *Ann. Neurol.* **14,** 55 (1983).

Regulation of Cytochrome P450 Gene Transcription by Phenobarbital

BYRON KEMPER

Department of Molecular and Integrative Physiology
College of Medicine at Urbana-Champaign
University of Illinois at Urbana-Champaign
Urbana, Illinois 61801

I. Introduction	26
II. Phenobarbital Induction of Drug Metabolism	27
A. Pleiotropic Effects	27
B. Induction of P450s by Inducers Other Than Phenobarbital	28
C. Approaches to Studies of Phenobarbital Induction of P450 Gene Transcription	28
III. Induction of P450s in *Bacillus megaterium*	30
A. The Repressor, BM3R1	30
B. Role of the Barbie Box Sequence and Positively Acting Proteins	32
C. Model for P450 Induction in *Bacillus megaterium*	33
D. Relevance of the Bacterial Mechanism to Mammalian Genes	34
IV. Proximal Promoter	35
A. Proposed Phenobarbital-Responsive Elements	35
B. Core Promoter Activity as the Function for Proximal Promoter Elements	38
V. Distal Phenobarbital-Responsive Enhancer	41
A. Chicken *CYPH1* Gene	41
B. Transgenic Analysis of the *CYP2B2* Gene	42
C. DNase I Hypersensitivity of the *CYP2B1/2* Genes	43
D. Localization of a Distal Enhancer of the *CYP2B2* Gene by Transfection into Primary Hepatocytes	44
E. Analysis of *CYP2B2* by Transfection of Liver *in Situ*	45
F. Comparison of the *Cyp2b9* and *Cyp2b10* PBRUs by *in Situ* Transfection	51
G. Analysis of the *Cyp2b9* and *Cyp2b10* PBRUs in Transfected Primary Hepatocytes	53
VI. Perspectives and Models of Phenobarbital Induction of P450 Genes	55
References	62

The ability of phenobarbital to induce levels of drug metabolism in mammals has been known for over 40 years. However, the molecular mechanisms underly-

ing increased expression of the genes of the key enzyme in drug metabolism, cytochrome P450, have not been elucidated, primarily because *in vitro* model systems in which the induction could be studied were not available. Transfected primary cultured hepatocytes, transfection of liver *in situ*, and transgenic mice now provide suitable models for phenobarbital induction. In this review, progress toward understanding the mechanism of phenobarbital induction of gene expression is discussed with an emphasis on the mammalian genes, *CYP2B1*, *CYP2B2*, and *Cyp2b10*, which are most highly inducible by phenobarbital. Barbiturate induction of P450s in *Bacillus megaterium*, which is the system best understood, and its relevance to mammalian mechanisms of induction are also discussed. In *B. megaterium*, the binding of a repressor to several motifs is reversed by direct effects of barbiturates and by induction of positively acting factors. One of the repressor binding sites, the barbie box, is present in many mammalian phenobarbital-inducible genes, including the promimal promoter regions of *CYP2B1*, *CYP2B2*, and *Cyp2B10*. In the mammalian P450 genes, evidence has been proposed for phenobarbital-regulated elements both in the proximal promoter region and in a distal enhancer region. The role of the proximal region is controversial. A positively acting element that overlaps the barbie box sequence and a negative element have been proposed to mediate induction of *CYP2B1/2*, based primarily on protein binding and cell-free transcription assays. In contrast, other investigators have not found differences in phenobarbital-dependent protein binding in the proximal promoter region nor mediation of phenobarbital induction by this region. A distal gene fragment, at about −2000 kb in *CYP2B1*, *CYP2B2*, and *Cyp2b10*, has been shown to be a phenobarbital-responsive enhancer independent of proximal promoter elements. This fragment contains several binding sites for proteins and several functional elements, including an NF-1 site, and, therefore, has been designated as a phenobarbital-responsive unit. Possible models are presented in which phenobarbital treatment induces altered chromatin structure, which allows the binding of positively acting factors, or activates factors already bound, to the distal enhancer and the proximal promoter.
© 1998 Academic Press

I. Introduction

Cytochromes P450 (P450, referring to proteins, or *CYP*, referring to genes) comprise a superfamily of enzymes with remarkably diverse functions (*1*). P450s are present in representatives of all classes of living organisms, from bacteria to humans (*2*). They have major roles in the detoxification or inactivation of bioactive compounds in the body and the synthesis and catabolism of many endogenous compounds, such as steroids and fatty acids. Although generally considered part of a detoxification system, P450s also may activate compounds in the body, sometimes detrimentally, as in the case of carcinogens and mutagens, and sometimes beneficially, as in the case of prodrugs such as cyclophosphamide, an anticancer agent. Nearly 500 P450s have been identified in all species, and within a single mammalian species more than 40 P450s have been characterized, and it is likely that 100 or more are present (1, 2). P450s in eukaryotes are membrane-bound proteins present in either the endoplasmic reticulum (microsomes) or in the mitochondria. Xenobiotic-metabolizing forms are predominantly in the microsomes. P450 is the key enzyme in the microsomal drug-metabolizing system. It is the terminal oxidase in a small electron transfer chain, binds to oxygen and an organic substrate, and catalyzes the incorporation of one atom of oxygen into the substrate and one into water (3). Although the prototypical reaction is the incorporation of a hydroxyl group into the substrate, over 40 distinct reactions utilizing thousands of substrates have been demonstrated for P450s. These reactions require the donation of two electrons to P450 from NADPH, which are transferred to P450 via P450 reductase, a flavoprotein, in the case of microsomal P450s. In the case of mitochondrial and bacterial P450s, electrons are transferred first to ferredoxin reductases, flavoproteins, then to ferredoxins, iron–sulfur proteins, before transfer to P450. This remarkable family of nearly ubiquitous enzymes plays key roles in metabolism underlying normal physiology and the interactions of organisms with the environment.

An important property of the drug-metabolizing system is increases in its activity in response to xenobiotics, in some cases the substrates of the enzymes but in other cases chemicals that are not substrates. These inducers include barbiturates, polycyclic hydrocarbons, peroxisomal proliferators, glucocorticoids, and ethanol (4). Generally only a subset of the P450s will be induced by individual compounds; for example, barbiturates induce P450s different from the P450s induced by polycyclic aromatic hydrocarbons and so on. In nearly every case, the increase in P450 activity has been shown to be due mostly to an increase in gene transcription which results in increased amounts of P450 in the cells. Ethanol and related compounds are an exception in which stabilization of the protein plays an important role (5).

II. Phenobarbital Induction of Drug Metabolism

A. Pleiotropic Effects

The ability of phenobarbital to modulate the metabolism of drugs and other xenobiotics in mammals has been known for nearly 40 years (6). In addition to the induction of P450s, phenobarbital also induces other drug-metabolizing enzymes and has dramatic effects on the hepatocyte: proliferation of endoplasmic reticulum membranes, stabilization of microsomal proteins, and tumor formation (7). The amounts of 29 cDNAs were either increased or decreased in RNA display analysis after phenobarbital treatment of chick embryos for 48 hr (8). The induction characteristics of P450s are also heterogeneous. Rat *CYP2B1/2* and mouse *Cyp2b10* are highly induced by phenobarbital whereas members of the *CYP2C* and *CYP3A* subfamilies are modestly induced. In primary cultures, the time courses and dose–response dependencies of induction of P450s and effects of cycloheximide vary (9–11). These pleiotropic effects of phenobarbital and heterogeneous responses suggest either that there are multiple mechanisms of phenobarbital induction or that the mechanism is sufficiently complex to allow for gene-specific and cell-specific variation in the induction. In addition, the pleiotropic effects require that primary effects on P450 induction be distinguished from secondary effects of the induction. The proliferation of the microsomal membrane, for example, may be secondary to the synthesis and insertion of P450 into the membrane (12).

B. Induction of P450s by Inducers Other Than Phenobarbital

Substantial progress toward understanding the basis for the induction of P450 gene expression has been made, but an understanding of the molecular mechanisms underlying phenobarbital regulation has lagged. The Ah receptor and associated proteins, Arnt and Hsp70, have been shown to be the receptor complex activated by the polycyclic aromatic hydrocarbons (13). The binding sites for the Ah receptor have been identified and the changes in the chromatin structure associated with activation of the P450 gene have been detected. Clofibrate and similar drugs, which promote the proliferation of peroxisomes, also induce P450s through the peroxisomal proliferator-activated receptor, and defined binding sites for this receptor are present in P450 genes (14). Although not part of the drug-metabolizing P450s, the genetic elements mediating regulation by cAMP of P450s involved in steroid biogenesis have been identified (15).

C. Approaches to Studies of Phenobarbital Induction of P450 Gene Transcription

In contrast, work on phenobarbital induction in mammals has progressed only slightly beyond the observations that the increase in P450 was primarily mediated by an increase in gene expression. The major roadblocks in understanding the phenobarbital effects have been the inability to identify a receptor and the lack of a suitable *in vitro* model for studying the regulation of gene expression. Phenobarbital requires nearly millimolar concentrations for maximum responses, which is prohibitive for identifying a receptor. Analogs with effects at micromolar concentrations have been identified, but specific interacting proteins remain elusive (7). Further confounding the search is the lack of obvious conformational similarity among a large number of "phenobarbital-like" inducers.

In view of these difficulties, an obvious approach is to work from the other end of the process, the identification of the gene elements and regulatory factors mediating the induction of the P450 genes. This approach, too, has proved difficult because the most propitious model system for this type of analysis, continuously cultured cells (7), cannot be used. The history of our own studies is representative of the frustrations in this field. After rather straightforward isolation of the cDNAs and genes for the phenobarbital-inducible CYP2C genes in the early to mid-1980s, attempts to analyze the genetic basis of the phenobarbital induction floundered on the lack of a model system. None of the immortalized liver cell lines expresses significant amounts of phenobarbital-induced P450s nor has proved to be a valid model for the *in vivo* induction of P450 by phenobarbital, although there are reports of modest responses to phenobarbital in cultured cells (16). Primary cultures of liver cells grown under normal conditions cease expression of the phenobarbital-inducible forms of P450s in the first couple of days of culture. Most of the work on phenobarbital induction up to 1995 was based on cell-free assay systems, the *in vivo* significance of which remains unproved. The development of methods for the culture of primary hepatocytes, which resulted in continued expression and response to phenobarbital of P450s, was an important advance in providing a possible system for analysis of transfected P450 genes (17–19). In 1995, a phenobarbital-responsive element was identified using this system (20). In 1996, a method of direct injection of DNA into rat liver, effectively an *in situ* transient transfection, was also shown to be an effective assay for phenobarbital induction (21). These advances, together with earlier studies demonstrating that rat P450 genes in transgenic mice responded to phenobarbital (22), now provide the tools to unravel the mysteries of the effects of the barbiturates on gene expression.

In this review, progress toward understanding the mechanism of phenobarbital induction of gene expression is discussed with an emphasis on the mammalian genes, CYP2B1, CYP2B2, and Cyp2b10, which are most highly inducible by phenobarbital (7). The review focuses on the molecular mechanisms at the level of gene transcription, with some emphasis on our own work. The signal transduction system involved in phenobarbital induction is largely unknown, but there is a substantial body of literature in this area that will not be covered. These studies suggest that phosphorylation is a component of the mechanism and that interactions with several hormones may modulate the phenobarbital response. Generally, studies on the mechanism of transcriptional activation by phenobarbital are in an early stage with no definitive regulatory factor identified as the target of phenobarbital action and a complex array of ubiquitous factors that seem to contribute to the induction. In only a few cases have the generally accepted minimal requirements yet been met to define a specific regulatory element, that is, inactivation of such an element by mutation, conferring responsiveness on a heterologous promoter by the proposed regulatory element, and identification of a regulatory factor that binds to the element. Except for the results in transgenic animals, the induction by phenobarbital in the *in vitro* assays has been much less than that observed *in vivo*. The review begins with a discussion of phenobarbital induction in *Bacillus megaterium* and an analysis of the relevance of the bacterial mechanism to that in the mammalian system, discusses the conflicting proposals about the role of the proximal promoter in the induction process, reviews the evidence for a role of distal sequences in the induction, and ends with speculation about the mechanism and future directions.

III. Induction of P450s in *Bacillus megaterium*

A. The Repressor, BM3R1

One of the fascinating aspects of barbiturate induction is the ubiquity of the response that occurs in many species, ranging from bacteria to humans. The regulation of barbiturate induction of P450 is understood best in bacteria, and some evidence implies that aspects of the bacterial induction mechanism are conserved in mammals. Two P450 genes in *B. megaterium* that are induced by barbiturates are designated P450s BM1 (*CYP106*) and BM3 (*CYP102*) (*23*). Regulation of the BM1 and BM3 genes has been analyzed in detail, and a complex coordinated mechanism of regulation was revealed that involves derepression of the expression of these genes by barbiturates (*245*) (Fig. 1). A key component of the regulation mechanism in *B. megaterium* is a repressor protein, BM3R1, the coding sequence for which is immediately upstream of the BM3 gene (*25*). A G39E mutation in this protein resulted in the constitutive expression of both BM1 and BM3 and complementation of the mutant bacteria with wild-type BM3R1 repressed the expression, demonstrating that the gene products acted in trans. The protein sequence of the repressor showed that it had a helix–turn–helix motif characteristic of DNA-binding proteins. The repressor binds to a palindromic operator sequence upstream of the BM3R1 gene located near a strong promoter that is responsible for expression of both BM3R1 and BM3. In *in vitro* protein binding assays, addition of barbiturates greatly reduced the binding of BM3R1 to the palindromic operator sequence (*26*). These genetic and biochemical studies

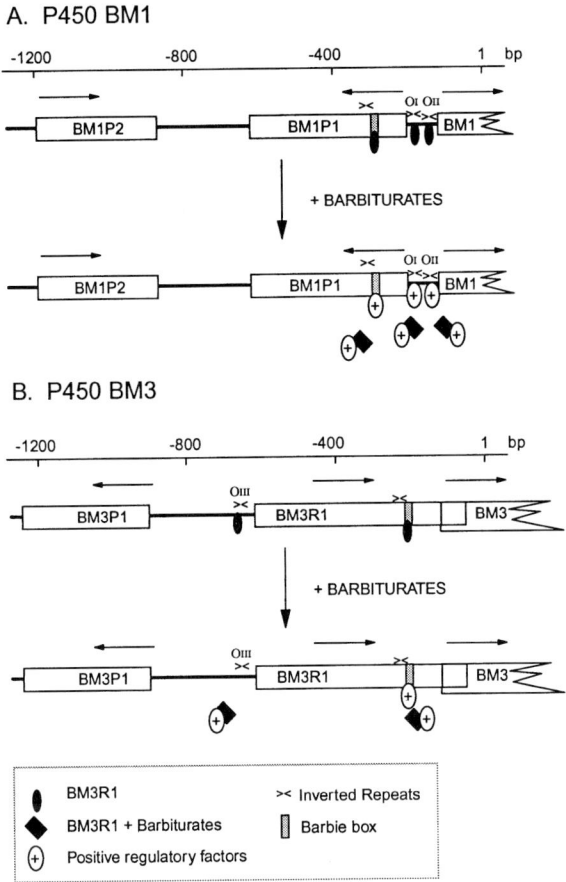

FIG. 1. Models of the induction of P450 BM1 and P450 BM3 by barbiturates in *Bacillus megaterium*. (A) Repression of the P450 BM1 gene in the absence of barbiturates involves the binding of the repressor BM3R1 to two palindromic sequences in the operator region (O_I and O_{II}) and to the barbie box sequence within the coding region for BM1P1. In the presence of barbiturates, BM3R1 can still bind to these sequences *in vitro*, but the binding is inhibited by positive factors, including BM1P1 and BM1P2, probably both by direct interaction with BM3R1 and by competition for the binding sites. (B) Similarly, repression of the P450 BM3 gene in the absence of barbiturates involves the binding of BM3R1 to the palindromic sequence in O_{III} and the barbie box sequence in the coding region for BM3R1. After incubation with barbiturates, binding to the operator site is directly inhibited and binding to the barbie box is inhibited by interactions of positive factors with BM3R1 and by competition for binding to a site adjacent to the barbie box. This model has been proposed by Fulco and colleagues (24, 29)

demonstrated that a major effect of barbiturates was to inhibit the binding of BM3R1 to the operator sequence, directly resulting in derepression of expression of its own synthesis and that of BM3. The BM1 gene promoter region did not have an analogous palindromic sequence, suggesting the presence of a second sequence to which BM3R1 binds.

B. Role of the Barbie Box Sequence and Positively Acting Proteins

A search for similar sequences in the BM1 and BM3 regulatory regions revealed a 17-bp sequence that was present at −318 to −302 in the BM1 gene (within the protein-coding region of BM1P1, a regulatory protein described below) and −243 to −227 in the BM3 gene (within the BM3R1 protein coding region upstream of the BM3 gene), which was termed the barbie box (27) (Fig. 1). A single protein–DNA complex was detected when DNA containing the barbie box sequence was incubated with proteins partially purified from *B. megaterium* extracts and the binding was inhibited if the bacteria had been previously incubated with a barbiturate or if the protein was incubated *in vitro* for 1 hr at 37°C with phenobarbital. The inhibition in binding was retained even if the phenobarbital was removed by dialysis before the DNA-binding reaction was carried out. In *B. megaterium* transformed with deletion fragments of the 5′ flanking region of the BM1 gene, deletion of the barbie box region resulted in a dramatic increase in the expression of the gene in control cells to levels equal to that seen in cells incubated with 4 m*M* pentobarbital (27). Confirming this result, point mutations within the barbie box of BM1 resulted in high-level constitutive expression of BM1 (28). In contrast, mutation of the BM3 barbie box resulted in little effect on basal expression, but increased expression after barbiturate induction. The barbie box mutations did not affect the relative increase in expression of BM1P1 or BM3R1 induced by barbiturates, but did increase both control and treated expression of BM1P1 by about 10-fold. More detailed deletions indicated that the region immediately upstream of the barbie sequence reduced but did not eliminate barbiturate responsiveness, that deletion of half of the barbie box eliminated induction without effects on the basal level, and that only deletion of the entire barbie box resulted in elevated basal expression and no barbiturate response (28). These results indicate the barbie box region is a complex element with positive regulatory elements at the 5′ end and negative ones at the 3′ end, probably overlapping each other.

Several proteins bind to the barbie box. BM3R1 was shown to bind to the barbie boxes in both the BM1 and BM3 genes (28). The mutations described above in the barbie sequence reduced but did not eliminate the binding of BM3R1. The lower binding affinity of the BM3R1 explains in part the effects of the mutations: constitutive expression of BM1, increases in expression of

Bm1P1, and barbiturate-mediated expression of BM3 and Bm3P1. Footprinting analysis showed that BM3R1 protected a site partially overlapping the BM3 barbie box and extending to 3′ flanking nucleotides, and the entire BM1 barbie box and extending to 3′ flanking nucleotides (29). Binding of BM3R1 to two palindromic sequences in the BM1 promoter region was also detected. In addition to binding of BM3R1 to the barbie box, two DNA–protein complexes were observed when the whole cell extracts from *B. megaterium* were incubated with barbie sequences. In contrast to the results with the partially purified proteins described above (27), the binding of both complexes with the whole-cell extracts were increased by barbiturate treatment of the cells (28). BM3R1 was not in these protein–DNA complexes because similar binding was observed with cells expressing the G39E mutant of BM3R1, which does not bind to the barbie box. In footprinting experiments, the positive factors in the whole cell extracts protected a palindromic sequence that included the BM1 barbie box and overlapped the binding site of the repressor, BM3R1 (28).

The nature of the positive binding factors remains unknown. Proteins encoded by two open reading frames upstream of the BM1 gene, designated BM1P1 and BM1P2, have been characterized (30). Both are induced by phenobarbital and overexpression of BM1P1 in *B. megaterium* results in greatly increased expression of BM1, indicating that it is a positively acting factor. Binding of both proteins to a fragment that contained the BM3R1 inverted repeat binding sites in the BM1 promoter was detected, and either protein blocked the binding of BM3R1. A fragment containing the barbie box from BM1 did not bind the two proteins, but both proteins were still capable of inhibiting the binding of BM3R1 to the barbie box. This result suggests that these proteins may interact directly with BM3R1 to inhibit binding although in the BM1 operator region direct competition for binding to the inverted repeats remains possible. The inability of these factors to bind to the barbie sequence indicates they are different from the positive acting factors detected in whole-cell extracts (28) or that additional proteins are required for their binding to the barbie box.

C. Model for P450 Induction in *Bacillus megaterium*

This series of elegant studies exploiting bacterial genetics provides insight into a complex regulatory system underlying phenobarbital induction even in a one-cell organism. The most likely mechanism, illustrated in Fig. 1, involves derepression both by direct effects of barbiturates on a repressor and by competition for DNA binding between the barbiturate-regulated repressor and activator proteins (24). BM3R1 binds to multiple sites, including palindromic sites in the operators for BM3 and BM1 and to barbie box sequences 5′ of both genes. Barbiturates directly inhibit the binding of the repressor to the operator sequence in the BM3 gene. Barbiturate-inducible positively acting factors compete for BM3R1 binding by binding to sites that overlap the other BM3R1 sites and possibly interact directly with BM3R1 to reduce its binding. The positive factors probably include BM1P1 and BM1P2 and other unidentified positive factors because binding characteristics of bacterial extracts differ from those of the purified BM1P1 and BM2P2.

D. Relevance of the Bacterial Mechanism to Mammalian Genes

In a general sense this regulation of gene expression by barbiturates in a bacterium provides a guide for examining similar regulation in eukaryotes. A most remarkable result reported in these studies, however, implied a direct relationship between the regulatory mechanisms in the two species. Sequences were detected in the mammalian *CYP2B1/2* genes similar to the barbie box sequence in bacteria. Proteins from either rat liver extracts or from bacterial cell extracts could bind to either the mammalian or bacterial sequences and the binding was barbiturate dependent (27). Binding of the bacterial protein was inhibited by barbiturates whereas the binding of the rat proteins was stimulated by treatment of rats with phenobarbital. As described above, positively acting bacterial proteins, bound to the barbie box region, were later detected, which might be more analogous to the mammalian proteins. Similar barbie box sequences were found in a variety of phenobarbital-inducible genes from several species, providing further evidence for the role of the barbie box in phenobarbital induction (28). Finally, in a cell-free system, induction of α_1-glycoprotein by phenobarbital was dependent on a barbie box sequence (31). These considerations suggest a remarkable evolutionary conservation of the mechanism of barbiturate induction.

As intriguing as such a hypothesis is, caution is required. There is little direct evidence for the role of the barbie box in mammalian regulation. The studies on α_1-glycoprotein cited above involved only a 50% increase in response to phenobarbital in a cell-free system, and although consistent with the hypothesis, are not compelling. The bacterial positively acting factors, although binding near the barbie box, do not appear to require the sequence for binding because they bind to mutated barbie sequences. The presence of similar sequences in the mammalian and bacterial genes may, therefore, be coincidental. As described below, researchers in several laboratories have analyzed the proximal promoter regions of *CYP2B1/2*, including the barbie box sequence, and have not been able to detect effects of phenobarbital on protein binding in this region. In three different transcription assays, constructions containing these proximal promoters did not respond to phenobarbital. In addition, peroxisomal proliferators have been shown to be potent inducers of BM3 by a mechanism similar to that of the barbiturates (32). The effectiveness as inducers was related to the presence of aliphatic side chains in the barbiturates. Because the peroxisomal proliferators are a class of mammalian inducers different from the barbiturates, the validity of the comparison of the bacterial and mammalian mechanisms for phenobarbital induction is reduced. Finally, the structure–activity relationships of barbiturates are different in the mammalian and bacterial systems, and the dose–response relationship is saturable in mammals, but not saturable to the limits of the solubility of the barbiturates in bacteria (7). Therefore, a role for the barbie sequence in phenobarbital induction in the mammalian genes remains problematic.

IV. Proximal Promoter

A. Proposed Phenobarbital-Responsive Elements

Investigations by Padmanaban and colleagues (33–37) have also provided evidence for phenobarbital-responsive sequences in the proximal promoter, some of which overlap with the barbie box described above. Studies in a cell-free transcription system derived from freeze-thawed rat liver nuclei demonstrated that transcription of a fragment of the *CYP2B1/2* gene (the sequence of *CYP2B2* and *CYP2B1* are identical in this region) from -179 to $+181$ was increased by treatment of the rats with phenobarbital (33). Similarly, transcriptional activity of this promoter fragment was dramatically increased in livers of animals treated with phenobarbital when the DNA was targeted to the liver by being complexed with asialoglycoprotein/poly(lysine) (34). Increased binding of the -179 to $+181$ fragment to rat liver nuclear proteins with an altered mobility was observed by gel retardation (33). Binding of an 85- to 95-kDa protein to the fragment detected by southwestern analysis was also increased after phenobarbital treatment. Footprint analysis indicated that the region from -54 to -89 was protected from DNase I digestion. The segment of DNA binding the phenobarbital-responsive protein was within -69 to -98 as determined by competitive gel-shift assays, and proteins of 42 and 39 kDa that bound to this fragment were detected by UV cross-linking (35). Deletion analysis indicated that progressive deletion from -116 to -75 decreased transcription in the cell-free system with nuclei from phenobarbital-treated rats. In standard cell-free transcription assays, using liver nuclei extracts, similar results were obtained and addition of oligonucleotides containing the -69 to -98 sequence decreased the amount of transcription product. In these last experiments, eight or more rather evenly spaced bands for transcription products were observed and the effects of phenobarbital were small so that interpretation was difficult. Nevertheless, the data are consistent with a positive element that mediates a phenobarbital response in the -69 to -98 region.

In similar studies, evidence was obtained suggesting that the region −160 to −127 contained a negative element (36). In freeze-thawed nuclei transcription studies, 5′ deletions from −160 to −116 appeared to increase activity after phenobarbital treatment. In transcription assays with nuclear extracts, transcription of the −179 to +25 template was increased when excess oligonucleotide containing the −160 to −127 region was added, consistent with titering of a negative factor. Protein binding to this oligonucleotide was also decreased in nuclei from phenobarbital-treated rats, and proteins of 68 and 44 kDa were detected by UV cross-linking.

Additional studies have been carried out to purify the proteins that bind to the positive and negative elements. By affinity chromatography to DNA containing the positive element, a 26- to 28-kDa protein was isolated from nuclear extracts from rats (34). In these studies three complexes were observed in binding studies with nuclear extracts, compared to two in earlier studies, and the purified factor formed a complex with the mobility of the fastest moving third complex. Interestingly, the same purified factor also bound to the negative element as determined by direct binding and competition for binding by the negative element. Treatment of the purified factor with kinase or phosphatase increased binding to the positive element, whereas treatment with kinase or phosphatase decreased or increased, respectively, binding to the negative element. The 26- to 28-kDa proteins could be phosphorylated by protein kinase A, and in nuclei from phenobarbital-treated rats, the 26- to 28-kDa proteins were phosphorylated more than in control nuclei. Addition of the purified proteins to cell-free transcription systems increased transcription from the −179 to +181 template and proteins from phenobarbital-treated animals were about twofold more active. A second factor of 65 kDa, which was isolated by affinity both to the positive element and heme, also increased activity in a cell-free transcription assay containing heme-depleted extracts from rat liver nuclei (37). This protein primarily formed the second of the three protein–DNA complexes observed with the positive element.

These studies have led to the proposal that the 26- to 28-kDa protein may bind to either the negative or positive element, with positive element binding favored by phosphorylation of the factor and negative element binding favored by dephosphorylation (34) (Fig. 2B). The model for regulation would then involve phenobarbital-mediated phosphorylation of the factor, which favors dissociation from the negative element and binding to the positive element. Further positive activity is mediated by binding of the 65-kDa protein to the positive element, and this binding is modulated by heme binding, which would be consistent with the decrease in response to phenobarbital observed in heme-depleted animals. A 94-kDa protein also interacts with the positive element in response to phenobarbital treatment and this increase is

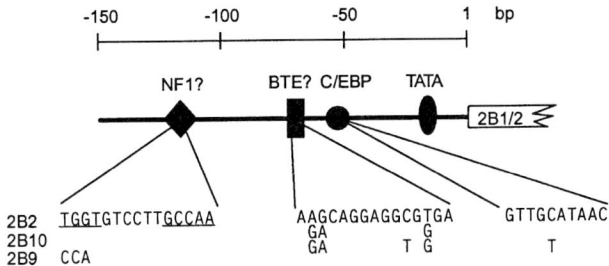
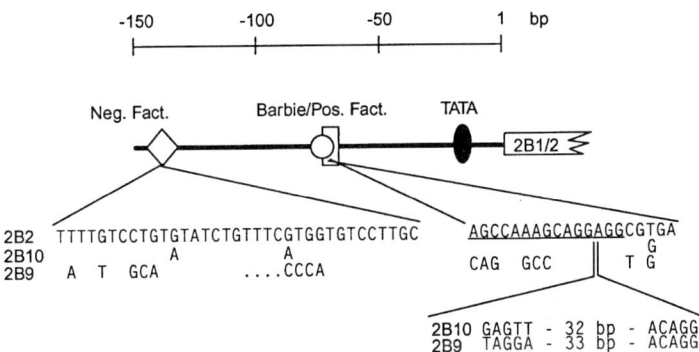

Fig. 2. Two views of the proximal promoter region of *CYP2B1/2*. Schematic diagrams of the promoter regions and possible regulatory factors are shown and the sequences of each element or region are shown for *CYP2B2*; differences in the sequences of *Cyp2b10* and *Cyp2b9* are indicated below the *CYP2B2* sequences. (A) Work from several laboratories (*38–41*) provides evidence for positively acting constitutive transcriptional factors; of those shown, only the C/EBP protein has been identified conclusively, and functional data exist for the basal transcription element (BTE) and C/EBP sites. These factors are proposed to form the core promoter necessary for basal and phenobarbital-induced expression, but are not directly responsive to phenobarbital treatment. (B) Work from the Fulco (*24*) and Padmanaban (*34*) groups supports a model in which factors in the proximal promoter mediate the phenobarbital regulation of *CYP2B1/2*. The barbie box sequence has been implicated in the regulation of P450s in *Bacillus megaterium* by genetic and functional studies, but its role in mammalian cells is largely based on sequence similarity. Because of a repeated sequence at the beginning and end of the 42-bp insertion in *Cyp2b9* and *Cyp2b10* relative to *CYP2B2*, the site of the insertion can be located as drawn, or in the middle of the barbie box, as reported (*43*). Because of the repeat the barbie box is largely conserved at the 5' end of the insertion as shown in B and the G-rich core of the BTE sequence is conserved at the 3' end of the insertion as shown in A. Reciprocal binding, mediated by phenobarbital, of the same factor to the negative-acting element and positive-acting element has been proposed to be the basis of phenobarbital induction of *CYP2B1/2* (*34*).

blocked if heme is depleted by administration of $CoCl_2$. This model is consistent with the data from the Padmanaban lab, but is largely based on cell-free transcription assays for the functional data, in which phenobarbital effects are small, and on protein binding studies. Neither is able to provide conclusive support for the model and there are several studies from other labs, described below, in which effects of phenobarbital on this region of the *CYP2B1/2* genes were not detected. The ability to direct the *CYP2B1/2* gene fragment to the liver by complexing it to asialyated glycoproteins and thus obtain dramatic stimulation by phenobarbital treatment (34) provides the means to obtain definitive evidence for this system. Specific mutagenesis of the regulatory elements, which eliminates the binding of the factors that have been identified, should correlate with decreased function of the proposed phenobarbital-responsive region.

B. Core Promoter Activity as the Function for Proximal Promoter Elements

There have been several other studies of proteins binding to the *CYP2B1/2* proximal promoter. Shephard *et al.* (38) analyzed binding by DNase I footprinting to the region from -1 to about -368. Within the region from -150 to -368, multiple regions of protein binding were observed, -330 to -336, -316 to -327, -289 to -308, -207 to -213, -173 to -189, and -156 to -174, which includes nearly 70% of the sequence in this region. The nature of the proteins that bind to these regions and the functional significance of the binding are unknown but the extensive protein binding may reflect the developmental, tissue-specific, or phenobarbital regulation of these genes. Of these sites, an oligonucleotide of the region from -183 to -199 was able to reduce cell-free transcription of the *CYP2B2* promoter by 30%, suggesting that a positive-acting factor may bind to this region. In both DNase I footprinting and gel-shift assays, protein binding to several of these sites was stronger with extracts from phenobarbital-treated animals, which suggests that they might play a role in the phenobarbital induction of the gene. The major concern with this conclusion is that rats were treated for 1 to 4 days with phenobarbital, so that it is not clear whether the changes observed are primary effects on gene induction or secondary effects. Transcription from the *CYP2B1/2* gene is increased within 6 hr after treatment with phenobarbital (9), so that any primary event mediating induction should occur within that time frame.

Regarding the *CYP2B2* proximal promoter region from -1 to -160, descriptions of results of protein binding studies from four groups differ from the results described above. Shephard *et al.* (38) described three footprints from -117 to -142, -48 to -66, and over the TATA region from about -10

to -35. Similar protein binding sites were reported by Luc *et al.* (*39*) at -116 to -129, -45 to -61, and -8 to -36; by Park and Kemper (*40*) at -119 to -138 and -45 to -64; and by Sommer *et al.* (*41*) at -116 to -143 and -47 to -61. Footprinting in the TATA region was not reported in the last two studies. The protein that binds to the -120 to -140 region has not been identified, but a sequence similar to the binding site of NF-1 is present in this region and this region competes weakly, compared to consensus binding sites, for C/EBP binding (*39*) (Fig. 2A). Similarly, the proteins binding to the TATA region have not been directly characterized but most likely are related to the TATA-binding protein and its associated proteins. Oligonucleotides containing sequences in these two regions competitively reduced transcription of the *CYP2B2* promoter in cell-free systems (*39*). Mutations in the NF-1-like element also reduced transcription.

The protein binding to the -45 to -65 region has been identified as a member of the C/EBP family of transcription factors. Oligonucleotides containing consensus sequences to the C/EBP binding site competed for the binding to this region in both footprint and gel-shift analyses (*39, 40*) and addition of either C/EBPα or C/EBPβ antisera resulted in supershifts of the DNA–protein complexes. Consistent with the fact that the liver contains multiple forms of C/EBP, which can form heterodimers (*42*), a complicated pattern of up to six protein–DNA complexes was observed. Not all were supershifted by the antisera to the α and β forms, thus other C\EBP forms in the liver presumably bind to this site as well (*40*). Oligonucleotides containing the -31 to -85 region increased cell-free transcription of the *CYP2B2* gene (*38*). Specific mutation of this region also decreased the transcription from *CYP2B2* promoters that were transfected into HepG2 and FGC4 hepatoma cells by 70 and 80%, respectively (*39, 40*). Coexpression of C/EBPα or C/EBPβ markedly stimulated the expression of the *CYP2B2* promoter in HepG2 and C33 cells and the increase was much smaller for promoters with mutations in the C/EBP binding site (*39, 40*). These experiments provide definitive evidence that the -45 to -68 region is a functional C/EBP site that increases transcription of the gene. In three of the studies in which phenobarbital treatments of animals were for 4 or 6 hr, no changes in the footprints in this region were observed (*39–41*). In one study, after treatment for 1 to 4 days, binding to the C/EBP site was increased (*38*). Again, because induction of the gene occurs by 6 hr, it is not clear whether the changes observed are primary effects of phenobarbital treatment. The complexity of C/EBP forms provides the potential for phenobarbital-dependent modifications of these factors that could affect transcription. There are, however, no direct functional data supporting a role for this factor in phenobarbital induction and, as discussed below, the available functional data are not consistent with a role for this element in phenobarbital induction.

In these studies, DNase footprint analysis did not reveal binding of proteins in the region containing the barbie box (-73 to -89) and the proposed phenobarbital-dependent positive element (-69 to -98) with either extracts from control or from phenobarbital-treated animals (38–41). In some studies, there were hypersensitive sites in these regions, which might indicate some protein binding (38, 39, 41). The inconsistencies among the data may be explained by different procedures used to isolate the nuclear factors and the conditions of the protein binding assays. Negative results are inherently weaker than positive results, but have been reported in several recent studies in contrast to the earlier reports of phenobarbital-dependent protein binding to the barbie box or within the positive element region, as discussed above.

Similar studies of the proximal promoter of the mouse *Cyp2b10* proximal promoter revealed protein footprints at -45 to -64, -154 to -180, and -215 to -235 (43). Correcting for the 42-bp insertion at about -80 in this gene relative to *CYP2B2*, the last two binding sites correspond to the C/EBP and NF-1 sites of *CYP2B2*. The sequences of these sites are well conserved in the two genes (Fig. 2). As in the rat gene studies, no binding to the barbie box region was observed and no differences in binding were observed with extracts from untreated or phenobarbital-treated mice.

There is general agreement that there is a positive element within the region -67 to -89. Cell-free transcription studies of progressive deletion mutations suggested that this region contains a positive element (35, 44). Similarly, deletion of sequences from -110 to -57 resulted in 90% reduction of transcription of the *CYP2B1/2* genes in HepG2 cells (40). The reduction was partially due to the deletion of the C/EBP site described above. A second candidate for a regulatory element in the region is a G-rich sequence that is similar to the basal transcription element (BTE) described for *CYP1A1* (45). In *CYP1A1*, this element binds Sp1-like proteins, including a specific protein, BTEP (46). Mutation of the BTE element reduced transcriptional activity of *CYP2B1* and *CYP2B2* in transfected HepG2 cells or by *in situ* transient transfection about 70%, which was about the same as the decrease when the C/EBP site was mutated (21, 40). In contrast, two independent mutations of the barbie box did not reduce expression. Although binding to the BTE was not detected by DNase I footprinting, protein binding to an oligonucleotide containing the BTE-like sequence was observed in gel-shift assays (40). These results indicate that the major positive element in the -69 to -98 region is the BTE element.

There are, therefore, two views of the role of the proximal promoter (Fig. 2). In one case, evidence from protein binding studies and mainly cell-free transcription functional assays suggests that elements in the proximal promoter region play a primary role in the response to phenobarbital. Reciprocal binding of proteins to either positive or negative elements mediated by phenobarbital is proposed to underlie the regulation. The second view is that the elements in the proximal promoter are mainly involved in the basal expression of the gene and bind constitutive factors that are present in the liver. These elements may be required for activation of the gene, i.e., for any expression, but do not directly mediate the phenobarbital response. Comparison of sequences between species, which might reveal conservation of critical sequences, is not very informative. All of the proposed elements are well conserved in the highly inducible *CYP2B2* and *Cyp2b10*. There is a 42-bp insertion in *Cyp2b10* and *Cyp2b9* compared to *CYP2B2* within the proposed phenobarbital-dependent positive element. However, this insertion contains a 4-bp repeat at the beginning and end so that the barbie box is largely conserved at the 5′ end of the insertion and the BTE region is conserved at the 3′ end. The less highly induced (47), or uninducible (48), *Cyp2b9* has substantial differences in the NF-1 site, the barbie box, and both the positive and negative phenobarbital-dependent regions. Thus, only the BTE core and the C/EBP sequence are conserved in all three genes. Most recent data from functional analysis of the genes favor the elements in the proximal promoter being part of the core promoter, but a role for the proximal promoter in phenobarbital induction has not been conclusively eliminated.

V. Distal Phenobarbital-Responsive Enhancer

A. Chicken *CYPH1* Gene

In contrast to the controversy concerning the proximal promoter, there is general agreement that sequences 5′ distal to the proximal promoter contribute to phenobarbital responsiveness. The initial evidence for upstream phenobarbital-responsive elements came from studies of the transcription of the chicken *CYP2H1* gene in chicken embryo hepatocytes in primary culture (49). In these experiments, about twofold induction by phenobarbital was observed with a *CYP2H1*–CAT construct that contained 8.9 kb of 5′ flanking region. Progressive deletion from −4.7 kb to −1.1 kb or deletion of the region from −5.9 to −1.1 eliminated the induction. Further, the region from −5.9 to −1.1 was able to confer phenobarbital responsiveness to a heterologous SV40 promoter and inductions of 14-fold were observed with the *CYP2H1* fragment in either orientation relative to the promoter. Further characterization of the specific elements involved have not been reported. In this chicken gene, thus, the proximal promoter region to −1.1 kb was not able to mediate phenobarbital induction and distal elements were required. The capability of the distal sequence to confer phenobarbital induction on a heterologous promoter provides strong evidence that this region was a primary target of phenobarbital action and not a general enhancer that was required to amplify phenobarbital effects mediated at proximal promoter elements.

B. Transgenic Analysis of the *CYP2B2* Gene

Studies in transgenic mice have also indicated that distal elements are required for phenobarbital induction of the *CYP2B2* gene (22). In these experiments, the transgenes contained the entire structural gene for *CYP2B2* and either 800 bp or 19 kb of 5' flanking region. Two transgenic colonies were founded for each construction. Expression was detected by either Northern analysis or RNA slot blotting. The constructions with only 800 bp of 5' flanking region were expressed at relatively high levels in the transgenic mice equivalent to 50 to 300% of endogenous expression of *CYP2B2* in rats when normalized to 18S ribosomal RNA. Expression was not increased by phenobarbital treatment. Expression of this construction was highest in the kidney, about twice that in the liver, and very low expression was observed in the testes. In contrast, expression of *Cyp2b9*, a mouse gene closely related to *CYP2B2*, was detected primarily in the liver, with low levels in the lung, after phenobarbital treatment. In the two transgenic lines containing 19 kb of 5' flanking sequence, expression was restricted to the liver as detected by RNA slot blots of 10 tissues. Expression of the transgenes in untreated mice was undetectable in one transgenic line and was detected at a low level in the other line. In phenobarbital-treated mice these transgenes were expressed at levels about 50% of the endogenous gene expression in rats. The high level of expression of these transgenes, similar to levels of the endogenous rat gene, provides confidence that the regulation of the genes is not dependent primarily on the site of insertion and neighboring regulatory elements but is controlled by the *CYP2B2* regulatory elements. The results establish that the proximal promoter region is not sufficient for phenobarbital induction of the *CYP2B2* gene nor for absolute tissue-specific expression in the liver. However, this region does provide substantial tissue specificity, because in two independent transgenic strains, it was expressed at high levels only in the liver and kidney and not in eight other tissues. Distal elements, upstream of -800, are required to restrict expression to the liver and for a response to phenobarbital.

These transgenic results are most consistent with the model for gene regulation in which there is a distal phenobarbital-responsive element and the proximal promoter elements are part of the core promoter required for expression of the gene at either basal or induced levels. There are some caveats to this conclusion. The most critical is that the distal element could be a general enhancer region required for high level expression, but that the phenobarbital switch or responsive element is in the proximal promoter. Another concern is that the copy numbers of the −800-bp constructions were 20 and 8 compared to 1 and 3 for the larger constructions. High copy numbers of genes may titrate critical regulatory factors so that improper regulation of the gene is observed. The small 5' flanking region in the −800 constructions may provide less "protection" against neighboring mouse genome enhancers, which could override the regulation by elements in the *CYP2B2* transgene. Such position effects are evident even in the two larger constructions in which expression in the untreated animals differs. The very similar properties of two transgenic lines for each construction argue against major position effects and the levels of expression near that of the endogenous gene argue against effects dependent on copy number, but not conclusively. To distinguish a general enhancing effect from a specific response to phenobarbital for the distal enhancer requires experiments utilizing the distal element with a heterologous promoter that does not respond to phenobarbital. As described above, this has been demonstrated with the *CYP2H1* gene and studies described below establish that a distal element of *CYP2B2* can confer phenobarbital inducibility to a heterologous promoter.

C. DNase I Hypersensitivity of the *CYP2B1/2* Genes

The transgenic studies described above indicate only that a phenobarbital-responsive element is between −800 and −19,000 bp. A study of the DNase I hypersensitivity of the *CYP2B2* gene in nuclei provided indirect evidence that defined more precisely a potential position for a distal phenobarbital-responsive region (39). Regions of hypersensitivity usually reflect binding of regulatory proteins to the region and disruption of the normal chromatin structure (50). These experiments utilized indirect end-labeling Southern techniques with the probe in the first exon. In spite of the facts that this probe cross-hybridized with the *CYP2B1* and *CYP2B2* genes and that there are multiple copies of these genes (51), two very distinct hypersensitive sites were observed in both untreated and phenobarbital-treated samples. The first was in the proximal promoter region and the second was a distal region from −2.2 to −2.3 kb. The distal region appeared to be slightly more sensitive to DNase I by being more prominent at lower concentrations of DNase I treatment. At the highest concentrations of DNase I reported, the proximal site was more prominent than the distal site in phenobarbital-treated samples and the reverse was true for the untreated sample. From this, it was concluded that the sensitivity increased for the proximal region after phenobarbital treatment and the distal site became more resistant. However, this is not apparent from the patterns at lower concentrations of DNase, and the decrease in the distal site fragments at higher DNase concentrations may be due to increased cleavage at the proximal sites. Cleavage at both distal and proximal regions in one DNA molecule would be detected only as the shorter fragment cleaved in the proximal region. These data show as expected that the hypersensitivity of the proximal promoter is related to the transcriptional activity of the gene and that hypersensitivity of the distal region may or may not be increased in response to phenobarbital.

Two very interesting conclusions may be drawn from these experiments. The first is that changes in the hypersensitive sites occur in response to phenobarbital, and the second is that under conditions in which the genes are not transcriptionally active, the chromatin structure of both the proximal region and the distal region is already disrupted. Presumably regulatory proteins are bound to these regions but are in a transcriptionally inactive form. Alternatively, the hypersensitive sites in the untreated samples may be provided by members of the *CYP2B* family that are consitititutively expressed (*52*). Interestingly, two hypersensitive sites at similar positions were detected in *CYP2C* genes that respond to phenobarbital, but the distal site was absent in *CYP2C3*, which does not respond to phenobarbital (*53*). The functional significance of the *CYP2C* hypersensitive regions has not been determined.

D. Localization of a Distal Enhancer of the *CYP2B2* Gene by Transfection into Primary Hepatocytes

A major advance in the studies of phenobarbital regulation of P450 genes was the development of culture conditions that maintained *CYP2B1/2* expression and response to phenobarbital (*17–19*). Prior to this advance, cell-free transcriptional assays had been the principal functional assay for expression of P450s, and these assays are notoriously difficult for the study of regulation by distal elements. Transfection of primary cultures of rat hepatocytes with about 4.5 kb of *CYP2B2* 5' flanking region fused to the CAT reporter resulted in about a fivefold induction after treatment of the cells for 48 hr with phenobarbital (*20*). Deletion to -2.5 kb did not decrease the induction, but deletion to -2.0 kb eliminated the induction. Internal deletion of the -2.5 to -1.7 fragment also eliminated the induction; however, induction was retained when the -1.7 to -0.2 kb region was deleted. A fragment from -2318 to -2155 was able to confer inducibility of 3.5-fold on the promoter-containing sequence to -1.7 and the induction was about doubled when two copies of the small fragment were used. In a construction containing the viral thymidine kinase promoter, the fragment from -2318 to -2155 conferred phenobarbital responsiveness if fused immediately upstream in either orientation or if fused at the 3' end of the CAT reporter gene in either direction. These important experiments provided the first functional data supporting a role for a distal element in the induction of *CYP* genes by phenobarbital. The presence of the functional element in the region -2.2 to -2.3 kb, which coincides with the DNase I-hypersensitive site in the gene, provides strong evidence that this region contains a phenobarbital-responsive unit (PBRU). The term "unit" is used rather than "element" because several elements are involved in the response (see below).

In other studies in primary cultures, results consistent with induction mediated by the PBRU have been reported. In rat hepatocyte cultures, a three- to fourfold induction of *CYP2B1* gene was observed when a 7-kb 5' flanking region fused to the chloramphenicol acetyltransferase gene was used (54). In rat hepatocytes cocultured with liver epithelial cells, 3.5- to 4.5-kb 5' flanking regions of the *CYP2B1* and *CYP2B2* genes did not respond to phenobarbital, but the 163-bp PBRU conferred induction of two- to threefold (55).

E. Analysis of *CYP2B2* by Transfection of Liver *in Situ*

A second method for assay of *CYP* regulatory elements was suggested by the demonstration that DNA injected directly into the liver of a rat *in situ* was taken up into hepatocytes and expressed (56). Although such direct injection techniques had been successful for tissues such as muscle, the liver was more resistant to uptake of DNA (57). Successful results of expression in liver of vectors containing luciferase reporters and either promoters from cytomegalovirus (CMV) or Rous sarcoma virus (RSV) were shown. Maximal expression was observed within 24 hr of injection, and by 3 days activity was very low but stable. The injected DNA remained in a low-molecular-weight state and was not detected in genomic DNA. This transient expression is similar to that obtained by transfection of mammalian cells and, therefore, may be referred to as *in situ* transient transfection. To obtain high levels of expression in these studies, rats were treated with dexamethasone (56). The rationale for using dexamethasone was that histological studies showed that acute inflammatory infiltrates were present along the needle tracks at the injection site in the liver, which might be detrimental for the cells in the area that had taken up DNA. Treatment with dexamethasone greatly decreased the inflammation and increased transcriptional activity by about 10-fold. However, dexamethasone treatment increased the transcription of the RSV and CMV promoter by four- to sevenfold in transfected cultured primary hepatocytes as well. The effects of dexamethasone then, at least partly, are direct effects on transcription and the antiinflammatory action may contribute to the increased expression, but this is not clearly established.

In situ transient transfection has been successfully used to assay transcriptional activity of *CYP2B* and *CYP2C* promoters and phenobarbital induction (21). Because this assay is not a standard method of promoter analysis, it will be described in more detail. Male Sprague–Dawley rats weighing 250–350 g were used. At 24 hr before surgery, rats were injected subcutaneously with 1 mg/kg body weight dexamethasone. The rats were anesthetized with ether by inhalation and a ventral midline incision was made;

the incision is large enough so that the largest lobe of the liver can be externalized. Between 350 and 400 µg of plasmid DNA was then injected directly into three sites in the liver in a volume of 1.5 ml of culture media with a 25-gauge, $\frac{5}{8}$-inch needle. The needle was inserted in the central part of the lobe and toward the margin of the lobe. It was left in place for about 1 min after each injection, and bleeding, if any, was controlled by direct pressure. The liver lobe was then internalized and the incision was sutured. The rats were then given a second injection of dexamethasone. Phenobarbital was administered intraperitoneally at this time as well. After 24 hr, the rats were euthanized by CO_2 inhalation and the livers were exposed. The injection sites on the injected lobe are visible and the part of the lobe from the central side of the needle marks to the edge of the lobes was excised. The liver fragment was minced and homogenized with a tight pestle in a Dounce homogenizer in 1.5 ml of buffer containing 1% Triton X-100. The homogenate was centrifuged in a microfuge and luciferase activity in the supernatant was assayed.

The *in situ* transfection assay has the advantage of assaying transcription of genes in hepatocytes, which are within the normal architecture of the liver and, thus, have the normal interactions with other liver cells and the extracellular matrix. The importance of this is illustrated by the rapid loss of expression of phenobarbital-inducible genes in primary cultures of hepatocytes that are grown on standard plastic dishes and medium (58). Either specialized medium or the addition of extracellular matrix is required for the persistent expression of these P450 genes and for a response to phenobarbital (17–19). Although primary hepatocyte cultures are effective models for phenobarbital induction, the possibility remains that some of the cellular interactions within the intact liver may not be accurately reproduced in the primary hepatocytes, which might affect the concentrations or activities of some regulatory factors. It is clear, for example, that the relative importance of regulatory elements in genes assayed in cultured cells is often different from that determined for the same gene in transgenic animals (59). The *in situ* injection assay offers an intermediate between these two types of assays. It is not yet clear whether the *in situ* assay system will provide information different from transfection of cultured cells, because both are transient transfections, in contrast to the transgenic animals. The limited amounts of data available from the liver *in situ* assay have generally been consistent with the results of the transfection of primary cultures of hepatocytes.

There are also disadvantages to the *in situ* assay system. The transfection is very inefficient, so that 300–500 µg of DNA is required for injection into a single rat liver. In addition, the processes of injection, homogenization, and determination of luciferase activity are relatively crude, leading to substantial variation in the results, i.e., variations of twofold under the best conditions. The variability requires that a number of animals be injected for each construction assayed or for each treatment protocol. A minimum of three animals is required for each treatment group to allow statistical analysis. Generally six to nine animals per treatment group are used, which permits reliable detection of changes of two- to threefold or greater. This requires on the order of 5 mg of plasmid DNA for each experiment. Because of the low efficiency of the transfection, only the luciferase assay is sensitive enough to detect expression. Thus, a second reporter cannot be used as an internal standard to reduce the variation. Substantial increases in the efficiency of the transfection would greatly improve this assay. It is possible that promoters that are highly active in liver, combined with liver-specific enhancers, may provide expression so that a second reporter can be used as a control. A second approach may be to increase the uptake of the DNA by coinjection of viral particles. Replication-defective adenovirus or adenovirus coats have been shown to increase the transfection efficiency of cultured cells by 100-fold or more (60, 61). Coadministration of DNA and defective adenovirus into salivary glands resulted in over an 100-fold increase in uptake and expression of the DNA (62). Chemiluminescent assays of β-galactosidase have been reported to be similar in sensitivity to the luciferase assays (63). There is, therefore, promise that more efficient *in situ* transfection assays can be developed so that less DNA is required and variability can be reduced by inclusion of an internal standard.

The *in situ* assay has been used to analyze the response of *CYP2C1* and *CYP2B2* genes to phenobarbital (21). Injection of DNA containing up to -3500 bp of *CYP2C1* and -1400 bp of *CYP2B2* fused to the luciferase promoter resulted in expression at a low level, but no response to treatment for 24 hr with 100 mg of phenobarbital/kg body weight. Expression was slightly inhibited by phenobarbital treatment, an effect that has been seen consistently. The activity of the *CYP2B2* promoter was about threefold greater than that of *CYP2C1*. The results of the *CYP2C1* promoter were disappointing because this gene has a hypersensitive site at about -2.5 kb, as was observed in the *CYP2B2* region. To determine if the *CYP2B2* fragment from -2318 to -2155, which mediated a phenobarbital response in primary hepatocytes, could do the same in the *in situ* injection assay, one to three copies of this region were inserted in front of the proximal promoters of either *CYP2C1* (-272 to $+1$) or *CYP2B2* (-110 to $+1$). For the heterologous proximal *CYP2C1* promoter, one copy of the PBRU resulted in about a 5-fold induction and three copies resulted in about 15-fold induction. Likewise with the homologous promoter, a single copy of the PBRU resulted in a 2.5-fold increase and two or three copies resulted in 4- to 5-fold increases. These results confirmed the studies in the primary hepatocyte cultures that this sequence distal from the promoter could mediate phenobarbital induction.

The ability of the PBRU to confer phenobarbital induction on heterologous promoters established that this region was a phenobarbital-responsive element and not just a general enhancer amplifying phenobarbital-dependent activation by elements in the proximal promoter. Nevertheless, because a role in phenobarbital induction had been proposed for proximal promoter elements, it was important to determine if the PBRU effects were dependent on any of the proximal promoter elements in *CYP2B1*. Mutations were made in the positively acting C/EBP and BTE elements (*40*), as well as the barbie box (*28*), and responsiveness to phenobarbital was assessed in constructions containing three copies of the PBRU fused to the mutated proximal promoters. Mutation of the BTE and C/EBP regions each reduced the expression of both the basal expression and the expression after phenobarbital treatment by about 70 to 75%, so that approximately the same fold induction was retained. The decrease in activity was similar to the decreased transcriptional activity of these mutated promoters when assayed by transfection into HepG2 cells (*40*). The elements, thus, behave as positively acting elements of the core promoter independent of phenobarbital treatment. When the first barbie box mutation unexpectedly resulted in increased expression of two- to threefold, a second mutation was made to eliminate the possibility that a positively acting element had been introduced by the mutation. Both mutations behaved similarly, with a small increase in transcription in untreated animals and responses to phenobarbital similar to that of wild type. These results are most consistent with the distal PBRU playing the primary role in phenobarbital induction independent of proximal promoter elements. Within the resolution of this assay, the proximal promoter elements appear to have a negligible role in the response to phenobarbital.

To better identify which sequences in the 163-bp PBRU were critical for the phenobarbital induction, a series of deletion mutations were analyzed (*63a*). If single copies of the mutated PBRU regions were analyzed, deletion of about 60 bp from the 5' end to an *Nco*I site at -2258 did not substantially reduce induction by phenobarbital, but deletion to -2231 did eliminate induction (Fig. 3). From the 3' end, deletion of 15 bp to -2170 did not reduce phenobarbital induction, but deletion to -2194 did eliminate induction. A fragment from -2258 to -2170 responded normally to phenobarbital treatment. From these studies with single copies of the PBRUs, this 88-bp fragment appears to represent the minimal phenobarbital-responsive unit. However, when three copies of the deleted fragments were analyzed, fragments, deleted to -2231 from the 5' side or -2194 from the 3' side, exhibited a response to phenobarbital only slightly reduced from wild type. These unexpected results suggest that a 37-bp segment from -2231 to -2194 retained elements required for the response to phenobarbital. However, a fragment containing three copies of the 37-bp fragment did not confer phenobarbital induction. This central region, therefore, requires an additional sequence

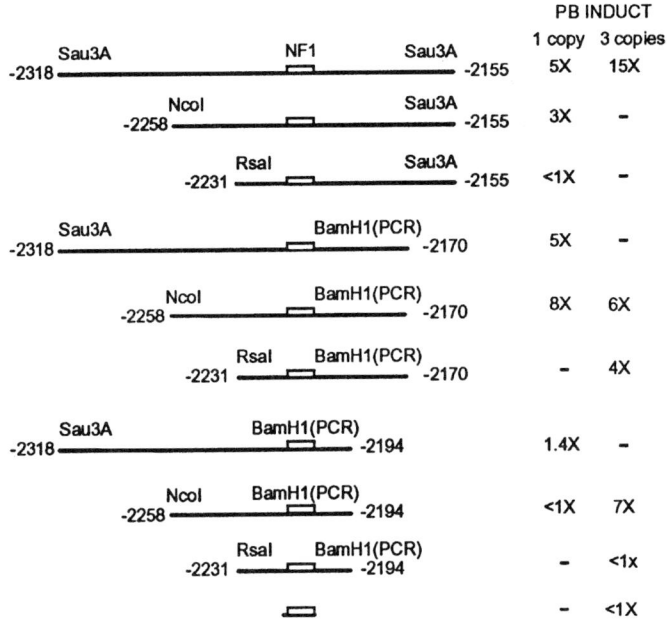

FIG. 3. Analysis of deletions of the *CYP2B2* PBRU by *in situ* transient transfections. Fragments of the 163-bp *Sau*3A fragment were fused as single copies or triple copies to the basal *CYP2C1* promoter and assayed by direct injection into rat liver *in situ*. Basal activities were similar for all constructions and the fold induction after phenobarbital treatment is indicated. Restriction sites used to create the deletions are indicated; the 3' *Bam*HI sites were created with primers for PCR amplification of the fragments.

from either side for phenobarbital induction, suggesting there are redundant elements in the sequence on either side. The explanation for the differences in the results with the single and triple copies is not clear, but may be the result of redundant elements in which deletion of one element required for activity of a single copy of the PBRU may be compensated for by multiple copies of a remaining element when three copies of the PBRU are assayed.

To further define specific regions important for the induction by phenobarbital, a linker scanning mutagenesis analysis, in which a *Kpn*I site was substituted for 6-bp blocks, was performed across the fragment from −2258 to −2170, which was the minimal active fragment in the assay using single copies of the mutated PBRUs (Fig. 4). An NF-1 site is present in the *CYP2B2* gene, which is the major protein-binding region of the segment by both DNase I footprinting and gel-shift assays (20; B. Kemper, unpublished). In this region, rather than linker scanning mutants, each of the two bipartite binding sequences and the intervening sequence were mutated separately

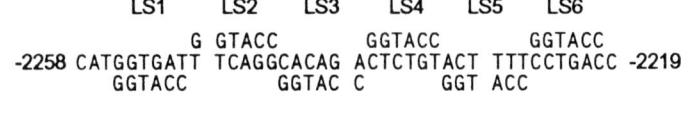
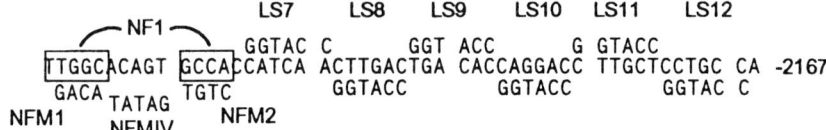

FIG. 4. Mutations introduced into the *CYP2B2* PBRU. *Kpn*I sites were introduced in a linker scanning analysis from −2258 to −2167, except for the NF-1 region, in which specific mutations were made in the two bipartite recognition sites (NFM1 and NFM2) and the intervening sequence (NFMIV). The mutants were assayed in the *in situ* transfection assay and the results are given in Table I.

(Fig. 4). Each of the mutants was assayed as a triple copy fused to the *CYP2C1* proximal promoter. Perhaps the most surprising result of this analysis is that none of the mutations eliminated the response to phenobarbital completely, although two regions clearly reduced the response (Table I). One of these was linker scanning mutants 4 and, particularly, 5, which reduced expression in phenobarbital-treated animals. The other region was the NF-1 binding site mutants. In addition, somewhat higher basal levels, but little effect on expression after phenobarbital treatment, were observed for linker scanning mutant 9 and activity in phenobarbital-treated animals was reduced in mutants 8, 10, and 12. The continued phenobarbital responsiveness for all of the mutants may result from incomplete inactivation of the critical element, however, the minimum number of changes was four for each 6-bp linker scanning mutant. Alternatively, there may be redundant constitutive and phenobarbital-responsive elements that can compensate for each other when three copies of the mutated PBRUs are assayed, the same explanation as for the effect of copy number on deletion mutants. There are several candidates for repeated sequences in the PBRU. As indicated in Fig. 5, the steroid hormone half-site AGGTCA is repeated three times and a 12-bp segment in the critical NF-1 region is repeated.

Mutation of either or both of the bipartite sites of the NF-1 motif substantially reduced the expression in phenobarbital-treated rats (Table I). In contrast, deletion of the sequence between the two sites, which is not specific for NF-1 binding, did not affect expression. The strongest binding of the proteins to the PBRU occurs at the NF-1-like sequence. Consensus NF-1 sites were able to compete for the binding of liver nuclear proteins to the frag-

TABLE I
Effects of Mutations in the PBRU Region
on Induction by Phenobarbital

Mutation[a]	Luciferase activity[b]		Fold induction
	Basal	Phenobarbital	
LS1	3.5 ± 0.6	15.6 ± 4.9	4×
LS2	2.1 ± 0.4	25.8 ± 7.8	12×
LS3	4.1 ± 1.0	37.0 ± 8.6	9×
LS4	3.1 ± 0.7	12.0 ± 2.8	4×
LS5	1.8 ± 0.3	8.9 ± 1.3	4×
LS6	3.9 ± 0.5	27.0 ± 7.4	7×
LS7	4.9 ± 0.8	18.0 ± 4.0	4×
LS8	4.1 ± 1.4	13.8 ± 3.4	3×
LS9	7.8 ± 2.0	28.8 ± 10.6	5×
LS10	2.0 ± 0.6	13.9 ± 3.0	5×
LS11	3.8 ± 0.7	20.5 ± 5.0	7×
LS12	1.6 ± 0.2	14.0 ± 2.1	9×
NFM1	1.8 ± 0.3	7.6 ± 1.4	4×
NFM2	2.4 ± 0.5	8.9 ± 1.5	4×
NFMIV	1.6 ± 0.3	23.7 ± 5.9	15×

[a]Either linker scanning (LS) mutations or specific mutations in the NF-1 region (as illustrated in Fig. 4) were introduced into the 163-bp fragment containing the PBRU.
[b]Total luciferase activity in arbitrary light units per injected liver × 10^{-3}. Three copies of each mutanted PBRU were inserted before the *CYP2C1* proximal promoter and transcriptional activity was determined by the *in situ* transient transfection assay.

ment, and antisera to NF-1, but not C/EBP, supershifted the protein–DNA complexes (63a). These NF-1 elements are, therefore, capable of binding NF-1. These data indicate that NF-1 plays a key role in the amplitude of the phenobarbital response. NF-1 is a constitutive liver factor, so it probably is not directly responsible for the response, which is consistent with the continued phenobarbital response when it is deleted. There are four NF-1 genes (64) and multiple variants produced from the genes (65), so it is possible that subtle differences in the sequences flanking the consensus NF-1 site might affect the relative binding of the different forms and that binding or activity of the favored factor is affected by phenobarbital treatment.

F. Comparison of the *Cyp2b9* and *Cyp2b10* PBRUs by *in Situ* Transfection

In mice, the gene *Cyp2b10* is orthologous to the rat *CYP2B1/2* genes, and a second closely related gene, *Cyp2b9*, is present. Both of these genes are responsive to phenobarbital, with *Cyp2b10* having the larger response (47) in

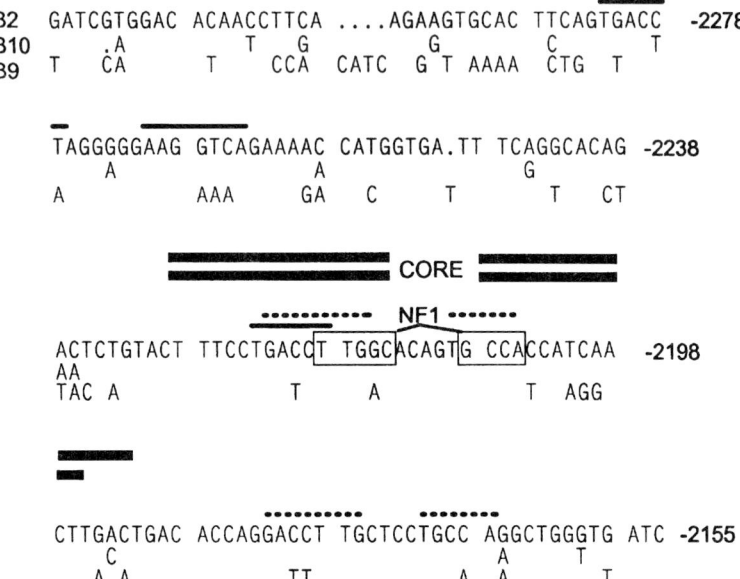

FIG. 5. Comparison of the PBRU sequences of CYP2B2, Cyp2b10, and Cyp2b9. The sequence of CYP2B2 (2B2) is shown at the top and differences in the sequences of Cyp2b10 (2B10) and Cyp2b9 (2B9) are shown below. The position of the NF-1 site is indicated and repeats of the AGGTCA sequence are indicated by a solid line above the sequences; a repeat of the critical sequence near the NF-1 site is indicated by the broken line. The 33-bp region identified by Honkakoski and Negishi (67) as the core phenobarbital-responsive region in Cyp2b10 is indicated by the lower very heavy lines extending from CORE; the upper very heavy lines demarcate the limits by 5' and 3' deletions of CYP2B2 for phenobarbital activity in the *in situ* transfection assay (see Fig. 3). Although the core structure is necessary for phenobarbital induction, it is not sufficient because multiple copies of the core regions do not confer inducibility. The sequences of Cyp2b9 and Cyp2b10 were derived from fragments isolated by PCR (I. Rivera-Rivera, H. Li, and B. Kemper, unpublished). The Cyp2b10 sequence is identical to that reported by Honkakoski and Negishi (67). The Cyp2b9 sequence differs from that reported in the GenBank database (M60267) at the positions numbered according to the CYP2B2 sequence [−2301]T, [−2246]A, and [−2129]delete T; and from the sequence reported by Honkakoski and Negishi (67), at [−2301]T, [−2273]A, [−2254]T, [−2252]C, [−2246]A, [−2243]G, and [−2242]T.

C57BL/6 mice. In some strains of mice Cyp2b9 is not responsive to phenobarbital (48, 66). To determine if a PBRU analogous to that in the rat gene was present in the mouse genes, the corresponding regions were isolated from mouse genomic DNA by polymerase chain reaction (PCR) (I. Rivera-Rivera, H. Li, and B. Kemper, unpublished). The Cyp2b9 PBRU sequence is at about −800 because there is a large deletion in the 5' flanking region relative to *CYP2B2*, and the *Cyp2b10* sequence is about −2.2 kb, as in *CYP2B2*. The *Cyp2b9* sequence is 83% similar to the *CYP2B2* PBRU region compared to 60 to 70% in other 5′ flanking regions, and the *Cyp2b10* sequence is 91% similar compared to less than 85% in neighboring 5′ flanking region (67; I. Rivera-Rivera, H. Li, and B. Kemper, unpublished). Three copies of the *Cyp2b9* sequence, equivalent to the 163-bp PBRU fragment of *CYP2B2*, were fused to the *CYP2C1* proximal promoter. Little or no response to phenobarbital treatment was observed. In contrast, three copies of the *Cyp2b10* sequence responded to phenobarbital to the same extent as the *CYP2B2* sequence. The *Cyp2b10* sequence has only three differences in the region equivalent to −2258 to −2170 of *CYP2B2;* the NF-1 site, the linker scanning 4/5 region, and the linker scanning 8/9 region, which contains an AP1-like site, are identical (Fig. 5). In contrast, the *Cyp2b9* sequence has critical changes in one of the NF-1 binding sites, immediately 5′ and 3′ of the NF-1 site, within the linker scanning 4 region and linker scanning 8/9 region. The NF-1 change and one or more of the other changes must contribute to the loss of phenobarbital responsiveness.

G. Analysis of the *Cyp2b9* and *Cyp2b10* PBRUs in Transfected Primary Hepatocytes

Negishi and colleagues have characterized phenobarbital induction of these two mouse genes using primary hepatocyte cultures for functional assays (43, 67). In initial studies, for *Cyp2b10*, two- to threefold induction was observed with 5′ flanking regions extending to −1400 and −4300 (43). Deletion of additional sequence to −971 resulted in loss of response to phenobarbital with little effect on the basal activity. The argument that the region contains a phenobarbital-responsive element was reduced by the inability of a fragment from −104 to −971 to confer phenobarbital inducibility to a heterologous promoter. In addition, the mouse sequence in this region is almost identical to the rat *CYP2B2* sequence and both contain an AGGTCA steroid half-binding site, proposed as a critical element (43). Constructions containing this region in *CYP2B2* did not exhibit phenobarbital responsiveness in primary hepatocytes or by *in situ* transfection (20, 21), although phenobarbital induction by a −1400 *CYP2B2* fragment was reported in transfected Fao cells (16). The mediation of phenobarbital induction by this region was confirmed in a later study and similar protein complexes were formed with this region and the *Cyp2b10* PBRU equivalent, as discussed below (67). This region may contribute to the overall induction of the *Cyp2b10* gene, but evidence for its role in the induction of *CYP2B2* is less clear. The lack of phenobarbital induction with *Cyp2b10* fragments to −971, which include the proposed phenobarbital-responsive elements in the proximal promoter, is also consistent with the studies on the *CYP2B2* gene in which the proximal promoter did not confer phenobarbital responsiveness.

In later studies, Honkakoski and Negishi (67) observed that deletion of the 5' flanking sequence of *Cyp2b10* from −4.4 to −2.4 kb resulted in an increase in induction from about three- to eightfold by TCPOBOP, a phenobarbital analog (67). In this case, further deletion to −1850 resulted in reduction of induction to about threefold. The deleted sequence between −2.4 and −1.8 kb contains the *Cyp2b10* sequence, analogous to the *CYP2B2* PBRU. DNase I footprinting was used to define potential regulatory protein-binding sites. In the region from −2428 to −2250, three weakly protected regions and three relatively strongly protected regions were observed and no differences between phenobarbital and control extracts were detected. A 177-bp fragment, −2426 to −2250, similar to the 163-bp fragment originally characterized for *CYP2B2*, was able to confer phenobarbital induction and various fragments of this 177-bp sequence were assayed for their activity. Fragments with 5' end points of −2364 or greater retained phenobarbital induction but the induction was reduced compared to fragments beginning at −2394. Fragments with 3' end points of −2297 or smaller retained phenobarbital induction, although the induction was reduced compared to those with end points at −2265. These results indicated that a 67-bp region from −2364 to −2297 was the minimal fragment retaining phenobarbital induction, in good agreement with the 88-bp *CYP2B2* fragment that retained activity when single copies of PBRU fragments were assayed by *in situ* transfection.

As described above, the *Cyp2b9* region corresponding to the *CYP2B2* PBRU differs in sequence at several potentially critical regions of the PBRU. As with the *in situ* injection assay, the *Cyp2b9* region was not capable of conferring phenobarbital inducibility to the tk promoter (67). The region from −2332 to −2300, which was shown in the deletion analysis to be critical for phenobarbital induction, contains six differences between the *Cyp2b9* and *Cyp2b10* sequences. When these changes were introduced into the *Cyp2b10* gene within a fragment from −2397 to −2265, induction was completely eliminated. Within the 33-bp region is an NF-1 consensus sequence and a steroid hormone receptor response element half-site. The sequence of the NF-1 site in *Cyp2b10* is identical to that in the rat *CYP2B2* gene, therefore the studies described above establishing that NF-1 proteins bind to this region in *CYP2B2* establish the *Cyp2b10* sequence as a functional NF-1 site as well. The *Cyp2b9* gene has differences in both the NF-1 and steroid hormone receptor motifs compared to the *Cyp2b10* gene.

Mutation of the NF-1 site reduced basal activity to about 65% and largely abolished the induction from 2.9-fold to 1.4-fold (67). Mutation of the steroid hormone binding motif increased the basal activity by about 2-fold and eliminated the induction. These same mutations also altered binding to the region assessed by DNase I footprinting and gel-shift assays. Mutation of the NF-1 site reduced binding substantially in the region so that only the sequence containing the steroid hormone-binding site was still protected and mutation of the steroid hormone-binding site resulted in reduced protection in the region of the steroid hormone half-site. Mutation of putative regulatory sites to the 5′ side in the −2364 to −2334 region did not substantially reduce induction, although basal activity was reduced. These results have led to the proposal that the 33-bp fragment containing the NF-1-like site is the core phenobarbital-responsive unit. Consistent with this conclusion is the presence of NF-1 elements and steroid hormone receptor half-sites in sequences of other phenobarbital-inducible genes, including near the DNase I-hypersensitive site present in *CYP2C1*, as noted (67).

The 33-bp *Cyp2b10* sequence proposed as the core phenobarbital element agrees very well with the central region of the *CYP2B2* element defined by deletions from the 5′ and 3′ side (Fig. 3). However, neither the 33- nor 37-bp fragment alone is able to confer phenobarbital responsiveness, even in multiple copies. Additional sequence from either side can confer responsiveness, thus there are redundant regulatory factors in the two flanking sequences that are required. There is general agreement between our studies using the *in situ* assay for *CYP2B2* and the primary hepatocytes for *Cyp2b10* that the NF-1 site within the core sequence plays a critical role in the amplitude of the induction. However, with multiple copies of the *CYP2B2* PBRU containing NF-1 site mutations, substantial induction with phenobarbital was retained, thus other sequences are important for the induction. Mutation of the steroid hormone receptor site overlapping with the NF-1 site in the *CYP2B2* gene, from AGGTCA to AGGGTA, did not detectably affect induction in the *in situ* assay, in contrast to the increase in basal activity and loss of induction in the primary hepatocyte assay when the sequence was changed to ACCTCA. In the primary hepatocytes, a single mutant construction of the AGGTCA sequence was made, so there is a slight possibility that a new positive regulatory element was inadvertently created by the mutation, which would account for the increased basal activity with this mutant. Thus, it is clear that the NF-1 site plays an important role in the phenobarbital induction, and mutations in the 2 bp immediately adjacent to the NF-1 site derepress the *Cyp2b10* gene. Whether this latter effect is due to altering the AGGTCA steroid hormone half-site or to changes in the specificity of the NF-1 site remains unclear.

VI. Perspectives and Models of Phenobarbital Induction of P450 Genes

The emerging consensus of the work described herein is that the primary regulation of the *CYP2B2* and *Cyp2b10* genes by phenobarbital is through elements distal to the promoter rather than within the proximal promoter elements. Although the core of the barbie box in the proximal promoter is retained in *Cyp2b10*, a 42-bp insertion relative to *CYP2B2* is present in this region (Fig. 2). Most investigators have found no differences in protein binding in the proximal region. Functional assays in transgenic mice, in transfected primary hepatocytes, and by direct injection into rat liver are all in agreement that proximal promoter sequences up to at least -800 cannot mediate phenobarbital induction alone. In contrast, the distal PBRU functions as an phenobarbital-responsive enhancer that can confer phenobarbital induction on heterologous promoters. Mutation of the barbie box motif does not reduce the induction mediated by the distal PBRU. These results are most consistent with the distal enhancer unit as the primary mediator of phenobarbital induction and the proximal elements as positive elements that comprise the core promoter activity, which, of course, is required for the phenobarbital induction. There remains a possibility that the proximal promoter does contribute to the phenobarbital induction, but that the assays in which induction is not observed are too insensitive to detect the effect. Clearly the large fold inductions observed for these genes *in vivo* are not reproduced in the *in vitro* assays and possibly position effects could be masking the induction in the transgenic mice. Even with these caveats, the contributions, if any, of the proximal promoter are likely to be less than those of the distal enhancer.

Within the resolution of the assays employed, deletion of the PBRU region eliminates the responsiveness of the *CYP2B2* gene to phenobarbital. However, as noted above, the induction observed in the transient transfection systems is much less than that observed for the genes *in vivo*. Thus, it is possible that there are other regions in the 5' flanking region or elsewhere that can contribute to the induction. Weak contributions from the proximal promoter have not been completely ruled out. Portions of the genes far upstream or downstream of the transcription start site have not been tested. In *CYP1* genes, which are induced by the Ah receptor, multiple sites for Ah receptor binding are common (*13*). In the *Cyp2b10* gene, there may be a second region downstream of the PBRU, which confers some phenobarbital responsiveness (*67*), and fragments of *CYP2B2*, which are missing the PBRU, have been reported to respond to phenobarbital induction in cultured cells (*16*). In addition, the PBRU region of *Cyp2b9* was incapable of conferring detectable phenobarbital induction in the primary hepatocyte and *in situ* transfection assays. *Cyp2b9* has been reported to be induced by phenobarbital *in vivo* (*47*), but has also been reported to be unresponsive to phenobarbital (*48*). The differences are related to strain differences (*66*). If it is inducible, is the inability of the *Cyp2b9* PBRU to confer phenobarbital responsiveness because it is a weaker enhancer below the detection level of the assays? Are there other regions in *Cyp2b9* that may mediate the response? Or are the two *Cyp2b9* PBRUs analyzed from constitutively expressed or unexpressed genes among the many highly related *Cyp2b9*-like genes in different species? Assay by transient transfection can only partially answer these questions about the number of phenobarbital-responsive regions in a gene, because the influence of distal elements is often underestimated in these assays when compared to *in vivo* transgenic models. It is likely that specific mutations of the PBRU in transgenes of *CYP2B2* or even targeted disruption by homologous recombination of the PBRU in its natural genomic context will be required to answer these questions definitively.

The possibility that there may be multiple mechanisms for phenobarbital induction is suggested by the observations that the time courses of induction and the dose dependencies of phenobarbital-inducible P450s differ and P450s can be crudely segregated into highly inducible and moderately inducible genes. The PBRU appears to be a complex unit containing several elements, probably including ubiquitous, as well as phenobarbital-responsive, ones. Possible combinations of the elements in the *CYP2B2* PBRU in other phenobarbital-inducible genes have been pointed out (67). Various combinations of these elements in the PBRUs of P450s and other phenobarbital-responsive genes could provide the complexity necessary to explain the gene-specific and tissue-specific variations in the characteristics of induction, providing a basis for the observed differences. Analysis of the PBRUs of other phenobarbital-inducible genes should provide insight into whether distinct mechanisms or simply variations on the same theme are involved.

The transient transfection or cell-free assays uniformly exhibit increases in transcriptional activity that are substantially less than that seen *in vivo*, 2- to 8-fold versus more than 100-fold. Part of this deficiency might be due to the absence of some elements that contribute to induction, but it seems likely that the state of the transfected DNA in transient assays plays a large role. *CYP2B1* expression is undetectable in untreated animals *in vivo*, yet expression of the transfected gene is detected in HepG2 cells, primary hepatocytes, and hepatocytes transfected *in situ* without phenobarbital treatment. As a general rule, binding of histones to DNA to form chromatin results in repression of expression. In the transient transfections, such chromatin structure is missing so that the normal silencing of the gene may be absent. Alternatively, there may be specific silencer sequences in the genes that are missing from the DNA gene fragments used in the assays. In either case, repression must be invoked to explain how the proximal promoter, which contains several positively acting regulator elements, can be essentially silent *in vivo*. It will be important to determine the binding of regulatory factors in these regions in native chromatin to determine if chromatin structure defines factor binding. Changes in the sensitivity of the gene region to DNase I, the sensitivity of hypersensitive sites, and the association with nuclear matrix after phenobarbital treatment suggest that changes in chromatin structure and binding may occur (68).

The studies on phenobarbital induction of P450 in bacteria have led to a model in which depression is a key feature (Fig. 1). As illustrated for P450 BM3, the binding of the repressor BM3R1 to a palindromic site in the operator region for the BM3R1 and BM3 operon and to barbie box sequences near a weaker promoter for BM3 represses the expression of both BM3 and BM3R1. In the presence of phenobarbital treatment, the binding activity of BM3R1 for the palindromic sequence is reduced. At the same time the synthesis of two or more positively acting factors is induced and these factors compete for the binding of the repressor at the barbie box region. The exact nature of the positively acting factors needs to be established, but such a model in general terms is well established.

The model for the induction of the mammalian P450s is less well defined. The bacterial depression model serves as a guide for the mammalian mechanism, but is probably not directly analogous. The similarity of barbie box sequence in bacteria and mammals may be suggestive of a common mechanism, but the inability of several laboratories to detect binding to this element suggest that the presence of similar sequences in bacterial and mammalian genes is largely coincidental. In addition, correlation of the aliphatic side chain of the barbiturates with induction in bacteria, and the strong induction by peroxisomal proliferators, are contrary to induction of phenobarbital-inducible P450s in mammals. Parenthetically, if the AGGTCA sequence in the PBRU is critical for phenobarbital induction, there may be an evolutionary connection between the two, because the peroxisomal proliferator-activated receptor binds to a repeat of the AGGTCA sequence. There are also substantial differences in the structure of the genes *in vivo* in bacteria and mammalian cells. Mammalian genes are complexed with histones, in a chromatin structure that depresses transcriptional activity, whereas bacterial DNA has a less complex structure and specific repressors such as BM3R1 repress the activity. Activation of the mammalian genes probably involves structural changes in the chromatin as well as activation of transcription, and is likely to be significantly more complex than the bacterial mechanism. Changes in the chromatin structure of the *CYP2B2/1* genes after phenobarbital treatment have also been detected by general DNase I sensitivity (68). Approximately a twofold increase in the sensitivity of the genes was observed. In addition, the amount of the genes associated with the nuclear matrix was increased by phenobarbital treatment, which would be consistent with structural changes in these genes.

The mammalian mechanism probably involves a derepression system. Before speculating on the mechanism, it is useful to review what is known. Several positively acting elements have been identified in the proximal promoter, redundant positive elements are present in the PBRU region, and a

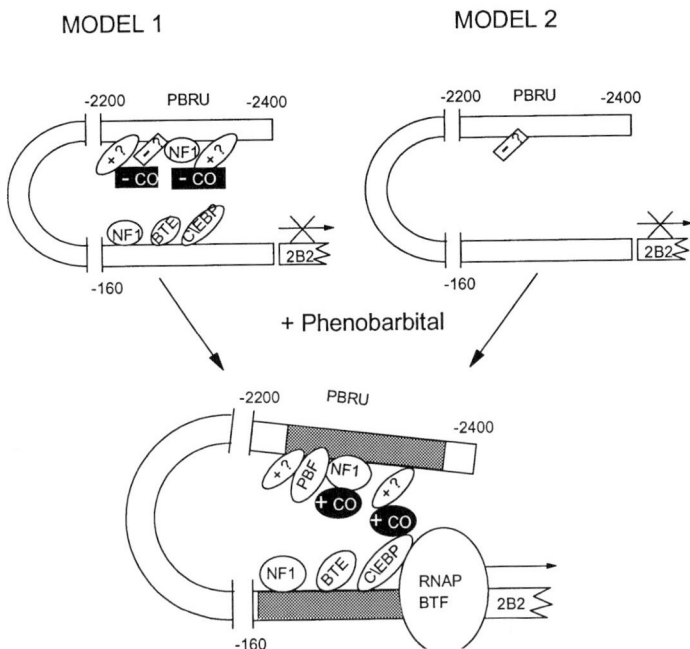

FIG. 6. Possible models for the activation of the *CYP2B2* promoter by phenobarbital. Multiple positive elements have been demonstrated in the PBRU and proximal promoters, but factors have not been characterized except for C/EBP in the proximal promoter and NF-1 in the PBRU. Factors other than these are hypothetical. Positive factors are indicated by circles or ovals and negative factors by rectangles. CO, Coactivators that do not bind directly to the DNA. The complex of basal transcription factors (BTF) and RNA polymerase II (RNAP) is indicated by the large oval. In model 1 most of the factors involved are bound to the promoter and PBRU in the untreated animals, and activity is repressed by suppressors or negative factors or chromatin structure. Phenobarbital treatment has little effect on the DNA-binding proteins, but reverses the repression so that the positively acting factors are now transcriptionally competent, which may involve binding different coactivators and alteration of the chromatin structure. In model 2, the chromatin structure in the untreated animal is not conducive to binding of the positive factors and possibly supports binding of negative elements. Phenobarbital treatment induces a change in the chromatin structure at the PBRU, which allows the binding of positive factors and possibly cofactors, resulting in a change in the chromatin structure at the proximal promoter and binding of positive factors there as well. This model is based on the mechanisms of activation of the MMTV promoter by glucocorticoids and the *CYP1A1* gene by aromatic hydrocarbons (69, 70). The presence of DNase I hypersensitivity regions at both the proximal promoter and PBRU in untreated animals favors model 1, assuming that the hypersensitivity actually reflects the disruption of chromatin structure in repressed *CYP2B2* genes.

repressor element may be present adjacent to the NF-1 site in the PBRU. The factors that bind to these elements that have been defined are NF-1 in the PBRU and C/EBP in the proximal promoter. Conjecture about the nature of all other factors and coregulator proteins shown in Fig. 6 is speculative. It is clear that the PBRU is a complex responsive unit containing several regulatory elements, one or more of which mediates the phenobarbital action either directly or indirectly.

The presence of the NF-1 site in the PBRU is particularly interesting. NF-1, although a constitutive factor, has been implicated in the ligand-dependent regulation of the MMTV promoter and *CYP1A1* (69, 70). In both cases, activation of the respective genes by the ligand results in binding of NF-1 to its site, presumably because of changes in chromatin structure. NF-1 binding is greatly reduced when DNA is in a normal nucleosomal structure (71). In addition, as many as 12 variants of NF-1 derived from four genes have been detected (64, 65), thus phenobarbital induction could in principle modulate a specific NF-1 that binds preferentially to the PBRU because of sequences flanking the NF-1 site. Alternatively, phenobarbital might result in modifications of NF-1 or adjacent binding proteins, resulting in the binding of a different NF-1 isoform. In addition to its positively acting function, NF-1 has been shown to be a negatively acting factor by competition with an overlapping site (72). Because mutation of the NF-1 site in *CYP2B2* and *Cyp2b10* does not increase transcription in the untreated animals, NF-1 in this case appears not to be a repressor but acts in a positive manner.

Two possible models for phenobarbital induction of *CYP2B2* are illustrated in Fig. 6. Ligand-mediated increases in gene transcription usually involve two steps, a change in the chromatin structure that allows binding of regulatory factors or alters their activity and activation or recruitment of the basal transcription factors and the RNA polymerase II complex to the RNA initiation site. Emerging general models of ligand-mediated activation suggest that chromatin remodeling is accomplished by acetylation of histones as an important modification in the activation and deacetylation for repression. Several activating transcription factors have acetylase activity, and suppressor coregulator factors with deacetylase activity have been described (73–75). Activation by factors may involve an allosterically induced structural change in one of the components of the basal transcription apparatus or recruitment of this apparatus to the promoter (76). The two models for phenobarbital induction have in common a change in the chromatin structure, possibly mediated by acetylation, and the speculation that coregulators may be present.

Model 1 is based on postulated mechanisms for the activation of gene expression by ligands for which the receptor binds to its binding site even in the absence of the ligand (77). In the untreated animal all of the factors are bound to their respective sites, which might include a negative factor in the PBRU, and the suppression is mediated by negative coregulators that maintain the chromatin in a transcriptionally inactive state. Phenobarbital acts by altering the binding of a protein to the PBRU, or its activity, such that the chromatin structure is altered to one that supports recruitment of the basal transcription machinery to the promoter, and transcription proceeds. This process might be mediated by positive coregulators. This model is consistent with the presence of DNase I hypersensitivity of the PBRU and proximal promoter even in untreated animals. It requires that any PB binding factor that binds would mimic binding of a protein in the untreated animal or that phenobarbital treatment alters the activity but not binding of a factor, because no difference in binding *in vitro* is seen with extracts from untreated or treated animals (67).

Model 2 is based on the induction of the MMTV promoter by glucocorticoids and of *CYP1A1* by polycyclic aromatic hydrocarbons (69, 70). In the model, the chromatin structure is closed to binding of factors in both the proximal and the distal regions, possibly because of a repressor factor, as shown, or deacetylation of the histones. In this case, the primary effect of phenobarbital is to cause the binding of a factor that alters the chromatin structure so that NF-1 and other positively acting factors can bind. Again, positively acting coregulators may contribute to this change. This is followed by the recruitment of the basal transcriptional machinery. This model is consistent with the lack of changes in binding to the DNA *in vitro* after phenobarbital treatment, because chromatin structure modulates the binding. It is not consistent with the presence of DNase I hypersensitivity of the PBRU observed in the untreated animals, unless this result does not reflect the state of the inactive gene, for example, if the hypersensitivity is due to closely related constitutively expressed genes.

It is likely the experiments in the next few months and years will distinguish among these possibilities. Work was presented at the Twelfth International Symposium on Microsomes and Drug Oxidations in 1996 by A. Anderson's group (Université Laval, Québec, Canada) on dissection of the functional elements of the *CYP2B2* PBRU assayed in primary hepatocytes (78) and by C. Omiecinski's group (University of Washington, Seattle) on the regulation of transgenes, containing *CYP2B2* 5′ flanking regions, which respond to phenobarbital. Completion of these studies will considerably enhance our understanding of the phenobarbital response. Studies of the chromatin structure of the PBRU, the proteins that bind under native chromatin structure, and the complexes of proteins that interact with the PBRU DNA-binding proteins should provide considerable insight. Such characterizations hold great promise to identify the regulatory protein that is the target of phenobarbital action. Functional studies in transgenic mice will provide the necessary functional studies *in vivo* to establish the quantitative contribution of the PBRU to the phenobarbital response. After many frustrating years, the tools are now available to address these questions conclusively, and the answers are just around the corner.

Acknowledgments

Research from our lab described in this paper was supported by NIH Grant GM39360. I thank M. Negishi for providing a prepublication copy of his paper.

References

1. M. J. Coon, A. D. N. Vaz, and L. L. Bestervelt, *FASEB J.* **10,** 428 (1996).
2. D. R. Nelson, L. Koymans, T. Kamataki, J. J. Stegeman, R. Feyereisen, D. J. Waxman, M. R. Waterman, O. Gotoh, M. J. Coon, R. W. Estabrook, I. C. Gunsalus, and D. W. Nebert, *Pharmacogenetics* **6,** 1 (1996).
3. J. T. Groves and Y.-Z. Han, *in* "Cytochrome P450: Structure, Mechanism, and Biochemistry" (P. R. Ortiz de Montellano, ed.), p. 3. Plenum, New York, 1995.
4. F. J. Gonzalez, *Pharmacol. Ther.* **45,** 1 (1990).
5. B. J. Song, R. L. Veech, S. S. Park, H. V. Gelboin, and F. J. Gonzalez, *J. Biol. Chem.* **264,** 3568 (1989).
6. H. Remmer, *Naturwissenshaften* **45,** 189 (1958).
7. D. J. Waxman and L. Azaroff, *Biochem. J.* **281,** 577 (1992).
8. F. W. Frueh, U. M. Zanger, and U. A. Meyer, *Mol. Pharmacol.* **51,** 363 (1997).
9. J. P. Hardwick, F. J. Gonzalez, and C. B. Kasper, *J. Biol. Chem.* **258,** 8081 (1983).
10. T. A. Kocarek, E. G. Schuetz, and P. S. Guzelian, *Mol. Pharmacol.* **38,** 440 (1990).
11. H. Burger, E. G. Schuetz, J. D. Schuetz, and P. S. Guzelian, *Arch. Biochem. Biophys.* **281,** 204 (1990).
12. R. Menzel, E. Kärgel, F. Vogel, C. Böttcher, and W. Schunck, *Arch. Biochem. Biophys.* **330,** 97 (1996).
13. J. P. Whitlock Jr., S. T. Okino, L. Dong, H. P. Ko, R. Clarke-Katzenberg, Q. Ma, and H. Li, *FASEB J.* **10,** 809 (1996).
14. E. F. Johnson, C. N. A. Palmer, K. J. Griffin, and M.-H. Hsu, *FASEB J.* **10,** 1241 (1996).
15. M. R. Waterman and L. J. Dischof, *Endocrinol. Res.* **22,** 615 (1996).
16. P. M. Shaw, M. Adesnik, M. C. Weiss, and L. Corcos, *Mol. Pharmacol.* **44,** 775 (1993).
17. E. G. Schuetz, D. Li, C. J. Omiecinski, U. Muller-Eberhard, H. K. Kleinman, B. Elswick, and P. S. Guzelian, *J. Cell. Physiol.* **134,** 309 (1988).
18. P. R. Sinclair, E. F. Schuetz, W. J. Bement, S. A. Haugen, J. F. Sinclair, B. K. May, D. Li, and P. S. Guzelian, *Arch. Biochem. Biophys.* **282,** 386 (1990).
19. D. J. Waxman, J. J. Morrissey, S. Naik, and H. O. Jauregui, *Biochem. J.* **271,** 113 (1990).
20. E. Trottier, A. Belzil, C. Stoltz, and A. Anderson, *Gene* **158,** 263 (1995).
21. Y. Park, H. Li, and B. Kemper, *J. Biol. Chem.* **271,** 23725 (1996).
22. R. Ramsden, K. M. Sommer, and C. J. Omiecinski, *J. Biol. Chem.* **268,** 21722 (1993).
23. A. J. Fulco, *Annu. Rev. Pharmacol. Toxicol* **31,** 177 (1991).
24. A. J. Fulco, J.-S. He, and Q. Liang, *in* "Cytochrome P450. 8th International Conference" (M. C. Lechner, ed.), p. 37. John Libbey Eurotext, Paris, 1994.
25. G.-C. Shaw and A. J. Fulco, *J. Biol. Chem.* **267,** 5515 (1992).
26. G.-C. Shaw and A. J. Fulco, *J. Biol. Chem.* **268,** 2997 (1993).
27. J.-S. He and A. J. Fulco, *J. Biol. Chem.* **266,** 7864 (1991).
28. Q. Liang, J.-S. He, and A. J. Fulco, *J. Biol. Chem.* **270,** 4438 (1995).
29. Q. Liang and A. J. Fulco, *J. Biol. Chem.* **270,** 18606 (1995).
30. J.-S. He, Q. Liang, and A. J. Fulco, *J. Biol. Chem.* **270,** 18615 (1995).

31. T. Fournier, N. Mejdoubi, C. Lapoumeroulie, J. Hamelin, J. Elion, G. Durand, and D. Porquet, *J. Biol. Chem.* **269,** 27175 (1994).
32. N. English, V. Hughes, and C. R. Wolf, *J. Biol. Chem.* **269,** 26836 (1994).
33. P. N. Rangarajan and G. Padmanaban, *Proc. Natl. Acad. Sci. U.S.A.* **86,** 3963 (1989).
34. L. Prabhu, P. Upadhya, N. Ram, C. S. Nirodi, S. Sultana, P. G. Vatsala, S. A. Mani, P. N. Rangarajan, A. Surolia, and G. Padmanaban, *Proc. Natl. Acad. Sci. U.S.A.* **92,** 9628 (1995).
35. P. Upadhya, M. V. Rao, V. Venkateswar, P. N. Rangarajan, and G. Padmanaban, *Nucleic Acid Res.* **20,** 557 (1992).
36. N. Ram, M. V. Rao, L. Prabhu, C. S. Nirodi, S. Sultana, P. G. Vatsala, and G. Padmanaban, *Arch. Biochem. Biophys.* **317,** 39 (1995).
37. S. Sultana, C. S. Nirodi, N. Ram, L. Prabhu, and G. Padmanaban, *J. Biol. Chem.* **272,** 8895 (1997).
38. E. A. Shephard, L. A. Forrest, A. Shervington, L. M. Fernandez, G. Ciaramella, and I. R. Phillips, *DNA Cell Biol.* **13,** 793 (1994).
39. P.-V. Luc, M. Adesnik, S. Ganguly, and P. M. Shaw, *Biochem. Pharmacol.* **51,** 345 (1996).
40. Y. Park and B. Kemper, *DNA Cell Biol.* **15,** 693 (1996).
41. K. M. Sommer, R. Ramsden, J. Sidhu, P. Costa, and C. J. Omiecinski, *Pharmacogenetics* **6,** 369 (1996).
42. S. C. Williams, C. A. Cantwell, and P. F. Johnson, *Genes Dev.* **5,** 1553 (1991).
43. P. Honkakoski, R. Moore, J. Gynther, and M. Negishi, *J. Biol. Chem.* **271,** 9746 (1996).
44. M. Hoffmann, W. H. Mager, B. J. Scholte, A. Civil, and R. J. Planta, *Gene Express.* **2,** 353 (1992).
45. A. Yanagida, K. Sogawa, K.-I. Yasumoto, and Y. Fujii-Kuriyama, *Mol. Cell. Biol.* **10,** 1470 (1990).
46. H. Imataka, K. Sogawa, K.-I. Yasumoto, Y. Kikuchi, K. Sasano, A. Kobayashi, M. Hayami, and Y. Fujii-Kuriyama, *EMBO J.* **11,** 3663 (1992).
47. N. Nemoto and J. Sakurai, *Arch. Biochem. Biophys.* **319,** 286 (1995).
48. P. Honkakoski, A. Kojo, and M. A. Lang, *Biochem. J.* **285,** 979 (1992).
49. C. N. Hahn, A. J. Hansen, and B. K. May, *J. Biol. Chem.* **266,** 17031 (1991).
50. S. C. R. Elgin, *J. Biol. Chem.* **263,** 19259 (1988).
51. M. Adesnik and M. Atchison, *CRC Crit. Rev. Biochem.* **19,** 247 (1985).
52. Labbe, A. Jean, and A. Anderson, *DNA Cell Biol.* **7,** 253 (1988).
53. J. Kim and B. Kemper, *Biochemistry* **30,** 10287 (1991).
54. N. Murayama, M. Shimada, Y. Yamazoe, K. Sogawa, K. Nakayama, Y. Fujii-Kuriyama, and R. Kato, *Arch. Biochem. Biophys.* **328,** 184 (1996).
55. C. Lerche, A. Fautrel, P. M. Shaw, D. Glaise, F. Ballet, A. Guillouzo, and L. Corcos, *Eur. J. Biochem.* **244,** 98 (1997).
56. R. W. Malone, M. A. Hickman, K. Lehmann-Bruinsma, T. R. Sih, R. Walzem, D. M. Carlson, and J. S. Powell, *J. Biol. Chem.* **269,** 29903 (1994).
57. J. A. Wolff, P. Williams, G. Acsadi, S. S. Jiao, A. Jani, and C. Wang, *Biotechniques* **11,** 474 (1991).
58. S. Newman and P. S. Guzelian, *Proc. Natl. Acad. Sci. U.S.A.* **79,** 2922 (1982).
59. M. M. McGrane, J. S. Yun, A. F. M. Moorman, W. H. Lameres, G. K. Hendrick, B. M. Arafah, E. A. Park, T. E. Wagner, and R. W. Hanson, *J. Biol. Chem.* **265,** 22371 (1990).
60. K. Yoshimura, M. A. Rosenfeld, P. Seth, and R. G. Crystal, *J. Biol. Chem.* **268,** 2300 (1993).
61. P. Fender, R. W. H. Ruigrok, E. Gout, S. Buffet, and J. Chroboczek, *Nature Biotechnol.* **15,** 52 (1997).
62. B. C. O'Connell, K. G. T. Hagen, K. W. Lazowski, L. A. Tabak, and B. J. Baum, *Am. J. Physiol.* **268,** G1074 (1995).
63. V. Jain and I. Magrath, *Anal. Biochem.* **199,** 119 (1991).

63a. S. Liu, Y. Park, I. Rivera-Rivera, H. Li, and B. Kemper, *DNA Cell Biol.*, in press (1998).
64. C. Santoro, N. Mermod, P. C. Andrews, and R. Tjian, *Nature (London)* **334,** 218 (1988).
65. U. Kruse and A. E. Sippel, *J. Mol. Biol.* **238,** 860 (1994).
66. M. Damon, A. Fautrel, A. Guillouzo, and L. Corcos, *Biochem. J.* **317,** 481 (1996).
67. P. Honkakoski and M. Negishi, *J. Biol. Chem.* **272,** 14943 (1997).
68. C. J. Stroop, A. A. Dutmer, and G. J. Horbach, *J. Biochem. Toxicol.* **11,** 59 (1996).
69. H. Richard-Foy and G. L. Hager, *EMBO J.* **6,** 2321 (1987).
70. L. Wu and J. P. Whitlock Jr., *Proc. Natl. Acad. Sci. U.S.A.* **89,** 4811 (1992).
71. P. Blomquist, Q. Li, and Ö. Wrange, *J. Biol. Chem.* **271,** 153 (1996).
72. D. Bernier, H. Thomassin, D. Allard, M. Guertin, D. Hamel, M. Blaquiere, H. Beauchemin, H. LaRue, M. Estable-Puig, and L. Belanger, *Mol. Cell. Biol.* **13,** 1619 (1993).
73. P. A. Wade, D. Pruss, and A. P. Wolffe, *Trends Biochem. Sci.* **22,** 128 (1997).
74. T. Heinzel, R. M. Lavinsky, T.-M. Mullen, M. Söderström, C. D. Laherty, J. Torchia, W.-M. Yang, G. Brard, S. D. Ngo, J. R. Davie, E. Seto, R. N. Eisenman, D. W. Rose, C. K. Glass, and M. G. Rosenfeld, *Nature (London)* **387,** 43 (1997).
75. L. Allend, R. Muhle, H. J. Hou, J. Potes, L. Chin, N. Schreiber-Agus, and R. A. DePinho, *Nature (London)* **387,** 49 (1997).
76. M. Ptashne and A. Gann, *Nature (London)* **386,** 569 (1997).
77. D. J. Mangelsdorf and R. M. Evans, *Cell* **83,** 841 (1995).
78. C. Stolz, M. H. Vachon, E. Trottier, S. Dubois, Y. Paquet, and A. Anderson, *J. Biol Chem.* **273,** 8528 (1998).

RNA and Protein Interactions Modulated by Protein Arginine Methylation

JONATHAN D. GARY AND
STEVEN CLARKE

*The Molecular Biology Institute
and The Department of Chemistry and
 Biochemistry
University of California, Los Angeles
Los Angeles, California 90095*

I. Introduction	66
II. Background	68
A. Guanidino-Methylated Arginine Residues in Proteins	68
B. Protein Arginine N-Methyltransferases	70
III. Identification of Protein Arginine N-Methyltransferase Genes	78
A. Mammalian PRMT1	78
B. *Saccharomyces cerevisiae* RMT1	83
C. Reconciling the Purification and Cloning Data	86
IV. Methyl-Accepting Substrates for Protein Arginine N-Methyltransferases	88
A. General	88
B. Asymmetric Dimethylarginine-Containing Substrates Identified *in Vivo*	89
C. Symmetric Dimethylarginine-Containing *in Vivo* Substrates—Myelin Basic Protein	99
V. Determination of Arginine Methyltransferase Substrates	100
A. Sequence Analysis of *in Vivo* Asymmetric Methylation Sites	100
B. RNA-Binding Proteins May Be the Major Group of Substrates for the Type I Protein Arginine N-Methyltransferases	103
C. Analysis of the *in Vivo* Symmetric Methylation Site of Myelin Basic Protein	109
D. Other Protein Arginine Methyltransferase Consensus Methylation Sites	110
VI. Functional Roles for the Protein Arginine N-Methyltransferase and Arginine Methylation	111
A. Asymmetric Methylation	111
B. Symmetric Arginine Dimethylation of Myelin Basic Protein	118
VII. Future Directions	123
References	124

This review summarizes the current status of protein arginine N-methylation reactions. These covalent modifications of proteins are now recognized in a number of eukaryotic proteins and their functional significance is beginning to be understood. Genes that encode those methyltransferases specific for catalyzing the formation of asymmetric dimethylarginine have been identified. The enzyme modifies a number of generally nuclear or nucleolar proteins that interact with nucleic acids, particularly RNA. Postulated roles for these reactions include signal transduction, nuclear transport, or a direct modulation of nucleic acid interactions. A second methyltransferase activity that symmetrically dimethylates an arginine residue in myelin basic protein, a major component of the axon sheath, has also been characterized. However, a gene encoding this activity has not been identified to date and the cellular function for this methylation reaction has not been clearly established. From the analysis of the sequences surrounding known arginine methylation sites, we have determined consensus methyl-accepting sequences that may be useful in identifying novel substrates for these enzymes and may shed further light on their physiological role. © 1998 Academic Press

I. Introduction

It has become increasingly clear that protein function is dependent on the covalent posttranslational modification of the 20 amino acid residues normally incorporated by ribosomes during protein synthesis. Some of these modifications are reversible, such as protein phosphorylation reactions, whereas others are apparently irreversible and can effectively create new types of amino acids to broaden the chemical diversity of polypeptides. In this latter group of modifications are a number of S-adenosylmethionine (AdoMet)-dependent methylation reactions that occur on the side-chain nitrogen atoms of lysine, arginine, and histidine residues (*1*). Although we understand some of the chemistry and enzymology of these reactions, less progress has been made in deciphering their functional importance.

In recent years, however, a number of advances have allowed us to get a clearer picture of the physiological role of one type of these reactions, the methylation of the guanidino group nitrogen atoms of protein arginine residues. Protein arginine methylation is found in most cells of higher eukaryotic species (*2*) as well as in lower eukaryotes, such as the trypanosomes (*3, 4*), but to date has not been found in prokaryotic organisms. As described below, it now appears that there are at least two distinct classes of protein arginine *N*-methyltranferases. The Type I enzyme catalyzes the formation of N^G-monomethylarginine and asymmetric N^G,N^G-dimethylarginine residues; whereas the Type II enzyme catalyzes the formation of N^G-monomethylarginine and symmetric N^G,N'^G-dimethylarginine residues (Fig. 1). Several *in vivo* substrates have been identified for the Type I (asymmetrically methylating) enzyme, including hnRNP A1, fibrillarin, and nucleolin. Interesting-

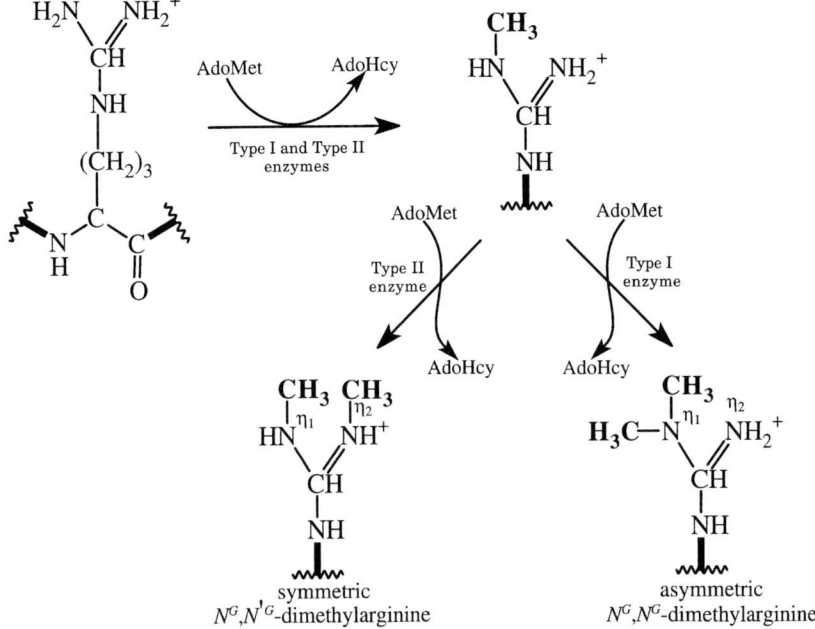

FIG. 1. The structures and stepwise synthesis of methylated arginine derivatives found in proteins. AdoMet serves as the methyl donor for two consecutive methylation events on the guanidino nitrogens of arginine residues. At least two protein arginine N-methyltransferases catalyze these reactions with both substrate and product specificity. One of the arginine methyltransferase activities (Type I) mono- and asymmetrically dimethylates hnRNP A1 as well as several other proteins that interact with RNA, including nucleolin and fibrillarin. The other known enzyme activity (Type II) mono- and symmetrically dimethylates myelin basic protein.

ly, all of these methyl-accepting proteins interact with RNA species, suggesting that the methylation of arginine residues might directly or indirectly modulate this interaction. In contrast, only a single *in vivo* substrate for the Type II (symmetrically methylating) methyltransferase is known: a major protein component of the myelin sheath, myelin basic protein.

A gene coding for a Type I methyltransferase has been identified in rat and yeast DNA (5–7) and apparent homologs have been found in human (8), mouse, sea urchin, *Caenorhabditis elegans, Pisolithus tinctorus*, rice, and *Arabidopsis thaliana* (9) DNA. The specific importance of asymmetric or symmetric protein arginine methylation in terms of a cellular function remains to be elucidated and current progress in this direction is addressed in this review. It is important to keep in mind that the metabolic cost of methylation is high (a net utilization of 12 ATPs is associated with the formation of a methylated product) (*10*); therefore, the ultimate benefit of this reaction must outweigh its energetic cost for it to be retained through evolution.

II. Background

A. Guanidino-Methylated Arginine Residues in Proteins

Methylated derivatives of arginine residues were first identified as radiolabeled acid hydrolysis products of *in vitro* incubations containing calf thymus nuclei and S-adenosyl-[^{14}C-*methyl*]-L-methionine ([^{14}C]AdoMet) (*11, 12*). These products, initially termed "unknown I" and "unknown II," eluted from a cation-exchange amino acid analysis column just before and just after the elution position of free arginine (*11, 12*). In subsequent work, unknowns I and II were clearly shown to result from an enzyme activity termed "protein methylase I" that transfers methyl groups from AdoMet to the endogenous "sulfated histone" fraction of the nuclei (*12*). These products were tentatively identified as guanidino-methylated derivatives of arginine based on the production of [^{14}C]methylurea after treatment with 2 N NaOH and the inability of arginase to release urea from them (*12*). Unknown I was determined to be methylated only at the guanidino group, whereas unknown II was suggested to be methylated at both the guanidino and α-amino groups (however, see below) (*12*).

Similar conclusions were reached from an examination of the products of *in vitro* methylation reactions containing [^{14}C]AdoMet, calf thymus histones, and soluble proteins from rat uterine cell cytosol (*13*). Cation-exchange amino acid analysis of the acid hydrolysis products from the histone methylation reaction also revealed two major radioactive species, one eluting prior to and the second coeluting with the free arginine standard (*13*). Identically eluting peaks were also detected when poly(L-arginine) was used as the methyl acceptor (*13*).

In parallel studies, evidence for methylated arginine residues in proteins was obtained from the analysis of urine for novel amino acids (*14*). These species can result from the inability of cells to metabolize fully certain amino acid residues in proteins (*15*). Chemical degradation analysis, elemental composition, and nuclear magnetic resonance (NMR) spectra identified N^G,N^G-dimethylarginine (asymmetric, DMA) and N^G,N'^G-dimethylarginine (symmetric, DMA') (Fig. 1) in human urine (*14*). The amounts of these amino acids in urine did not vary with dietary intake, suggesting that they resulted from the hydrolysis of cellular proteins (*14*). The existence of these amino acids as residues in cellular proteins *in vivo* was then confirmed from the direct analysis of acid hydrolysates of bovine brain proteins as well as protein hydrolysates from various rat organs (16, 17). Synthetic standards (14), whose structures were determined by elemental analysis, NMR, and infrared (IR) spectroscopy (16) as well as various chromatographic techniques (17), were used for chromatographic comparison with the compounds isolated from protein hydrolysates. In rats, the levels of protein-incorporated asymmetric dimethylarginine ranged from 0.01 μmol/g of protein in blood to 1.67 μmol/g of protein in spleen; the amounts of the protein-incorporated symmetric dimethylarginine varied from 0.01 μmol/g of protein in blood to 0.31 μmol/g of protein in brain (16). Overall, the concentration of the asymmetric derivative was always greater than the symmetric, with the highest levels of both found in the small intestine, spleen, lung, liver, and brain (2, 16). In addition to asymmetric and symmetric dimethylarginine, N^G-monomethylarginine (the presumed biosynthetic intermediate) was also found in the protein hydrolysates from bovine brain (16).

Finally, the identities of the hydrolysis products from *in vitro* methylation reactions containing rat brain or liver proteins, with [^{14}C]AdoMet as the methyl donor, were then unambiguously identified as mono- as well as asymmetric and symmetric dimethylarginine by comparison to standards using various chromatographic techniques (17). The α-amino and δ-imino methylated derivatives previously suggested (12) were ruled out as potential *in vivo* products after alkaline hydrolysis and comparison of their chromatographic properties with standards (16). Taken together, these data indicate that the peak originally identified as unknown I (11, 12) was probably a mixture of asymmetric and symmetric dimethylarginine (17), whereas unknown II (11, 12) was most likely monomethylarginine (17).

The isolation and quantitation of methylarginine derivatives have been typically accomplished through the use of paper electrophoresis, chemical degradation, high-pressure liquid chromatography (HPLC) analysis, and cation-exchange chromotography (2, 5, 6, 14, 16, 18, 19, 20, 21). It is also possible to determine these residues directly in polypeptides by automated Edman sequencing (22). This latter method not only allows the positive determination of the methylated residue but also gives the position of the methylated residue within the sequence. Although it has not yet been used extensively in this area, electrospray mass spectroscopy is a potentially powerful tool to determine rapidly whether a given protein may be covalently modified, as well as the site and type of the modification. It is also possible that antibodies specific for one or more types of methylated arginine residues could be developed as general analytical probes, much as antiphosphotyrosine antibodies have been useful in identifying phosphotyrosine-containing proteins (23). For example, an antibody has been found that recognizes arginine methylated Npl3 isolated from yeast but not the unmodified recombinant Npl3 expressed in *Escherichia coli* (24). Subjecting the bacterially expressed Npl3 to an *in vitro* protein arginine methylation reaction allowed antibody recognition of the protein. These results suggest that the antibody was specifically interacting with the methylated version of the protein, although it is not clear if part of the recognition epitope includes determinants aside from the methylarginine residues.

B. Protein Arginine N-Methyltransferases

An enzymatic activity initially termed "protein methylase I" and now generally referred to as protein arginine N-methyltransferase (EC 2.1.1.23) was first described in calf thymus (11, 12). This broad class of enzyme catalyzes the transfer of methyl groups from AdoMet to the terminal nitrogens ($\eta 1$ and $\eta 2$) of protein-incorporated arginine residues to generate monomethylarginine as well as symmetric and asymmetric dimethylarginine (Fig. 1). After the identification of these three methylated arginine derivatives in proteins *in vivo* and *in vitro* (16, 17), two central questions concerning the enzymatic activities remained: (1) how many protein arginine N-methyltransferases are responsible for creating the three derivatives and (2) what is the range of possible methyl-accepting substrates acted on by each enzyme?

To answer these questions, a number of groups have pursued the purification of these enzyme activities using specific methyl-accepting proteins in *in vitro* assays. Since the initial partial purification (12), there have been at least 15 reports concerning the purification of these methyltransferases from various tissues (Table I).

Because the sulfated histone fraction from calf thymus was the first identified methyl-accepting substrate for a protein arginine N-methyltransferase (12), most of the initial purifications used histone fractions to track the activity at each step. Histone fractions $f2a_1$, f2b, and f3 as well as the purified histones H4, H2A, and H2B were also found to be good *in vitro* substrates for arginine N-methyltransferase activities isolated from rat organs and human placenta (13, 25, 26). For the histone preparations methylated by the rat enzyme, the exact methylated arginine derivative formed was not determined, only that it was dimethylarginine (13, 25). The human placental arginine methyltransferase, on the other hand, methylated a histone preparation (calf thymus type IIAS; Sigma) to give all three derivatives, asymmetric and symmetric dimethylarginine as well as monomethylarginine, in a ratio of 50:7:44, respectively (26). Several other proteins tested as potential substrates, including soy bean trypsin inhibitor, bovine serum albumin, lysozyme, cytochrome *c*, hemoglobin, RNase, and fibrinogen, were not found to be methyl acceptors (13, 25, 26).

In 1971, myelin basic protein was found to be a good *in vitro* methyl-accepting substrate for a protein arginine N-methyltransferase in guinea pig brain extracts (27), and subsequent work confirmed this in mouse spinal cord extracts (28) and rat brain extracts (29). The products of these reactions were mono- and symmetric dimethylarginine (30). These *in vitro* methylation experiments were based on the results of the sequence and amino acid analyses of myelin basic protein from several species. Specifically, position 107 of bovine brain myelin basic protein is occupied by monomethylarginine (40–80%) and symmetric dimethylarginine (20%) (18, 31, 32). Sequence analysis also revealed the presence of mono- and symmetric dimethylarginine in endogenous myelin basic protein from rat brain (33) and chicken brain (34). Amino acid analysis of native myelin basic protein acid hydrolysates from several additional species, including bovine, frog, rabbit, monkey, and turtle, also showed the presence of mono- and symmetric dimethylarginine, indicating the general nature of this modification (21, 35). The presence of two methyl-accepting proteins with nonidentical but overlapping subtypes of methylated arginine residues suggested that cells may contain more than one type of protein arginine N-methyltransferase.

Biochemical evidence for the presence of more than one enzyme initially came from the differential ammonium sulfate fractionation of two protein arginine N-methyltransferase activities from rat brain that were more specific for either histones (now Type I) or myelin basic protein (now Type II) as methyl-accepting substrates (36). However, neither methyltransferase was completely separated from the other activity in this study. Interestingly, a second group (37) was not able to repeat this ammonium sulfate fractionation (36). They could not resolve the myelin basic protein methylating activity from the histone activity after a single chromatographic step that resulted in an 11-fold purification (37). From these data and the fact that myelin basic protein and histones both competitively inhibited the methyltransferase activity (29), they suggested that one methyltransferase recognized both substrates (37).

In later work, however, these two methyltransferase preparations (36) were further purified by gel filtration (30). Assays of acid-hydrolyzed myelin basic protein methylated by the gel filtration-purified Type II enzyme and [^{14}C]AdoMet indicated the presence of only radioactive monomethylarginine and symmetric dimethylarginine (30). In similar assays, the gel filtration-purified Type I activity only catalyzed the formation of radioactive monomethylarginine and asymmetric dimethylarginine in histone methyl-accepting substrates (30). This histone-methylating activity was also apparently free of any myelin basic protein methylation activity (30). It should be noted, however, that the Type II activity could still methylate histones, but the identity of the methylated arginine residue in these histones was not determined (30), allowing for the possibility of contamination by a Type I enzyme.

Additional evidence in favor of the multiple enzyme theory was provid-

TABLE I
PURIFICATIONS OF PROTEIN ARGININE N-METHYLTRANSFERASES

Year of purification	Ref.	Enzyme source	Fold purification	Substrate[a]	AdoMet K_m (μM)	pH optimum	Molecular mass[b]	Methylated residues[c]
Types I and II								
1968	12	Calf thymus	34	Histones	21	7.4	ND[d]	Guanidino-methylated arginine
1971	13	Rat organs	ND	Histones/(poly)Arg	10	8.5	ND	Methylated arginine
1971	25	Rat thymus	35	Histones	ND	8.1	150kDa	Guanidino-methylated arginine
1975	30	Rat brain	250	Histones	ND	ND	ND	DMA, MMA
1978	52	Krebs II ascites cells	150	Histones	2.5	ND	500	DMA, MMA
1986	39	Bovine brain	52	Histones	ND	ND	ND	DMA, DMA′, MMA
1988	40	Calf brain	ND	Histone	8	ND	275 kDa	ND
1991	26	Human placenta	400	Histones	5	7.4–7.6	ND	DMA, DMA′, MMA
1994	48	Rat liver	195	hnRNP A1	6.3	7.6	450 kDa	DMA, MMA
1995	49	HeLa cells	550	hnRNP A1	5.8	7–7.4	450 kDa	DMA, MMA
1971	27	Guinea pig brain	ND	Myelin basic protein	ND	ND	ND	DMA, MMA

Year	Ref	Source	Column4	Substrate	Col6	Col7	Native MW	Products
1974	37	Rat brain	6–11	Histones/myelin basic protein	ND	ND	ND	DMA, MMA
1975	30	Rat brain	170	Myelin basic protein	ND	ND	ND	DMA, MMA
1977	38	Calf brain	119	Histones/myelin basic protein	7.6	7.2	ND	DMA, DMA′, MMA
1988	40	Calf brain	404	Myelin basic protein	4.4	ND	500 kDa	DMA, MMA
1988	41	Bovine brain	ND	Myelin basic protein	7.6	ND	72 kDa[e]	ND
Type III								
1982	50	Wheat germ	90	Histone	5.7	9	ND	DMA, MMA
Type IV								
1985	53	*Euglena gracilis*	52	Cytochrome *c*	40	7	36 kDa	MMA

[a]Best substrates are listed.
[b]Determined by gel filtration.
[c]DMA, N^G,N^G-dimethylarginine; DMA′, N^G,N'^G-dimethylarginine; MMA, monomethylarginine.
[d]ND, not determined.
[e]Determined by SDS-PAGE.

ed by the partial purification of the Type I arginine methyltransferase from calf brain (38). For each active methyltransferase pool generated during the purification, the relative efficiency of histone and myelin basic protein methylation was also monitored at each step of the purification. In addition, the amount of radioactivity incorporated into each of the three methylated derivatives of arginine was determined using only histones as the methyl-accepting substrate. Throughout the purification of the Type I enzyme, the ratio of histone to myelin basic protein methylation efficiency increased from 1.6 to 8.9 (38), indicating the existence of at least two enzyme activities. In this same purification, the ratios of the methylated arginine products in histones did not vary greatly (DMA:DMA':MMA = 40:5:55). From these data, the authors therefore suggested that although there are different methyltransferases recognizing different substrates, a single protein arginine N-methyltransferase could form all three methylated arginine derivatives. Presumably the type of posttranslationally modified arginine formed would be dependent on the local context (sequence or structure) of the arginine residue within the substrate protein, or perhaps on the presence of regulatory subunits associated with the methyltransferase.

Perhaps the best evidence for the existence of at least two different types of protein arginine methyltransferases in a single tissue came from the partial purification of a Type I methyltransferase from bovine brain capable of methylating histones and not myelin basic protein (39), and from the separate purifications of a myelin basic protein-specific and a histone-specific methyltransferase from calf brain (40) (Table I). The Type I and Type II enzymes isolated from calf brain were found to have differential stabilities to treatment with heat, p-chloromercuribenzoate, and guanidine-HCl (40); these results clearly distinguish the two enzymes from each other. Furthermore, the Type I enzyme did not methylate myelin basic protein (40). However, the Type II enzymes from two separate preparations (40, 41) had some activity toward histones, but the K_m was three orders of magnitude greater for histones than for myelin basic protein and the V_{max} was, interestingly, twofold higher for histones than for myelin basic protein as methyl acceptors (40). It is also possible that it is solely the affinity of substrate binding that leads to the specificity of methylation by the Type II protein arginine N-methyltransferase and that this enzyme is able to methylate Type I substrates at a low efficiency. Surprisingly, in this study the identities of methylated arginines incorporated into the histones by either enzyme were not determined (40). However, the myelin basic protein substrate was found to be specifically mono- and symmetrically dimethylated (80:20) by the Type II enzyme, and the asymmetric dimethylarginine derivative was not detected (40) (Table II).

The purified calf brain Type I methyltransferase has a native molecular

TABLE II
Presence of Various Isoforms of Methylated Arginine in Several Substrates from Different Enzyme Preparations

Year of purification	Ref.	Substrate	Asymmetric dimethylarginine	Symmetric dimethylarginine	Monomethylarginine
1975	30	Histones	+[a]	ND[b]	+
1977	38	Histones	32%	5%	63%
1978	52	Histones	+	ND	+
1982	50	Histones	ND	+	+
1986	39	Histones	19%	13%	69%
1991	26	Histones	50%	7%	44%
1994	48	Histones	51%	ND	49%
		hnRNP A1	41%	ND	59%
1994	5	hnRNP A1	+	ND	+
1994	6	Histones	+	ND	+
		hnRNP A1	+	ND	+
1995	49	hnRNP A1	+	ND	+
1971	27	Myelin basic protein	ND	+	+
1975	30	Myelin basic protein	ND	+	+
1988	40	Myelin basic protein	ND	20%	80%
1990	45	Myelin basic protein	?[c]	?[c]	+

[a] Presence of residue, not quantitated in reference.
[b] ND, none detected.
[c] Only dimethylarginine determined.

mass of 275 kDa, estimated by gel-filtration chromotography, and this enzyme activity is associated with two major polypeptides of 110 and 75 kDa by sodium dodecyl sulfate polyacrylamide gel electrophoresis (SDS-PAGE) (40). The purified calf brain Type II myelin basic protein-specific enzyme was determined to be about 500 kDa by gel-filtration analysis, and includes two polypeptides of 100 and 72 kDa by SDS-PAGE (40). However, another purification of the Type II enzyme suggested a single polypeptide with a molecular mass between 71 and 74.5 kDa (41). These data are consistent with both methyltransferases being multimers, although the subunit structures of this complex are far from clear at this point. The recently cloned mammalian Type I protein arginine N-methyltransferase gene codes for a polypeptide of approximately 40 kDa that has enzymatic activity by itself (6). This raises the possibility that the higher molecular weight species identified in both preparations by SDS-PAGE were contaminants, although these larger polypeptides may have regulatory roles similar to those seen in other enzymes (42).

In 1977, protein components of the 40S ribonuclear complex were also found to contain methylated arginine residues *in vivo* (*43, 44*). The polypeptides, heterogeneous nuclear ribonuclear proteins (hnRNPs) A1 and A2, contain almost exclusively the asymmetric dimethylarginine derivative with a minor amount of monomethylarginine, whereas none of the symmetric derivative was detected (*43, 44*). Bovine brain Type I and Type II arginine methyltransferase preparations were unable to methylate recombinant hnRNP A1; therefore, this methylation was at first considered to be of a third class (*45*). However, Type I methyltransferase preparations from rat liver and calf brain were later shown to methylate recombinant hnRNP A1 much more efficiently than histones *in vitro* (*46–48*). The K_m for hnRNP A1 is 100-fold lower and the V_{max} is at least sevenfold higher than for histones (*47*). Recombinant hnRNP A1 overexpressed in *E. coli* is a much better substrate than native hnRNP A1 protein, presumably because the latter is already nearly fully methylated (*46*). Analysis of the arginine methylation products in the recombinant hnRNP A1 after an *in vitro* reaction showed the presence of only mono- and asymmetric dimethylarginine (*48*). The fact that the methylated arginine residues found in hnRNP A1 *in vivo* and after *in vitro* reactions with a Type I methyltransferase preparation were the same as those identified from *in vitro* reactions with histones led to the conclusion that the Type I histone and hnRNP A1 methyltransferases were the same (*46, 47*). This substrate has since been used in two additional purifications of protein arginine *N*-methyltransferases (*48, 49*).

The Type I rat liver hnRNP A1 methyltransferase has an estimated molecular mass of 450 kDa by gel filtration, and SDS-PAGE analysis of the purest fraction shows an apparent single polypeptide at 110 kDa (*48*). The enzyme has maximal activity toward recombinant hnRNP A1, with relative activities of only 23 and 11% for myelin basic protein and histones, respectively (*48*). Although myelin basic protein was apparently a substrate for the purified material, amino acid analysis of the hnRNP A1 and histone substrate demonstrated the presence of only monomethylarginine and asymmetric dimethylarginine, with none of the symmetric form (*48*). These methylated products are consistent with the activity of the Type I enzyme, because myelin basic protein does not contain any asymmetric dimethylarginine (*31, 32*). It is surprising that amino acid analysis or tryptic peptide microsequencing was not done on the methylated myelin basic protein in these experiments to distinguish between the possibilities that the preparation was contaminated with a Type II methyltransferase, that the myelin basic protein was contaminated with a substrate for the Type I enzyme, or that myelin basic protein could be methylated at a site distinct from Arg-107 by the Type I enzyme (*48*).

Subsequently, a more purified form of an hnRNP A1 methyltransferase from human ovarian carcinoma HeLa cells was prepared and its specificity for methylating recombinant hnRNP A1, histones, but not myelin basic protein was determined (49). This methyltransferase was apparently not purified to homogeneity because the purest fraction contained at least eight polypeptide bands by silver-stained SDS-PAGE analysis (49). The two most prominent bands were at 100 and 45 kDa (49). As with the rat liver enzyme (48), the HeLa cell methyltransferase also chromatographed as a 450-kDa polypeptide on gel filtration (49). Again, it is not clear which, if any, of these polypeptides observed in the final stages of the purifications are actually part of the native enzyme.

The above discussion has focused on the two major mammalian classes of identified protein arginine N-methyltransferase activities: the Type I asymmetric, hnRNP A1/histone-methylating enzyme and the Type II symmetric, myelin basic protein-specific enzyme. Other activities have been characterized from nonmammalian sources. For example, a distinct histone H4-specific methyltransferase has been characterized in plants (50, 51). A partially purified wheat germ protein arginine N-methyltransferase preparation was found to methylate endogenous histones, giving solely mono- and symmetric dimethylarginine residues (50). In comparison, the Type I enzyme generally catalyzes the formation of only monomethylarginine and asymmetric dimethylarginine residues when incubated with histones (52) (Table II). The wheat germ enzyme was also incapable of methylating myelin basic protein whose only methylated arginine residue is found as either monomethylarginine or symmetric dimethylarginine (50). Together with the fact that endogenous histones from wheat germ but not those from calf thymus contain symmetric dimethylarginine residues (50) and given the difference in the nature of the *in vitro*-methylated reaction products, these results support the conclusion that the wheat germ enzyme is distinct from its mammalian counterpart, and we designate it Type III.

A fourth type of enzyme was purified from the photosynthetic protozoan *Euglena gracilis* (53). This methyltransferase specifically monomethylated a single arginine residue at position 38 in cytochrome *c* from horse heart (53). The enzyme did not methylate histones to a great extent, but did methylate myoglobin at over half the velocity seen with cytochrome *c* (53). The native size of this Type IV enzyme is estimated to be 36 kDa by gel filtration (53) and is therefore much smaller than the native mammalian enzymes. The endogenous methyl-accepting substrates for this enzyme have not been characterized.

In summary, at least four types of protein arginine methyltransferases have been identified in various eukaryotic species, ranging from the protists, such as *Leishmania* sp. (3), to higher organisms, including humans (26). Protein arginine methylation has, however, never been observed in bacterial species. Liu and Dreyfuss (*49*) could not detect any Type I activity from *E. coli* using recombinant hnRNP A1 as the methyl-accepting substrate, although they also could not detect the activity in *Saccharomyces cerevisiae*, which was subsequently shown to have a Type I methyltransferase (*5, 7*). Additional evidence for the lack of a Type I protein arginine *N*-methyltransferase in *E. coli* comes from the fact that recombinant hnRNP A1 is a better substrate than its endogenous form, presumably because it is fully unmethylated (*46*). Despite the number of purifications published on this activity from eukaryotic organisms, no peptide sequence results have been presented. Therefore definitive proof of two or more distinct activities, either in substrate or methylated product specificity, awaits the expression of recombinant forms of all of these enzymes with purified substrates and amino acid analysis of the reaction products.

III. Identification of Protein Arginine *N*-Methyltransferase Genes

A. Mammalian PRMT1

The first mammalian gene encoding a protein arginine *N*-methyltransferase was identified from a yeast two-hybrid screen (*54, 55*) designed to identify proteins that could interact with murine TIS21, an immediate-early/primary response gene product (*56–58*), and the closely related murine BTG1 gene product (*59, 60*). From a rat cDNA library, a single clone, 3G, was found to interact specifically with both bait proteins in the two-hybrid screen as well as in *in vitro* binding experiments (*6*). A basic local alignment search tool (BLAST) search through the National Center for Biotechnology Information (NCBI) (http://www.ncbi.nlm.nih.gov/) (*61*) queried with the 3G sequence against the nonredundant protein database identified the *E. coli* ribosomal protein L11 methyltransferase as the highest scoring alignment for a protein with a known function (*6*). This bacterial methyltransferase is thought to methylate an amino-terminal alanine residue as well as an internal lysine residue (*62, 63*). The region of similarity between this protein and 3G is relatively short (29 amino acids) but encompasses two highly conserved regions common among a majority of soluble methyltransferases and AdoMet-binding proteins (*64*). These regions, I and post I, are thought to be involved in the binding of AdoMet, and recent crystallographic evidence on a catechol *O*-methyltransferase and two DNA methyltransferases demonstrates a direct interaction between these regions and AdoMet (*65*). In addition to region I and post I, 3G also was found to have two additional motifs common to methyltransferases (II and III) (*64*)(Fig. 2). From this sequence homology data, it was likely that 3G was also a methyltransferase.

```
  1 M A - - - - - - A A E A A N C I M E V S C G Q A E S S E K P N A E D M T S K D Y Y F D S Y A H F      PRMT1 (3G) rat
  1 M E N F V A T L A N G M S L Q P P L E E V S C G Q A E S S E K P N A E D M T S K D Y Y F D S Y A H F  PRMT1 (HCP1/IR1B4) human
  1 M S K - - - - - - - - - - - - - T A V K D S A T E K T K L S E - - S E Q H Y F N S Y D H Y            Rmt1 (Odp1/Hmt1) yeast

                        post I
 43 G I H E E M L K D E V R T L T Y R N S M F H N R H L F K D K V V L D V G S G T G I L C M F A A K A G  PRMT1 (3G) rat
 51 G I H E E M L K D E V R T L T Y R N S M F H N R H L F K D K V V L D V G S G T G I L C M F A A K A G  PRMT1 (HCP1/IR1B4) human
 31 G I H E E M L Q D T V R T L S Y R N A I I Q N K D L F K D K I V L D V G C G T G I L S M F A A K H G  Rmt1 (Odp1/Hmt1) yeast

                                                                                        II
 93 A R K V I G I E C S S I S D Y A V K I V K A N K L D H V V T I I K G K V E E V E L P V E K V D I I T  PRMT1 (3G) rat
101 A R K V I G I V C S S I S D Y A V K I V K A N K L D H V V T I I K G K V E E V E L P V E K V D I I T  PRMT1 (HCP1/IR1B4) human
 81 A K H V I G V D M S S I T E M A K E L V E L N G F S D K I T L L R G K L E D V H L F P P K V D I I I  Rmt1 (Odp1/Hmt1) yeast

                     III
143 S E W M G Y C L F Y E S M L N T V L H A R D K W L A P D G L I F P D R A T L Y V T A I E D R Q Y K D  PRMT1 (3G) rat
151 S E W M G Y C L F Y E S M L N T V L Y A R D K W L A P D G L I F P D R A T L Y V T A I E D R Q Y K D  PRMT1 (HCP1/IR1B4) human
131 S E W M G Y F L L Y E S M M D T V L A R D H Y L V E G G L I F P D K C S I H L A G L E D S Q Y K D    Rmt1 (Odp1/Hmt1) yeast

                            post III
193 Y K I H W E N V Y G F D M S C I K D V A I K E P L V D V V D P K Q L V T N A C L I K E V D I Y T V    PRMT1 (3G) rat
201 Y K I H W W E N V Y G F D M S C I K D V A I K E P L V D V V D P K Q L V T N A C L I K E V D I Y T V  PRMT1 (HCP1/IR1B4) human
181 E K L N Y W Q D V Y G F D Y S P F V P L V L H E P I V T V E R N N V N T T S D K L I E F D L N T V    Rmt1 (Odp1/Hmt1) yeast

243 K V E D L T F T S P F C L Q V K R N D Y V H A L V A Y F N I E F T - - R C H K R T G F S T S P E S P  PRMT1 (3G) rat
251 K V E D L T F T S P F C L Q V K R N D Y V H A L V A Y F N I E F T - - R C H K R T G F S T S P E S P  PRMT1 (HCP1/IR1B4) human
231 K I S D L A F K S N F K L T A K R Q D M I N G I V T W F D I V F P A P K G K R P V E F S T I G P H A P Rmt1 (Odp1/Hmt1) yeast

291 Y T H W K Q T V F Y M E D Y L T V K T G E E I F G T I G M R P N A K N N R D L D F T I D L D F K G Q   PRMT1 (3G) rat
299 Y T H W K Q T V F Y M E D Y L T V K T G E E I F G T I G M R P N A K N N R D L D F T I D L D F K G Q   PRMT1 (HCP1/IR1B4) human
281 Y T H W K Q T I F Y F P D D L D A E T G D T I E G E L V C S P N E K N N R D L N I K I S Y K F E S N   Rmt1 (Odp1/Hmt1) yeast

341 L C E L S C S T - - - - D Y R M R                                                                    PRMT1 (3G) rat
349 L C E L S C S T - - - - D Y R M R                                                                    PRMT1 (HCP1/IR1B4) human
331 G I D G N S R S R K N E G S Y L M H                                                                  Rmt1 (Odp1/Hmt1) yeast
```

FIG. 2. Protein sequence alignment of three Type I protein arginine N-methyltransferases identified from the rat (GB U60882), human (GB D66904, corrected, and Ref. 8), and yeast (GB Z35903). Those residues identical to the rat PRMT1 protein sequence are boxed. The conserved AdoMet binding regions (I, post I, II, III, and post III) common to the majority of soluble methyltransferases and AdoMet-binding proteins (64) are indicated by an overline.

Using an amino-terminal glutathione S-transferase fusion (GST) to 3G, various protein substrates were tested as possible substrates (6). Although the fusion protein was not able to methylate isoaspartyl-containing peptides (66) or the catalytic subunit of protein phosphatase 2A (67, 68), it was able to methylate peptide R1 (69), which contains a single methylation site for the Type I protein arginine N-methyltransferase (Section V,A) (6). Although the peptide was only monomethylated on the arginine residue, analysis of *in vitro*-methylated histones and recombinant hnRNP A1 demonstrated the presence of both mono- and asymmetric dimethylarginine (6). These results prove that the 3G cDNA, derived from a rat mRNA, encodes a Type I protein arginine N-methyltransferase, or at least its catalytic subunit. The suggestion that the hnRNP A1 and the histone arginine methyltransferases are identical is also strongly supported by these results: the identical methylated arginine products are formed from a methylation reaction with recombinantly expressed and purified GST–3G and either of the substrates. However, the affinity of the methyltransferase fusion protein for hnRNP A1 is much greater than for crude histones (6). In addition, myelin basic protein and cytochrome *c* were not substrates for the fusion protein. The gene encoding this protein has been termed PRMT1 (Protein aRginine N-Methyl Transferase. The PRMT1 protein has a calculated polypeptide molecular mass of 40.5 kDa and elutes as a native 180-kDa polypeptide species by gel-filtration chromotography (6). Recent results with the yeast two-hybrid system indicate that PRMT1 is able to at least dimerize with itself (J. Tang and H. Herschman, personal communication) and it is possible that the native enzyme is a homotetramer, although more complex subunit structures cannot be eliminated at this point. Northern analysis of the PRMT1 transcript shows that it is present in all tissues tested, and in RAT1 fibroblast cells it is constitutively expressed and its mRNA levels are not up-regulated after mitogen stimulation (6).

Human homologs of PRMT1 have recently been identified in two separate studies (8, 9). The human HCP1 (highly conserved protein) was identified as a multicopy cDNA suppressor of a yeast *ire15* mutant (9). This mutation causes a decrease in the expression of the inositol 1-phosphate synthase and inositol transporter genes (70). Suppressors of the *ire15* mutation allowed growth on inositol-free minimal agar plates (9). The published sequence of the human HCP1 is very similar to the rat sequence, except for a region between amino acids 147 and 175 that could be the result of a frameshift from sequencing errors (8). Indeed, an insertion of an A at nucleotide position 468 and an insertion of the dinucleotide GC at position 549 establishes a new open reading frame that is 96% identical in amino acid sequence to the rat PRMT1 protein (Fig. 2). The possible activity of HCP1 as a methyltranferase, however, was not assessed in this report (9).

A second human PRMT1 homolog was identified in a yeast two-hybrid screen designed to find proteins that interact with the intracytoplasmic domain of the interferon-α,β cytokine receptor (8). This human PRMT1 homolog was designated IR1B4 and is identical in sequence to the corrected HCP1 sequence (Figure 2)(8). The methyltransferase activity of human PRMT1 (IR1B4) was demonstrated by the fact that a histone-methylating activity could be coimmunoprecipitated using antibody to the interferon receptor IFNAR1 (8). In addition, extracts from a human myeloma cell line grown in the presence of an antisense construct of human PRMT1 (IR1B4) had an inhibitory effect on the level of endogenous protein arginine N-methyltransferase activity (8).

The relationship of the PRMT1 methyltransferase to the previously partially purified Type I histone and hnRNP A1 methyltransferases shown in Table I is not clear at this point. The native enzyme size of the PRMT1 enzyme appears to be smaller (180 kDa versus 275–450 kDa) than that observed in the previous preparations, and the catalytic unit has a lower molecular mass (40.5 kDa) than even the smallest of the major polypeptide species identified in any of the purifications (Section II,B).

In addition to the human PRMT1 homolog experimentally identified by both Nikawa *et al.* (9) and Abramovich *et al.* (8), searches of the database of expressed sequence tags (dbEST at NCBI) revealed three potential splicing variants of the human PRMT1, represented by human cDNA clones R99195, H87819, and H71767 (Fig. 3). We obtained and fully sequenced the insert within each of these clones (J. Gary, unpublished). With the exception of the amino terminus, the sequences of the human clones are nearly identical to the rat PRMT1 sequence. In fact, the differences seen may represent polymorphisms or errors in library construction. For example, clone H87819 (identical to the human sequences HCP1 and IR1B4) encodes a Met instead of Phe at amino acid position 77, whereas clone R99195 has a Val replacing a Leu at position 322, and clone H71767 has two Asp residues replacing two Glu residues at positions 335 and 348. The most significant differences between the human variants and the rat PRMT1 are found at the amino terminus (Fig. 3). From Fig. 3, the relatedness of R99195 to H87819 by alternative splicing is apparent. H87819 can be derived by splicing out the 121-bp segment of R99195, but the origin of the 5' fragment of H71767 is less evident.

We have expressed all three variants as GST fusion proteins. All are able to methylate an artificial substrate, GST–GAR (J. Gary, unpublished). This substrate was constructed by fusing the coding region of GST (from pGEX-2T; Pharmacia) in-frame with the amino-terminal region of human fibrillarin (amino acids 1–148) (Section IV,B,1) (R. Kagan and J. Gary, unpublished). This region of fibrillarin is known to contain asymmetric dimethylarginine *in*

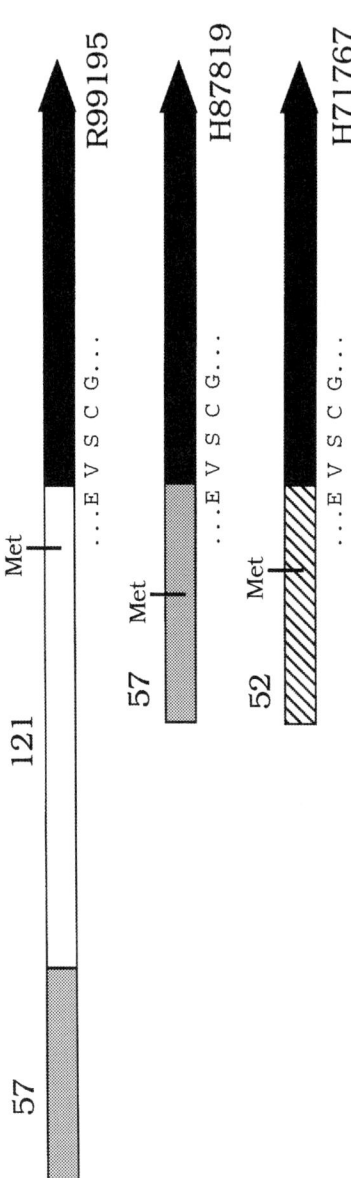

```
1 M V G V A - - - - - - - - - - - - E V S C G Q A E S S E K P N A E D M T S K    R99195 protein
1 M E N F V A T L A N G M S L Q P P L E E V S C G Q A E S S E K P N A E D M T S K    H87819 (HCP1 & IR1B4) protein
1 M A A A E A A N C I M - - - - - E V S C G Q A E S S E K P N A E D M T S K    H71767 protein
```

FIG. 3. Three alternative splicing variants of the human Type I protein arginine N-methyltransferase identified in the database of expressed sequence tags. The upper panel is a schematic diagram of the different mRNA species; regions of identity are similarly shaded. The proposed initiator methionines (Met) were assigned because they are the most 5'-AUG codons that are in-frame with the methyltransferase sequence that follows. A translation of the sequence near the splice junction where the three forms are identical is shown under each mRNA species. The lower panel is a protein sequence alignment of the translated amino termini from the three splice variants. The H87819 spliced form has also been identified in two separate screens as IR1B4 and HCP1 (8, 9).

vivo (*71*) and is therefore a substrate for the Type I protein arginine *N*-methyltransferase PRMT1. The reason for the three differently spliced forms of the human PRMT1 *in vivo* is not clear. Experiments using hypomethylated RAT1 fibroblast extracts (AdOx treated) (*6*) together with the different forms of the methyltransferase have not yet revealed any variation in methyl-acceptor substrate specificity (J. Tang and J. Gary, unpublished). Perhaps the different 5′ untranslated regions of each of the different mRNA transcripts play an important role.

In theory, the EST database should reflect the abundance of certain transcripts *in vivo:* common mRNAs should be randomly cloned proportionately more than rarer transcripts. Therefore, it is interesting to note that the H87819 splice variant has been identified twice in the literature (as HCP1 and IR1B4) in screens for specific functions but the other two clones have not, because the H87819 and R99195 amino termini are unique in dbEST and the H71767 form is present in at least five other clones.

Another study identified a rat gene encoding a protein with sequence homology to the rat PRMT1 between the methyltransferase regions I and III, but that was divergent outside of these regions (*248*). The gene is currently designated PRMT3. A GST–PRMT3 fusion enzyme can methylate GST–GAR (*248*), which classifies it as another Type I protein arginine *N*-methyltransferase.

An additional potential human arginine *N*-methyltransferase gene has been proposed based on sequence similarity alone and has been designated hHMT2 and PRMT2 (GB X99209, H. Scott, unpublished; GB U80213, N. Katsanis, unpublished). We had also identified this gene from BLAST searches of dbEST as human clones T75034 and T77642 (J. Gary, unpublished). The sequence of this potential methyltransferase (GB X99209 and U80213) is more distantly related to the rat PRMT1 than is the rat PRMT3 (27% amino acid identity between PRMT1 and PRMT2; 44% identity between PRMT1 and PRMT3), but the sequence within the conserved methyltransferase regions is very similar. The hHMT2/PRMT2 gene product has an additional interesting sequence feature: the amino-terminal portion has an SH3 domain, which is present in many signal transduction proteins and is known to bind a polyproline motif (*72*). We have expressed this protein as a GST fusion and have not detected any activity toward the synthetic Type I methyltransferase substrate GST–GAR, proteins present in the *rmt1* "hypomethylated" mutant yeast cytosol, nor myelin basic protein, the Type II substrate (A. Frankel and J. Gary, unpublished). Therefore it seems premature to designate this species as a protein arginine *N*-methyltransferase.

B. *Saccharomyces cerevisiae RMT1*

On identification of the rat PRMT1 as a Type I protein arginine *N*-methyltransferase, BLAST searches against the nonredundant protein database revealed a homolog in the yeast *S. cerevisiae* with 45% amino acid identity, the hypothetical *ODP1* gene product (5). *ODP1* was originally identified as an open reading frame downstream of the gene *PDX3* (73), and then was later resequenced through the yeast genome sequencing effort (yeast open reading frame *YBR0320*) (74) (Fig. 2). It encodes a potential protein product with a calculated molecular mass of 39.8 kDa and contains the conserved methyltransferase regions I, post I, II, and III, as well as the post III region that is not found in PRMT1 or its mammalian homologs. This sequence similarity, combined with biochemical data on the rat PRMT1 gene product, strongly implicated the *ODP1* gene product as a Type I protein arginine *N*-methyltransferase as well. A previous analysis of yeast ribosomal proteins indicated the presence of methylated arginines *in vivo* (75). However, a recent attempt to detect a Type I arginine *N*-methyltransferase activity in yeast was unsuccessful (49).

The ability of a GST–Odp1 fusion enzyme to methylate the synthetic R1 peptide substrate (69) was the first experimental indication that Odp1 is a protein arginine *N*-methyltransferase (5). The fusion protein was also able to mono- and asymmetrically dimethylate histones, hnRNP A1, and to a lesser degree cytochrome *c* and myoglobin, but did not methylate myelin basic protein at all (5) (J. Gary and M. Yang, unpublished). We thus renamed *ODP1*, calling it *RMT1*, for protein a<u>R</u>ginine *N*-<u>M</u>ethyl <u>T</u>ransferase, to signify its ability to mono- and asymmetrically dimethylate a broad range of substrates. *In vivo* and *in vitro* experiments using a strain in which the chromosomal copy of *RMT1* was disrupted showed that the mutant strain was viable but highly defective in the ability to form *N*-methylated arginine derivatives, whereas the parent strain produced only mono- and asymmetric dimethylarginine (5).

The *in vitro* methylation of cytosolic extracts from both the parent and mutant cells with [^{3}H]AdoMet allowed at least five proteins lacking these arginine derivatives to be visualized by SDS-PAGE analysis and fluorography (5). In addition, incubating the GST–Rmt1 fusion with cytosol from this mutant strain resulted in the specific mono- and asymmetric dimethylation of at least 13 polypeptides (5). Similar experiments with *in vivo*-labeled extract from each strain show that at least four polypeptide species (55, 42, 38, and 29 kDa) are no longer methylated in *rmt1* mutants (J. Gary, unpublished). This is a rather broad substrate specificity compared to the mammalian GST–PRMT1 fusion protein that recognizes only a single 55-kDa species in the mutant yeast extract (presumably Npl3 by SDS-PAGE migration) as well as 55- and 65-kDa polypeptides in untreated RAT1 fibroblast cytosolic fractions (5,6). This difference in substrate specificity might be explained if mammalian cells have multiple Type I methyltransferases with varying specificities instead of one enzyme that recognizes the majority of the substrates (6).

Rmt1 is responsible for over 80% of the *in vivo* and *in vitro* asymmetric arginine dimethylating activity in yeast (5). Approximately 65% of the *in vivo* monomethylating activity can be attributed to Rmt1 (5). We are currently attributing this significant *in vitro* Rmt1-independent activity to a less abundant protein arginine N-methyltransferase enzyme that may be released during homogenization from a cellular compartment, where it would not normally encounter such high concentrations of hypomethylated proteins (5). The protein responsible for this activity has not been identified.

RMT1 was also identified in a screen for mutants that caused synthetic lethality with a temperature-sensitive mutant allele of *NPL3* in yeast (7). The Npl3 protein has been implicated in various processes (24), including nuclear protein import (76), pre-rRNA processing (77), as well as binding and shuttling mRNA out of the nucleus (78, 79). Null mutants of *NPL3* are not viable (76); therefore, a partially impaired temperature-sensitive mutant was used to screen for mutants that might further exacerbate the phenotype of the *npl3* temperature-sensitive mutant, indicating either that the proteins have some physical interaction or are present in the same or overlapping pathways (7). The screen yielded four mutants, all in a single complementation group designated *SLN1* (synthetically lethal with *NPL3*). The *SLN1* gene was identified by complementation and found to be identical to *RMT1* (7). Importantly, a null disruption mutant of *RMT1/SLN1* is viable and does not have any of the phenotypes associated with a *npl3* mutant (7). It was shown that immunoprecipitated yeast Rmt1/Sln1 as well as recombinant Rmt1/Sln1–Myc methylated Npl3 and recombinant hnRNP A1 *in vitro* (7). With this information, the authors renamed *SLN1*, calling it *HMT1* (hnRNP methyltransferase) (7), although it seems more reasonable to now identify it as simply *RMT1* because of its broad substrate specificity. Despite the fact that the type of methylated residue formed by Hmt1 was not determined, analysis of the CNBr cleavage products of methylated Npl3 indicated that the site of methylation was in the carboxy-terminal region of Npl3 (7, 24). This region of Npl3 has a glycine- and arginine-rich (GAR) domain (80) containing the consensus methylation site for Type I protein arginine N-methyltransferases (Section V,A).

The subcellular localization of the Rmt1/Hmt1–Myc construct was found to be nuclear, as determined by indirect immunofluorescence, and is consistent with the presence of a weak bipartite nuclear localization signal (7). Preliminary immunofluorescence data indicate that the rat PRMT1 is present in the nucleus, as well (J. Tang and H. Herschman, personal communication). This result is interesting because PRMT1 was identified by its association with a protein, TIS21, that is completely cytosolic (58), and IR1B4/hPRMT1 was identified by its interaction with the cytoplasmic tail of a plasma membrane receptor (8). Furthermore, some of the early purification studies indicated a cytosolic localization (13, 37).

C. Reconciling the Purification and Cloning Data

The past three decades of research concerning the identities of protein arginine N-methyltransferases in different mammalian tissues have provided a database that is often conflicting and convoluted. From the purification data alone, there appear to be at least two major mammalian enzymes (Section II,B). However, there are discrepancies among the results of the partially purified methyltransferase preparations, and it is also unclear how the recently constructed GST fusion enzymes are related to these activities.

The Type I asymmetric arginine methyltransferase activity originally corresponded to a histone methyl-accepting activity and was also considered by some to be distinct from the Type II symmetric myelin basic protein-specific methyltransferase based on this substrate preference alone (30, 38, 40). However, data from various purifications of the Type I histone methyltransferase showed that arginines in histones were methylated to all three derivatives (Table II). This could be due to the substrate directing different types of methylation from the same enzyme due to the sequence or structure surrounding the methylation site (45). But in light of the evidence from experiments with more highly purified enzyme and substrate (5, 6, 49), the presence of symmetric dimethylarginine in histones may be due instead to contamination with the Type II (symmetric) myelin basic protein methyltransferase in each of these partial purifications and the subsequent action of both enzymes on histones. Histones do contain arginine residues in several different contexts as potential methylation sites and may allow varying degrees of catalysis by both types of enzymes. Myelin basic protein-specific methyltransferase preparations do methylate histones *in vitro*, but always to a lesser degree (40, 41). It will be very important in future work to determine whether histones are actually being methylated by the Type II enzyme and if these histones contain mostly symmetric or asymmetric dimethylarginine residues. From these results, it may be possible to distinguish between the presence of a contaminating Type I methyltransferase in Type II preparations or whether the Type II enzyme can also recognize symmetric methylation sites within histones.

Both the Type I asymmetric (RNA-binding protein-specific) and Type II symmetric (myelin basic protein-specific) arginine methyltransferases appear to exist in larger complexes than their apparent constituent polypeptide chains would suggest. By gel-filtration analysis, the masses for the enzymes range from 150 to 500 kDa (Table I). In our identification of the rat PRMT1 enzyme from RAT1 fibroblast cells, we found that the native methyltransferase eluted from a Superdex S200 gel-filtration column as a 180-kDa species (6). The native yeast enzyme, Rmt1, eluted at an approximate molecular mass of 74 kDa by Superdex S200 gel-filtration analysis (J. Gary, unpublished), suggesting it may be a homodimer.

At this point we cannot resolve the discrepancies seen in the native molecular masses of the mammalian enzymes. No attempts have been made so far to compare directly the gel filtration behaviors of the different enzyme preparations shown in Table I, and there is enough uncertainty in the experimental descriptions that it is difficult to say whether the enzymes are of similar size. For example, Gallwitz (25) did not mention what protein standards were used in order to make the molecular mass calculation of 275 kDa for the rat thymus enzyme. Ghosh *et al.* (40) did not run all of the standards simultaneously, but rather over several column runs, and stated that the runs were variable by as many as three fractions. Liu and Dreyfuss (49) did not show the data for their determination of the molecular mass (450 kDa), and Rawal *et al.* (48) did not run the standards with the sample. Finally, we used endogenous rat proteins for the column calibration and therefore ran the sample and the standards simultaneously (6). We, however, did not include a standard over 161 kDa and may have therefore underestimated the mass (6).

Another problem has been attempting to assign those polypeptides identified in the mammalian purifications with the protein arginine N-methyltransferase gene products that have been recently characterized. The molecular masses of the polypeptides remaining in the active pool after purification range from 45 to 110 kDa by SDS-PAGE analysis (40, 48, 49); however, both the mammalian and yeast proteins migrate on SDS-PAGE as approximately 40-kDa species (5, 6). It is clear from the methylation experiments using the GST fusion enzymes with homogeneous substrates such as recombinant hnRNP A1 or GST–GAR that PRMT1 and Rmt1 are the catalytic units of a Type I protein arginine N-methyltransferase.

The additional higher molecular weight polypeptides seen in many of the purifications may be contaminating proteins and/or regulatory or additional catalytic subunits. As previously mentioned, our gel-filtration and SDS-PAGE analysis data are consistent with the native rat PRMT1 being a homotetramer (homodimer for yeast), but the larger, more complex structures suggested by the results of several purifications cannot be excluded until experiments such as quantitative native immunoprecipitation of the methyltransferase can be done. It is also possible that the situation is more complex. The PRMT1 product may be a minor species within cells, and the genes encoding the actual catalytic subunits of the Type I and Type II enzymes identified in the purifications (Table I) may have not yet, in fact, been identified. This seems unlikely given that the substrate specificities of the GST–PRMT1 fusion protein and the Type I enzyme appear to be very similar, but further work is needed in this area. Even the designations Type I and Type II are tenuous, because they are based solely on substrate and reaction product specificity. These classes are not mutually exclusive. In fact, the Type I and Type II methyltransferase complexes could feasibly differ only in their subunit organization. A similar scenario has been suggested for the phosphatase PP2A, which has a complex and interchanging subunit structure, multiple substrates and seems to affect various processes at different points in the cell cycle (81–83). Although much of the current progress in this area has been through the identification of new arginine methyltransferases from database searches or yeast two-hybrid analyses, it may be more practical in the future to follow the trail backward from newly or previously identified substrates.

IV. Methyl-Accepting Substrates for Protein Arginine N-Methyltransferases

A. General

The first *in vitro* methyl-accepting substrates for protein arginine *N*-methyltransferases were crude as well as fractionated histones from calf and rat thymus (12, 13, 25). However, *in vitro* methylation reactions with endogenous rat brain proteins showed that although the majority of the incorporation was into methylarginine derivatives, only 20% of the incorporation was into the crude histone fraction (17, 84). These results indicate that histones, although apparently they can be methylated *in vitro*, are not necessarily the major *in vivo* substrates.

The identification of endogenous arginine methylation substrates is not straightforward, because the *in vivo* methyl-accepting sites may already be partially or fully occupied in tissue extracts and in purified proteins, reducing the incorporation of radioactivity from radiolabeled AdoMet in the *in vitro* assays. For instance, the myelin basic protein from carp, which is not methylated *in vivo* (35), is a fivefold better substrate than myelin basic protein from cows and is 17-fold better than myelin basic protein from rabbit in terms of the [^{14}C]methyl group incorporation into the protein from [^{14}C]AdoMet (45). Nonetheless, specific polypeptide substrates can be detected from cell lysates by fluorography after *in vitro* labeling with radiolabeled AdoMet and a partially purified methyltransferase preparation. For example, several substrates from human placental cell cytosol and nuclei were identified using a Type I protein arginine *N*-methyltransferase partially purified from the same cells (26).

By incubating rat pheochromocytoma PC12 cells with adenosine dialdehyde, a specific inhibitor of S-adenosyl-L-homocysteine hydrolase, it was possible to increase the proportion of unmethylated substrates within the cells by causing an increase in the intracellular concentration of S-adenosylhomocysteine, a product inhibitor of all AdoMet-dependent methyltransferases (85). About 72% of the endogenously incorporated methyl groups from an adenosine dialdehyde-treated homogenate incubated with [^{3}H]AdoMet could be accounted for by asymmetric dimethylarginine alone (85). Similarly, specific substrates such as hnRNP A1 were identified in adenosine dialdehyde-treated RAT1 fibroblast cells incubated with [^{3}H]AdoMet and the GST–PRMT1 fusion (6). In this experiment, hnRNP A1 was immunoprecipitated from the methylation reaction containing adenosine dialdehyde-treated cytosol by either a specific monoclonal antibody or a control antibody. Analysis of protein in the immunoprecipitated pellet by SDS-PAGE and fluorography indicated that a methylated band corresponding to the migration position of hnRNP A1 was present only in the lane using the anti-hnRNP A1 antibody and not the control antibody (6). Rather than using adenosine dialdehyde with yeast, it is possible to obtain hypomethylated cytosol through genetic manipulations; yeast *rmt1* null mutants are specifically devoid of a majority of asymmetric dimethylarginine derivatized proteins (5). Therefore, cytosol from these mutant cells would be expected to contain the *in vivo* substrates in their unmethylated forms. Experiments using this hypomethylated (*rmt1*) yeast cytosol allowed the identification of approximately 13 radiolabeled polypeptide substrates using the GST–Rmt1 fusion protein and [^{3}H]AdoMet (5). The actual identities of these polypeptides have not been determined. However, a list of candidate gene products is described in Section V,B,1.

In the next section, we present the identities of known *in vivo* substrates, determined directly by sequencing or amino acid analysis, for both types of protein arginine N-methyltransferases. The methylation consensus sequence for the two general types of substrates parallels the distinction between their two respective methyltransferases. From this compilation of methylation sites, it is also possible to predict other potential substrates from existing protein databases that contain both characterized as well as hypothetical proteins from various genomic sequencing projects.

B. Asymmetric Dimethylarginine-Containing Substrates Identified *in Vivo*

1. FIBRILLARIN

Fibrillarin is a 34-kDa polypeptide component of a nucleolar small nuclear ribonuclear protein involved in the first processing step of preribosomal RNAs (86). It is associated with the U3, U8, and U13 small nuclear RNAs (87). Amino acid analysis of the rat protein indicated that approximately 13

TABLE III
List of Positively Determined *in Vivo* Substrates of Arginine
Methylation and Their Homologs from Other Species

Definition	Organism	ID[a]
Fibrillarin		
	Homo sapiens	g182591
	Mus musculus	g296547
	Xenopus laevis	g214143
	Drosophila melanogaster	g929566
	Caenorhabditis elegans	g1495012
	Schizosaccharomyces pombe	g544285
	Giardia intestinalis	g520824
	Leishmania major	g544283
	Tetrahymena thermophila[b]	g578561
Nucleolin		
	Homo sapiens	g189305
	Rattus norvegicus	g205790
	Mus musculus	g53453
	Cricetulus griseus	g387050
	Gallus gallus	g212412
	Xenopus laevis	g64936
hnRNP A1		
	Homo sapiens	g32344
	Macaca mulatta	g422724
	Rattus norvegicus	g133255
	Mus musculus	g1507693
	Xenopus laevis	g133258
	Drosophila melanogaster	g133253
Basic fibroblast growth factor	*Homo sapiens*	g183083
Myelin basic protein		
	Homo sapeins	g187408
	Rattus norvegicus	g126804
	Mus musculus	g387414
	Gallus gallus	g126798
	Bos taurus	g126796
	Cavia porcellus	g126797
	Ovis aries	g223882
	Xenopus laevis	g1816437
	Raja erinacea[c]	g1177855
	Squalus acanthias[c]	g1177857
	Heterodontus francisci[c]	g126799

[a] Unique identification number of GenBank retrieval.
[b] No GAR or RGG.
[c] No R at 107.

mol of asymmetric dimethylarginine are present per mole of fibrillarin, with no mono- or symmetric dimethylarginine (71). Amino-terminal sequence analysis identified six asymmetric dimethylarginine residues within the first 31 residues, and no unmodified arginine residues were detected. All of the methylated arginines were present in a glycine- and arginine-rich (GAR) domain (71) (Section V,A). Rat fibrillarin homologs are present in many different species, ranging from the *Physarum polycephalum* B-36 protein (88) to human fibrillarin, although their methylated arginine content has not been determined experimentally (Table III).

2. NUCLEOLIN

Nucleolin, a 110-kDa polypeptide, is one of two major phosphoproteins from the nucleolus (89). It has also been implicated in several steps of ribosome biogenesis (90): regulation of RNA polymerase I transcription, modulation of chromatin conformation, binding to nascent rRNAs (91, 92), and possible shuttling between the nucleus and cytoplasm to facilitate transport of ribosomal proteins (93). Rat nucleolin, originally designated C23, also contains asymmetric dimethylarginine (approximately 9 mol per mole of nucleolin) and trace amounts of monomethylarginine *in vivo* (94). Sequence analysis of a 53-amino acid tryptic peptide derived from the carboxy-terminal region of the protein identified 10 asymmetric arginine residues, but no mono- or symmetric dimethylarginine, nor any unmodified arginine residues (20).

3. HETEROGENEOUS NUCLEAR RIBONUCLEOPROTEIN A1

Sequence analysis of a carboxy-terminal tryptic peptide fragment of calf thymus unwinding protein 1 (UP1) indicated the presence of a single asymmetric dimethylarginine residue at position 193 (194 if an initiator Met is included) out of 195 residues, and amino acid composition analysis also revealed one residue of asymmetric dimethylarginine per polypeptide (95). The UP1 used in this study may not have been full length, however, because its rat homolog, helix-destabilizing protein (HDP) (96), has an additional 124 residues at the carboxy terminus. This region in HDP contains the GAR domain, suggesting that the one residue of methylated arginine per polypeptide in UP1 may underestimate the total asymmetric dimethylarginine present in the full-length protein. Subsequently, it was determined that HDP is identical to hnRNP A1 (97, 98). Kumar *et al.* (97) determined that endogenous hnRNP A1 from calf thymus contained 3.1 asymmetric dimethylarginine residues per protein. The earlier determination of 0.31 residues of asymmetric dimethylarginine per hnRNP A1 molecule (43) cannot be explained assuming that full-length hnRNP A1 was analyzed.

The hnRNP A1 protein is a 34-kDa polypeptide and a major component of the 40S ribonucleoprotein particle (43). A role for hnRNP A1 has been established in the differential selection of proximal or distal 5' splice sites (99, 100). In vitro, hnRNP A1 can promote alternative exon skipping in certain constructs (99). As in the previous examples of in vivo substrates, hnRNP A1 is also nuclear localized and may shuttle to the cytoplasm as well (101, 102).

Recombinant hnRNP A1 has the distinction that it is the only in vivo substrate that has been extensively used as a methyl-accepting substrate in vitro. It has been used to monitor enzyme activity in several purifications of the Type I methyltransferase (Section II,B) (Table I). Analysis of the in vitro-methylated hnRNP A1 with various endo- and exoprotein hydrolases, including V8 protease, trypsin, and carboxypeptidase B, suggested that the site of methylation is the same arginine (position 194; SSQ**R**GRS) that was determined to be the methylated residue in UP1 (47). Arginine residue 196 is not methylated; carboxypeptidase B treatment of the tryptic peptide KQEMASASSSQRGR releases only free arginine, not a methylated derivative. The fact that the bond at Arg-194 is not cleaved by trypsin and that after the carboxypeptidase B treatment the radioactivity is still associated with the peptide suggests that the methylation of the only remaining arginine occurs at position 194 (47). The methylation status of the other four arginines in the GAR region could not be absolutely determined, but rather was suggested only by the methylation of smaller tryptic peptide fragments derived from the region of the protein carboxy-terminal to residue 196 (47). In a separate study, sequence analysis of a tryptic fragment corresponding to amino acids 197–215 shows an unidentified amino acid at position 206 that is encoded as an arginine residue, possibly indicating the presence of an asymmetric dimethylarginine residue at that position (47, 97).

Although hnRNP A1 has been the most extensively studied, the 40S ribonucleoprotein particle is composed of at least 11 additional polypeptides (44). In addition to hnRNP A1, five other unidentified polypeptide components of this particle were found to contain asymmetric dimethylarginine in vivo (44). Amino acid analysis of bulk ribonuclear proteins from the 40S core particle showed that 0.5 mol% of the amino acids was asymmetric dimethylarginine, and that specifically hnRNP A2 contained 1.4 asymmetric dimethylarginine residues per polypeptide (43).

4. BASIC FIBROBLAST GROWTH FACTOR

The mammalian mitogenic protein, basic fibroblast growth factor (bFGF), can stimulate growth in a wide range of mesodermal and neuroectodermal cell types (103, 104) as well as vascular endothelial cells (105). Although this protein is a potent mitogen, an active mechanism for the secretion of bFGF from cells in vivo has remained elusive (106). The protein has no secretion signal sequences and has the general features of a cytosolic protein (107). In fact, evidence exists to support the idea that bFGF is actually only released from cells after cell death and/or cell lysis (108–111). Therefore, although the determined mitogenic action of bFGF seems to separate it from the other methylated RNA-binding substrates mentioned above, the true function of this polypeptide intracellularly remains to be determined.

This polypeptide has been determined to be an 18-kDa species by the isolation of human bFGF (112) and sequence analysis of a human cDNA clone (113, 114). Amino-terminally extended forms of bFGF have also been isolated, ranging from just two additional amino acid residues (115) to an additional 53 residues, creating a 25-kDa protein (114). Interestingly, these high-molecular-weight forms of bFGF are translated from alternative CUG start codons (116). Sequence analysis of tryptic fragments of the 25-kDa bFGF isolated from guinea pig brains suggested the presence of at least three sites of arginine methylation in the context of a GAR domain (114). The putative methylated phenylthiohydantoin–amino acid derivative eluted slightly earlier than a symmetric dimethylarginine derivative from myelin basic protein (117), suggesting that the methylated residue is an asymmetric dimethylarginine.

Although the 18-kDa form of bFGF is predominantly cytoplasmic, the larger forms of bFGF are specifically localized to the nucleus (106, 118), and it has been suggested that nuclear localization is dependent on the methylation of the peptide (106). When NIH 3T3 cells were incubated with a general inhibitor of AdoMet-dependent methyltransferases, 5′-methylthioadenosine, the larger forms of bFGF were not posttranslationally modified, as determined by differential electrophorectic mobility on SDS-PAGE (106). This presumably unmethylated bFGF did not accumulate in the nucleus as was found in control cells. This result was not due to a decrease in the stability of the unmodified form (106). In fact, treatment with methylthioadenosine actually stabilized all forms of bFGF in pulse-chase experiments and specifically increased the expression of the high-molecular-weight forms (106).

5. Histones?

It is important to note that there are uncertainties about the methylation of histone fractions, the original methyl-accepting substrate of the Type I enzyme in *in vitro* studies. For instance, it is still not certain exactly what types of methylated arginine residues are present in the *in vitro*-methylated preparations and whether the methyl acceptors are indeed histone polypeptides! Amino acid analysis of *in vitro* methylation reactions containing histones, [^{3}H]AdoMet, and a partially purified preparation of Type I methyltransferase indicates the presence of a small amount of symmetric dimethylarginine residues (26, 38, 39) (Table II). However, four additional studies did not detect the symmetric form in similar methylation assays with their Type I methyltransferase preparations (*30, 48, 119*) or with a recombinant GST–PRMT1 methyltransferase fusion purified from *E. coli* (*6*) (Table II). The radiolabeled symmetric dimethylarginine residues detected in the former three studies may be due to a contaminating Type II arginine methyltransferase that methylates the exogenous histones.

Several fractionated histone preparations as well as individually purified histones have been used as substrates for the asymmetric protein arginine *N*-methyltransferase (*12, 13, 25, 26, 38, 52*). The "histone" substrates used include the "sulfated histone fraction," the arginine-rich fraction, the lysine-rich fraction, calf thymus fractions IIb–III and IV, f2a1(IV), f1, f2a, f2b, f3, and chicken fraction IV, as well as pure histone H1, H2A, H2B, and H4. The histones in these protein preparations have been assumed generally to be uncontaminated. Therefore, the experimental determination of arginine-specific methylation involved only the bulk precipitation and quantitation of the radioactivity present in the total protein of the reaction. Quantitation often involved direct determination of radiolabeled methylated arginine content by amino acid analysis of protein acid hydrolysates. However, it was also common to just count the total acid precipitable counts from the reaction and assume they corresponded to the amount of radiolabeled methylated arginine.

Given the difference in the *in vitro* activity of the methyltransferase on hnRNP A1 and histones (0.49 µg of hnRNP A1 gives an activity similar to 100 µg of histones) (*5*), it is possible that the presumed histone methylation is due to the preparation having a small amount of contamination with a true substrate and that histones are not, in fact, methyl-accepting substrates. None of the studies, except for those dealing with the Type III wheat germ histone methyltransferase (*50, 51*), analyzed tryptic fragments or performed amino acid analysis that would prove the actual arginine methylation of histones *in vitro*.

By incubating calf thymus histone fraction IIAS from Sigma, a histone preparation common to the arginine methylation literature, with the purified rat Type I methyltransferase fusion enzyme (GST–PRMT1) or yeast fusion enzyme (GST–Rmt1) and [^{3}H]AdoMet, we found only a single, major methylated polypeptide (Fig. 4). We were able to visualize the histones by separating the components of the reaction on a detergent–acid–urea gel system (*120, 121*) and Coomassie staining the gel (Fig. 4). Fluorography indicated that the radiolabeled polypeptide corresponded to the elution position of histone H2B (*120*). Another band, which is 5- to 10-fold less intense, was also observed on the fluorograph and migrates at a position similar to histone H2A or H3 (Fig. 4). Significantly, histones H1 and H4, which previously had been suggested to be substrates (*26, 38*), are present but are not methylated under the conditions used here (Fig. 4). The protein sequences of histones H2B, H2A, and H3 do not contain sequences similar to the glycine- and argi-

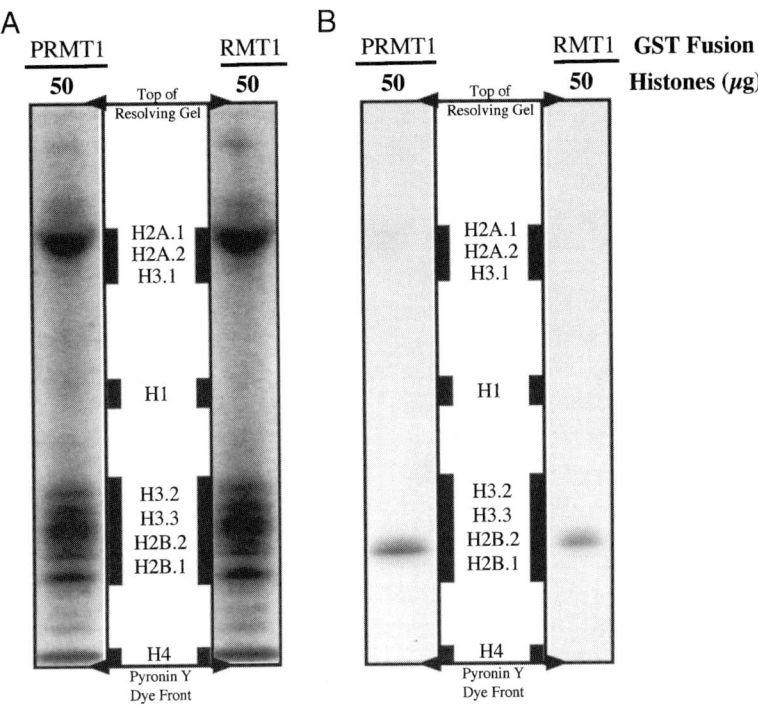

FIG. 4. Histone H2B is a candidate methyl-accepting substrate from *in vitro* arginine-methylated calf thymus histone fraction IIAS. The rat and the yeast Type I protein arginine N-methyltransferase fusions, GST-PRMT1 and GST-Rmt1, respectively, were incubated with 50 μg of calf thymus histone fraction IIAS (Sigma) and [^{3}H]AdoMet in a reaction identical to those previously described (5, 6). The reactions were stopped by the addition of sample loading buffer and boiling for 5 min. The samples were then separated on a detergent–acid–urea gel (120, 121) to resolve the various histone components. A photograph of the stained gel is shown (A). The gel was then prepared for fluorography as described previously (5, 6). The fluorograph was exposed for 3 days at −80°C (B). A species running similarly to the published migration position of H2B (120) was the major methyl-accepting species. A minor methylated species was also seen in the area where H2A and H3 are known to run (120).

nine-rich region or the consensus methylation sequence that is found in *in vivo* substrates (Section V,A). However, the histones do contain many arginine residues and they may become artifactual sites of methylation when present at high enough concentrations. In fact, even poly(arginine) has been identified as a substrate when as much as 3 mg was used per reaction (13, 38). As with the other reports of histone methylation *in vitro*, we cannot rule out the possibility that contaminating methyl-accepting proteins run similarly to the labeled bands and are the actual methyl-accepting substrates in the reaction. Indeed, the known *in vivo* substrates for the Type I protein arginine

N-methyltransferase are very abundant in the nucleus and may be ubiquitous contaminants of histone preparations. At the least, we can state that under our conditions, using recombinant methyltransferase fusion proteins and histone fraction IIAS, we observed the methylation of a single major and one minor polypeptide *in vitro*, and not the multiple histone species suggested by the literature (*12, 13, 25, 26, 38, 52*).

From the data described above, it is not clear which histones, if any, are methylated *in vitro* and evidence for the presence of methylated arginines *in vivo* is even more conflicting. Methylated arginines have been suggested to be present in histones *in vivo* from labeling experiments with rat tissues and mammalian cell culture using [*methyl*-^{14}C]Met (*122–124*). Unfortunately, the identities of the radiolabeled peaks, supposedly representing methylated arginine species, from the amino acid-analysis of the acid hydrolyzed proteins were not directly compared to free methylated arginine standards. It was also reported that histones isolated from rat liver nuclei contained methylated arginines, whereas those from calf thymus did not (*125*). However, in additional studies, no methylated arginines were detected in histones isolated from rat organs (*126, 127*). It is possible that, as mentioned above, the histones isolated in the experiments in which methylarginine was detected were contaminated with other nuclear or nucleolar proteins that are actual *in vivo* substrates. In fact, it was an additional purification step after the usual acid extraction of "crude histones" that demonstrated that a histone preparation could be contaminated with nonhistone proteins containing asymmetric dimethylarginine (*127*).

The situation in nonmammalian cells may be different, however. The most compelling evidence for the *in vivo* methylation of histones is a report that the methylation pattern in *Drosophila* histone H3 shifts from lysine to arginine after a heat-shock treatment at 37°C (*128*). In this case, thin-layer chromatography detected both symmetric and asymmetric dimethylarginine in acid hydrolysates of histone H3 isolated from *Drosophila* Kc III cells after heat shock (*128*). A final determination as to whether histones contain methylarginine derivatives *in vivo* awaits the application of the mass spectroscopy technique either described by Edmonds *et al.* (*129*) or used by Kouach *et al.* (*130*) on a variety of histones isolated under different conditions and from different organisms. It is certainly possible that histones are methylated in some species (flies and plants) (*51*) and not in others (mammalian).

6. OTHER POTENTIAL *in Vivo* GAR DOMAIN-CONTAINING SUBSTRATES

Two additional proteins have been suggested to contain asymmetric dimethylarginine *in vivo:* ICP27 from the herpes simplex virus and Npl3 from yeast. ICP27 (infected cell protein) is an immediate-early gene product produced by the herpes simplex virus soon after infection of a cell (*131*). The protein is essential for virus growth and it is specifically required for the viral transition to the late phase of infection (*132*), as well as the general stimulation of viral DNA replication (*133*). A direct mechanism for its mode of action has not been determined. However, it is thought to be posttranslational through the altering of the pre-mRNA processing machinery (*134, 135*). Consistent with these functions, ICP27 is found in the nucleus and also accumulates in the nucleolus (*136, 137*). The *in vivo* methylation of ICP27 was demonstrated in African green monkey kidney Vero cells labeled with [*methyl*-^{3}H]Met in the presence of protein synthesis inhibitors (*138*). The methylation of the protein was shown to be dependent on the presence of consensus asymmetric arginine dimethylation sites in the amino-terminal third of ICP27 (*138*). However, direct amino acid analysis or protein sequencing of ICP27 has not been done to confirm the identity of the methylated residues as asymmetric dimethylarginine.

Similarly, the evidence supporting the *in vivo* arginine methylation of yeast Npl3 is based on the presence of a carboxy-terminal consensus sequence for asymmetric arginine methylation (*7*). CNBr cleavage of Npl3 isolated from yeast labeled *in vivo* with [*methyl*-^{3}H]Met versus [^{35}S]Met indicated that methylation occurs in the carboxy-terminal region of the protein (*7*). Additional evidence for *in vivo* methylation came from the discovery of an antibody that was specific for Npl3 isolated from yeast, but not from *E. coli* (*24*). It was subsequently found that the antibody recognizes the bacterial recombinant Npl3 if it is incubated with yeast extracts and AdoMet. Binding was competitively inhibited by peptides corresponding to the consensus asymmetric arginine dimethylation site (R1 or R3) (*24, 69*). The authors concluded that the epitope recognized by the antibody included the asymmetrically dimethylated arginines of Npl3 (*24*). The *NPL3* gene was originally isolated because a mutant of this gene was defective in nuclear protein import (*76*). However, it is currently believed that the protein is also involved in the binding and export of mRNA from the nucleus (*139*). Temperature-sensitive mutants of *NPL3* accumulate mRNA in the nucleus (*79*), and the wild-type protein shuttles between the nucleus and the cytoplasm (*78*).

Although neither ICP27 nor Npl3 have been proved to contain asymmetric dimethylarginine *in vivo*, we have included them in this section because of the strong indirect evidence.

7. Candidate *in Vivo* Substrates

Several other possible methyl-accepting substrates have been identified *in vivo*, including the high-mobility group proteins 1 and 2 from calf thymus, heat-shock proteins HSP70A and HSP70B from chicken fibroblasts, as well as myosin from rats. The mono- and asymmetric dimethylation of these proteins was determined by amino acid analysis of acid hydrolysates derived from each of the endogenously isolated proteins. These studies, however, do not address the possibility of contamination in their preparations of these putative methyl-accepting substrates. Therefore, the methylated arginines in these isolates could conceivably be due to contamination with a small amount of a highly methylated Type I protein arginine *N*-methyltransferase substrate, such as fibrillarin or hnRNP A1. It will be necessary to obtain amino acid sequence data from these possible methyl-accepting substrates in order not only to identify the type of methylated residue but also to positively determine the identity of the substrate protein.

   i. High-mobility group (HMG) proteins are nonhistone proteins associated with chromosomal DNA (*140*). Although the functions of HMG1 and HMG2 remain unclear, they are known to bind DNA in the minor groove of AT tracts (*141*). It has also been shown that the proteins induce DNA bending and may promote cis-acting protein–protein interactions or be important in the compaction of DNA (*142*). Direct amino acid analysis of endogenous HMG1 and HMG2 isolated from calf thymus indicated the presence of asymmetric but not mono- or symmetric dimethylarginine (*143*).
   ii. Two of the major species in the 70-kDa heat-shock protein superfamily have been shown to be methylated on arginine residues *in vivo* in chicken embryo fibroblasts (*144*) and in mouse 3T3 fibroblasts (*145*). These proteins act as molecular chaperones to ensure proper protein folding under stressed and unstressed conditions. Protein methylation in both systems is tightly linked to protein synthesis (*145*). The level of monomethylarginine in the HSC70 homolog is markedly reduced in 3T3 cells but not in chicken cells on arsenite treatment that induces the heat-shock response. The level of monomethylarginine is reduced, however, in the chicken HSP70B protein after such treatment. Significantly, peptide binding and ATPase activities are similar in purified preparations of recombinant rat HSC70 (presumably unmethylated) and bovine brain HSC70 (presumably methylated), suggesting that methylation may not affect these properties of the proteins (*146*).
   iii. There is a limited literature describing the methylation of myosin on arginine residues. Only the asymmetric derivative was found in the myosin of cultured rat muscle cells labeled with [*methyl*-^{14}C]Met (*147*). Arginine-methylated myosin was also detected directly in tissues from rat fetuses and pups younger than 7 days old. Adult rats had no detectable methylated arginine residues in myosin (*147, 148*). Besides the lack of protein sequence data, no attempts have been made to determine which polypeptide chain of myosin is actually methy-

lated. This developmental regulation of methylarginine content in myosin has not been reported for the other constitutively methylated asymmetric substrates such as fibrillarin (71), which makes it more unlikely that the methylated arginines in the myosin preparations are derived from a contaminating protein.

C. Symmetric Dimethylarginine-Containing *in Vivo* Substrates—Myelin Basic Protein

Several studies have identified only the mono- and symmetrically dimethylated arginine, at position 107, corresponding to the bovine sequence, in myelin basic protein from human, monkey, bovine, rabbit, guinea pig, chicken, turtle, and frog species; however, there was no mono- or symmetric dimethylarginine found in carp myelin basic protein (18, 21, 22, 31, 32, 34, 35). Only two studies have identified asymmetric dimethylarginine in myelin basic protein (19, 149); these preparations may have been contaminated, however, with other proteins containing this residue (34, 35).

Mammalian myelin basic protein is one of the major protein components of the myelin sheath (150). It is located in compact myelin and interacts with opposed cytoplasmic membrane surfaces, perhaps playing an essential role in maintaining this structure (151–153) (see Ref. 154 for review). The protein exists in two isoforms, 18.5 and 14 kDa, generated from the same gene by differential mRNA splicing (155). In the rat, the smaller form of myelin basic protein is identical to the larger form, except for a 40-amino acid internal deletion, and it is also methylated on the corresponding arginine (33). The two isoforms can be further posttranslationally modified to create one of eight charge isomers (155). Other posttranslational modifications of this protein include phosphorylation, ADP-ribosylation, and conversion of arginine residues to citrulline residues (155).

Myelin basic protein is the only example of a protein known to contain symmetric dimethylarginine *in vivo*. To date, all of the Type II arginine methyltransferases have been isolated from the brain tissue from various species (Table I), matching the localization of the only known substrate. Methylation experiments with hypomethylated, adenosine dialdehyde-treated rat PC12 cells (Section IV,A) also suggest that there may be far fewer symmetric dimethyl-accepting arginine substrates than asymmetric ones (85). Amino acid analysis of these PC12 extracts after the addition of [^{3}H]AdoMet showed that only 2.9% of [^{3}H]methyl group incorporation was into symmetric dimethylarginine, compared to 71.5% for asymmetric dimthylarginine (85). This result should be viewed with some caution, because this type of analysis necessitates that the protein methyltransferases can recognize their substrates after completion of protein synthesis and folding. It may be that the enzymes responsible for the asymmetric versus symmetric dimethylation of arginine residues are differentially able to methylate methyl-accepting substrates that have been released from the ribosome. In fact, Type II methyltransferase activity has been found in other tissues besides brain (29), suggesting that other nonmyelin basic protein substrates may exist for this methyltransferase.

V. Determination of Arginine Methyltransferase Substrates

A. Sequence Analysis of *in Vivo* Asymmetric Methylation Sites

From the four amino acid sequences in which the actual methylated residue(s) were directly determined, a total of 20 asymmetric dimethylarginine residues have been identified (Fig. 5A). Because the methylated arginine residues of fibrillarin and bFGF are found near the amino terminus and those of nucleolin and hnRNP A1 are near the carboxy terminus, it appears that the methylation site is independent of its location within the polypeptide. This is also supported by the observation that the GST–GAR fusion protein is a substrate (Section III,A) (J. Gary, unpublished).

By compiling data on the residues surrounding the methylated arginine residue, it is possible to generate a consensus sequence for asymmetric dimethylarginine formation: (G/F)GGRGG(G/F) (Fig. 6). Only the arginine and glycine at positions 0 and +1 are found in all methylation sites (Fig. 6). Whether the glycine residue at position +1 is essential for methylation is not known. We chose to limit our comparison to positions −3 to +3 among these sequences, because the percentage of residue conservation had dropped to below 50% by the third position away from the arginine (Fig. 6). For these proteins, the methylation occurs at an RGG-methylation site that is within the larger context of the GAR domain (Fig. 6). The term "RGG" is used to describe the actual site of asymmetric methylation; the Arg-Gly sequence is completely conserved for known Type I methyltransferase substrates, and the final Gly is included to distinguish this site from the myelin basic protein symmetric methylation site Arg-Gly-Leu (Section V,C)

In order to determine the general nature of the consensus generated here, we also separately compiled the residues surrounding the arginine residues in the GAR regions from the homologs of the known *in vivo* substrates from different species (Fig. 6 and Table III). Interestingly, the known *in vivo* methylation site for hnRNP A1 is not a particularly good match to the consensus (compare Figs. 5A and 6). In fact, a peptide identical to a portion of this sequence (SSQRG) is a very poor substrate for a partially purified Type

A

rat Fibrillarin	FSP**R**GGGFGG**R**GGFGD**R**GG**R**GGG**R**GG**R** (GGF)
rat Nucleolin	FGG**R**GG**R**GGF GG**R**GGG**R**GG**R**GGFGG**R**G**R**GGFGG**R**GGF**R**GG**R**GGG
guinea pig bFGF (25 kDa-form)	VGG**R**G**R**G**R**GTA
rat UP1/hnRNP A1	SSQ**R**GR(S)

B

```
                    203┐                                    ┌236
```

rat hnRNPA1	GGG**R**GGGFGGNDNFG**R**GGNFSG**R**GGFGGS**R**GGG
monkey hnRNPA1	GGG**R**GGGFGGNDNFG**R**GGNFSG**R**GGFGGS**R**GGG
mouse hnRNPA1	GGG**R**GGGFGGNDNFGGFGGS**R**GGG
Xenopus hnRNPA1	FGG**R**GGNFGGN**R**GGGGGFGN**R**GYG
Drosophila hnRNPA1	GGG**R**GGPGG**R**AGGN**R**GNMGGGNYGNQNGGGNWNNGGNNWGNN**R**GGN

FIG. 5. Sequences of proteins containing Type I protein arginine N-methyltransferase methylation sites. (A) The sequences surrounding arginine residues that have been directly determined to be methylated *in vivo*. These asymmetrically dimethylated arginine residues are boldfaced and underlined. Residues in parentheses were not reported in the original studies but were taken from the corresponding GenBank entry and are included here for the compilation of the consensus sequence. (B) The GAR domains of hnRNP A1 proteins from various organisms.

I protein arginine N-methyltransferase from rat liver and calf brain, but a peptide (GNFGGGRGGGFGG) corresponding to a portion of the GAR motif six residues away is a much better substrate (*156*). For these reasons, in the expanded compilation we included only those arginine residues that were clearly in the GAR domain (Figs. 5B and 6). A surprising feature is that the fibrillarin from *Tetrahymena thermophila* does not contain a GAR domain and is therefore presumably not arginine methylated *in vivo*. In the end, the results of this expanded comparison are not drastically different from those obtained with just the known substrates (Fig. 6).

The structure of the GAR domain has not been determined. This domain, however, could adopt higher order structures akin to other proteins that have high percentages of glycine residues, e.g., silk and collagen. The structure of the larger GAR domain may not necessarily be crucial for methylation. This short RGG consensus noted above seems to be sufficient to direct asymmetric arginine methylation, as seen in the methylation of the synthetic peptide R1 (GGFGGRGGFG) by a Type I protein arginine N-methyltransferase from rat PC12 cells (*69*) as well as the rat GST–PRMT1 and yeast GST–Rmt1 fusion proteins (*5, 6*).

From this consensus sequence, the likelihood that histones, specifically H2B, H2A, and H3 (possible histone substrates identified in Fig. 4), are *in vivo* substrates of the same enzyme that methylates hnRNP A1 and fibrillar-

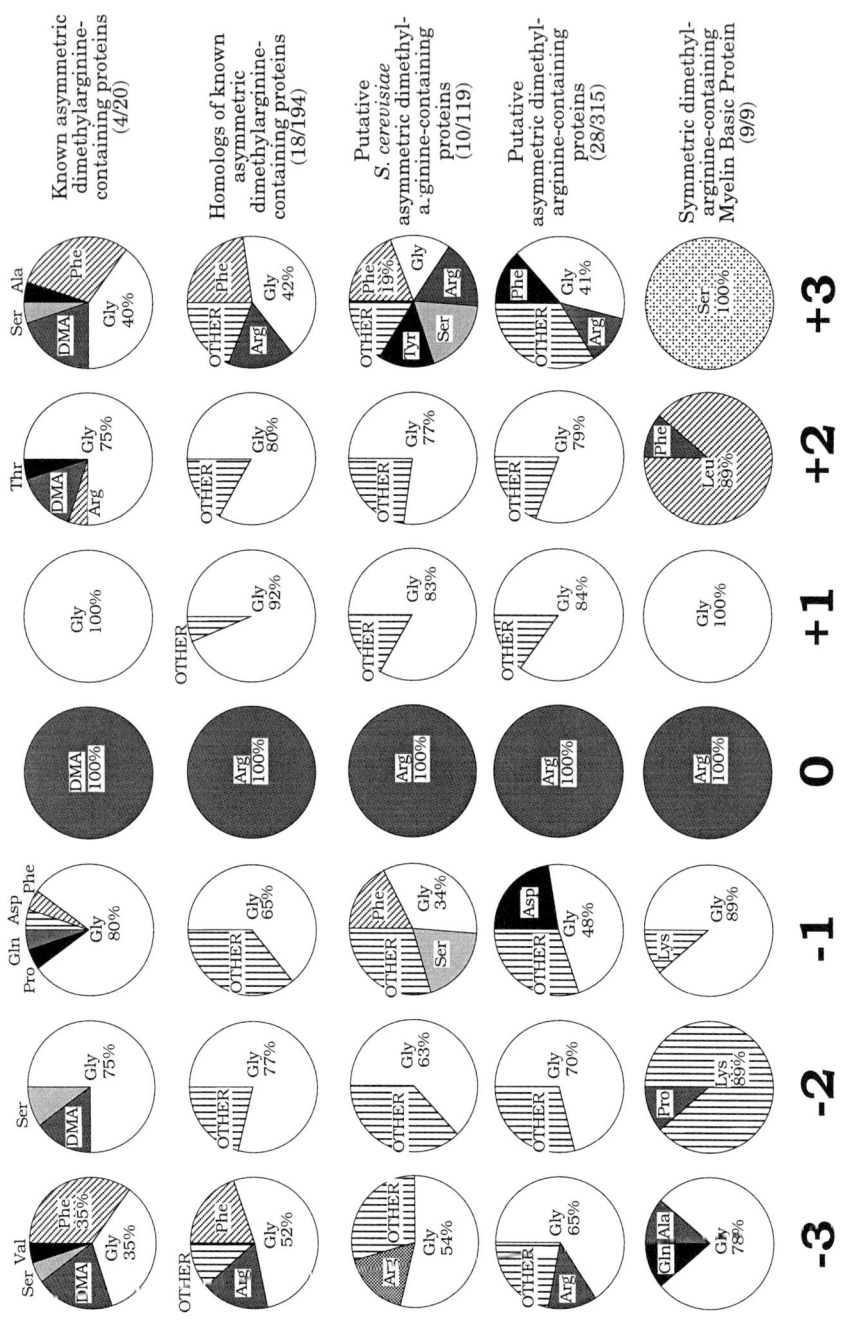

in is low. Neither of the histones contain sequences resembling the RGG consensus sequence in a GAR domain. The *in vitro* methylation of these histones may be due to an artifactual *in vitro* activity of the Type I methyltransferase on the histones or to the presence of a contaminating Type I methyl-accepting substrate in the histone preparation.

Of the other potentially arginine-methylated proteins described in Section IV,B,7, only the HMG1 and HMG2 proteins have directly corresponding protein sequences in the NCBI database that can be examined for the methylation consensus. Neither the bovine HMG1 (GB X12796) nor the pig HMG2 (GB J02895) protein contains RGG-methylation consensus sequences. For the other proteins, without experimental protein sequence data it is not clear which of the protein sequences present in the protein database correspond to the methylated HSP70A and HSP70B (*145, 157*); furthermore, the methylation site within myosin was not specifically determined to be on the light or heavy chain (*147, 148*). In these latter cases, examination of several HSP70 sequences as well as myosin heavy- and light-chain sequences revealed that these proteins also do not contain the methylation sites. It is always possible, however, that an additional methyltransferase is responsible for the activity. Interestingly, however, the myosin heavy chain 1B from the amoeba does contain consensus asymmetric arginine dimethylation sites. In either case, myosin distinguishes itself as a possible methyl acceptor for the Type I protein arginine *N*-methyltransferase because it is one of only a few potential substrates whose primary function does not involve binding to RNA.

B. RNA-Binding Proteins May Be the Major Group of Substrates for the Type I Protein Arginine *N*-Methyltransferases

1. POTENTIAL *Saccharomyces cerevisiae* SUBSTRATES

Using the methyl-accepting consensus sequence derived above for the Type I protein arginine *N*-methyltransferase, it is possible to identify other candidate proteins through protein database searches. Such a search was per-

---

FIG. 6. Frequency of amino acid residues that are found at positions surrounding known or proposed methylated arginine residues. For each row of pie charts, the number of sequences and all of their known or proposed methylated arginines (indicated in parentheses, respectively) were compiled in the subgroups denoted in the text on the right side of the figure. The methylated arginine residue is designated to be at position 0 and the three flanking amino acids on either side are also shown. A percentage for the amino acid that is most frequently present at a particular position is given under the amino acid abbreviation. Amino acids that are present less than 15% of the time at a particular position were included in the OTHER category.

formed by Najbauer *et al.* (69) with a single methylation consensus sequence (FGGRGGF) against all of the proteins in the GenPept database (release 72.0, November, 1992). This search identified the members of the nucleolin, fibrillarin, and hnRNP A1 families of known *in vivo* substrates. Four putative *S. cerevisiae* methyl-accepting substrates were also listed: Gar1, Npl3, Ssb1 (now Sbp1), and Nsr1.

Using the consensus sequence defined above, we chose to search for potential substrates in the yeast *S. cerevisiae*, because not only is the sequence of the entire genome now available, but also because a gene encoding a Type I protein arginine *N*-methyltransferase catalytic subunit (*RMT1*) was identified in this organism (5, 7). Our original analysis of the *rmt1* mutant yeast strain indicated that there were at least five polypeptides that were endogenously methylated *in vitro* in the parent versus the mutant strain (5). Additional *in vivo* labeling experiments with these two strains using [^{3}H]AdoMet show that at least four individual peptides are methylated in an *RMT1*-dependant manner (J. Gary, unpublished). Furthermore, the addition of GST–Rmt1 to *rmt1* cytosol caused the methylation of approximately 13 distinct polypeptides (5), suggesting that a number of different methyl-accepting substrates are present in yeast cells.

Initial searches of the *S. cerevisiae* genome were performed on the BLAST server at the Stanford Genome Database (http://genome-www.stanford.edu/Saccharomyces/) using a string of 21 amino acids composed of three repeats of the methylation site consensus determined above. Because all of the full-length *in vivo* substrates for the Type I methyltransferase contain multiple methylated arginine residues, we used this longer search sequence to eliminate putative positives that contained only a single arginine in an RGG context. To make the search exhaustive, each of the GAR domains identified in this first tier of the search was used to search the database again. The regions of similarity between the consensus search sequence and those returned by the BLAST server sometimes did not contain arginine residues in the proper context determined above, most likely due to the fact that the search string was relatively short and highly repetitive; for instance, poly(Gly) sequences containing no arginine residues were common false positives. Therefore the sequences returned by the server had to be examined manually to remove those that did not contain a GAR region or any RGG consensus sites. From these searches, 10 proteins were identified as possible substrates for the Type I protein arginine *N*-methyltransferase (Table IV).

As might be expected, the compilation of the amino acid sequences surrounding the putative arginine modification sites in the GAR domains of these potential yeast substrates shows a consensus similar to the one derived from the known substrates (Fig. 6). However, two differences were found that should be noted. At position -1, Ser and Phe residues are found with Gly as

TABLE IV
LIST OF PUTATIVE TYPE I ARGININE N-METHYLTRANSFERASE SUBSTRATES
FROM Saccharomyces cerevisiae

ORF name	Common names	ID[a]	Localization[b]	Nucleic acid binding	Ref.
YDR432w	NPL3 /NOP3 / MTS1 /MTR13	g172051	Nu/Cy	+	77, 139
YDL014w	NOP1	g1430978	No	+	161, 247
YHR089c	GAR1	g487935	No	+	80
YHL034c	SBP1 /SSBR1	g508771	No/Nu	+	177, 179
YNL112w	DBP2	g1302034	ND	ND	186
YGR159c	NSR1 /SHE5	g1323271	No	+	167, 172
YLR398c	SKI2	g410320	No	ND	181
YCL011c	RLF6 /GBP2	g1907116	ND	+	174
YOL123w	HRP1 / NAB4/ NAB5	g1420003	ND	ND	160
YGL122c	NAB2	g1322681	Nu	+	176

[a]Unique identification number for GenBank retrieval.
[b]Nu, Nuclear; No, nucleolar; Cy, cytoplasmic; ND, not determined.

the predominant residues. At position +3, a more diverse group of amino acids was found than would have been predicted from the known substrates.

One of the substrates identified by the search of the yeast genome database is the NPL3 gene product (also known as the genes NOP3, NAB1, MTS1, and MTR13), for nuclear protein localization (76). Npl3 has been implicated in a wide variety of cellular processes: nuclear protein import (76), rRNA processing (77), association with and export of poly(A+) RNA from the nucleus (78, 79, 139), mitochondrial protein import (158), and involvement in mating-type silencing (159). Current models propose that Npl3 shuttles between the nucleus and the cytoplasm as a carrier of mRNA (78, 139) and also may be involved in mRNA splicing (160; C. Siebel and C. Guthrie, unpublished). As described earlier, the yeast Type I protein arginine N-methyltransferase was identified in a synthetic lethal screen to identify genes that could exacerbate the phenotype of an npl3 temperature-sensitive allele (7). In addition, Npl3 was found to be in vivo and in vitro methylated in its carboxy-terminal region, where the GAR domain is also located (7, 24).

Of the other nine substrates identified by the homology searches, six (Nop1, Gar1, Sbp1, Nsr1, Rlf6, and Nab2) have been experimentally shown to bind nucleic acids, and the remaining three (Dbp2, Ski2, and Hrp1) display sequence similarity to nucleic acid-binding species as well (Table IV). Three of these potential methyl-accepting substrates (Nop1, Gar1, and Nsr1) appear to be nucleolar and are involved in ribosome biogenesis. Nop1 is the yeast homolog of fibrillarin (*161–163*), and is also found to be associated with small nucleolar RNA (*164*). This nucleolar protein is essential for yeast viability and various temperature-sensitive mutants of *NOP1* are impaired in pre-rRNA maturation and processing (*165*). Gar1 is also an essential nucleolar protein that is required for rRNA processing (*80, 166*). The gene, in yeast, was identified on a Southern blot by its hybridization to a cDNA probe derived from the GAR domain of *Xenopus* fibrillarin; as an aside, the putative methyl-accepting substrates *NOP1* and *SSB1* (now *SBP1*) were also identified in this screen (*80*). Nsr1 is a nucleolar protein (*167*) that was originally identified as the protein p67, which could specifically bind to the nuclear localization sequence of histone H2B (*168*). Sequence analysis revealed that *NSR1* is related to mammalian nucleolin (*169, 170*). Null mutants of this gene have a slow-growth phenotype and are deficient in pre-rRNA processing (*171*). The *NSR1* gene was then subsequently cloned under two separate circumstances: as a cold-shock inducible gene (*170*), as well as from a yeast genomic screen using a human hnRNP A1 cDNA as the probe (*172*). *NSR1* was also recently cloned in a screen designed to detect proteins with the ability to bind to a single-stranded telomeric repeat *in vitro;* another possible methyl-accepting substrate, Rlf6, was also found during this screen (*173*). Although the localization of Rlf6 has not been determined, its ability to bind telomeric repeats *in vitro* as well as its requirement for the proper distribution of the Rap1 telomerase-associated protein within the nucleus suggest that its localization is at least nuclear (*174*). The importance of nuclear Rap1 protein localization, and therefore Rlf6, is underscored by the fact that this protein has been shown to be responsible for telomere length regulation in the protein-counting mechanism established in yeast (*175*).

Other yeast candidates for asymmetric methylation at arginine residues are also localized to the nucleus and interact with RNA. The essential *NAB2* gene was cloned in a screen identifying proteins that interact with polyadenylated mRNA (*176*). Nab2 is a nuclear protein that can be purified from the polyadenylated mRNA fraction of yeast cells in which proteins have been cross-linked to mRNA *in vivo* by UV irradiation (*176*). Sbp1 is a nuclear/nucleolar protein that was originally identified as the single-stranded nucleic acid-binding protein Ssb1(*177–179*). Sbp1 is not essential for viability and seems to not be required for promoting mRNA splicing *in vitro*, but it may compete for mRNA binding with true splicing factors or be involved in some aspect of RNA metabolism (*179, 180*). *HRP1* was found because mutations in this essential gene suppressed the phenotype of an *npl3* temperature-sensitive allele (*160*). Hrp1 has sequence homology to the hnRNP A and hnRNP B families and is similarly localized to the nucleus (*160*).

The last two potential yeast methyl-accepting substrates identified are the putative RNA helicases Ski2 and Dbp2. One of the functions of the nucleolar protein Ski2 (*181*) is to act as an antiviral protein *in vivo* by blocking the translation of viral mRNAs that are not capped or polyadenylated (*182*). Under normal conditions, it may inhibit the translation of fragmented mRNAs, which may also lack these 5' or 3' features (*183–185*). DBP2 was identified in a two-hybrid screen for proteins that interact with Upf1, a protein involved in the selective degradation of nonsense codon-containing mRNAs (*186*).

All of the proposed endogenous substrates of the yeast Type I protein arginine *N*-methyltransferase appear to be nuclear or even nucleolar. All of the known substrates of the mammalian enzyme are also nuclear. From the fact that the known substrates as well as all of the putative methyl-accepting proteins in yeast interact with RNA, it is tempting to speculate that methylation of arginine residues can modulate the nucleic acid interactions of these proteins.

2. POTENTIAL SUBSTRATES FROM OTHER ORGANISMS

We also performed database searches on the entire NCBI nonredundant protein database (February, 1997) in order to identify additional substrates in organisms other than *S. cerevisiae*. The search was done in a fashion similar to that described above for the yeast example, although the second tier of searches was not done. Non-GAR-containing sequences were again manually excluded. The 28 highest scoring sequences are compiled in Table V. The consensus sequence obtained for these proteins (Fig. 6) is very similar to that of the known substrates for the Type I arginine methyltransferase. The glycine residue at position -1 is less frequently observed, similar to the situation seen with the yeast candidate protein substrates (Fig. 6).

Several of the substrates in Table V are homologs or family members of known substrate proteins or proposed yeast substrates for the Type I protein arginine *N*-methyltransferase: hnRNPs (U and A3), helicases, nucleolar proteins (Nopp44/46 and the "nucleolin homolog" C27H5.3), and the *Schizosaccharomyces pombe* Gar2 nucleolar protein involved in ribosome biogenesis. As previously noted, the common feature of almost all of these proteins is that they have either been directly shown, or are thought on the basis of sequence similarity, to bind nucleic acids, mostly RNA. Only one protein on the list, myosin heavy chain 1B, is not known to bind nucleic acids as a primary function.

We should note that many of the same proteins were identified in a previous database search for substrates of the Type I protein arginine *N*-methyltransferase (*69*). We have, however, excluded two of the sequences originally identified in that report, based on our criteria for selecting putative substrates. Cytochrome P450 and the mouse replication protein A were not included here because the RGG-methylation consensus of these proteins was not present in the context of multiple repeating methylation sites. However,

TABLE V
LIST OF PUTATIVE TYPE I ARGININE N-METHYLTRANSFERASE SUBSTRATES
FROM VARIOUS ORGANISMS

Definition	Organism	ID[a]	Nucleic acid binding[b]
RBP56	*Homo sapiens*	g1613775	By homology
EWS	*Homo sapiens*	g1666067	+
hTAFII68	*Homo sapiens*	g1628403	+
hnRNP A2	*Homo sapiens*	g500638	+
FUS/TLS RNA-binding protein	*Homo sapiens*	P35637[c]	By homology
hnRNP U	*Homo sapiens*	g32358	+
Pigpen	*Bos taurus*	Ref. 219[d]	By homology
Ribosomal protein S2	*Rattus rattus*	g57717	By homology
FLI2	*Mus musculus*	g193323	+
hnRNP A3	*Xenopus laevis*	P51968[c]	By homology
GCR 101	*Drosophila melanogaster*	S49193[e]	?
caz RNA-binding protein	*Drosophila melanogaster*	S54729[e]	+
p62 putatative ATP-dependent RNA helicase	*Drosophila melanogaster*	P1909[c]	By homology
T01C3.7	*Caenorhabditis elegans*	g1495012	By homology
ORF from yk76b9.5	*Caenorhabditis elegans*	g1255346	?
F58G11.2	*Caenorhabditis elegans*	g1742993	By homology
C27H5.3	*Caenorhabditis elegans*	g540269	By homology
GAR2	*Schizosaccharomyces pombe*	P41891[c]	+
DBP2 p68 RNA helicase	*Schizosaccharomyces pombe*	g173419	+
GAR1	*Schizosaccharomyces pombe*	Q06975[c]	+
GBP1	*Chlamydamonas rheinhardi*	g520518	+
AtGRP2b	*Arabidopsis thaliana*	g1063684	By homology
Bmsqd-1	*Bombyx mori*	g784909	By homology
Cutinase negative-acting protein	*Fusarium solani*	g1438951	+
RNA helicase, PRH75	*Spinacia oleracea*	g1488647	By homology
Nopp44/46	*Trypanosoma brucei*	g1314705	+
Glycine-rich protein	*Solanum lycopersicum*	S14985[e]	?
Glycine-rich RNA-binding protein	*Daucus carota*	Q03878[c]	By homology
EBNA-1	Epstein–Barr virus	g1334880	+
ICP27	Herpes simplex virus	P36295[c]	?
Myosin heavy chain 1B	*Acanthamoeba castellanii*	MWAXIB[e]	−

[a]Unique identification number for GenBank retrieval.
[b]As listed in the reference (ID) provided in column 3.
[c]Unique identification number for Swiss-Prot retrieval.
[d]See reference list.
[e]Unique identification number for PIR retrieval.

we want to stress the fact that the presence of a single RGG-methylation consensus in a protein without a larger GAR domain does not exclude proteins from being *in vitro* substrates and therefore possible *in vivo* methyl-accepting substrates as well. In fact, three proteins that were not identified from the protein database search are methylated *in vitro*. Expressed forms of the fragile X protein, FMRP, as well as *Drosophila* PSI and SXL, can be methylated by the GST fusion of the yeast Type I protein arginine N-methyltransferase *in vitro* (J. Gary and S. Warren, unpublished; J. Gary and D. Rio, unpublished). Interestingly, these three proteins are also nucleic acid-binding proteins *(138, 187, 188)*. In addition, PSI and SXL, like hnRNP A1, are regulators of mRNA splicing *(188–191)*. Sequence analysis of the SXL and PSI proteins indicate that each contains one possible asymmetric arginine methylation site, and they are not optimal consensus sequences compared to the consensus proposed above. FMRP, on the other hand, has up to seven arginines in a small GAR domain *(131)*. Perhaps only two or three of these are in an RGG context that is within the consensus parameters.

C. Analysis of the *in Vivo* Symmetric Methylation Site of Myelin Basic Protein

Biochemical evidence supports the idea that the mammalian Type II myelin basic protein-specific symmetric protein arginine N-methyltransferase is distinct from the Type I asymmetric methyltransferase (6). This conclusion is also supported by the identification of a consensus methylation site that is similar to but distinct from the asymmetric methylation site (Fig. 6). The Type II consensus site was determined by aligning the known as well as the analogous postulated methylation sites within myelin basic proteins from various organisms (Table III) in a manner similar to that noted previously. Although the arginine residue, like the Type I methyltransferase substrates, is still generally flanked on both sides by a glycine residue, additional glycine residues are apparently absent from positions -2 and $+2$. Instead, these positions generally contain lysine and leucine residues, respectively. Interestingly, myelin basic proteins from the marine organisms shark, skate, and dogfish do not have an arginine at the analogous position (107) to the bovine protein. These proteins have not been tested in an *in vitro* methylation reaction with a Type II methyltransferase preparation in order to determine if another arginine residue may be methylated. On the other hand, the myelin basic protein from carp is known not to be methylated *in vivo* (35), but it does seem to contain a methylatable arginine residue because it was found to be the best substrate for a bovine brain Type II protein arginine N-methyltransferase (45). The exact position of the *in vitro*-methylated arginine in the myelin basic protein from carp, however, was not determined.

Direct analysis of the methyl-accepting site for the Type II methyltransferase has been approached via synthetic peptide substrates. Rawal *et al.* (156) performed activity assays with a partially purified calf brain preparation of the Type II protein arginine N-methyltransferase, using various peptides based on the myelin basic protein sequence surrounding Arg-107 as methyl-accepting substrates. They found that the peptide matching residues 104–109 (GKG**R**GL) was a good methyl acceptor, but peptides where the flanking glycine residues on either side of Arg-107 were changed independently to several other amino acids, including histidine, phenylalanine, glutamine, leucine, and aspartate, were not substrates (*156*). It is of particular interest to note that the circularized peptide (by disulfide formation) CGKG**R**GLC was found to be a better substrate than its linear version (*156*). This may be an important indication that tertiary structure also plays a role in substrate recognition by the Type II methyltransferase. In native full-length myelin basic protein, Arg-107 is only five residues away from a triproline sequence, thought to impose a sharp bend in the polypeptide (*155*, *192*), and this feature may have been mimicked by the synthetically circularized peptide.

D. Other Protein Arginine Methyltransferase Consensus Methylation Sites

1. Plant Type III Histone H4-Specific Arginine Methyltransferase

The Type III methyltransferase was isolated from wheat germ extract and specifically mono- and symmetrically dimethylates endogenous histones as well as calf thymus histone H4 (*50*). In a subsequent study, the site of methylation was determined by amino acid analysis of peptide fragments from calf histone H4 incubated with [^{14}C]AdoMet and a preparation of the wheat germ Type III methyltransferase (*51*). A single peptide fragment corresponding to residues 24–35 was radioactive and contained only monomethylarginine (*51*). The site of methylation could be established because this peptide contains only one arginine residue at position 35. The amino acid context surrounding this methylation site (PAI**R**RLA) is different from all the other arginine methyltransferases and supports the recognition of this enzyme as a separate subtype.

2. Type IV Cytochrome *c*-Specific Arginine Methyltransferase

An activity able to monomethylate an arginine residue in cytochrome *c* *in vitro* has been identified in the protist *Euglena gracilis* (*53*) and designated the Type IV enzyme. The arginine methylation site in cytochrome *c* was determined to be at position 38 in the horse heart protein that was presumably used because the endogenous *E. gracilis* cytochrome *c* is fully methylated *in vivo* (*53*). The arginine is in a context distinct from the asymmetric or myelin basic protein consensus sequences. Using cytochrome *c* sequences from horse, human, yeast, mouse, and rat cells, the consensus amino acids surrounding the analogous Arg-38 are (I/L)FG<u>R</u>(K/H)(S/T)G. It would seem from the consensus sequence data that if the modification does occur *in vivo*, the cytochrome *c* methyltransferase may also be a distinct enzyme. Interestingly, the yeast arginine methyltransferase fusion protein (GST–Rmt1), known to have a broader substrate specificity, is able to poorly methylate cytochrome *c*, whereas the mammalian GST–PRMT1 does not (5, 6). Despite the activity of the GST–Rmt1 fusion protein toward cytochrome *c*, it is unable to methylate myelin basic protein (5), which would seem to have a methylation site closer to its own consensus sequence.

VI. Functional Roles for the Protein Arginine *N*-Methyltransferase and Arginine Methylation

A. Asymmetric Methylation

The importance of asymmetric protein arginine *N*-methylation is underscored by the number of possible substrates involved in important cellular processes (Table V) and the presence of the activity in many different organisms and tissues (2). By attempting to understand aspects of both the asymmetric protein arginine *N*-methyltransferase and the substrates it acts on, we may begin to ascertain the significance of this posttranslational modification in the context of a cell. In the following sections we describe possible functions for asymmetric arginine methylation.

1. SIGNAL TRANSDUCTION

Perhaps the clearest evidence to support a role for the Type I protein arginine *N*-methyltransferase in signal transduction can be seen from the ways in which the mammalian methyltransferase genes have been cloned. Rat PRMT1 was found because of its interaction with TIS21 and BTG1 in a yeast two-hybrid system (6). TIS21 is an early-response/immediate-early gene product, the expression of which is up-regulated independently of protein synthesis in response to a mitogenic signal such as hormones, serum, or phorbol esters (56, 58). Its closely related family member BTG1 was originally identified because of its proximity to a breakpoint found in leukemias; however, it was found that BTG1 expression could also be induced by epidermal growth factor in RAT1 fibroblast cells (6). Interestingly, in dissociated brain cells from embryonic mice, a Type I arginine methyltransferase activity is stimulated 120 and 130% by the β-adrenergic agonist (−)isoproterenol and by epinephrine, respectively (*193*).

The first human homolog of rat PRMT1, HCP1, was identified in a screen for genes that allowed *ire15* mutant yeast cells to overcome their inability to grow without inositol. HCP1 expression in the *ire15* yeast cells caused an increase in the transcription of *INO1* mRNA, encoding inositol 1-phosphate synthase, which is greatly reduced in the mutant cells (9, 70). In a third study, the human homolog of PRMT1 was also isolated as a protein that specifically interacts with the intracytoplasmic domain of the cytokine receptor INFAR1 (8). This domain of the receptor provides a docking site for proteins already known to be involved in signal transduction, such as tyrosine and serine/threonine kinases (194, 195).

Another role for Type I arginine methylation in signal transduction may be in the induction of the heat-shock response, at least in nonmammalian systems. After a heat shock or arsenite treatment, the methylation states of arginine residues within histone H3 and two of the 70-kDa heat-shock proteins are changed. Histone H3 in *Drosophila* shows an increase in methylarginine content (mono- and asymmetric dimethylarginine) (128) and HSP70A and HSP70B from chicken fibroblasts have decreased levels of monomethylarginine (144, 145). It is not clear whether the decrease in methylation of HSP70A and HSP70B is due to a true decrease in methylation or an artifact of the inability of the endogenous methyltransferase to cope adequately with the rapid overexpression of these proteins in response to a heat shock. In either case, the function of this change in the state of methylation is not understood. However, it is known that a large portion of HSP70 becomes localized to the nucleus after a heat shock (145).

There have been several reports that correlate signal transduction with possible changes in arginine methylation in unidentified proteins. For example, lipopolysaccharide stimulation of pre-B cells results in increased membrane protein methylation (196), whereas epidermal growth factor and nerve growth factor stimulate methylation within PC12 cells (197). However, there is no evidence that the increases in protein methylation are associated with arginine residues. Experiments designed to look specifically at the results of inhibiting both of the protein arginine N-methyltransferases found that various S-adenosylhomocysteine analogs, such as 5'-deoxy-5'-S-(2-methylpropyl)-5'-thioadenosine (SIBA), were good inhibitors of the Type I and Type II methyltransferases, but did not affect the protein lysine methyltransferase activity (198). These same analogs also inhibited foci formation in chick embryo fibroblasts transformed by the Rous sarcoma virus, circumstantially linking arginine methylation and the process of viral transformation (198). Sinefungin and SIBA both cause a decrease in all three methylated arginine derivatives *in vivo* when administered to murine splenic lymphocytes and inhibit the normal cellular lymphoproliferative response to lipopolysaccharide, again implicating the involvement of an arginine methyltransferase in a signaling pathway (199).

PROTEIN ARGININE METHYLATION   113

In fact, arginine methylation may even cooperate with protein phosphorylation events to control activity. A second posttranslational modification of the S. cerevisiae methyl-accepting substrate Npl3 has been proposed, whereby the serine residues adjacent to the proposed methylated arginine residues are phosphorylated (C. Siebel and C. Guthrie, unpublished). Having this phosphorylated SR domain (24, 200) overlap with an arginine-methylated GAR domain could give rise to multiple modes of regulation of Npl3 activity (24).

2. NUCLEAR LOCALIZATION

There are several examples whereby the GAR domain and even the arginine methylation event have been directly implicated in promoting nuclear import. Almost all of the known asymmetric dimethylarginine-containing proteins are localized to the nucleus and even more specifically the nucleolus. A majority of the proposed substrates from yeast and other organisms are similarly localized (Table IV).

Many studies have been done characterizing the signal for nuclear import. Most of the nuclear proteins that are actively imported contain basic amino acid stretches that are termed nuclear localization sequences (NLSs) (201). Those destined for the nucleolus often contain additional sequences for this targeting (201). The GAR domains of many of the proposed substrates for the Type I protein arginine N-methyltransferase were assessed for the ability to promote nuclear or nucleolar targeting. A yeast *npl3* mutant missing the carboxy-terminal portion of the protein containing the GAR domain fails to localize to the nucleus (78). Import of nucleolin into the nucleus requires a NLS, but its further localization into the nucleolus requires the GAR domain and the adjacent RNA-binding domains (202, 203). The yeast Nsr1 protein also requires both the GAR domain and the RNA-binding domains for proper nucleolar import (167). Interestingly, a mutant form of the herpes simplex virus protein ICP27, lacking the RGG sites, is nuclear but not nucleolar localized (131, 132, 249). Conflicting results were reported when similar experiments were performed with both hnRNP A1 and Gar1. The nuclear localization signal within hnRNP A1 is contained within a 40-amino acid domain in the carboxy-terminal region, 30 amino acids downstream from the GAR domain (204). With Gar1, it was also found that neither of the GAR domains was sufficient for nucleolar localization, and they did not seem to be required because the central domain and an adjacent basic domain alone were sufficient for proper localization (201). However, having a nuclear localization signal in close proximity to or overlapping a nucleotide-binding region (Section VI,A,4) suggests a possible mechanism for the shuttling of certain proteins between the nucleus and the cytoplasm, such as with hnRNP A1 and Npl3 (78, 101, 102, 250). When the proteins are imported into the nucleus, they encounter a high concentration of mRNA substrate to bind.

The binding of mRNA with or near to the GAR domain could partially block the localization function of the domain, allowing the protein and its "cargo" to exit into the cytoplasm. The bound RNA species could then dissociate because of the lowered cytosolic concentration, and the empty shuttle protein would then be actively targeted back to the nucleus.

Although all of the above examples are still consistent with the GAR domain playing an accessory role in nuclear or nucleolar import, a direct role for not only the GAR domain but also the asymmetric dimethylation of arginines within the GAR domain was shown for the nuclear import of the high-molecular-weight forms of basic fibroblast growth factor. The 18-kDa form of bFGF is predominantly cytoplasmic; however, the higher molecular weight forms are preferentially found in the nucleus and contain asymmetrically dimethylated arginine residues (Section IV,B,4) (*114, 117, 118, 205*). These higher molecular weight forms are transcribed from alternative upstream CUG start codons, and the amino-terminal extension contains a GAR domain (*116*). After the addition of the methyltransferase inhibitor 5′-methylthioadenosine to 3T3 cells, the high molecular weight forms of bFGF were no longer preferentially localized to the nucleus as in the control cells (*106*). The higher molecular weight forms of bFGF from methylthioadenosine-treated 3T3 cells demonstrate increased electrophoretic mobility compared to when they are immunoprecipitated from untreated cells (*106*). This change in mobility was assumed to be due to methylation, because the shift is dependent on the presence of methionine, although no direct measurement of the amount of methylation in these higher molecular weight bFGF species from methylthioadenosine-treated cells was reported (*106*). Furthermore, methylthioadenosine did not increase the turnover rate of either form of bFGF and it actually enhanced the level of expression of the higher molecular weight form (*106*).

3. Nucleic Acid Binding

The most common feature of the proteins that contain GAR domains is their ability to bind single-stranded nucleic acids (Tables IV and V). For example, the GAR domain of the yeast Nab2 protein, when *in vitro* transcribed and translated in various protein contexts, binds poly(G), poly(U), and single-stranded DNA linked to a solid support resin (*176*). A GAR domain from the herpes simplex virus protein, ICP27, can also confer RNA-binding abilities when fused to heterologous proteins (*138*). ICP27 is critical for several viral functions, all of which affect cellular mRNA processing; perhaps the ability of the ICP27 GAR domain to bind nucleic acids may explain why RGG-deficient forms of ICP27 do not allow efficient replication of the herpes simplex virus in Vero cells (*131, 138*). In Epstein–Barr-virus infected cells, a similar decrease in transcriptional activiation was observed when a mutant EBNA-1 protein with altered GAR domains was expressed compared to a wild-type version of the protein (206, 207).

These examples are, however, in proteins that have not been positively confirmed to contain methylated arginine residues. Therefore, the study of the function of the GAR domain in hnRNP proteins, where methylation is established, may be more relevant to understanding its role in RNA binding. The carboxy-terminal domain of hnRNP A1 (residues 203–236 in rat hnRNP A1; see Fig. 5B), containing the GAR domain, provides cooperative binding to single-stranded DNA cellulose and may be responsible for the protein–protein interactions important in this cooperativity (208). The carboxy-terminal domain is also necessary for the alternative splicing activity, the stable RNA binding, and the optimal annealing activity of hnRNP A1 (100, 208). Interestingly, bacterially produced hnRNP A1 constructs that have had their carboxy-terminal GAR domain removed are not functional in their ability to bind RNA or promote alternative splicing (100). Similarly, in hnRNP U, loss of the GAR domain abolishes single-stranded DNA binding activity *in vitro*, but an expressed form of the GAR domain of this protein alone is able to bind single-stranded DNA (209). It is important to note that in all of the *in vitro* studies performed with hnRNP A1, the protein was isolated from recombinant *E. coli* cells that do not have arginine methyltransferase activity. To examine the effect of methylation of the GAR domain, the nucleic acid-binding abilities of methylated versus unmethylated recombinant hnRNP A1 were compared (210). It was found that the methylated form of the protein eluted earlier in a salt gradient from a single-stranded DNA–cellulose column, indicating a weaker binding compared to the unmethylated form (210). The general association of the GAR domain with single-stranded nucleic acids is thought to be through the interaction of the positive charge on the arginine residues and the negative charge on the phosphate backbone (211). More specific interactions may arise through specific hydrogen bonds that are possible with the guanidino nitrogens and the nucleic acids. The methylation of these nitrogens does not affect the gross charge of the side chain, but the addition of two methyl groups onto a single nitrogen may disrupt the more specific hydrogen bond interactions (211). This hypothesis is consistent with the experimental data showing a decreased affinity for single-stranded DNA by the methylated form of hnRNP A1. Because only 1.45 arginines per hnRNP A1 molecule (out of a total of six potential sites) were shown to be methylated *in vitro* (210), the difference in binding affinity may be more substantial in more fully methylated proteins *in vivo*. A "less-specific" form of hnRNP A1 may be advantageous in light of the fact that the protein is responsible for differential 5′ splice site selection (99, 212).

4. Developmental Aspects

The Type I methyltransferase activity seems to be developmentally regulated. Arginine methyltransferase activity from either the whole homogenate or the cytosolic fraction is high in rat fetal brain and then declines rapidly after birth (213). Although the arginine methyltransferase activity is reduced by half in the first 30 days after birth (213), monomethylarginine residues in rat brain increase with a sigmoidal curve over the first 180 days (2). The initial increase begins at day 15, and the half-maximal content is reached at 45 days (2). During this same period, the content of asymmetric dimethylarginine residues did not vary (2).

Similar experiments with synchronized HeLa S-3 cells showed that Type I methyltransferase activity gradually increases after the initiation of mitosis and reaches a maximum 18 hr later in late G_2 (214). In a separate experiment, levels of methylated endogenous substrates, the 40S and 60S ribosomal subunits, were examined throughout the cell cycle (215). The levels of arginine-methylated 40S subunit were highest in late G_1, and the arginine-methylated 60S subunit was most abundant in early S phase (215).

5. Arginine Methylation with Unknown Consequences

There are many other substrates that are potentially asymmetrically methylated *in vivo* (see Table V and Section IV,B,7), of which several do not have a function or do not appear to be related to nucleic acid-binding proteins. Determining the function of these proteins and trying to assess the role of the arginine methylation is an obvious course of action. Among these proteins are histones, which may be methylated *in vivo* in response to a heat shock, as well as HMG1 and HMG2. The interactions of both of these proteins with DNA may be modulated by arginine methylation. A histone methyltransferase has even been suggested to be involved in the oddly shaped chromosomal structures seen in cells with chronic erythremic myelosis (216, 217).

Methylated arginine residues are resistant to proteolytic degradation by trypsin (20, 27, 31). This has led to the hypothesis that the modification may be involved in protecting certain proteins from proteolytic degradation. This resulting stabilization may be beneficial to some of the other uncharacterized substrates, including the heat-shock proteins HSP70A and HSP70B, cytochrome *c*, and myosin.

6. Conclusions

The roles discussed above for asymmetric protein arginine *N*-methylation are not mutually exclusive and they can be brought together under a rather general model described below. It is intriguing that proteins that are respon-

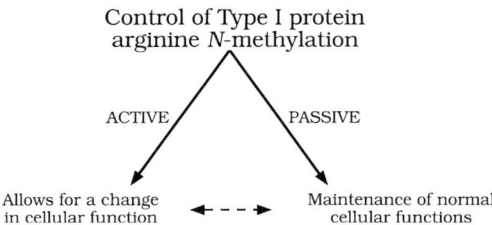

FIG. 7. A schematic of the two possible modes of *in vivo* Type I protein arginine N-methylation.

sible for binding single-stranded nucleic acids may be localized to the correct organelle by the same sequence that aids in the cooperative binding of the nucleic acid substrate. In addition, this system provides a way for the cell to receive extracellular signal inputs and transmit them directly to the nucleus.

In its simplest form, the model proposes that asymmetric dimethylation of arginine residues can occur in two modes, the passive and the active (Fig. 7). Passive methylation is proposed to be constitutive and occurs typically in unstimulated cells, under conditions of "normal" metabolism. For example, proteins such as fibrillarin and nucleolin have been found only in their fully methylated forms *in vivo* (20, 71), and these proteins as well as others may require this complete derivatization for proper localization and/or function. However, under certain conditions, "active" methylation of new, normally under- or unmethylated (or even unexpressed) substrates is required. In response to an external stimulus (cytokine, hormone, heat, etc.) the methyltransferase activity may be modulated to change substrate specificity and therefore affect cellular processes. For example, the reported levels of asymmetric dimethylarginine in hnRNP A1 *in vivo* vary (43, 97); therefore, the methylation state of hnRNP A1 may be changing under certain conditions to nonspecifically modify splicing patterns in mRNA transcripts. In these studies, hnRNP A1 was isolated from HeLa cells grown to cell densities that varied by as much as three orders of magnitude (43, 97). Perhaps the difference in the density of cell–cell contacts could trigger a change in the arginine methylation levels.

The conditions under which the asymmetric methyltransferase activity shifts from passive to active may be as general as a response to a positive growth signal, as seen in the up-regulation of methyltransferase activity in regenerating hepatic tissue in rats (218). This pathway could include interactions with the methyltransferase regulators TIS21 and BTG1 to effect the desired result. It is also possible that this branch of the active pathway is what causes the induction of *INO1* mRNA in yeast *ire15* mutants. The active pathway may also be induced under conditions that induce an inhibition of growth, as in the treatment of human myeloma U266S cells with interferon-β (8). This active pathway is blurred with the passive arm, however, because during either type of response it is possible that the expression of passively methylated substrates, which need to be methylated for proper function, is also up-regulated. Perhaps this increase in Type I methyltransferase activity is to cope with the enhanced expression of growth-induced methyl-accepting substrates such as the pigpen protein, whose cellular levels increase during the growth and differentiation of epithelial cell during angiogenesis (*219*). Active up-regulation of the methyltransferase activity toward similar substrates would constitute the crossover between these two pathways (Fig. 7).

As with other posttranslational modifications involved in signal transduction, whether it be phosphorylation or methylation, it is important to reset the system back to the resting state by removing the modification. It is at this point that the model proposed in Fig. 7 and the role of asymmetric dimethylation of arginines in signal transduction are problematic. No demethylase has been identified for asymmetrically dimethylated arginine residues, although one has been suggested for the myelin basic protein symmetrically dimethylated arginine (*220*). Experiments with *in vivo*-labeled histones from rat tissues indicated that the turnover of the methylated arginines coincided with the degradation of the histones (*122–124*). If the mechanism for the removal of asymmetrically dimethylated proteins is specific proteolysis, it would be a very costly method of signal transduction in terms of cellular resources.

B. Symmetric Arginine Dimethylation of Myelin Basic Protein

1. BACKGROUND

The relative and absolute content of mono- and symmetric dimethylarginine in myelin basic protein reported in the literature varies widely among species (*19, 21, 22, 31, 32, 34, 35*). Levels of symmetric dimethylarginine residues range from 0.033–0.6 mol per mole of myelin basic protein when isolated from human, monkey, bovine, chicken, guinea pig, rabbit, rat, frog, and turtle tissues (*19, 21, 35*). In these same myelin basic protein isolates, monomethylated arginine levels ranged from 0.17 to 0.48 mol per mole of myelin basic protein (*21, 35*). These studies also determined the *in vivo* ratios of symmetric dimethylarginine to monomethylarginine at position 107, which varied as much as from 0.14 to 1.95 (*19, 22, 32, 34, 35*). In 1977, it was hypothesized that this difference may not be due to different experimental techniques, but rather to a difference in the *in vivo* ages of the myelin basic protein (*221*). This idea is supported by the fact that monomethylated myelin basic protein from 2-day-old chickens has a half-life of 40 days, and the symmetrically dimethylated species has an apparent half-life of less than 50 hr (*34*), whereas newly formed myelin basic protein from mouse brain, labeled with [^{3}H]leucine, has a half-life of approximately 36 days (*222*).

Further evidence indicating a difference in the extent of arginine methylation of myelin basic protein *in vivo* with age was obtained from experiments with rabbits, rats, and mice of different ages (12 days and 2 years; 8 days and 10 months; and 10 days and 6 months, respectively). Importantly, myelination in mice begins 10 days after birth and reaches a maximum at 30 days, followed by a steady decline of accumulation (*222*), whereas in rats total myelin protein begins to accumulate at 10–15 days after birth as well; however, it reaches a maximum level at 100 days that is sustained beyond 1 year after birth (*221*). In both animals, the expression of myelin basic protein also follows the general trend of myelination. From the rabbits, rats, and mice, as much as a 9-fold increase in symmetric myelin basic protein-derived dimethylarginine residues and a 14-fold increase in myelin basic protein derived-monomethylarginine were observed in the older animals versus the younger ones (*149*). Interestingly, this large difference in the absolute amounts of the two arginine derivatives is not reflected in the ratios of the two, which did not vary greatly between the age groups (*149*).

Although an explanation for the increase of arginine methylation in older animals is not obvious, it may at least suggest a functional relationship between the extent of methylation and myelination. In support of this hypothesis, the Type II methyltransferase activity from mice is maximally active toward exogenous bovine myelin basic protein at 17–40 days after birth, coinciding with the peak of myelination (*222, 223*). As expected, the levels of methylated arginine derivatives also increased during this period (*149, 221*). Furthermore, the methylation of myelin basic protein during myelination may be under hormonal control, because Type II activity from cultured mouse brain cells is stimulated by the administration of thyroid hormone (*224*).

In addition to these studies, analyses were also conducted in mutant strains of dysmyelinating mice. Jimpy mice produce normal levels of myelin basic protein, but fail to incorporate it into the myelin sheath (*225*). Type II protein arginine *N*-methyltransferase activity was reduced by an average of 70% in brain extracts from the 21-day-old homozygous jimpy mice compared to the normal controls; in the heterozygous mice activity levels remained similar (90%) to the control level (*225*). Earlier in development (12–15 days), significant differences in enzymatic activity could not be detected among the three genotypes (*225*). The myelin basic protein from 23-day-old heterozygous and homozygous jimpy mice contain 23 and 16%, respectively, of the symmetric arginine derivative compared to the mother control, as well as 32 and 7%, respectively, for the monomethylarginine species (*149*). No explanation has been offered for the fact that heterozygous jimpy mice are phenotypically normal and have almost wild-type levels of Type II methyltransferase activity, but have significantly reduced levels of myelin basic protein-derived methylated arginine derivatives (*149, 225*). Alternatively, for the 12-day-old shiverer dysmyelinating mutant mice, the Type II protein arginine *N*-methyltransferase activity from brain extracts was greater than twofold higher than the normal controls, and fell to near normal levels by day 21 (*226*). The amounts of the specific methylated arginine derivatives were not assayed for the shiverer mice.

2. FUNCTIONAL MODELS

Attempts to understand how the methylation of the single arginine residue could effect myelination began with the modeling of the myelin basic protein structure (*192*). Myelin basic protein is quite refractory to crystallization techniques; however, early molecular modeling algorithms predicted high levels of β-sheet that could not be confirmed experimentally (*192*). The results of these modeling exercises was that a methylated versus an unmethylated arginine would make different contacts with nearby residues. Specifically, the methylated form would have favorable interactions with Pro97, stabilizing the hairpin loop created by the triproline sequence (*192*). A more recent analysis of the structure of myelin basic protein incorporated data from three-dimensional reconstructions of electron microscope analyses (*155*). This revised structure is based on the β-sheet backbone proposed earlier (*192*); however, the myelin basic protein structure model could now be constrained to fit the electron microscopy reconstruction data (*155*). In this new model, myelin basic protein has a C-shaped configuration and is still composed mainly of β-sheets. The area containing the triproline sequence and the arginine methylation site is suggested to be in a position where posttranslational modifications may affect the global structure of the protein (*155*). Interestingly, however, the methylation of the arginine was not mentioned in the study (*155*) as one of the possible *in vivo* modifications to myelin basic protein.

Exogenously added myelin basic protein can induce the aggregation of unilamellar myelin vesicles *in vitro*, which can be monitored by light-scattering changes (*227*). Experiments comparing unmethylated carp myelin basic protein with the presumably *in vivo*-methylated bovine protein suggested that the arginine methylation was more efficient at promoting aggregation by 0.93 kcal/mol, whereas the unmethylated myelin basic from carp was found to be no more efficient than other cationic proteins tested, such as lysozyme or poly(histidine) (*227*). An additional increase (0.13 kcal/mol) in efficiency for the bovine myelin basic protein was observed after it had been further methylated *in vitro* by a preparation of the Type II protein arginine N-methyltransferase from bovine brain (227). Interestingly, the myelin basic protein from carp was not arginine methylated *in vitro* and assayed for vesicle aggregation (227). The authors concluded that the effect of arginine methylation in the vesicle assay could not be attributed solely to the increased hydrophobicity of the methylated arginine residue, but included additional gains in free energy from more favorable ionic interactions (227).

Furthermore, incubation of mouse embryonic brain cells with the methyltransferase inhibitor sinefungin lead to an inhibition of the formation of compact myelin without decreasing the gross amount of myelin basic protein (228). From these results, it was suggested that the methylation of Arg-107 played a key role in the compaction of the myelin sheath. However, analysis of various myelin fractions at different degrees of compaction showed that the highest methyl acceptability was found in the most compact fraction of multilayered myelin, suggesting that it was the least, and not the most, methylated fraction (229). It was proposed that the methylated myelin basic protein is excluded from the compact myelin because it is less cationic than the unmethylated form, and the extent of positive charge of myelin basic protein is thought to be critical for the association of the two opposing negatively charged phopholipid membranes (229). However, in guanidine-HCl the nature of the charge on methylated myelin basic protein was not different from the unmethylated form; in fact, only after the removal of the denaturant was any difference seen (229), although this may be attributable to improper refolding *in vitro* rather than to any real difference *in vivo*.

A solution to this apparent paradox established by the vesicle aggregation experiments compared to assays on compact myelin is not at hand. It is possible that the arginine methylation of myelin basic protein is a dynamic rather than a static process. One may speculate that the methylation of Arg-107 occurs during or soon after its synthesis to prevent its rapid complexation with the membrane and allow for proper folding in a fashion similar to the proposed role of the phosphorylation of Thr-99 and Ser-166 (*192*). This scenario would cause a reevaluation of the data indicating that methylation induced favorable interactions with the membrane (227). The data may be viewed as artifactual if the role of the methylation is only critical soon after the protein is synthesized. Once the derivatized myelin basic protein is properly folded, it could associate with the membrane and lose its arginine methylation (or demethylation could precede membrane association). Indeed, as mentioned above, the protein may only go from symmetric dimethylarginine to monomethylarginine, as is observed in the experiment with young versus old animals (*149*). This aspect of the model is supported by the discovery in a crude preparation of the bovine brain myelin basic protein methyltransferase of a demethylase activity that could not be attributed to a loss of myelin basic protein through general proteolysis (*220*).

3. MYELIN BASIC PROTEIN METHYLATION AND THE RELATIONSHIP TO DISEASE

Defects in myelination have been associated with alterations in methyl group metabolism. For example, a deficiency of vitamin B_{12} (a necessary cofactor for the synthesis of methionine, a precursor to the methyl-group donor AdoMet) is associated with malformed myelin sheaths in subacute combined degeneration disease (SCD) (*27*). Another inhibitor of AdoMet synthesis, 1-aminocyclopentane carboxylic acid (cycloleucine), can induce myelin lesions similar to those seen in SCD when injected into chickens (*230*) or when given to mice (*231, 232*). Finally, nitrous oxide (an inhibitor of methionine synthetase) also causes these lesions in myelin, and the neurological as well as the histological phenotypes can be fully reversed by supplementing the diet with methionine (*233*). However, whether these effects are mediated in whole or in part via arginine methylation in myelin basic protein is not clear. Administration of cycloleucine to rats does cause a 26% decrease in myelin basic protein methylation; however, no effect was found with nitrous oxide treatment of rats, nor did a depletion of vitamin B_{12} in bats by dietary measures cause a significant decline (*234*).

The methylation of arginine residues and their eventual metabolism may also reflect on pathological conditions. Levels of asymmetric dimethylarginine in the urine of 12 patients with muscular distrophy were initially found to be 2-fold higher than in 10 normal children, and the average ratio of the asymmetric to the symmetric derivative increased from 1.4 to 2.6 (*235*). Later it was observed that both types of dimethylarginine residue levels were almost 10-fold higher in a single patient affected with muscular dystrophy compared to 9 unaffected individuals, and the ratio of the two residues was 1.6 (*236*). The discrepancy between these two studies has yet to be explained; however, it may be due to differences in sample size or to the fact that children (5–16 years old) were specifically used in the earlier study (*235*), while presumably only a single adult was used in the later one, although the age of the individual is not specifically stated (*236*). This change in the absolute excretion levels of dimethylarginine may still be of use for the early diagnosis of this disease.

The levels of these dimethylated arginine derivatives in normal human urine (16–71 mmol/mg creatinine) are comparable to the levels of arginine (6–40 mmol/mg creatinine), but are significantly lower than for other methylated amino acids, such as 3-methylhistidine (120–385 mmol/mg creatinine) (*235, 236*). This difference suggests that there is an active mechanism for the metabolism of the proteolytic products of arginine-methylated proteins

In fact, several enzymes have been characterized that may be involved in the catabolism of these free methylated arginines in rats (237–241). Some of the by-products of these reactions include N^G-methylagmatine, ornithine, the corresponding α-keto acids, and citrulline. The importance of these catabolic enzymes may become relevant in light of the fact that methylguanidine may be related to several uremic disorders (242). Additionally, free monomethylarginine (243) is known to inhibit nitric oxide synthase *in vitro* (244, 245), and perhaps the dimethylarginine derivatives can inhibit this enzyme as well.

VII. Future Directions

In this review, we have provided an overview of protein arginine methylation, a field in in which there is considerable activity at this time. Current excitement comes from the identification of new protein arginine N-methyltransferase genes and the analysis of new potential methyl-accepting substrates.

At present, at least two mammalian genes are known that encode catalytically active arginine methyltransferase subunits, PRMT1 and PRMT3. Three splice variants of the PRMT1 gene in humans have also been described. All can be classified as Type I methyltransferases based on their activity with the synthetic substrate GST–GAR, but they still methylate only a single 55-kDa polypeptide in *rmt1* yeast extracts, compared to the 13 substrates recognized by the GST–Rmt1 yeast fusion Type I protein arginine N-methyltransferase (5) (J. Tang and J. Gary, unpublished results). There are also five "orphan" mammalian substrates (HMG1, HMG2, HSP70A, HSP70B, and myosin) that have been tentatively shown to contain methylated arginines *in vivo* but do not contain the asymmetric methylation consensus. It is unclear whether distinct gene products are required for their methylation. A gene for the Type II arginine methyltransferase has also not yet been identified. Together, these findings suggest the existence of potentially many more protein arginine N-methyltransferases in higher organisms with different substrate specificities. The finding of additional protein arginine N-methyltransferase substrates is critical for discovering new methyltransferases and determining the complexity of this family of enzymes and ultimately its role in various cellular processes. At this stage, it will be important to reinvestigate those potential substrates identified only by amino acid analysis with more purified forms of the enzymes and more sensitive means of analysis, such as mass spectroscopy, to verify the original claims. Having a larger pool of identified substrates, recombinantly expressed, gives us the ability to either purify novel methyltransferases from cellular extracts or assess the activity of the methyltransferase fusion enzymes already constructed.

The most crucial and rewarding work will be done by those whose protein of interest is a substrate for a protein arginine N-methyltransferase. The ability to obtain purified methylated or unmethylated protein from recombinant bacterial cells coexpressing a protein arginine N-methyltransferase for use in an *in vitro* assay will open the door to understanding the function of this methylation event. The coexpression of the methyltransferase and the substrate is suggested over *in vitro* methylation based on our results attempting to stoichiometrically methylate recombinant Npl3 with GST–Rmt1. Under all of the conditions tested, such as high AdoMet concentrations and thrombin removal of the GST moiety, we were unable to achieve stoichiometric methylation (C. Seibel and J. Gary, unpublished). This result supports earlier hypotheses that the methylation event was either cotranslational or very tightly coupled to it (*246*).

As more of the proteins listed in Table V are characterized, perhaps additional *in vitro* assays will be developed and used to probe the functional advantage of arginine methylation. In the yeast system, these types of functional assays can be performed *in vivo* using an *RMT1* versus an *rmt1* strain in combination with mutations in the gene of choice. Indeed, this is how the *RMT1* gene was identified (*7*). Using this type of approach, it might be possible to ascertain what portions of the substrate proteins act in conjunction with methylation, causing an interesting phenotype that may reveal its role in nucleotide binding, signal transduction, or localization. For example, are point mutations in the conserved RNA recognition motifs more deleterious in an *rmt1* strain, or can a GAR domain deletion be rescued by including a heterologous nuclear localization signal? Making synthetic substrates by the addition of either Type I or Type II arginine methylation sites to certain proteins may also be instructive.

The past 30 years of literature concerning the symmetric and asymmetric dimethylation of arginine residues is large and often contradictory. As the specific methyltransferase genes are cloned and expressed and methods of detecting and analyzing the derivatized residues improve, it will be of great importance to revisit many of the questions raised here with the hope of unifying the record and providing a physiological rationale for this posttranslational modification reaction.

References

1. S. Clarke, *Curr. Biol.* **5,** 977 (1993).
2. Y. Kakimoto, Y. Matsuoka, M. Miyake, and H. Konishi, *J. Neurochem.* **24,** 893 (1975).

3. P. Paolantonacci, F. Lawrence, F. Lederer, and M. Robert-Gero, *Mol. Biochem. Parasitol.* **21**, 47 (1986).
4. N. Yarlett, A. Quamina, and C. J. Bacchi, *J. Gen. Microbiol.* **137**, 717 (1991).
5. J. D. Gary, W. J. Lin, M. C. Yang, H. R. Herschman, and S. Clarke, *J. Biol. Chem.* **271**, 12585 (1996).
6. W. J. Lin, J. D. Gary, M. C. Yang, S. Clarke, and H. R. Herschman, *J. Biol. Chem.* **271**, 15034 (1996).
7. M. F. Henry and P. A. Silver, *Mol. Cell. Biol.* **16**, 3668 (1996).
8. C. Abramovich, B. Yakobson, J. Chebath, and M. Revel, *EMBO J.* **16**, 260 (1997).
9. J. Nikawa, H. Nakano, and N. Ohi, *Gene* **171**, 107 (1996).
10. D. E. Atkinson, "Cellular Energy Metabolism and its Regulation." Academic Press, New York, 1977.
11. W. K. Paik and S. Kim, *Biochem. Biophys. Res. Commun.* **29**, 14 (1967).
12. W. K. Paik and S. Kim, *J. Biol. Chem.* **243**, 2108 (1968).
13. A. M. Kaye and D. Sheratzky, *Biochim. Biophys. Acta* **190**, 527 (1969).
14. Y. Kakimoto and S. Akazawa, *J. Biol. Chem.* **245**, 5751 (1970).
15. T. Teerlink and B. E. de, *J. Chromatog.* **4 91**, 418 (1989).
16. T. Nakajima, Y. Matsuoka, and Y. Kakimoto, *Biochim. Biophys. Acta* **230**, 212 (1971).
17. Y. Kakimoto, *Biochim. Biophys. Acta* **243**, 31 (1971).
18. E. H. Eylar, S. Brostoff, G. Hashim, J. Caccam, and P. Burnett, *J. Biol. Chem.* **246**, 5770 (1971).
19. S. W. Brostoff, A. Rosegay, and W. J. Vandenheuvel, *Arch. Biochem. Biophys.* **148**, 156 (1972).
20. M. A. Lischwe, R. G. Cook, Y. S. Ahn, L. C. Yeoman, and H. Busch, *Biochemistry* **24**, 6025 (1985).
21. G. E. Deibler and R. E. Martenson, *J. Biol. Chem.* **248**, 2392 (1973).
22. P. R. Young and F. Grynspan, *J. Chromatog.* **421**, 130 (1987).
23. M. Ohtsuka, S. Ihara, R. Ogawa, T. Watanabe, and Y. Watanabe, *Int. J. Cancer* **34**, 855 (1984).
24. C. W. Siebel and C. Guthrie, *Proc. Natl. Acad. Sci. U.S.A.* **93**, 13641 (1996).
25. D. Gallwitz, *Arch. Biochem. Biophys.* **145**, 650 (1971).
26. M. K. Paik, K. H. Lee, S. S. Hson, I. M. Park, J. H. Hong, and B. D. Hwang, *Int. J. Biochem.* **23**, 939 (1991).
27. G. S. Baldwin and P. R. Carnegie, *Science* **171**, 579 (1971).
28. A. J. Crang and W. Jacobson, *Biochem. Soc. Trans.* **8**, 611 (1980).
29. N. Sundarraj and S. E. Pfeiffer, *Biochem. Biophys. Res. Commun.* **52**, 1039 (1973).
30. M. Miyake, *J. Neurochem.* **24**, 909 (1975).
31. S. Brostoff and E. H. Eylar, *Proc. Natl. Acad. Sci. U.S.A.* **68**, 765 (1971).
32. G. S. Baldwin and P. R. Carnegie, *Biochem. J.* **123**, 69 (1971).
33. P. R. Dunkley and P. R. Carnegie, *Biochem. J.* **141**, 243 (1974).
34. D. H. Small and P. R. Carnegie, *J. Neurochem.* **38**, 184 (1982).
35. G. E. Deibler and R. E. Martenson, *J. Biol. Chem.* **248**, 2387 (1973).
36. M. Miyake and Y. Kakimoto, *J. Neurochem.* **20**, 859 (1973).
37. G. M. Jones and P. R. Carnegie, *J. Neurochem.* **23**, 1231 (1974).
38. H. W. Lee, S. Kim, and W. K. Paik, *Biochemistry* **16**, 78 (1977).
39. G. H. Park, L. P. Chanderkar, W. K. Paik, and S. Kim, *Biochim. Biophys. Acta* **874**, 30 (1986).
40. S. K. Ghosh, W. K. Paik, and S. Kim, *J. Biol. Chem.* **263**, 19024 (1988).
41. P. R. Young and C. M. Waickus, *Biochem. J.* **250**, 221 (1988).
42. B. McCright, A. M. Rivers, S. Audlin, and D. M. Virshup, *J. Biol. Chem.* **271**, 22081 (1996).

43. A. L. Beyer, M. E. Christensen, B. W. Walker, and W. M. LeStourgeon, *Cell* **11,** 127 (1977).
44. J. Karn, G. Vidali, L. C. Boffa, and V. G. Allfrey, *J. Biol. Chem.* **252,** 7307 (1977).
45. S. K. Ghosh, S. K. Syed, S. Jung, W. K. Paik, and S. Kim, *Biochim. Biophys. Acta* **1039,** 142 (1990).
46. R. Rajpurohit, W. K. Paik, and S. Kim, *Biochim. Biophys. Acta* **1122,** 183 (1992).
47. R. Rajpurohit, S. O. Lee, J. O. Park, W. K. Paik, and S. Kim, *J. Biol. Chem.* **269,** 1075 (1994).
48. N. Rawal, R. Rajpurohit, W. K. Paik, and S. Kim, *Biochem. J.* **300,** 483 (1994).
49. Q. Liu and G. Dreyfuss, *Mol. Cell. Biol.* **15,** 2800 (1995).
50. A. Gupta, D. Jensen, S. Kim, and W. K. Paik, *J. Biol. Chem.* **257,** 9677 (1982).
51. S. G. Disa, A. Gupta, S. Kim, and W. K. Paik, *Biochemistry* **25,** 2443 (1986).
52. P. Casellas and P. Jeanteur, *Biochim. Biophys. Acta* **519,** 243 (1978).
53. J. Z. Farooqui, M. Tuck, and W. K. Paik, *J. Biol. Chem.* **260,** 537 (1985).
54. S. Fields and O. Song, *Nature (London)* **340,** 245 (1989).
55. S. Fields and R. Sternglanz, *Trends Genet.* **10,** 286 (1994).
56. B. S. Fletcher, R. W. Lim, B. C. Varnum, D. A. Kujubu, R. A. Koski, and H. R. Herschman, *J. Biol. Chem.* **266,** 14511 (1991).
57. A. Bradbury, R. Possenti, E. M. Shooter, and F. Tirone, *Proc. Natl. Acad. Sci. U.S.A.* **88,** 3353 (1991).
58. B. C. Varnum, S. T. Reddy, R. A. Koski, and H. R. Herschman, *J. Cell. Physiol.* **158,** 205 (1994).
59. R. Rimokh, F. Berger, P. Cornillet, K. Wahbi, J. P. Rouault, M. Ffrench, P. A. Bryon, M. Gadoux, O. Gentilhomme, D. Germain *et al.*, *Genes Chromosom. Cancer* **2,** 223 (1990).
60. J. P. Rouault, C. Samarut, L. Duret, C. Tessa, J. Samarut, and J. P. Magaud, *Gene* **129,** 303 (1993).
61. S. F. Altschul, W. Gish, W. Miller, E. W. Myers, and D. J. Lipman, *J. Mol. Biol.* **215,** 403 (1990).
62. M. J. Dognin and B. Wittmann-Liebold, *Hoppe-Seylers Z. Physiol. Chem.* **361,** 1697 (1980).
63. A. Vanet, J. A. Plumbridge, and J. H. Alix, *J. Bacteriol.* **175,** 7178 (1993).
64. R. M. Kagan and S. Clarke, *Arch. Biochem. Biophys.* **310,** 417 (1994).
65. G. Schluckebier, M. O'Gara, W. Saenger, and X. Cheng, *J. Mol. Biol.* **247,** 16 (1995).
66. J. D. Lowenson and S. Clarke, in "Deamidation and Isoaspartate Formation in Peptides and Proteins" (D. W. Aswad, ed.), p. 47. CRC Press, Boca Raton, Florida, 1995.
67. H. Xie and S. Clarke, *J. Biol. Chem.* **268,** 13364 (1993).
68. H. Xie and S. Clarke, *J. Biol. Chem.* **269,** 1981 (1994).
69. J. Najbauer, B. A. Johnson, A. L. Young, and D. W. Aswad, *J. Biol. Chem.* **268,** 10501 (1993).
70. J. Nikawa, *Gene* **149,** 367 (1994).
71. M. A. Lischwe, R. L. Ochs, R. Reddy, R. G. Cook, L. C. Yeoman, E. M. Tan, M. Reichlin, and H. Busch, *J. Biol. Chem.* **260,** 14304 (1985).
72. N. Katsanis, M. L. Yaspo, and E. M. Fisher, *Mammal. Genome* **8,** 526 (1997).
73. A. Loubbardi, C. Marcireau, F. Karst, and M. Guilloton, *J. Bacteriol.* **177,** 1817 (1995).
74. H. Feldmann, M. Aigle, G. Aljinovic, B., andre, M. C. Baclet, C. Barthe, A. Baur, A. M. Becam, N. Biteau, E. Boles *et al.*, *EMBO J.* **13,** 5795 (1994).
75. T. Kruiswijk, A. Kunst, R. J. Planta, and W. H. Mager, *Biochem. J.* **175,** 221 (1978).
76. M. A. Bossie, C. DeHoratius, G. Barcelo, and P. Silver, *Mol. Biol. Cell* **3,** 875 (1992).
77. I. D. Russell and D. Tollervey, *J. Cell Biol.* **119,** 737 (1992).
78. J. Flach, M. Bossie, J. Vogel, A. Corbett, T. Jinks, D. A. Willins, and P. A. Silver, *Mol. Cell. Biol.* **14,** 8399 (1994).

79. I. Russell and D. Tollervey, *Eur. J. Cell Biol.* **66,** 293 (1995).
80. J. P. Girard, H. Lehtonen, M. Caizergues-Ferrer, F. Amalric, D. Tollervey, and B. Lapeyre, *EMBO J.* **11,** 673 (1992).
81. S. Zolnierowicz, C. Csortos, J. Bondor, A. Verin, M. C. Mumby, and A. A. DePaoli-Roach, *Biochemistry* **33,** 11858 (1994).
82. C. Kamibayashi, R. Estes, R. L. Lickteig, S. I. Yang, C. Craft, and M. C. Mumby, *J. Biol. Chem.* **269,** 20139 (1994).
83. P. Turowski, A. Fernandez, B. Favre, N. J. Lamb, and B. A. Hemmings, *J. Cell Biol.* **129,** 397 (1995).
84. W. K. Paik and S. Kim, *J. Neurochem.* **16,** 1257 (1969).
85. J. Najbauer and D. W. Aswad, *J. Biol. Chem.* **265,** 12717 (1990).
86. J. P. Aris and G. Blobel, *Proc. Natl. Acad. Sci. U.S.A.* **88,** 931 (1991).
87. K. Tyc and J. A. Steitz, *EMBO J.* **8,** 3113 (1989).
88. M. E. Christensen and K. P. Fuxa, *Biochem. Biophys. Res. Commun.* **155,** 1278 (1988).
89. A. W. Prestayko, M. O. Olson, and H. Busch, *FEBS Lett.* **44,** 131 (1974).
90. G. Jordan, *Nature (London)* **329,** 489 (1987).
91. B. Lapeyre, H. Bourbon, and F. Amalric, *Proc. Natl. Acad. Sci. U.S.A.* **84,** 1472 (1987).
92. M. Caizergues-Ferrer, P. Mariottini, C. Curie, B. Lapeyre, N. Gas, F. Amalric, and F. Amaldi, *Genes. Dev.* **3,** 324 (1989).
93. R. A. Borer, C. F. Lehner, H. M. Eppenberger, and E. A. Nigg, *Cell* **56,** 379 (1989).
94. M. A. Lischwe, K. D. Roberts, L. C. Yeoman, and H. Busch, *J. Biol. Chem.* **257,** 14600 (1982).
95. K. R. Williams, K. L. Stone, M. B. LoPresti, B. M. Merrill, and S. R. Planck, *Proc. Natl. Acad. Sci. U.S.A.* **82,** 5666 (1985).
96. F. Cobianchi, D. N. SenGupta, B. Z. Zmudzka, and S. H. Wilson, *J. Biol. Chem.* **261,** 3536 (1986).
97. A. Kumar, K. R. Williams, and W. Szer, *J. Biol. Chem.* **261,** 11266 (1986).
98. F. Cobianchi, R. L. Karpel, K. R. Williams, V. Notario, and S. H. Wilson, *J. Biol. Chem.* **263,** 1063 (1988).
99. A. Mayeda, D. M. Helfman, and A. R. Krainer, *Mol. Cell. Biol.* **13,** 2993 (1993).
100. A. Mayeda, S. H. Munroe, J. F. Caceres, and A. R. Krainer, *EMBO J.* **13,** 5483 (1994).
101. S. Pinol-Roma and G. Dreyfuss, *Nature (London)* **355,** 730 (1992).
102. S. Pinol-Roma and G. Dreyfuss, *Mol. Cell. Biol.* **13,** 5762 (1993).
103. D. Gospodarowicz, G. Neufeld, and L. Schweigerer, *Cell Diff.* **19,** 1 (1986).
104. D. Gospodarowicz, A. Baird, J. Cheng, G. M. Lui, F. Esch, and P. Bohlen, *Endocrinology* **118,** 82 (1986).
105. R. Montesano, J. D. Vassalli, A. Baird, R. Guillemin, and L. Orci, *Proc. Natl. Acad. Sci. U.S.A.* **83,** 7297 (1986).
106. G. Pintucci, N. Quarto, and D. B. Rifkin, *Mol. Biol. Cell* **7,** 1249 (1996).
107. C. Basilico and D. Moscatelli, *Adv. Cancer Res.* **59,** 115 (1992).
108. A. Jackson, S. Friedman, X. Zhan, K. A. Engleka, R. Forough, and T. Maciag, *Proc. Natl. Acad. Sci. U.S.A.* **89,** 10691 (1992).
109. S. R. Opalenik, J. T. Shin, J. N. Wehby, V. K. Mahesh, and J. A. Thompson, *J. Biol. Chem.* **270,** 17457 (1995).
110. P. L. McNeil, L. Muthukrishnan, E. Warder, and P. A. D'Amore, *J. Cell Biol.* **109,** 811 (1989).
111. P. T. Ku and P. A. D'Amore, *J. Cell. Biochem.* **58,** 328 (1995).
112. M. T. Story, F. Esch, S. Shimasaki, J. Sasse, S. C. Jacobs, and R. K. Lawson, *Biochem. Biophys. Res. Commun.* **142,** 702 (1987).

113. J. A. Abraham, J. L. Whang, A. Tumolo, A. Mergia, J. Friedman, D. Gospodarowicz, and J. C. Fiddes, *EMBO J.* **5,** 2523 (1986).
114. A. Sommer, D. Moscatelli, and D. B. Rifkin, *Biochem. Biophys. Res. Commun.* **160,** 1267 (1989).
115. A. Sommer, M. T. Brewer, R. C. Thompson, D. Moscatelli, M. Presta, and D. B. Rifkin, *Biochem. Biophys. Res. Commun.* **144,** 543 (1987).
116. H. Prats, M. Kaghad, A. C. Prats, M. Klagsbrun, J. M. Lelias, P. Liauzun, P. Chalon, J. P. Tauber, F. Amalric, J. A. Smith, *et al.*, *Proc. Natl. Acad. Sci. U.S.A.* **86,** 1836 (1989).
117. W. H. Burgess, J. Bizik, T. Mehlman, N. Quarto, and D. B. Rifkin, *Cell Reg.* **2,** 87 (1991).
118. M. Renko, N. Quarto, T. Morimoto, and D. B. Rifkin, *J. Cell. Physiol.* **144,** 108 (1990).
119. P. Casellas and P. Jeanteur, *Biochim. Biophys. Acta* **519,** 255 (1978).
120. W. M. Bonner, M. H. West, and J. D. Stedman, *Eur. J. Biochem.* **109,** 17 (1980).
121. A. Zweidler, *Meth. Cell Biol.* **17,** 223 (1978).
122. P. Byvoet, *Biochim. Biophys. Acta* **238,** 375a-376b (1971).
123. P. Byvoet, G. R. Shepherd, J. M. Hardin, and B. J. Noland, *Arch. Biochem. Biophys.* **148,** 558 (1972).
124. P. Byvoet, *Arch. Biochem. Biophys.* **152,** 887 (1972).
125. W. K. Paik and S. Kim, *J. Biol. Chem.* **245,** 88 (1970).
126. J. A. Duerre and S. Chakrabarty, *J. Biol. Chem.* **250,** 8457 (1975).
127. M. T. Tuck and R. Cox, *Carcinogenesis* **3,** 431 (1982).
128. R. Desrosiers and R. M. Tanguay, *J. Biol. Chem.* **263,** 4686 (1988).
129. C. G. Edmonds, J. A. Loo, R. D. Smith, A. F. Fuciarelli, B. D. Thrall, J. E. Morris, and D. L. Springer, *J. Toxicol. Environ. Health* **40,** 159 (1993).
130. M. Kouach, D. Belaiche, M. Jaquinod, M. Couppez, D. Kmiecik, G. Ricart, D. A. Van, P. Sautiere, and G. Briand, *Biol. Mass Spectrom.* **23,** 283 (1994).
131. W. E. Mears, V. Lam, and S. A. Rice, *J. Virol.* **69,** 935 (1995).
132. M. K. Hibbard and R. M. Sandri-Goldin, *J. Virol.* **69,** 4656 (1995).
133. A. M. McCarthy, L. McMahan, and P. A. Schaffer, *J. Virol.* **63,** 18 (1989).
134. W. R. Hardy and R. M. Sandri-Goldin, *J. Virol.* **68,** 7790 (1994).
135. R. M. Sandri-Goldin, M. K. Hibbard, and M. A. Hardwicke, *J. Virol.* **69,** 6063 (1995).
136. D. M. Knipe, D. Senechek, S. A. Rice, and J. L. Smith, *J. Virol.* **61,** 276 (1987).
137. S. A. Rice, L. S. Su, and D. M. Knipe, *J. Virol.* **63,** 3399 (1989).
138. W. E. Mears and S. A. Rice, *J. Virol.* **70,** 7445 (1996).
139. M. S. Lee, M. Henry, and P. A. Silver, *Genes Dev.* **10,** 1233 (1996).
140. G. H. Goodwin and E. W. Johns, *Eur. J. Biochem.* **40,** 215 (1973).
141. J. F. Maher and D. Nathans, *Proc. Natl. Acad. Sci. U.S.A.* **93,** 6716 (1996).
142. T. T. Paull, M. J. Haykinson, and R. C. Johnson, *Genes Dev.* **7,** 1521 (1993).
143. L. C. Boffa, R. Sterner, G. Vidali, and V. G. Allfrey, *Biochem. Biophys. Res. Commun.* **89,** 1322 (1979).
144. C. Wang, E. Lazarides, C. M. O'Connor, and S. Clarke, *J. Biol. Chem.* **257,** 8356 (1982).
145. C. Wang, J.-M. Lin, and E. Lazarides, *Arch. Biochem. Biophys.* **297,** 169 (1992).
146. C. Wang and M.-R. Lee, *Biochem. J.* **294,** 69 (1993).
147. M. Reporterer and J. L. Corbin, *Biochem. Biophys. Res. Commun.* **43,** 644 (1971).
148. R. Helm, Z. Deyl, and O. Vancikova, *Exp. Gerontol.* **12,** 245 (1977).
149. N. Rawal, Y. J. Lee, W. K. Paik, and S. Kim, *Biochem. J.* **287,** 929 (1992).
150. L. F. Eng, F. C. Chao, B. Gerstl, D. Pratt, and M. G. Tavaststjerna, *Biochemistry* **7,** 4455 (1968).
151. E. E. Golds and P. E. Braun, *J. Biol. Chem.* **251,** 4729 (1976).
152. F. X. Omlin, H. D. Webster, C. G. Palkovits, and S. R. Cohen, *J. Cell Biol.* **95,** 242 (1982).
153. J. F. Poduslo and P. E. Braun, *J. Biol. Chem.* **250,** 1099 (1975).

154. R. E. Martenson, in "Biochemistry of Brain" (S. Kumar, ed.), p. 49. Pergamon, Oxford, 1980.
155. R. A. Ridsdale, D. R. Beniac, T. A. Tompkins, M. A. Moscarello, and G. Harauz, *J. Biol. Chem.* **272,** 4269 (1997).
156. N. Rawal, R. Rajpurohit, M. A. Lischwe, K. R. Williams, W. K. Paik, and S. Kim, *Biochim. Biophys. Acta* **1248,** 11 (1995).
157. C. Wang, J. M. Lin, and E. Lazarides, *Arch. Biochem. Biophys.* **297,** 169 (1992).
158. E. M. Ellis and G. A. Reid, *Gene* **132,** 175 (1993).
159. S. Loo, P. Laurenson, M. Foss, A. Dillin, and J. Rine, *Genetics* **141,** 889 (1995).
160. M. Henry, C. Z. Borland, M. Bossie, and P. A. Silver, *Genetics* **142,** 103 (1996).
161. R. Henriquez, G. Blobel, and J. P. Aris, *J. Biol. Chem.* **265,** 2209 (1990).
162. D. Tollervey, H. Lehtonen, M. Carmo-Fonseca, and E. C. Hurt, *EMBO J.* **10,** 573 (1991).
163. R. P. Jansen, E. C. Hurt, H. Kern, H. Lehtonen, M. Carmo-Fonseca, B. Lapeyre, and D. Tollervey, *J. Cell Biol.* **113,** 715 (1991).
164. T. Schimmang, D. Tollervey, H. Kern, R. Frank, and E. C. Hurt, *EMBO J.* **8,** 4015 (1989).
165. D. Tollervey, H. Lehtonen, R. Jansen, H. Kern, and E. C. Hurt, *Cell* **72,** 443 (1993).
166. B. Lubben, P. Fabrizio, B. Kastner, and R. Luhrmann, *J. Biol. Chem.* **270,** 11549 (1995).
167. C. Yan and T. Melese, *J. Cell Biol.* **123,** 1081 (1993).
168. W. C. Lee, Z. X. Xue, and T. Melese, *J. Cell Biol.* **113,** 1 (1991).
169. K. Kondo and M. Inouye, *J. Biol. Chem.* **267,** 16252 (1992).
170. K. Kondo, L. R. Kowalski, and M. Inouye, *J. Biol. Chem.* **267,** 16259 (1992).
171. W. C. Lee, D. Zabetakis, and T. Melese, *Mol. Cell. Biol.* **12,** 3865 (1992).
172. C. Gamberi, G. Contreas, M. G. Romanelli, and C. Morandi, *Gene* **148,** 59 (1994).
173. J. J. Lin and V. A. Zakian, *Nucleic Acids Res.* **22,** 4906 (1994).
174. L. M. Konkel, S. Enomoto, E. M. Chamberlain, P. McCune-Zierath, S. J. Iyadurai, and J. Berman, *Proc. Natl. Acad. Sci. U.S.A.* **92,** 5558 (1995).
175. S. Marcand, E. Gilson, and D. Shore, *Science* **275,** 986 (1997).
176. J. T. Anderson, S. M. Wilson, K. V. Datar, and M. S. Swanson, *Mol. Cell. Biol.* **13,** 2730 (1993).
177. A. Y. Jong, R. Aebersold, and J. L. Campbell, *J. Biol. Chem.* **260,** 16367 (1985).
178. A. Y. Jong and J. L. Campbell, *Proc. Natl. Acad. Sci. U.S.A.* **83,** 877 (1986).
179. A. Y. Jong, M. W. Clark, M. Gilbert, A. Oehm, and J. L. Campbell, *Mol. Cell. Biol.* **7,** 2947 (1987).
180. M. E. Cusick, *Biochim. Biophys. Acta* **1171,** 176 (1992).
181. S. G. Lee, I. Lee, S. H. Park, C. Kang, and K. Song, *Genomics* **25,** 660 (1995).
182. D. C. Masison, A. Blanc, J. C. Ribas, K. Carroll, N. Sonenberg, and R. B. Wickner, *Mol. Cell. Biol.* **15,** 2763 (1995).
183. S. P. Ridley, S. S. Sommer, and R. B. Wickner, *Mol. Cell. Biol.* **4,** 761 (1984).
184. A. W. Johnson and R. D. Kolodner, *Mol. Cell. Biol.* **15,** 2719 (1995).
185. S. M. Noble and C. Guthrie, *Genetics* **143,** 67 (1996).
186. F. He and A. Jacobson, *Genes Dev.* **9,** 437 (1995).
187. E. Sakashita and H. Sakamoto, *J. Biochem.* **120,** 1028 (1996).
188. C. W. Siebel, A. Admon, and D. C. Rio, *Genes Dev.* **9,** 269 (1995).
189. M. D. Adams, R. S. Tarng, and D. C. Rio, *Genes Dev.* **11,** 129 (1997).
190. J. Wang and L. R. Bell, *Genes Dev.* **8,** 2072 (1994).
191. J. Valcarcel, R. Singh, P. D. Zamore, and M. R. Green, *Nature (London)* **362,** 171 (1993).
192. G. L. Stoner, *J. Neurochem.* **43,** 433 (1984).
193. S. G. Amur, G. Shanker, and R. A. Pieringer, *J. Neurosci. Res.* **16,** 377 (1986).
194. C. Abramovich, L. M. Shulman, E. Ratovitski, S. Harroch, M. Tovey, P. Eid, and M. Revel, *EMBO J.* **13,** 5871 (1994).

195. H. Yan, K. Krishnan, A. C. Greenlund, S. Gupta, J. T. Lim, R. D. Schreiber, C. W. Schindler, and J. J. Krolewski, *EMBO J.* **15**, 1064 (1996).
196. R. E. Law, J. B. Stimmel, M. A. Damore, C. Carter, S. Clarke, and R. Wall, *Mol. Cell. Biol.* **12**, 103 (1992).
197. D. A. Kujubu, J. B. Stimmel, R. E. Law, H. R. Herschman, and S. Clarke, *J. Neurosci. Res.* **36**, 58 (1993).
198. J. Enouf, F. Lawrence, C. Tempete, M. Robert-Gero, and E. Lederer, *Cancer Res.* **39**, 4497 (1979).
199. B. R. Kim and K. H. Yang, *Toxicol. Lett.* **59**, 109 (1991).
200. E. Birney, S. Kumar,, and A. R. Krainer, *Nucleic Acids Res.* **21**, 5803 (1993).
201. J. P. Girard, C. Bagni, M. Caizergues-Ferrer, F. Amalric, and B. Lapeyre, *J. Biol. Chem.* **269**, 18499 (1994).
202. M. S. Schmidt-Zachmann and E. A. Nigg, *J. Cell Sci.* **105**, 799 (1993).
203. B. Messmer and C. Dreyer, *Eur. J. Cell Biol.* **61**, 369 (1993).
204. H. Siomi and G. Dreyfuss, *J. Cell Biol.* **129**, 551 (1995).
205. R. Z. Florkiewicz, A. Baird, and A. M. Gonzalez, *Growth Factors* **4**, 265 (1991).
206. D. K. Snudden, J. Hearing, P. R. Smith, F. A. Grasser, and B. E. Griffin, *EMBO J.* **13**, 4840 (1994).
207. J. L. Yates and S. M. Camiolo, *Cancer Cells* **6**, 197 (1988).
208. S. G. Nadler, B. M. Merrill, W. J. Roberts, K. M. Keating, M. J. Lisbin, S. F. Barnett, S. H. Wilson, and K. R. Williams, *Biochemistry* **30**, 2968 (1991).
209. M. Kiledjian and G. Dreyfuss, *EMBO J.* **11**, 2655 (1992).
210. R. Rajpurohit, W. K. Paik, and S. Kim, *Biochem. J.* **304**, 903 (1994).
211. B. J. Calnan, B. Tidor, S. Biancalana, D. Hudson, and A. D. Frankel, *Science* **252**, 1167 (1991).
212. A. Mayeda and A. R. Krainer, *Cell* **68**, 365 (1992).
213. W. K. Paik, S. Kim, and H. W. Lee, *Biochem. Biophys. Res. Commun.* **46**, 933 (1972).
214. H. W. Lee, W. K. Paik, and T. W. Borun, *J. Biol. Chem.* **248**, 4194 (1973).
215. F. N. Chang, I. J. Navickas, C. Au, and C. Budzilowicz, *Biochim. Biophys. Acta* **518**, 89 (1978).
216. L. Kass and C. J. Zarafonetis, *Proc. Soc. Exp. Biol. Med.* **145**, 944 (1974).
217. L. Kass and D. Munster, *Biochim. Biophys. Acta* **524**, 497 (1978).
218. H. W. Lee and W. K. Paik, *Biochim. Biophys. Acta* **277**, 107 (1972).
219. M. C. Alliegro and M. A. Alliegro, *Dev. Biol.* **174**, 288 (1996).
220. P. R. Young and C. M. Waickus, *Biochem. Biophys. Res. Commun.* **142**, 200 (1987).
221. M. O. Aspillaga and J. R. McDermott, *J. Neurochem.* **28**, 1147 (1977).
222. L. P. Chanderkar, W. K. Paik, and S. Kim, *Biochem. J.* **240**, 471 (1986).
223. A. J. Crang and W. Jacobson, *J. Neurochem.* **39**, 244 (1982).
224. S. G. Amur, G. Shanker, and R. A. Pieringer, *J. Neurochem.* **43**, 494 (1984).
225. S. Kim, M. Tuck, M. Kim, A. T. Campagnoni, and W. K. Paik, *Biochem. Biophys. Res. Commun.* **123**, 468 (1984).
226. S. Kim, M. Tuck, L. L. Ho, A. T. Campagnoni, E. Barbarese, R. L. Knobler, F. D. Lublin, L. P. Chanderkar, and W. K. Paik, *J. Neurosci. Res.* **16**, 357 (1986).
227. P. R. Young, D. A. Vacante, and C. M. Waickus, *Biochem. Biophys. Res. Commun.* **145**, 1112 (1987).
228. S. G. Amur, G. Shanker, J. M. Cochran, H. S. Ved, and R. A. Pieringer, *J. Neurosci. Res.* **16**, 367 (1986).
229. S. K. Ghosh, N. Rawal, S. K. Syed, W. K. Paik, and S. D. Kim, *Biochem. J.* **275**, 381 (1991).
230. D. H. Small, P. R. Carnegie, and R. M. Anderson, *Neurosci. Lett.* **21**, 287 (1981).
231. W. Jacobson, G. Gandy, and R. L. Sidman, *J. Pathol.* **109**, 13 (1973).

232. G. Gandy, W. Jacobson, and R. Sidman, *J. Physiol.* **233,** 1P- (1973).
233. J. M. Scott, J. J. Dinn, P. Wilson, and D. G. Weir, *Lancet* **2,** 334 (1981).
234. R. Deacon, P. Purkiss, R. Green, M. Lumb, J. Perry, and I. Chanarin, *J. Neurol. Sci.* **72,** 113 (1986).
235. M. F. Lou, *Science* **203,** 668 (1979).
236. N. Rawal, Y. J. Lee, J. N. Whitaker, J. O. Park, W. K. Paik, and S. Kim, *J. Neurol. Sci.* **129,** 186 (1995).
237. W. K. Paik, M. Abou-Gharbia, D. Swern, P. Lotlikar, and S. Kim, *Can. J. Biochem. Cell Biol.* **61,** 850 (1983).
238. T. Ogawa, M. Kimoto, H. Watanabe, and K. Sasaoka, *Arch. Biochem. Biophys.* **252,** 526 (1987).
239. T. Ogawa, M. Kimoto, and K. Sasaoka, *Biochem. Biophys. Res. Commun.* **148,** 671 (1987).
240. T. Ogawa, M. Kimoto, and K. Sasaoka, *J. Biol. Chem.* **264,** 10205 (1989).
241. T. Ogawa, M. Kimoto, and K. Sasaoka, *J. Biol. Chem.* **265,** 20938 (1990).
242. P. P. DeDeyn, P. Robitaille, M. Vanasse, I. A. Qureshi, and B. Marescau, *Nephron* **69,** 411 (1995).
243. S. Ueno, A. Sano, K. Kotani, K. Kondoh, and Y. Kakimoto, *J. Neurochem.* **59,** 2012 (1992).
244. N. M. Olken and M. A. Marletta, *Biochemistry* **32,** 9677 (1993).
245. N. M. Olken, Y. Osawa, and M. A. Marletta, *Biochemistry* **33,** 14784 (1994).
246. M. Miyake and Y. Kakimoto, *Metab. Clin. Exp.* **25,** 885 (1976).
247. T. Dandekar and D. Tollervey, *Nucleic Acids Res.* **21,** 5386 (1993).
248. J. Tang, J. D. Gary, S. Clarke, and H. R. Herschman, *J. Biol. Chem.* **273,** in press (1998).
249. R. M. Sandri-Goldin, *Genes Develop.* **12,** 868 (1998).
250. E. C. Shen, M. F. Henry, V. H. Weiss, S. R. Valenti, P. A. Silver, and M. S. Lee, *Genes Develop.* **12,** 679 (1998).

Genetic Regulation of Phospholipid Metabolism: Yeast as a Model Eukaryote

SUSAN A. HENRY[1] AND
JANA L. PATTON-VOGT

Department of Biological Sciences
Carnegie Mellon University
Pittsburgh, Pennsylvania 15213

I. Introduction 134
   A. Yeast as a Model System 134
   B. Phospholipid Synthesis in Yeast: A Brief Overview 135
   C. Phospholipid Turnover in *Saccharomyces cerevisiae*: An Overview .. 137
II. Regulation of Phospholipid Metabolism in Yeast 141
   A. Genetic and Biochemical Studies 142
   B. Regulation of Phospholipid Biosynthesis in *Saccharomyces cerevisiae* 153
III. Model for the Regulation of Genes Containing UAS_{INO} 165
IV. Summary and Future Directions 172
   References 173

Baker's yeast, *Saccharomyces cerevisiae*, is an excellent and an increasingly important model for the study of fundamental questions in eukaryotic cell biology and genetic regulation. The fission yeast, *Schizosaccharomyces pombe*, although not as intensively studied as *S. cerevisiae*, also has many advantages as a model system. In this review, we discuss progress over the past several decades in biochemical and molecular genetic studies of the regulation of phospholipid metabolism in these two organisms and higher eukaryotes. In *S. cerevisiae*, following the recent completion of the yeast genome project, a very high percentage of the gene–enzyme relationships in phospholipid metabolism have been assigned and the remaining assignments are expected to be completed rapidly. Complex transcriptional regulation, sensitive to the availablity of phospholipid precursors, as well as growth phase, coordinates the expression of the structural genes encoding these enzymes in *S. cerevisiae*. In this article, this regulation is described, the mechanism by which the cell senses the ongoing metabolic activity in the pathways for phospholipid biosynthesis is discussed, and a model is presented. Recent information relating to the role of phosphatidylcholine turnover in *S. cerevisiae* and its relationship to the secretory pathway, as well as to the regulation of phospholipid metabolism, is also presented. Similarities in the role of phospholipase D-mediated phosphatidylcholine turnover in the secretory process in yeast and mammals lend further credence to yeast as a model system. © 1998 Academic Press

---

[1] To whom correspondence should be addressed.

I. Introduction

The questions our laboratory has studied for the past two decades involve regulation of the pathways controlling metabolism of the membrane phospholipids in two yeast species, *Saccharomyces cerevisiae* and *Schizosaccharomyces pombe*. In the evolution of living cells, the development of functional membranes is presumed to have been an early and essential step (*1*). Membranes provide structural integrity to the cell, separating the inside of the cell from its environment and, in eukaryotes, also defining the architectural boundaries of cellular subcompartments. Ester-linked glycerophospholipids, with their property of self-assembly into bilayers at the interface between aqueous environments, are a dominant feature of biological membranes in eukaryotes and eubacteria (*2*).

Viewed from the perspective of metabolic regulation in eukaryotic cells, the synthesis of biological membranes is enormously complex. A key challenge for the growing cell is the coordination of ongoing biosynthesis of phospholipids to ensure their delivery at the crucial times, the right locations, and in the correct proportions. This intricate choreography controls the proportional synthesis of hundreds of molecular species consisting of a variety of polar head groups and a wide array of combinations of fatty acid side chains of varying chain length and desaturation. These diverse molecules must be delivered to a multiplicity of cellular locations, a process requiring not only intricate communication between multiple membrane sites separated by aqueous environments, but also the regulated transport and exchange of lipids between the membrane compartments.

A. Yeast as a Model System

In the baker's yeast, *S. cerevisiae*, mechanisms by which the eukaryotic cell exerts control over the complex process of membrane biogenesis can be dissected using the unparalleled molecular genetics of this organism. *Saccharomyces cerevisiae* has long been the eukaryotic model organism of choice for studies of fundamental processes such as transcriptional regulation, recombination, secretion, protein sorting, and cell division. Its powerful and well-developed molecular genetics, well-characterized biochemistry, and ease of cultivation on chemically defined media are now coupled with the availablity of the complete sequence of its genome (*3*). Furthermore, a project now well underway (*4*) will result in construction and analysis of disruption mutants of all 5885 genes of the S. cerevisiae genome (*5*). The set of experimental tools now available for investigation of basic cellular processes in *S. cerevisiae* is unmatched in any other eukaryotic organism. However, *S. cerevisiae* differs from other eukaryotes even in some fundamental cellular processes. For example, *S. cerevisiae*, unlike most eukaryotes, divides by budding rather than fission. Nevertheless, many features of the control of the cell cycle of S. *cerevisiae* have proved to be homologous to other eukaryotes (6).

Whereas S. *cerevisiae* has 5885 genes and 12 Mb of DNA, humans have an estimated 50,000–100,000 genes and 3000 Mb of DNA (3, 7). However, virtually all of the genes of S. *cerevisiae* are expected to have homologs in other eukaryotic organisms (3). It is estimated that some 1200 million years have elapsed since the lineages of S. *cerevisiae* and other fungi separated from the eukaryotic lineages that led to multicellular animals and plants (8, 9). In terms of gene organization, S. *cerevisiae* has a much higher density of coding sequences than any other well-studied eukaryote, including the fission yeast, S. *pombe* (3, 9). *Saccharomyces cerevisiae* has one protein coding sequence per 2 kb on average in its genome as opposed to one potential coding sequence per 2.3 kb in S. *pombe* and one coding sequence per 30 kb in humans (3). Indeed, S. *cerevisiae* has many fewer introns than any other eukaryote that has been studied, including S. *pombe*, which has a higher density of coding sequences than other eukaryotes (3). Only about 4% of the genes in S. *cerevisiae* have introns, whereas approximately 40% of S. *pombe* genes have introns (3). These two yeasts are very distantly related and are believed to have diverged from one another at least 1000 million years ago (8). Presumably, those proteins that have homologs in both yeasts, and those features of cellular organization and regulation that they share in common, existed prior to the separation of their lineages from their last common ancestor and, very likely, from the eukaryotic lineages leading to multicellular organisms. We believe, therefore, that a comparative study of regulation in these two organisms may reveal fundamental mechanisms of metabolic regulation common to all eukaryotes.

B. Phospholipid Synthesis in Yeast: A Brief Overview

With respect to the pathways of lipid metabolism, S. *cerevisiae* and S. *pombe* are fairly typical eukaryotes. They both synthesize the inositol- and choline-containing phospholipids that are typical of all eukaryotes, via pathways (Fig. 1) that are, for the most part, similar to those in other eukaryotes (10–12). The two yeasts differ from each other, with respect to phospholipid metabolism, in one major aspect: S. *pombe* is unable to synthesize the precursor L-*myo*inositol 1-phosphate (I1-P) and is consequently an inositol auxotroph (13), whereas S. *cerevisiae*, like other eukaryotes, synthesizes I1-P from glucose 6-phosphate (Fig. 1) (14).

Yeast, like other eukaryotes, synthesizes phosphatidylinositol (PI)[2] from free inositol and cytidinediphosphate diacylglycerol (CDP-DG) (Fig. 1). PI is subsequently phosphorylated to form PIP and PIP_2 (Fig. 1). The existence of

---

[2] Abbreviations: C, choline; CDP-C, cytidinediphosphate choline; CDP-DG, cytidinediphosphate diacylglycerol; CDP-DME, cytidinediphosphate dimethylethanolamine; CDP-E, cytidinediphosphate ethanolamine; CDP-MME, cytidinediphosphate monomethylethanol-

PIP and PIP$_2$ in *S. cerevisiae* was first described by Lester and colleagues (15), who also demonstrated that those molecules exhibit an active metabolism of their phosphomonoester groups in response to cellular ATP levels (16, 17). This active metabolism can be reconstituted *in vitro* using an isolated plasma membrane fraction and exogenous ATP (18). Auger *et al.* (19) reported that phosphatidylinositol-3-phosphate is as abundant as phosphatidylinositol-4-phosphate in *S. cerevisiae*. A PI 4-kinase, which is essential for viability (20), is encoded by the *PIK1* gene (see Table I). Another potential PI 4-kinase gene, *STT4*, was identified by mutations conferring staurosporine sensitivity (21). Yoshida *et al.* (21) reported that the *STT4* deletion mutant is viable and that cell homogenates prepared from the deletion mutant exhibit decreased PI 4-kinase activity compared to wild type. However, PI 4-kinase activity has not been directly documented for the Stt4p, *in vitro*. Dennis Voelker and colleagues (personal communication) isolated mutants defective in transport of phosphatidylserine (PS) from its site of synthesis in the endoplasmic reticulum/mitochondrial-associated membrane compartments to the location of the *PSD2*-encoded PS decarboxylase in the Golgi/vacuole compartments (22). They found a mutant (*pstB1*) with reduced transport of PS *in vivo* that is an allele of *stt4*. The *pstB1* mutant has reduced PI 4-kinase activity and shows some accumulation of PS at the expense of phosphatidylethanolamine (PE). These results suggest that PI 4-kinase activity plays a role in lipid transport to the Golgi/vacuole (Dennis Voelker, personal communication). However, in contrast to the results of Yoshida *et al.* (21), the deletion mutation constructed by Dennis Voelker and colleagues is not viable.

---

amine; CL, cardiolipin; C-P, choline phosphate; CTP, cytidinetriphosphate; DAG, *sn*-1,2-diacylglycerol; DGPP, diacylglycerol pyrophosphate; DHAP, dihydroxyacetone phosphate; DME, dimethylethanolamine; DME-P, dimethylethanolamine phosphate; E, ethanolamine; E-P, ethanolamine phosphate; FA, fatty acid; Glu-6-P, glucose-6-phosphate; gly-3-P, glycerol-3-phosphate; GroPC, glycerophosphocholine; GroPIns, glycerophosphoinositol; I, inositol; lyso-PA, 1-acylglycerol-3-phosphate; MME, monomethylethanolamine; MME-P, monomethylethanolamine phosphate; PA, phosphatidic acid; PC, phosphatidylcholine; PDME, phosphatidylimethyethanolamine; PE, phosphatidylethanolamine; PG, phosphatidylglycerol; PGP, phosphatidylglycerol phosphate; PI, phosphatidylinositol; PIP, phosphatidylinositol phosphate; PIP$_2$, phosphatidylinositol biphosphate; PMME, phosphatidylmonomethylethanolamine; PS, phosphatidylserine; S, serine; TG, triacylglycerol. For gene abbreviations see Tables I–V.

FIG. 1. Pathways of phospholipid biosynthesis and turnover in *Saccharomyces cerevisiae*. This schematic details the metabolic steps and corresponding structural genes of phospholipid biosynthesis in yeast. Allele designations are alongside reactions for which mutations have been characterized on a biochemical level. The designations given are for the recessive mutant alleles rather than the wild-type structural genes. A more complete list of structural gene designations is given in Tables I–V. Mutations that cause Opi$^-$ or Opc$^-$ phenotypes are indicated. Metabolic steps that have been shown to be regulated by inositol and choline are indicated by a double line.

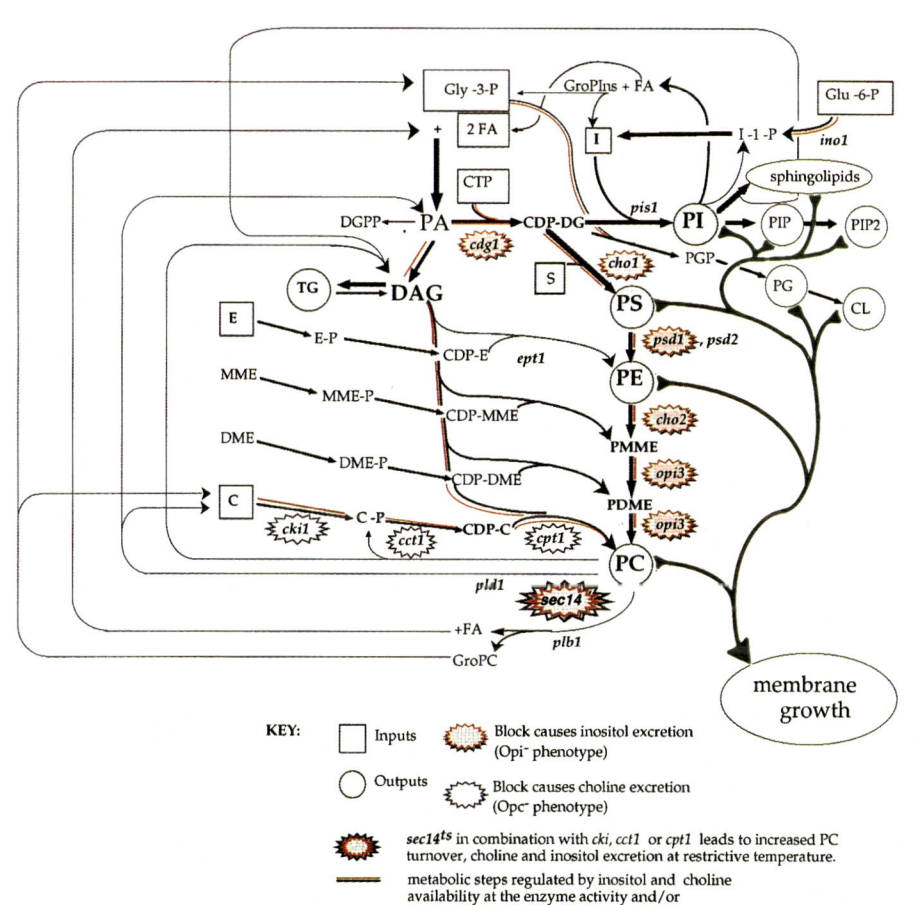

A PI 3-kinase that is essential for vacuolar protein sorting is encoded by the *VPS34* gene (23). The *TOR1* and *TOR2* genes, originally identified by mutations that confer resistance to rapamycin, encode proteins with homology to PI 3-kinase (24). Likewise, the *TEL1* gene, which plays a role in maintaining telomere length in yeast, encodes a product with homology to PI kinases (25, 26). *TEL1* also shows homology to the *MEC1* (*ESR1*) gene of yeast, which is involved in mitotic cell division. The *MEC1* gene product also exhibits homology to PI kinases, including those of *TOR1* and *TOR2* (27, 28) (Table I). However, none of these gene products (i.e., Tor1p, Tor2p, Tel1p, or Mec1p) has been demonstrated to have PI kinase activity.

In terms of glycerophospholipid synthesis, yeast cells differ from mammalian cells primarily with respect to the mechanism by which they synthesize phosphatidylserine and the predominant route by which they synthesize phosphatidylcholine (PC). Yeast synthesize PS from CDP-DG and serine (29) in a reaction catalyzed by the membrane-associated enzyme PS synthase (30) (Fig. 1), whereas, mammals synthesize PS via an exchange reaction with phosphatidylethanolamine and free serine (12). Yeast cells growing without choline supplementation synthesize PC primarily via a three-step methylation of PE (31–33), but can also synthesize PC from free choline via the CDP-choline pathway (Fig. 1), originally documented by Kennedy and Weiss (34). When yeast cells are growing in the absence of choline, the CDP-choline pathway serves primarily to recycle choline from lipid turnover (35). This pathway can serve as a major or predominant source of PC in yeast when choline is supplied exogenously, especially in mutants with defects in synthesis of PC via the methylation pathway (36–39). In mammals, however, the CDP-choline pathway is the predominant route of PC biosynthesis (12). The methylation pathway in mammals, first described by Bremer and Greenberg (40), functions primarily in the liver (41, 42).

C. Phospholipid Turnover in *Saccharomyces cerevisiae:* An Overview

In mammals, much has been learned in the past decade about the role of lipid turnover in signal transduction. Currently, with the excitement surrounding yeast as a model system, there has been much interest in learning whether the signal transduction pathways involving metabolites generated by lipid turnover, which act as second messengers in mammals, also occur in yeast. Homologs of the mammalian phospholipases B, C, and D (PLB, PLC, PLD) genes (Table II) have been detected in yeast, and gene disruptions have been made (43–48). All such gene disruption strains created to date are viable. For example, deletion of a phospholipase B gene (*PLB1*) results in no apparent phenotypic defect, although strains bearing that disruption have decreased lysophospholipase/phospholipase B activity (43).

TABLE I
GENES ENCODING BIOSYNTHETIC ENZYMES INVOLVED IN PHOSPHOLIPID SYNTHESIS

Enzyme	Gene designation	Ref.
α-subunit of fatty acid synthetase	FAS1	255–257
β-subunit of fatty acid synthetase	FAS2	255, 256
Acetyl-CoA carboxylase	ACC1/FAS3[a]	226, 258
Acyl-CoA desaturase	OLE1	259
Acyl-CoA synthetase	FAA1	260
Acyl-CoA synthetase	FAA2	261
Acyl-CoA synthetase	FAA3	261
Acyl-CoA synthetase	FAA4	262
Serine palmitoyltransferase subunit	LCB1	263
Serine palmitoyltransferase subunit	LCB2/SCS1[b]	264, 265
Fatty acyltransferase	SLC1	179
Inositol phosphorylceramide synthase	AUR1	266
Inositol-1-phosphate synthase	INO1	107, 121
CDP-DAG synthase	CDS1/CDG1[c]	130, 131
Phosphatidylserine decarboxylase	PSD1[c]	166, 167
Phosphatidylserine decarboxylase	PSD2	22
Phosphatidylserine synthase	CHO1/PSS1[c]	30, 159, 211, 212, 215
Phospholipid N-methyltransferase	PEM1/CHO2[c]	37, 147
Phospholipid N-methyltransferase	PEM2/OPI3[c]	129, 147
Choline kinase	CKI1[d]	170, 267
sn-1,2-Diacylglycerol ethanolaminephosphotransferase	EPT1	151, 171
sn-1,2-Diacylglycerol cholinephosphotransferase	CPT1[d]	152, 171
Choline phosphate:CTP cytidyltransferase	CCT1[d]	169
Cardiolipin synthase	CLS1	267a
Diacylglyceropyrophosphatase	DPP1	188, 188a
	DPP2/LPP1[e]	—
Phosphatidylinositol synthase	PIS1	109, 110, 112
Phosphatidylinositol 4-kinase	PIK1	20, 268
Phosphatidylinositol 4-kinase	STT4	21
Phosphatidylinositol 3-kinase	VPS34	23
Phosphatidylinositol 3-kinase homologs[e]	TOR1	24
	TOR2	—
	MEC1/ESR1	27, 28
	TEL1	25, 26

[a]Alternative names for genes are indicated by a slash (i.e., ACC1/FAS3).

[b]The SCS1 gene designation has been used separately by Zhao et al. (265) to identify the subunit of serine palmitoyl transferase (also known as LCB2) and by Hosaka et al. (222) to identify a high-copy suppressor of the ire15 mutation. The gene isolated by Hosaka et al. (222) is identical to the INO2 gene (220).

[c]Mutations in these genes lead to an Opi⁻ phenotype under certain growth conditions (see Fig. 1).

[d]Mutations in these genes lead to an Opc⁻ phenotype when combined with sec14^{ts} (Fig. 1).

[e]Based solely on homology data.

TABLE II
GENES ENCODING PHOSPHOLIPASES

Enzyme	Gene designation	Ref.
Phospholipase B	*PLB1*	43
Phospholipase C	*PLC1*	44
Phospholipase D	*PLD1/SPO14*	45–47, 49, 269

The *PLD1* gene (*45–47*), originally isolated as the *SPO14* gene (*49*), is required for meiosis, but is not required for viability. *PLD1* encodes a protein that appears to be primarily membrane-associated, with a predicted molecular mass of 195 kDa. Assays *in vitro* have shown the enzyme to be PC specific, Ca^{2+} independent, stimulated by phosphatidylinositol 4,5-bisphosphate, and inhibited by oleate (*45*), similar to the human (h) hPLD1 gene product (*50*). In fact, the hPLD1 gene was cloned based on its sequence homology to *S. cerevisiae PLD1*. A phospholipase D activity distinct from PLD1p was first detected by Paltauf and colleagues (*51*) and was subsequently confirmed by others (*52*). The second PLD activity in yeast is PS/PE specific, Ca^{2+} dependent, not stimulated by PIP_2 *in vitro*, and is probably structurally unrelated to PLD1p (*51*).

A phospholipase C gene (*PLC1*) has been identified in *S. cerevisiae*, and it bears closest resemblance to the γ class of PI-PLC enzymes. Strains carrying deletion mutations in *PLC1* are viable under normal laboratory conditions, but display temperature-sensitive growth, osmotic sensitivity, and aberrant growth on galactose, raffinose, and glycerol (*44*). Strains carrying mutations in *PLC1* also exhibit defective chromosome segregation (*53*). Although Plc1p was shown to cleave PI 4,5-P_2 *in vitro*, no *in vivo* PLC-mediated turnover of polyphosphoinositides has been documented under normal growth conditions in the presence of glucose. However, nitrogen starvation and subsequent refeeding does lead to the production of DAG and inositol 1,4,5-trisphosphate (IP_3), products of PLC-mediated turnover (*54*). Furthermore, Marini *et al.* (*54a*) reported a cell-cycle-dependent hydrolysis of PC to DAG and choline phosphate (C-P), that is stimulated by the Cdc28 protein kinase.

Turnover via a PLC-mediated mechanism is not the only, or even the major, route of phosphoinositide catabolism in *S. cerevisiae*. The complete deacylation of PI, presumably via a PLB or PLA and a lysophospholipase, to form extracellular glycerophosphoinositol (GroPIns) is a predominant PI turnover route (*55*). The polyphosphoinositides can also turn over via a deacylation event (*56*). The large, rapid, and reversible changes in the levels of

PIP and PIP_2 demonstrated by Lester and colleagues to occur when *S. cerevisiae* is alternately exposed to and starved for glucose were subsequently shown to occur via deacylation events. Hawkins *et al.* (56) demonstrated that the products of this polyphosphoinositide turnover are extracellular glycerophosphoinositol 4-phosphate and glycerophosphoinositol 4,5-bisphosphate. This suggests that the nutrient signal of glucose refeeding activates one or more phospholipases A or B, which act on PI, PI 4-P, and PI 4,5-P_2.

The production and subsequent reutilization of extracellular GroPIns produced by PI turnover is an interesting aspect of phospholipid metabolism in yeast. Studies of PI turnover performed by Angus and Lester (55) revealed the surprising fact that approximately half of the inositol and phosphorous lost from PI during growth in rich media occurs as GroPIns in the media. The remaining half occurs in the phosphoinositol sphingolipids and a small percentage is phosphorylated to form PIP and PIP_2. GroPIns accumulates in the culture media at a level of approximately 25% of the cellular PI, compared to much lower levels of extracellular glycerophosphocholine and glycerophosphoethanolamine (55). It was shown that this process requires an energy source, because removal of glucose from the media reduces the formation of extracellular GroPIns and increases the production of extracellular inositol (57). Exogenously added PI was quantitatively converted to GroPIns, with the probable intermediate formation of monoacylglycerophosphoinositol, leading the authors to speculate that PI deacylation occurs at the cell surface via phospholipases (57).

Paltauf *et al.* (58) showed that *Saccharomyces uvarum*, a near inositol auxotroph, also produces extracellular GroPIns. In addition, the authors found that on starvation for inositol, *S. uvarum* takes up [^{32}P]phosphate and [^{3}H]inositol from exogenous glycero[^{32}P]phospho[^{3}H]inositol at approximately equal rates. Ongoing studies in our laboratory have demonstrated that *S. cerevisiae* is also capable of reutilizing GroPIns, and that both the production and reutilization of this metabolite are affected by inositol availability (59). That is, extracellular GroPIns is produced at much greater rates when inositol is available in the media, and GroPIns is reutilized by the cells only when the levels of inositol in the media are low. In studies employing a strain in which both of the inositol permeases (Table III) have been deleted (i.e., an *itr1 itr2* mutant) (60), we have shown that GroPIns enters the cell via a transporter distinct from the inositol permeases (59). In addition, studies with [^{14}C]glycerophospho[^{3}H]inositol have indicated that, although GroPIns enters the cell intact, the inositol moiety but not the glycerol moiety is incorporated into lipids. A GroPIns transporter gene (*GIT1*) has now been cloned (*60a*) and is currently under study.

TABLE III
GENES ENCODING TRANSPORT PROTEINS INVOLVED IN
PHOSPHOLIPID METABOLISM

Function	Gene designation	Ref.
Inositol transport	ITR1	60, 73
Inositol transport	ITR2	60, 73
Choline transport	CTR1 (HNM1)	229, 270
Glycerophosphoinositol transport	GIT1	59, 60a
PI/PC transport	SEC14	84, 85

II. Regulation of Phospholipid Metabolism in Yeast

The focus of this article is primarily on the work on genetic regulaton of lipid metabolism in yeast. Our laboratory has been involved in the analysis of this regulation over the past 20 years and great progress has been made during this period, by us and by many other laboratories, in elucidating the mechanisms by which phospholipid biosynthetic pathways are regulated in yeast (10, 11, 61–66). A connection between the regulation of phospholipid biosynthesis and phospholipid turnover has also emerged (35, 59, 60a). Phospholipid metabolism in yeast is controlled on multiple levels, including biochemical control of enzymes in several parts of the pathway. These levels of control have been reviewed by Carman and Zeimetz (65). The regulation includes diverse mechanisms, including noncompetitive inhibition of PS synthase by inositol (67), regulation by sphingolipid bases (68, 69), regulation by cytidinetriphosphate (CTP) (70), phosphorylation (71, 72), degradation of the inositol transporters in response to inositol (73, 74), and regulation of choline phosphate:CTP cytidyltransferase by the PI/PC transporter (74a).

Although the subject of localization of the phospholipid biosynthetic enzymes has been extensively explored in yeast (10, 75–78), the role of the movement of the various lipids between compartments in the overall regulation of phospholipid metabolism is largely unexplored. However, the discovery of novel phenotypes related to organelle structure and secretory function associated wtih defects in lipid metabolism suggest that this area of exploration will be very fruitful (79). These novel phenotypes include aberrations in the structure of various membrane-bound organelles in mutants defective in the synthesis of long-chain fatty acids (79, 80) and vacuolar (81) and mitochondrial defects (36) associated with cho1 mutants defective in PS synthesis. Mitochondrial defects have been reported in a number of mutants defective in PC synthesis (82). In addition, a secretory defect is associated with mutations in the PI/PC transporter protein (Sec14p) (83–85). Moreover, mutations in the CDP-choline pathway for PC biosynthesis suppress the $sec14^{ts}$ defect (86, 87).

Much of the work in our laboratory has centered on transcriptional regulation, particularly the regulation of the *INO1* gene, the structural gene for inositol-1-phosphate synthase (IPS) (Table I). The *INO1* gene is the most highly regulated of the coordinately regulated yeast genes encoding enzymes of phospholipid biosynthesis. The *INO1* gene shows a complex pattern of transcriptional regulation that is found in a large number of genes encoding enzymes of phospholipid biosynthesis (10, 62, 64, 88). The native *INO1* transcript shows 30-fold or more derepression (88), whereas other coregulated enzymes show as little as two- to threefold derepression (89–92). Thus, various reporter constructs made from the *INO1* promoter region (93, 94) have become powerful tools for the analysis of the regulation controlling all the coregulated genes of phospholipid biosynthesis (35, 38, 95–99).

A. Genetic and Biochemical Studies

1. INOSITOL-1-PHOSPHATE SYNTHASE MUTANTS IN *Saccharomyces cerevisiae*

Given the high degree of regulation exhibited by the *INO1* gene, it was fortuitous that our studies on regulation of phospholipid biosynthesis in *S. cerevisiae* started with an analysis of the genetics and biochemistry of inositol biosynthesis. In yeast (100), as in other eukaryotic organisms (14, 101–104), I1-P is synthesized from glucose-6-phosphate (Fig. 1) in a complex reaction series that involves a coupled oxidation/reduction reaction, steric rearrangement, and ring closure and is catalyzed by the cytoplasmic enzyme inositol 1-phosphate synthase (103–107). I1-P is subsequently dephosphorylated by I1-P phosphatase (101, 108). Free inositol is used to synthesize PI in a reaction catalyzed by the membrane-associated enzyme phosphatidylinositol synthase (29, 109–112). PI synthase activity is influenced by the phospholipid composition of the cell membrane (110) and it is regulated in response to carbon source (113), but it is not regulated in response to inositol and choline availability (89, 110). In contrast, IPS activity is repressed over 30-fold in cells grown in the presence of inositol (100, 107). When yeast cells are grown in the presence of inositol, the rate of PI synthesis increases markedly over the rate of synthesis in the absence of exogenous inositol (65, 67). This increase in PI synthesis occurs at the expense of PS synthesis and is controlled by the availability of CDP-DG, as well as inositol (65, 67). The availability of inositol is, in turn, controlled by the expression of IPS, the product of the *INO1* gene (10, 11, 64, 114).

Yeast IPS is a 240,000-Da tetramer consisting of identical subunits of ap-

TABLE IV
Phospholipid Biosynthetic Regulatory Genes

Function	Gene designation	Ref.
Positive regulatory gene/high-copy suppressor of ire15 and BSD2-1 inositol auxotrophy	INO2/SCS1[a]	98, 122, 142, 220–222, 231
Positive regulatory gene	INO4	122, 217, 231
Negative regulatory gene	OPI1	89, 125, 219

[a] See footnote b in Table I concerning redundant use of the SCS1 gene designation. The INO2 designation has priority and, for clarity, SCS1 should not be used as a cognate of INO2.

proximately 60,000 Da (106, 107). The yeast (107), rat (105), and other forms of IPS proved to be remarkably similar in their biochemical characteristics (14). IPS was the first enzyme of phospholipid synthesis to be purified to homogeneity from yeast, and yeast IPS was the first form of the enzyme for which the structural gene was identified (107).

Genetic analysis of inositol biosynthesis in yeast revealed 10 genetic loci conferring inositol auxotrophy, most of which were represented by only a single allele (115, 116). All of these mutants lacked detectable IPS activity (117). Later, it became evident that mutations at many loci involved in general transcription regulation and other fundamental cellular processes interfere with IPS expression and confer inositol auxotrophy (see Table V for a summary). Mutants representing many of the functions shown in Table V were probably present among the mutants in the original collection of inositol auxotrophs (115, 117), but mutants represented by only a single allele were not fully characterized. In the initial genetic analysis, we focused our attention on those loci represented by more than one allele, of which there were only three: INO1, INO2, and INO4 (Tables I and IV). Among the inositol auxotrophs isolated in the two original screens, ino1 mutants accounted for approximately 70% of the total and ino2 and ino4 mutants each represented less than 10% (115, 116). The ino1 mutants showed a complex pattern of interallelic complementation (115, 118, 119), suggesting that the protein encoded by the INO1 locus had multiple subunits, as indeed, the subsequent purification and characterization of IPS confirmed (106, 107). Antibody raised in response to the purified protein revealed many ino1 mutants that produce protein that cross-reacts with this antibody (107, 119). On the basis of this evidence, the INO1 locus was identified as the structural gene for IPS (107). Later, cloning (120) and sequencing (121) of the INO1 gene confirmed that the protein encoded by the INO1 gene has an amino acid composition and N-terminal sequence that matches the purified subunit of IPS.

TABLE V
Other Genes Known to Affect *INO1* Expression

Function/effect on *INO1* expression	Gene	Ref.
RNA polymerase II subunit/mutations cause inositol auxotrophy	RPB1	132, 134
RNA polymerase II subunit/mutations cause inositol auxotrophy	RPB2	133
RNA polymerase II subunit/mutations cause inositol auxotrophy	RPB4	135
Global transcription factor/mutations cause inositol auxotrophy	SRB2	271
Global transcription factor/mutations cause inositol auxotrophy	SPT7	136
TATA binding protein/mutations cause inositol auxotrophy	SPT15	137
Global transcription factor/mutations cause inositol auxotrophy	SPT20/ADA5	138
Global regulator/*SIN1* deletion suppresses the inositol auxotrophy of *swi1*, *swi2*, *swi3* mutants	SIN1/SPT2	139, 272
Global regulator/required for *INO1* repression; a *sin3* deletion mutation causes Opi⁻ phenotype and suppresses the inositol auxotrophy of a *swi1* mutant	SIN3/SDI1/ UME4/RPD1	95, 96, 99, 273
Global regulator/required for *INO1* repression	UME6	99
Global transcription factor/mutations cause inositol auxotrophy	SWI1 (ADR6/GAM3)	246–248
Global transcription factor/mutations cause inositol auxotrophy	SWI2 (SNF2)	246–248
Global transcription factor/mutations cause inositol auxotrophy	SWI3	246–248
Histone H4/mutations (*hhf1*) suppress the inositol auxotrophy of *snf2* mutants	HHF1	249
Putative protein kinase involved in unfolded protein response pathway/mutations cause inositol auxotrophy	IRE1	143, 274, 253a
Transcription factor involved in unfolded protein response pathway/mutations cause inositol auxotrophy	IRE2/HAC1	143, 144, 253a
Unknown/mutations cause inositol auxotrophy	IRE15	275
Unknown/mutations suppress actin and secretory defects and cause inositol auxotrophy	SAC1	141, 245
Unknown/mutations suppress secretory defects and cause inositol auxotrophy	BSD2	142
Unknown/mutations cause dominant inositol auxotrophy	CSE1	153
Unknown/high-copy suppressor of *ire15* inositol auxotrophy	SSC2	275
Involved in protein degradation/mutations cause inositol auxotrophy	DOA4, UBC4, UBC5	276
Unknown/required for correct regulation of *INO1*	DEP1	240
Involved in release of transport vesicles from the ER/mutations cause an Opi⁻ phenotype	SEC13/DAM303	145
PI/PC transfer protein/mutations cause an Opi⁻ phenotype when in conjunction with a CDP-choline pathway mutation	SEC14	35, 84, 85
Structural genes in phospholipid biosynthesis/mutations cause an Opi⁻ phenotype	CDG1/CDS1, PSD1, CHO1, CHO2/PEM1, OPI3/PEM2	37, 38, 129–131, 158, 223, 225

a. Mutants Defective in the Expression of Inositol-1-Phosphate Synthase. Availability of specific antibody (*107*) to the IPS subunit made it possible to demonstrate that the enzyme subunit is virtually undetectable in extracts of cells grown in the presence of inositol (*107*). Moreover, the subunit is also absent in extracts of *ino2* and *ino4* mutant cells, suggesting that the wild-type *INO2* and *INO4* gene products are required for derepression and expression of the IPS subunit, the product of the *INO1* gene. On this basis, *INO2* and *INO4* genes were putatively identified as positive regulators of *INO1* expression (*107*). Following the cloning of the *INO1* gene, it was demonstrated that this regulation occurs at the level of *INO1* transcript abundance (*88*). Subsequently, it was shown that the *ino2* and *ino4* mutations affect not only IPS expression, but also expression of the phospholipid N-methyltransferases (PLMTs) leading to PC biosynthesis via methylation of PE (*92, 122*), as well as PS synthase (*90, 123*). Such pleiotropic phenotypes, common to many of the mutants to be discussed here, are due to the complexities of the underlying transcriptional regulation that coordinates the expression of many enzymes of phospholipid biosynthesis, a topic that will be explored in depth in this article.

b. Mutants Defective in the Repression of Inositol-1-Phosphate Synthase. To isolate mutants constitutive for IPS expression, we had initially hoped to expose yeast cells to inositol analogs in hopes of finding one or more compounds that would inhibit the growth of wild-type yeast, thus, allowing us to isolate resistant mutants, some of which might prove to harbor regulatory defects. This approach proved entirely unsuccessful because among the inositol analogs tested, none had any effect on the growth of will-type yeast (*124, 125*). Therefore, the classical approach of seeking regulatory mutants with resistance to a metabolic analog was not an available strategy. Instead, we turned to the approach of seeking mutants that overproduced inositol and excreted it into the growth medium. A bioassay for inositol excretion involving the use of an *ino1*, *ade1* mutant strain was devised. The strain actually employed for this work was a diploid that was homozygous for the *ino1* and *ade1* mutations (*124, 125*). We reasoned that the use of a diploid in our screen would negate any complication that could arise from mating between the tester strain and the colonies being screened. The *ade1* marker gave the tester strain a red colony phenotype and the *ino1* marker caused it to be auxotrophic for inositol. Thus, the tester strain will not grow in the absence of inositol, but, on inositol-free media, it forms a distinctive halo of red growth around colonies that are excreting inositol into the surrounding medium (*124–126*). Strains possessing the phenotype of inositol excretion are said to be Opi$^-$.

The Opi$^-$ (overproduction of inositol) phenotype is tested by spraying or spreading the tester strain onto inositol-free plates (*124–126*). The inositol-free growth medium used for testing the Opi$^-$ phenotype permits derepression of IPS in wild-type cells (*100, 107, 117*). Therefore, the Opi$^-$ phenotype as measured on inositol-free plates does not reflect constitutive *INO1* expression in the presence of inositol. Rather, the Opi$^-$ plate assay tests for overproduction of inositol, under a growth condition (inositol-free medium) where IPS is normally expressed (derepressing growth condition). Thus, we could not predict whether the Opi⁻ phenotype would also correlate to the inability to repress IPS in response to exogenous inositol. Virtually all Opi⁻ mutants identified, however, proved to overexpress IPS under both repressing and derepressing growth conditions (i.e., presence and absence of inositol) (*124, 125, 127*). Several independent *opi1* mutant alleles, which proved to be defective in a negative regulatory function controlling IPS (*127*), and many other enzymes of phospholipid biosynthesis (*89*) were identified among the original Opi⁻ mutants (*125*). The *opi3-3* mutant, which was later found to have a defect in a structural gene encoding a PLMT activity required for PC biosynthesis (*128, 129*), was also isolated in these initial screens for Opi⁻ mutants (*125*), as was the *cdg1* mutant, which has reduced cytidinediphosphate diacylglycerol synthase activity (*130*) (see Fig. 1 for the positions of structural gene lesions on this pathway). This very same *cdg1* mutant allele was confirmed to have a single base pair mutation in the structural gene, *CDS1*, encoding CDG-DG synthase (*131*).

c. Pleiotropic Phenotypes of Ino⁻ and Opi⁻ Mutants. The search for Opi⁻ mutants, like the search for Ino⁻ mutants, yielded both regulatory and structural gene mutants. Both the Opi⁻ and Ino⁻ phenotypes have proved to be associated with mutations defining a surprisingly wide range of loci (Tables IV & V). Many of the Opi⁻ and Ino⁻ mutants have very pleiotropic phenotypes and the Ino⁻ phenotype has proved to be quite common in mutants with defects in processes that are apparently quite removed from phospholipid biosynthesis. For example, mutants with a defect in components of the general transcription apparatus, such as the large subunit of RNA polymerase and the TATA binding protein (Spt15p), have an Ino⁻ phenotype (*132–139*), as do *sac1* and *BSD2-1* mutants, which suppress actin and secretory pathway defects (*140–142*). Mutations in the *IRE1* and *IRE2/HAC1* genes, which are involved in the unfolded protein response pathway, also confer inositol auxotrophy (*143, 144*). The Opi⁻ phenotype is associated both with regulatory and structural gene defects in phospholipid synthesis, but it is also found in mutants with defects in the secretory pathway (*35, 145, 146;* S. Kohlwein, personal communication) and in *ume6* and *sin3* mutants defective in global transcriptional regulators (*95, 96, 99*). A summary of mutants exhibiting Opi⁻ and Ino⁻ phenotypes is given in Table V. We now believe that our understanding of the regulatory control of *INO1* and other enzymes of phospholipid biosynthesis has progressed to the point where we can offer some speculation on the reasons for the occurrence of Opi⁻ and Ino⁻ phenotypes in such a diverse array of mutants. This is a topic to which we will return.

2. CHOLINE-REQUIRING MUTANTS OF *Saccharomyces cerevisiae*

Screens for mutants of S. *cerevisiae* potentially defining additional steps in phospholipid biosynthesis have been undertaken in a number of laboratories, including our own (36, 37, 129, 147–154), and the results, although successful, were often surprising. The success of a mutant screen depends, in large part, on the accuracy with which the investigator predicts the phenotype of a mutant that has lost a particular function. In the case of phospholipid biosynthetic mutants in yeast, many of the initial phenotype predictions were not entirely accurate, and the mutants isolated, such as the Opi⁻ and Ino⁻ mutants described above, often had very surprising and/or pleiotropic phenotypes. For example, based on the choline auxotrophic phenotype of *Neurospora crassa* mutants (155) defective in the three-step methylation that leads to PC from PE (Fig. 1), we and others expected that a search for choline auxotrophs in *S. cerevisiae* would yield mutants similar to the *Neurospora* mutants. It did not. Several independent searches for choline auxotrophs led to the isolation solely of *cho1* mutants (36, 148, 149, 156), which are not defective in PE methylation. The *cho1* mutants are defective in PS synthase (Fig. 1) and have an auxotrophic requirement satisfied either by ethanolamine or by choline (36, 149). Feeding the *cho1* mutants exogenous ethanolamine or choline restores PC biosynthesis by bypassing PS as an intermediate in PE and PC biosynthesis (Fig. 1). However, the *cho1* mutants are unable to make phosphatidylserine under any circumstance. The tightest of the *cho1* mutants, including *cho1* deletion mutants, make no detectable PS (36, 67, 90, 157–159). It has, however, been shown that the *cho1* mutants exhibit a growth deficit on rich medium, but grow as well as wild type in the presence of high concentrations of salt (5). They also have severe vacuolar (81) and mitochondrial defects (36, 82).

These observations concerning the growth requirements and phospholipid composition of *cho1* mutants led to the startling conclusion that S. *cerevisiae* cells can survive by making membranes essentially devoid of PS (36, 149, 157). The analysis of the *cho1* mutants also confirmed the earlier work of Steiner and Lester (29), who suggested that PS is synthesized in yeast from serine and CDP-DG in a reaction catalyzed by PS synthase. This reaction apparently does not occur in mammalian cells, in which PS is made via an exchange reaction with PE (12). However, bacteria do make PS from CDP-DG and serine in a reaction catalyzed by PS synthase, but the PS synthase enzymes in yeast and *E. coli* differ in a number of characteristics (30).

3. Mutants of *Saccharomyces cerevisiae* Defective in PE Methylation

Thus, the search for choline auxotrophs in S. *cerevisiae* resulted not in the expected isolation of mutants defective in PE methylation, but surprisingly in the isolation of mutants unconditionally defective in PS synthesis. Equally surprising, the S. *cerevisiae* mutants defective in PE methylation are not, in fact, strict choline auxotrophs (37, 129, 160). The first phospholipid methylation mutant to be isolated in S. *cerevisiae*, the original *opi3-3* mutant isolated by Greenberg *et al.* (*124, 135*), was selected on the basis of a regulatory phenotype, inositol overproduction, rather than on the basis of choline auxotrophy. Despite the lack of a growth requirement for choline, *opi3* mutants synthesize very low levels of PC, unless supplied with exogenous choline, and they accumulate very high levels of the precursor monomethylethanolamine (MME) (*128*). The *opi3* mutants were ultimately shown to be defective in the bifunctional enzyme that catalyzes the final two methylations of phosphatidylmonomethylethanolamine (PMME) to PC (*31, 129*) (Fig. 1).

The *opi3* mutants isolated in our laboratory proved to be allelic to mutants designated *pem2*, independently isolated by Kodaki and Yamashita (*147*). The *PEM2/OPI3* gene was identified as the structural gene for one of two phospholipid N-methyltransferase activities in yeast (*147*). The *OPI3/PEM2* gene disruption mutants have biochemical defects similar to the original (*opi3/pem2*) point mutants and are not choline auxotrophs (*129, 160*), although they grow more rapidly if supplied with choline. In addition, the growth of *opi3* mutants is inhibited at 37°C if exogenous MME is present in the growth medium (*129*), a phenotype that is presumably due to their aberrant lipid composition and the membrane-destabilizing effects of PMME.

The *S. cerevisiae cho2/pem1* mutants, which are defective in the first methylation step, PE → PMME (*31, 37, 147*), are, likewise, not stringent choline auxotrophs (*37, 160*), although their growth rate in the absence of choline is slower than that of the wild type (*37, 38, 160*). The original *cho2* mutant identified in our laboratory was isolated as a second-site mutation that altered the phenotype of a strain with a *cho1* mutation. The *cho1* mutants, as previously discussed, have an auxotrophic requirement satisfied by ethanolamine or choline. In a *cho1* genetic background, *cho2*, as a second-site mutation, imposes strict choline auxotrophy no longer satisfied by ethanolamine (*37, 161*). However, when crossed out of the *cho1* genetic background, the *cho2* mutant is prototrophic for choline and, like the *opi3* mutants, the *cho2* mutants have an Opi$^-$ phenotype (*37, 38*). Biochemical studies showed that the *cho2* mutants have very reduced ability to methylate PMME and, when grown in the absence of choline, PC constitutes 10% or less of the total phospholipid composition compared to 40% or greater in wild type. The *pem1* mutants isolated independently by Kodaki and Yamashita (*147*) have similar biochemical defects. The *PEM1* gene was identified as the structural gene for the PE methyltransferase, which catalyzes the initial methylation of PE to form PMME (*147*), and the *PEM1* and *CHO2* genes are identical (*37*).

One explanation for the ability of *cho2/pem1* and *opi3/pem2* mutants to grow in the absence of choline is the fact that the enzymes encoded by the *CHO2/PEM1* and *OPI3/PEM2* genes are at least partially functionally redundant (*37, 162*). Thus, *cho2* mutants retain some ability to methylate PE because the *OPI3/PEM2* gene product can methylate PE to a limited extent. Indeed, overexpression of the *OPI3* gene was shown to complement the *cho2* phenotype (*162*). However, functional redundancy of the *CHO2* and *OPI3* gene products is not the total explanation for the lack of choline auxotrophy in *cho2* and *opi3* mutants. The *cho1, cho2*, and *opi3* mutants all grow despite having very aberrant lipid compositions (*10*). The unexpected growth of the *cho1* mutants with membranes devoid of PS and the growth of *opi3* and *cho2* mutants with membranes containing very low levels of PC and with other major aberrations of lipid composition, suggest that S. *cerevisiae* cells have unusually plastic requirements for phospholipids. Indeed, other fungi, such as N. *crassa* and S. *pombe*, which share the same fundamental pathways for phospholipid biosynthesis, do not appear to have the same tolerance for perturbation of lipid composition. In these two organisms, the mutants biochemically analogous to the S. *cerevisiae cho2/pem1* and *opi3/pem2* mutants are strict choline auxotrophs (*39, 155, 163, 164, 164a*).

4. CHOLINE AUXOTROPHS OF *Schizosaccharomyces pombe*

So far, no other fungal organism has been studied as extensively as S. *cerevisiae*. However, for reasons discussed at the beginning of this article, the fission yeast, S. *pombe*, is viewed as a valuable eukaryotic model system and its genome is also being sequenced. Our laboratory has conducted studies of phospholipid biosynthesis in S. *pombe* confirming that the basic pathways of phospholipid biosynthesis in S. *pombe* are similar to those in S. *cerevisiae*, except for the absence of IPS (*13*) We have also conducted an analysis of S. *pombe* choline auxotrophs. The original set of choline requiring mutants was obtained from Professor Leupold and subsequently an additional series was isolated in our laboratory by S. Fernandez (*39, 163, 164, 164a*).

The search for choline auxotrophs of S. *pombe* yielded mutants representing two loci (*39, 163, 164, 164a*). The S. *pombe cho1*⁻ mutants have a biochemical defect analogous to the *opi3* mutants of S. *cerevisiae*. They are unable to carry out the last two methylations leading to PC biosynthesis and thus can methylate PE to form PMME but fail to form phosphatidyl-dimethylethanolamine (PDME) or PC. Unlike the S. *cerevisiae opi3/pem1* mutants, the S. *pombe cho1*⁻ mutants are stringent choline auxotrophs and cannot grow unless supplemented with choline or dimethylethanolamine (DME) (*39, 164, 164a*). The S. *pombe cho2*⁻ mutants are analogous to the S. *cerevisiae cho2* mutants in that they cannot carry out the methylation of PE → PMME. However, unlike the S. *cerevisiae cho2* mutants, they are strict choline auxotrophs. Thus, apparently S. *pombe*, like N. *crassa* (*155*), has more stringent requirements for methylated lipids than does S. *cerevisiae* (*164, 164a*). Furthermore, none of the searches for choline auxotrophs in S. *pombe* or N. *crassa* produced mutants with the properties of the S. *cerevisiae cho1* mutants, suggesting that a defect in PS synthesis may not be compatible with survival in these organisms, even when choline or ethanolamine is supplied exogenously, thereby permitting synthesis of the downstream lipids, PE and PC (*164, 164a*).

What might explain the difference between S. *cerevisiae* and these other fungi with respect to cellular phospholipid requirement? One clue may lie in the observation that S. *cerevisiae cho1* mutants and other mutants with severe alterations in the synthesis of phospholipids have a higher tendency than wild type to generate Rho$^-$ petites (*82*). *Saccharomyces cerevisiae* is a facultative anaerobe and can survive the complete loss of its mitochondrial genome and almost all of its mitochondrial functions. It may well be that mitochondria are more sensitive to perturbations of lipid composition than are other membranous organelles. Organisms that cannot withstand the loss of major mitochondrial functions may, therefore, be more sensitive than S. *cerevisiae* to changes in lipid composition.

Isolation and Characterization of the Schizosaccharomyces pombe cho1$^+$ and cho2$^+$ Genes. The S. *pombe cho1$^+$* gene was cloned by complementation of the S. *pombe cho1$^-$* mutant phenotype. It encodes a 204-amino acid protein with a predicted molecular mass of approximately 22.8 kDa (*164, 164a*). The S. *cerevisiae PEM2/OPI3* and the S. *pombe cho1$^+$* gene products exhibit considerable homology to each other (*39*) and to the mammalian enzyme encoded by the PEMT gene (*42*). The two fungal gene products share approximately 51% similarity to each other and 41% similarity to the rat/mouse PEMT gene product (*39, 164a*). The sequence similarity is distributed over the entire length of the gene products. However, a region of high sequence identity, concentrated in a stretch of 48 amino acids, located between amino acids 120 and 160 in the S. *pombe* sequence (*39*), suggests a possible conserved functional domain. Overall, there are approximately 66% conserved amino acids in this region among the three gene products (i.e., S. *cerevisiae PEM2/OPI3*, S. *pombe cho1$^+$*, and rat PEMT). This conserved domain has been identified as a putative binding site for the phospholipids PMME and PDME (*147, 165*). All three gene products are predicted by Kyte Doolittle profiles to be integral membrane proteins because they contain several membrane-spanning domains.

The S. *pombe cho2$^+$* gene was isolated by complementing the choline auxotrophy in an S. *pombe cho2$^-$* mutant (*164, 164a*). Translation of the open reading frame encoded by the S. *pombe cho2$^+$* gene product predicts a protein of 732 amino acids with a calculated molecular mass of 83,284 Da. The S. *pombe cho2$^+$* gene product shows 56% similarity overall compared to the S. *cerevisiae CHO2/PEM1* gene. The hydropathy profiles of the gene products encoded by the S. *cerevisiae PEM1/CHO2* and S. *pombe cho2$^+$* genes suggest that they are both integral membrane proteins (*39, 147*).

REGULATION OF PHOSPHOLIPID METABOLISM   151

A comparative analysis of the predicted amino acid sequences of the S. cerevisiae CHO2/PEM1 and S. pombe cho2$^+$ genes shows that there are conserved regions within each gene that share homology to each other, suggesting that these regions resulted from an internal duplication. These regions of duplication in the S. pombe gene exhibit approximately 64% similarity to the analogous regions in S. cerevisiae. The CHO2/PEM1 and cho2$^+$ genes share some similarity with the OPI3/PEM2 and cho1$^+$ genes (39, 147), suggesting the two classes of gene products may have evolved by a process of duplication and divergence from a common ancestral gene.

The gene product encoded by the rat liver PEMT cDNA (165) was found to be capable of complementing the S. pombe cho1$^-$ mutation as does the S. cerevisiae OPI3 gene (39, 164, 164a). However, whereas the OPI3 gene could partially complement the cho2 defect in S. cerevisiae, the S. pombe cho1$^+$ gene is not capable of complementing the S. pombe cho2$^-$ mutation. These results suggest that the cho1$^+$ and cho2$^+$ genes have specific and precise functional roles in S. pombe (164), whereas the OPI3/PEM2 and CHO2/PEM1 genes of S. cerevisiae have some functional redundancy (37, 162).

The regulation of the S. pombe methyltransferases is complex and the mechanism of regulation in response to soluble precursors may be quite different than in S. cerevisiae. This is, perhaps, not surprising given the fact that S. pombe is a natural inositol auxotroph (13). Schizosaccharomyces pombe wild-type cells labeled with [^{14}C]methyl-methionine accumulate over 40% of lipid-associated label into PC when grown in the absence of choline, but only 2% when grown in the presence of choline (64). This reduction in PC labeling in cells grown in the presence of exogenous choline is much greater than the reduction in labeling seen in S. cerevisiae grown under comparable conditions (122, 163, 164). This suggests that growth in the presence of exogenous choline represses the PE methylation pathway to a much greater degree in S. pombe than in S. cerevisiae (39, 163, 164). The reduction of methyltransferase activity in S. pombe cells grown in the presence of choline does not involve a reduction in the level of cho1$^+$ and cho2$^+$ mRNA abundance, suggesting that a posttranscriptional mechanism may be involved (164).

The S. pombe cho1$^+$ and cho2$^+$ transcripts are both repressed in stationary phase. As we will discuss subsequently, similar growth phase regulation occurs in S. cerevisiae. Thus, growth phase regulation of phospholipid metabolism may be similar in S. pombe and S. cerevisiae (164).

5. ADDITIONAL GENE–ENZYME RELATIONSHIPS IN
   PHOSPHOLIPID BIOSYNTHESIS IN Saccharomyces cerevisiae

In the past several years, S. cerevisiae mutants defining the last remaining steps in the pathway, PA → CDP-DG → PS → PE → → → PC, have been identified. The recent characterization of phosphatidylserine decarboxylase psd1 and psd2 mutants and cloning of the respective genes (22, 166, 167), and the identification of the *cdg1* mutant (*130*) as an allele of *CDS1*, which encodes CDP-DG synthase (*131, 168*), complete the assignment of gene–enzyme relationships in S. *cerevisiae* in the reaction sequence leading from PA to PC (Fig. 1; Table I). Mutants defining each step in the CDP-choline pathway (Kennedy pathway) for PC biosynthesis have also been identified and characterized (*151, 152, 169–174*). Table I shows the assessment of gene–enzyme relationships in phospholipid synthesis in S. *cerevisiae* and the position of these activities in the relevant pathways is shown graphically in Fig. 1. Figure 1 also shows the position in the pathway of structural gene mutants that have Opi$^-$ phenotypes.

The gene–enzyme relationships in the initial steps leading to the formation of PA remain, however, to be fully described. In eukaryotes, two routes of PA synthesis have been described (*175*). Glycerol-3-phosphate is acylated by acyl-CoA in a reaction catalyzed by glycerol-3-phosphate acyltransferase to form 1-acyl-glycerol-3-phosphate (lyso-PA). Alternatively, dihydroxyacetone phosphate (DHAP) can be acylated to form acyl-DHAP in a reaction catalyzed by DHAP acyltransferase. Acyl-DHAP reductase is then required to produce lyso-PA. The DHAP pathway has been detected in S. *cerevisiae* (*176*), but its role is not clear. The DHAP pathway occurs in organisms that produce ether lipids, but ether lipids have not been detected in S. *cerevisiae*. Furthermore, in S. *cerevisiae*, it has not been determined whether glycerol-3-phosphate acyltransferase and DHAP acyltransferase are one enzyme or two separate enzymes (*11*). Racenis *et. al.* (*176*) reported that the two activities had distinctly different biochemical characteristics, but Tillman and Bell (*154*) identified a mutant (*tta1*) in which both activities were affected.

The gene corresponding to the *tta1* mutation has yet to be cloned and characterized at the molecular level. A temperature-sensitive mutant (*DAM303*) originally identified as potentially defective in the early steps of phospholipid biosynthesis proved to be an allele of *sec13*, defective in the secretory pathway (*145, 177*). A third mutant (*slc1*, sphingolipid compensation), defining a gene that apparently encodes a fatty acyltransferase, was isolated as a suppressor (*178, 179*) of the *lcb1* mutant (*lcb1* mutants lack serine palmitoyl transferase, an essential enzyme in sphingolipid biosynthesis) (*180–183*). The *slc1* mutant was shown to make novel inositol glycerophospholipids, which apparently mimic sphingolipids by containing a very long-chain fatty acid and the typical sphingolipid head groups. The novel phospholipids made by the *slc1* mutant contain a C_{26} fatty acid attached to PI, mannosyl-PI, and inositol-P-mannosyl-PI. The *slc1* suppressor mutant is believed to encode an altered acyltransferase capable of adding a C_{26} fatty acid to the sn-2 position of a glycerophospholipid (*179*).

Because deletion of the *SLC1* gene is not lethal (*179*), other acyltransferases must exist. Zinser *et al.* (*76*) reported that the highest specific activity of glycerol-3-phosphate acyltransferase is present in the lipid particle fraction, a subcellular compartment consisting mainly of neutral lipid, but significant activity was also detected in the microsomal fraction (75). Athenstaedt and Daum (184) have studied the role of the *SLC1* and *TTA1* (also known as *GAT1*, for glycerol 3-phosphate acyltransferase) gene products by assaying phosphatidic acid (PA) biosynthesis in organelles from strains containing either the *slc1* or the *tta1/gat1* mutation. Both mutants make very little PA *in vitro* with lipid particles as an enzyme source, but the *slc1* mutant is able to make lyso-PA, suggesting that it is defective in the second acylation step, leading from lyso-PA to PA. When lipid particles are prepared separately from the *slc1* and *tta1/gat1* mutants and subsequently mixed, they complement each other *in vitro* and produce PA, suggesting that the *SLC1* and *TTA1/GAT1* gene products carry out sequential acylation reactions. However, the double mutant (*tta1/gat1, slc1*) is viable even though cell-free extracts of this strain support very reduced levels of PA biosynthesis *in vitro*. The residual glycerol-3-phosphate acyltransferase activity in the double mutant as well as in *slc1* and *tta1/gat1* mutants was localized to microsomes, suggesting that two independent mechanisms of PA synthesis may exist in yeast.

The gene–enzyme relationships governing the important interconversions of PA, DAG, and diacylglycerol pyrophosphate (DGPP) also have not been completely defined in yeast (65). On an enzymological level, DGPP phosphatase (185) and two forms of PA phosphatase (45 and 105 kDa) (186, 187) have been characterized in yeast. The PA phosphatase step appears to be highly regulated. Both forms of the enzyme are inhibited by sphingoid bases (69) and the two forms are differentially regulated by phosphorylation (71) and by inositol (187). The partitioning of PA between the competing reactions leading to DAG and CDP-DG may well be controlled by PA phosphatase, because both forms of the enzyme have a greater affinity for PA than does CDP-DG synthase (65, 186, 187). DGPP phosphatase catalyzes the dephosphorylation of DGPP to form PA and the yeast enzyme, *in vitro*, will also dephosphorylate PA to DAG but shows substrate preference for DGPP (185, 188, 188a). It can also use lyso-PA as a substrate. The yeast DGPP phosphatase resembles the mammalian PAP2 form of PA phosphatase (188, 188a). The roles of DGPP and DGPP phosphatase have not yet been fully elucidated in yeast, but there is considerable evidence in other organisms that they participate in signal transduction (188, 189).

B. Regulation of Phospholipid Biosynthesis in *Saccharomyces cerevisiae*

1. COORDINATION OF PHOSPHOLIPID BIOSYNTHESIS WITH ONGOING CELLULAR METABOLISM

Using inositol-requiring *ino1* strains, we examined the effects on cellular metabolism of an interruption in PI biosynthesis (190). These studies showed that inositol starvation in *S. cerevisiae* leads to a dramatic slowing of PI biosynthesis (within 30 min of removal of inositol). Employing these same strains, Hanson (*191*) detected a drop in the synthesis of the major cell wall carbohydrates, mannan and glycan, that occurred parallel with the slowing of PI biosynthesis. However, no coordinate cessation in synthesis of other phospholipids or slowing of overall macromolecular synthesis was found to be correlated with the slowing of PI synthesis (*161, 190, 192*). Indeed, the cells kept dividing for one generation, and even after cell division had stopped, macromolecular synthesis continued unabated for a period of time equivalent to another full generation (*190*). The result of such ongoing metabolism in cells that had ceased dividing was that the inositol-starved cells became very dense (*190*). Indeed, spheroplasts made from such cells were found to be osmotically unstable unless a much higher level of osmotic support was provided than that which is required for support of normal yeast spheroplasts (*193*).

The increased density of inositol-starved yeast cells allowed them to be separated from normal growing cells by centrifugation in a ludox gradient (*190*). This observation became the basis for a selection procedure used by Novick *et al.* (*194*) in the isolation of yeast mutants (Sec$^-$) defective in the secretory pathway. Likewise, Letts and Dawes (*145, 148*) used the selection of dense mutant cells on a ludox gradient to identify mutants defective in overall phospholipid metabolism. Notably, however, the temperature-sensitive mutant (DAM303) that Letts and Dawes (*145*) isolated proved to be an allele of *sec13*, isolated by Novick *et al.* (*194*) in an analogous screen. Letts and Dawes (*145*) reported that cessation of phospholipid synthesis in the *sec13* mutant was a very early event following the shift to the restrictive temperature. Furthermore, they reported that the *sec13* allele (DAM303) that they studied had an Opi$^-$ phenotype at the restrictive temperature. These observations suggest a link between phospholipid biosynthesis and the secretory pathway, but the observation has not been further pursued.

On the basis of our studies with inositol starvation in *ino1* mutants, we concluded that PI biosynthesis is not directly coordinated with other ongoing cellular processes, including synthesis of other lipids (*64, 190*). The yeast cell does not appear to have any direct mechanism for sensing a block in PI biosynthesis that enables it to respond by shutting off other metabolic processes. Rather, ongoing metabolism in the absence of PI biosynthesis leads to rapid, metabolism-dependent "inositol-less death" (*190, 192, 195*). A similar phenomenon was reported many years ago in *Neurospora* (*196*), suggesting that it may reflect a general feature of metabolic regulation in fungi.

In contrast, failure to provide *cho1* mutants with ethanolamine or choline, a growth condition that results in cessation of PE and PC biosynthesis, does not result in comparable cell death (*36, 158, 161*). The cessation of PC biosynthesis in ethanolamine-starved cells is accompanied by concurrent slowing of PI biosynthesis (*158*), and, indeed, limitation of PC biosynthesis leads to a concurrent slowing of the cell growth rate (*38*). However, under these conditions, IPS is constitutively expressed and *cho1* mutants exhibit an Opi$^-$ phenotype (*158*), a subject to which we will return.

2. COORDINATE REGULATION OF PHOSPHOLIPID SYNTHESIS IN *Saccharomyces cerevisiae*

The activity of the phospholipid *N*-methyltransferases that catalyze the formation of PC from PE were shown by Waechter and Lester (*32, 33, 197*) to be regulated in *S. cerevisiae* in response to availability of choline and/or the methylated intermediates, DME and MME. Likewise, Carson *et. al.* (*198*) showed that PS decarboxylase activity is repressed if choline is added to standard yeast media. These studies all involved addition of choline or one of the methylated intermediates, MME or DME, to standard yeast medium, which contains inositol. Yamashita and Oshima (*199*) reported that inositol also causes a reduction in PLMT activity when added to the medium of wild-type *S. cerevisiae* cells. Subsequently, it was demonstrated that the repression of the PLMT activity in response to the addition of choline occurs only in media already containing inositol (*61, 89, 122, 150*). Addition of choline to the growth medium in the absence of inositol has little or no effect on either the PLMT or PS synthase activity (*89, 90, 92, 200*). Inositol alone causes partial repression of the activity of these enzyme activities, and when choline is added in the presence of inositol, there is a further reduction in activity (*61, 89, 90, 92, 122*).

Prior to the isolation of the structural genes encoding enzymes of phospholipid biosynthesis, the list of enzymes demonstrated to exhibit repression in response to inositol and choline included PS decarboxylase, CDP-DG synthase, and phosphatidylglycerophosphate synthase (*122, 123, 198, 201–204*), as well as the PLMTs, as discussed above, and IPS, shown initially to be repressed at the level of enzyme activity in response to inositol (*100, 116, 117*). Some of these enzyme activities were also shown to be regulated in response to other precursors such as serine or ethanolamine, MME, or DME, but only if inositol is also present (*92, 200, 201, 203, 205*). Curiously, one form of phosphatidate phosphatase (*187, 206*), phosphatidylinositol ceramide phosphatidylinositol transferase (IPC synthase) (*207*), and I1-P phosphatase (*108*) are all induced severalfold by inositol.

With the cloning of the structural genes encoding enzymes of phospholipid synthesis, the regulation in response to inositol and choline was shown to occur at the transcriptional level and to include a much larger set of activities, including the enzymes of the CDP-choline pathway, as well as the choline and inositol transporters, and enzymes involved in fatty acid biosynthesis (*10, 11, 62, 64–66*).

The *ino2*, *ino4*, and *opi1* mutants, originally isolated on the basis of misregulation of IPS, were later found to have pleiotropic defects, including misregulation of the entire set of coordinately regulated enzymes described above (*10, 11, 63*). Thus, the entire set of enzymes subject to repression by inositol and choline is also coordinately controlled by the products of the *ino2*, *ino4*, and *opi1* genes. Based on the fact that *ino2*, *ino4*, and *opi1* are all "loss of function," recessive mutations, the *INO2* and *INO4* wild-type gene products are presumed to be positive regulators required for derepression (full expression) of the coordinately regulated enzymes. By similar reasoning, the *OPI1* gene product is a negative regulator, required for repression of these same activities (*63, 64, 89, 122, 125, 208*). Double mutants, containing an *opi1* mutation in combination with an *ino2* and *ino4* mutations are epistatic to *opi1*. This result suggests that the function of the *OPI1* gene product requires the presence of the *INO2* and *INO4* gene products. Thus, the *INO2* and *INO4* gene products presumably have a more direct effect than the *OPI1* gene product on expression of *INO1* and other coregulated genes (*210*).

3. Transcriptional Regulation of *INO1* and Coregulated Genes of Phospholipid Biosynthesis

The first structural gene encoding a yeast phospholipid enzyme to be isolated was *CHO1* (*211*). This gene [sometimes also referred to as *PSS1* (*212, 213*); Fig. 1] was independently isolated and characterized on a molecular level by a number of laboratories (*91, 159, 211–215*). The cloning and sequencing of the *INO1* structural gene for IPS was also soon reported(*120, 121*) followed by the *PEM1/CHO2* and *PEM2/OPI3* genes (*37, 129, 147*). The genes encoding the structural genes for a majority of the reactions shown in Fig. 1 are now identified (Table I). The *INO4* regulatory gene, which was isolated by complementation of an *ino4* mutant (*216*), was the first regulatory gene involved specifically in controlling phospholipid biosynthesis in yeast to be characterized (*217*). The *OPI1* gene was isolated (*218, 219*) on the basis of its proximity to the already cloned *SPO11* gene. The *INO2* gene was first isolated by complementation of the *ino2* mutant phenotype (*220*). *INO2* was identified a second time by its proximity to the *KIN1* protein kinase gene, which was reported to be a high-copy suppressor of *ino2* null mutants (*221*). It was also identified as a high-copy suppressor of the *ire15* mutation (*222*) (see Table V for *ire15* and the footnote in Table IV concerning nomenclature for *INO2*). *INO2* was isolated a fourth time as a high-copy suppressor of the inositol auxotrophic phenotype of the dominant *BSD2-1* mutation (Table V) (*142*).

The isolation of the structural and regulatory genes involved in phospholipid biosynthesis has enabled a molecular dissection of the coordinate regulation that was first observed at the level of enzyme activity. At the enzymatic level, IPS had been shown only to be repressed in response to inositol *(100, 107)*. At the level of *INO1* transcript abundance, it became clear that although the presence of inositol alone causes the major repression (10-fold or greater), the addition of choline causes an additional threefold repression if inositol is already present in the growth medium *(88, 218)*. However, choline when present alone in the growth medium has no effect on *INO1* expression. The pattern of regulation of the *CHO1* transcript is similar. When inositol is added to the growth medium, the *CHO1* transcript is repressed two- to threefold. Addition of choline in the presence of inositol leads to additional repression to a total of four- to fivefold overall. However, addition of choline alone has no effect *(90, 123)*. In the case of the *CHO1* gene, regulation of the transcript was shown to mirror precisely the presence of the enzyme subunit as detected immunologically *(200)*. The list of structural genes showing this pattern of transcriptional regulation continues to expand and now includes *OPI3/PEM2, CHO2/PEM1 (38, 91, 92, 99, 129), CDS1/CDG1 (131, 168, 223), CKI1 (224), CPT1 (174), CCT1 (173), PSD1 (225), ACC1* (the structural gene for acetyl Co-A carboxylase) *(226)*, and the choline and inositol transporters *(60, 73, 227–229)*. Notably, however, the genes showing this regulation exhibit widely differing levels of repression, ranging from 30-fold overall for *INO1 (88, 93, 94, 218)* to only about two- to threefold for *CHO2* and *OPI3 (38, 92)*.

The molecular characterization of the *INO1* gene revealed the presence of multiple copies of a repeated element in its promoter *(62, 93, 218)*. Initially, we and our collaborators identified this element as a 9-bp consensus (5′ ATGTGAAAT 3′), based on the sequence of the repeats in the *INO1* promoter *(62, 93, 218, 230)*. This element was also detected in the published sequences of the coregulated *CHO1, CHO2/PEM1,* and *OPI3/PEM2* genes *(62, 93, 214)*. Independently, Kodaki *et al. (91, 213)*, working on the *PEM1/OPI3, PEM1/CHO2,* and *PSS1/CHO1* genes, reported the presence of an 8-bp repeat (consensus 5′ CATGTGAA 3′). As pointed out by Paltauf *et al. (10)*, the 8-bp repeat of Kodaki *et al. (92, 213)* and the 9-bp element of Hirsch *(218)* and Lopes *et al. (93, 230)* are overlapping and together constitute a 10-bp element with a consensus of 5′ CATGTGAAAT 3′. Furthermore, contained within this 10-bp consensus sequence is the canonical binding site, CANNTG *(10, 98, 231)*, for DNA-binding proteins of the basic helix–loop–helix (bHLH) class *(232)*. The 10-bp element has now been reported in a very wide-ranging and growing list of genes *(10, 11)*, including all of the genes previously mentioned that show regulation in response to inositol and choline. We have designated this element (consensus 5′ CATGT-GAAAT 3′) as the inositol-sensitive upstream activating sequence, or UAS_{INO} *(10, 97, 231)*. The 10-bp consensus element itself is sufficient to drive fully regulated expression of a heterologous reporter gene when inserted into a construct lacking an upstream activating sequence, but containing an otherwise functional promoter (97, 233).

Specific DNA protein complexes, detectable by electrophoretic mobility band shift assays (EMSAs), form when fragments of DNA from the *INO1* promoter are incubated with cell-free extracts of wild-type yeast (230). Two classes of complexes are observed when the DNA used in the assay contains a functional copy of UAS$_{INO}$. One class of complex can be competed away using an excess of an oligonucleotide containing the original 9-bp repeat (5' ATGTGAAAT 3') with any base in the tenth position 5' (230). The proteins composing this complex, called the nonamer binding factor (NBF), have not yet been identified (210). However a second complex was shown to be dependent on the products of the *INO2* and *INO4* genes (230). This complex can be competed only by oligonucleotides containing the 10-bp element 5' CATGTGAAAT 3' 997, 98, 234). The *INO2* and *INO4* gene products cotranslated *in vitro* form a heterodimer, even in the absence of DNA, and this Ino2p/Ino4p complex is sufficient to bind DNA fragments containing UAS$_{INO}$ (231).

The *INO2* and *INO4* genes have been characterized at a molecular level (217, 220). Both encode gene products having homology to DNA-binding proteins containing the bHLH motif (232, 236) first described in mammalian oncogenes such as c-*Myc*, and, as mentioned above, UAS$_{INO}$ contains within it the bHLH consensus binding site, CANNTG. Curiously, the *INO2* and *INO4* genes contain UAS$_{INO}$ in their promoter sequences and autoregulation of *INO2* plays a role in the overall pattern of phospholipid regulation (99, 237). A number of *ino2* and *ino4* mutations have been characterized at the molecular level (98, 234, 238). Those mutations that lead to loss of binding the Ino2p/Ino4p complex to UAS$_{INO}$ all involve either point mutations in the helix–loop–helix motif or the preceding basic region or, alternatively, chain termination or frameshift mutations that lead to truncation of a major portion of the protein (98, 234). Complete deletion of either or both of these genes is compatible with cell viability. The deletion mutants are inositol auxotrophs similar to the original *ino2* and *ino4* point mutants (98, 210, 231).

The *OPI1* gene contains within it a leucine zipper motif and polyglutamine stretches (219). The promoter of the *OPI1* gene contains a UAS$_{INO}$ sequence and its transcript is repressed in stationary phase and in cells grown in the presence of inositol (210). The *OPI1* deletion mutant is similar in phenotype to the original *opi1* point mutants. It is viable, has a strong Opi$^-$ phenotype under all growth conditions, and exhibits completely constitutive overexpression of the coregulated genes containing UAS$_{INO}$ (97, 99, 210, 219). The *opi1* mutants also fail to repress UAS$_{INO}$-containing genes in stationary phase (114). The stationary phase repression of genes containing UAS$_{INO}$ will be discussed below.

A set of oligonucleotides representing a series of systematic substitutions at each position within the 10-bp UAS_{INO} element, and in the flanking bases at the 5' end, has been constructed (97). Mutational analysis of slightly different versions of this element have also been reported by Kodaki et al. (213) and Schüller et al. (239). Every mutation away from the consensus within the core bHLH binding domain of UAS_{INO} (i.e., CATGTG, the first six bases of the consensus) has a detrimental effect on the ability of the element to support regulated expression of a heterologous reporter element transformed into wild-type yeast strains (97). Mutations in the four bases flanking the 5' end of the element also influenced its ability to function as a UAS element. Mutations in the two bases 3' to the core bHLH consensus (i.e, bases 7 and 8 from the 5' end) also have substantial effects on the ability of the element to support expression, and mutations in the last two bases at the 3' end (i.e., bases 9 and 10) of the element have lesser effects. Every mutated UAS_{INO} element, however, that retains any measurable ability to support expression of the heterologous reporter gene exhibits regulation in response to inositol and choline, and each version of the element that is functional as a UAS element is also constitutive in the *opi1* mutant background. Furthermore, the ability of an element to bind the Ino2p/Ino4p complex as measured in a competition assay is also correlated to its strength *in vivo* as a UAS element. Thus, it appears that all of the regulatory functions controlled by UAS_{INO} are mediated by a single element and all of these functions exhibit identical sequence specificity (97).

4. EFFECT OF GROWTH PHASE ON EXPRESSION
   OF THE COORDINATELY REGULATED ENZYMES
   OF PHOSPHOLIPID BIOSYNTHESIS

The same set of enzymes that are repressed in the presence of inositol and choline is also regulated in a growth phase-dependent manner (205). The coordinately regulated enzymes are maximally expressed in logarithmic phase of growth in the absence of inositol and choline. In stationary phase, these same enzymes are repressed whether or not inositol is present (205) and the growth phase regulation, like the regulation in response to inositol and choline, occurs at the transcriptional level (204, 241). Furthermore, the *opi1* mutant, which fails to repress the expression of *INO1* and other coregulated genes in response to inositol and choline, also fails to repress them in stationary phase (114). The repression of the coregulated genes in stationary phase suggested to us that genes containing UAS_{INO} might be regulated in response to nutrient depletion. Thus, the effects of transient depletion of carbon, nitrogen, and phosphorous were tested and we found that transient depletion of any one of these nutrients leads to rapid repression of *INO1*. This response is also mediated by UAS_{INO} and is dependent on the *OPI1* gene product, because in *opi1* mutants, the *INO1* gene is constitutively expressed during nutrient deprivation (P. Griac and S. Henry, unpublished data). Thus, we believe that repression of *INO1* and coregulated genes in stationary phase, as well as the response to nutrient deprivation, are mediated by the same promoter element, namely UAS_{INO}, and the same regulatory factors, namely the *INO2, INO4,* and *OPI1* gene products, that mediate repression in response to inositol and choline (P. Griac and S. Henry, unpublished data).

5. Dependence of the Coordinate Regulation of Phospholipid Biosynthesis in *Saccharomyces cerevisiae* on Ongoing Phosphatidylcholine Biosynthesis

As previously discussed, the *opi3* mutant, which was isolated in the original screen for the Opi⁻ phenotype (*125*), proved to be defective in carrying out the final two methylations in PC biosynthesis (i.e., PMME → PDME → PC) (Fig. 1) (*128, 129*). The Opi⁻ phenotype of the *opi3* mutant is conditional and can be eliminated by including either choline or DME (but not MME) in the growth medium (*129*). Furthermore, in *opi3* mutants, the *INO1* gene and its product, IPS, are not repressed in response to inositol unless choline or DME is also present in the growth medium (*128, 129*). In contrast, in wild-type cells IPS is repressed over 10-fold in response to inositol alone (*107*) at the level of *INO1* transcription (*88*). The conditional Opi⁻ phenotype of *opi3* mutants is also distinct from the phenotype of *opi1* mutants in which the constitutive overexpression of IPS (and other coregulated genes of phospholipid biosynthesis) is unresponsive to any combination of inositol and other soluble precursors such as choline (*89*). The *cho2* mutants, which are defective in carrying out the first methylation of PE (Fig. 1), also have a conditional Opi⁻ phenotype (*37, 38*). In the case of *cho2* mutants, however, the Opi⁻ phenotype is eliminated and IPS regulation in response to inositol is restored if MME, DME, or choline is added to the growth medium, whereas in *opi3* mutants, only DME and choline have this effect (*129*).

Similar conditional Opi⁻ phenotypes are seen in all structural gene mutants in the reaction sequence CDP-DG → PS → PE → → → PC. For example, the *cho1* mutants, which are defective in the production of PS from serine and CDP-DG, fail to repress IPS when starved for choline (*158*). In the case of *cho1* mutants, however, regulation can also be restored by ethanolamine in addition to MME, DME, or choline. In the case of *cho1* mutants, ethanolamine enters the pathway downstream of the metabolic lesion in PS synthase, thus restoring PC biosynthesis (*158*). Griac (*225*) has shown that supplementation with ethanolamine also eliminates the Opi⁻ phenotype of *psd1* mutants, which are defective in PS decarboxylase (*166, 167*). Ethanolamine enters the pathway downstream of the *psd1*, as well as the *cho1*, metabolic lesion (Fig. 1). In both cases, ethanolamine is able to return PC biosynthesis to near wild-type levels and, in so doing, eliminates the Opi⁻ phenotype and restores *INO1* regulation (*158, 225*).

When we first observed the correlation between PC biosynthesis and *INO1* regulation in response to inositol, we deemed it "cross-pathway control" and we concluded that repression of IPS in response to inositol requires ongoing PC biosynthesis (*37, 129, 158*). This regulatory phenomenon affects IPS regulation at the level of transcription of the *INO1* gene (*38, 88, 129*). The change in the pattern of transcriptional response in the *cho2* mutants is not limited to the *INO1* gene and has been documented for the transcript of the coregulated *OPI3* gene (*38*). At the level of enzyme activity, a similar dependence on ongoing PC biosynthesis for regulation in response to inositol was reported for phosphatidylglycerophosphate synthase (*202*) and the PLMT activities encoded by *CHO2/PEM1* and *OPI3/PEM2* (*92*). Thus, the regulatory dependence on ongoing PC biosynthesis presumably affects expression of all the coregulated enzymes of phospholipid biosynthesis at the level of transcription of their structural genes (*38*).

6. Studies on the Nature of the Regulatory Signal Connecting Ongoing PC Biosynthesis and Regulation of *INO1* and Other Coregulated Genes

The association of regulatory phenotypes with mutations in structural genes suggests that the mechanism controlling repression and derepression of the coregulated genes responds to a signal emanating from the metabolism. If this is true, it should be possible to learn something about the nature of the signal, and the part of the pathway from which it originates, through the study of strains in which portions of the pathway are systematically blocked using combinations of mutations. The regulatory phenotypes associated with mutations in various steps in phospholipid biosynthesis are shown in Fig. 1. Inspection of this figure reveals that Opi⁻ phenotypes are limited to the series of mutants spanning the reaction series leading from PA and culminating in PC biosynthesis via the PE methylation pathway. Because every step in this series confers the Opi⁻ phenotype, we propose that the signal is generated by some metabolite (or general metabolic condition) that is created at the beginning or the end of the sequence. (If the signal lay in the middle of the sequence, there would presumably be distinctly different effects when a mutation lay just above versus just below the signal point.) Therefore, our attention was drawn to the two ends of the reaction series associated with these regulatory phenomena (i.e., PC at the end, and PA at the beginning).

Using strains carrying the *cho2* mutation in combination with mutations in the CDP-choline pathways for PC biosynthesis, we exhaustively explored the relationship between regulation in response to inositol and ongoing PC biosynthesis via the PE methylation and the CDP-choline pathways (*38*). These studies indicated that regulation in response to inositol does not require a functioning CDP-choline pathway so long as the PC biosynthesis via the PE methylation pathway is unimpaired. Likewise, the PE methylation pathway is unnecessary for the regulation if choline is supplied exogenously, so that PC can be synthesized via the CDP-choline pathway (*38*). No single, specific precursor or intermediate in either of the two routes of PC synthesis (i.e., PE methylation or CDP-choline pathways) was found to be correlated to, or required for, the restoration of the regulation in response to inositol. Likewise, free choline has no effect on the regulation unless it actually participates in PC biosynthesis. Finally, the actual proportion of PC in the phospholipid composition is not correlated to the regulatory response. The restoration of the regulation of *INO1* and other coregulated genes was found to be correlated only to restoration of PC biosynthesis and/or a wild-type growth rate (*38*). Similar conclusions have been reached in studies employing the *psd1* mutant (*225*).

Because the studies on the *cho2* and CDP-choline pathway mutants did not provide evidence linking the signal directly to any specific precursor of PC biosynthesis (*38*), we must consider the possibility that the signal is generated near the beginning of the series of reactions leading from PA to PC. The *cdg1* mutant (*130*) lies at the start of the series of structural gene mutants that possess Opi$^-$ phenotypes, defining the metabolic steps leading from PA to PC (Fig. 1). Unlike the *cho1, psd1, cho2,* and *opi3* mutants, the Opi$^-$ phenotype of the original *cdg1* mutant [now known to be an allele of the *CDS1* locus (*168*)] is not affected by choline (*242*). Furthermore, repression of *INO1* in the *cdg1* (*cds1*) mutant does not occur in response to any combination of inositol and choline (S. Kohlwein, personal communication). As previously discussed, the *cdg1* (*cds1*) mutant has a partial (but unconditional) block in the conversion of PA → CDP-DG, reducing the flow of substrate through this crucial reaction. Shen and Dowhan have shown that complete deletion of the *CDS1* gene encoding CDP-DG synthase, the enzyme that is defective in the *cdg1* mutant, is lethal (*168, 223*). However, they produced a conditional mutant by placing the *CDS1* gene under the control of the *GAL* promoter in a strain containing a *CDS1* deletion (*223*), and in this fashion showed that the *INO1, CHO1, CHO2,* and *OPI3* genes are all derepressed in response to declining levels of the *CDS1* gene product, CDP-DG synthase.

Thus, the signal responsible for regulation in response to inositol can be restored just as efficiently by relieving the metabolic impediment in the conversion to PA to CDP-DG at the beginning of the reaction series leading from PA to PC, as it can by supplying choline to an *opi3* or a *cho2* mutant, thereby directly restoring PC biosynthesis at the end of the sequence. Thus, we reason the signal could equally well be emanating from either end of the reaction series leading from PA to PC. Fortuitously, a crucial finding capable of distinguishing between these two alternative models was provided by the discovery of a relationship between PC turnover and *INO1* derepression (*35*).

7. Derepression of *INO1* in Response to Elevated PC Turnover

The discovery of a connection between PC turnover and regulation of *INO1* grew out of a serendipitous observation (*146, 243*) of a temperature-sensitive Opi$^-$ phenotype in strains carrying a temperature-sensitive *sec14^{ts}* allele. The *SEC14* gene encodes a PI/PC transporter (Sec14p) essential for viability and the secretory pathway (*84, 85*). The *sec14^{ts}* mutants raised to their restrictive temperature of 37°C arrest at the late Golgi stage of the secretory pathway (*83*). Cleves et al. (*86, 87*) subsequently reported that mutations in the CDP-choline pathway for PC biosynthesis suppress the growth and secretory defects of *sec14* mutants, including the null allele. Thus, strains carrying *sec14^{ts}* in combination with *cki1*, *cpt1*, or *cct1* can grow at a nearly wild-type rate at the *sec14^{ts}* restrictive temperature of 37°C. We have observed an Opi$^-$ phenotype, which is present only at the restrictive temperature in double mutants carrying *sec14^{ts}* in combination with a mutation in any one of the three steps in the CDP-choline pathway for PC biosynthesis (*cki1, cct1,* or *cpt1;* see Fig. 1) (*35*). Such strains exhibit a choline excretion phenotype, overproduction of choline (Opc$^-$), as well as an Opi$^-$ phenotype (*35*). The choline excretion phenotype, which was first observed by S. Kohlwein (personal communication), is detected using a plate assay similar to that used to visualize the Opi$^-$ phenotype. Specifically, the strains being tested are grown on petri plates lacking choline and are sprayed with an overlay consisting of a *cho2, opi3* strain, which is a choline auxotroph. The detector strain will grow as a halo around the colony being tested when choline is excreted (*35*).

Strains (Fig. 1) with blocks in the CDP-choline pathway (*cki1, cct1,* or *cpt1* mutants) exhibit a modest Opc$^-$ phenotype, but double-mutant strains carrying *sec14^{ts}*, in combination with one of the CDP-choline mutations, exhibit a much increased Opc$^-$ phenotype. Strains such as *sec14^{ts} cki1*, containing the *sec14^{ts}* allele in combination with a CDP-choline pathway mutation, excrete large amounts of free choline into the growth medium when shifted to the *sec14^{ts}* restrictive temperature (37°C). The choline excretion phenotype of such strains is the result of accelerated PC turnover via a phospholipase D (PLD)-mediated route (Fig. 1). In a *sec14^{ts} cki1* strain growing at 37°C, the increased turnover of PC results in excretion of more than 50% of the total PC-associated choline per generation (35). In CDP-choline pathway mutants such as *cki1*, at least 6–7% of cellular PC is degraded per generation via a PLD-mediated route. Thus, in the *sec14^{ts} cki1* strain at 37°C, the rate of PC degradation is elevated four- to fivefold compared to the *SEC14 cki1* strain (35).

PLD-mediated turnover of PC produces one molecule of PA (Fig. 1) for every molecule of choline produced. However, PA is rapidly reused in synthesis of other phospholipids, whereas, in cells carrying a CDP-choline pathway lesion, the choline cannot be readily reused and is excreted. We propose that the accelerated turnover of PC observed in *sec14^{ts} cki1* cells shifted to the *sec14^{ts}* restrictive temperature is related to the mechanism by which the *cki1* and other CDP-choline pathway mutations suppress the *sec14* secretory pathway block. When we introduced a disruption of *PLD1*, which encodes the major Ca^{2+}-independent, PIP_2-stimulated phospholipase D activity of the yeast cell, into a *sec14^{ts} cki1* strain, the triple mutant was unable to grow at 37°C. Furthermore, the *sec14^{ts} cki1 pld1* triple mutant no longer exhibited the inositol and choline excretion (Opi$^-$ and Opc$^-$) phenotypes observed in the *sec14^{ts} cki1* strain (243a). These results suggest that the PLD-mediated turnover of PC is required for suppression of the *sec14* defect by *cki1* and other CDP-choline pathway mutations. McGee *et al.* (244) reported that when *sec14^{ts}* is inactivated, the PC content of the Golgi membranes rises. However, Kearns *et al.* (245) have argued that DAG production is responsible for suppression of the *sec14^{ts}* phenotype in the *sac1* mutant background. DAG and PA are related to each other by a single phosphorylation/dephosphorylation step (Fig. 1). In order to distinguish between the relative roles of PA and DAG in the secretory pathway and *INO1* regulation, it will be necessary to have mutants in these specific phosphorylation/dephosphorylation reactions.

The Opc$^-$ phenotype of *sec14^{ts} cki1* strains and measurement of choline excretion in liquid culture indicate that degradation of PC via a PLD-mediated mechanism increases simultaneously with the shift to the restrictive temperature (35). In the *sec14^{ts}* mutants carrying a CDP-choline pathway lesion, the reduction in PC content, coupled with the excess PA produced during PLD-mediated turnover, may provide a localized environment in the Golgi conducive to the membrane-mediated events of vesicle budding and/or fusion required for the secretory process to proceed. However, in a *sec14^{ts}* mutant with an intact CDP-choline pathway, the choline liberated by increased turnover when Sec14p is inactivated is apparently immediately reused, setting up a futile cycle of PC degradation and resynthesis. When excess PA is produced by PLD-mediated turnover in the *sec14^{ts} cki1* strain, it leads to derepression of the *INO1* gene, which continues into stationary phase, the excess production of inositol may well stimulate PIP_2 production, leading to further stimulation of Pld1p, reinforcing the overall metabolic condition of enhanced PC turnover (35).

Derepression of *INO1* expression in the *sec14^{ts} cki1* strain in response to increased PC turnover requires a block (i.e., the *cki1* mutation) in choline reutilization in PC biosynthesis via the CDP-choline pathway (35). In the *cho1, cho2,* and *opi3* mutants, *INO1* derepression occurs when the mutant cells lack a supply of exogenous choline (or other intermediates) to make PC via an intact CDP-choline pathway (37, 38, 129, 158). Two distinctly different metabolic conditions, one involving increased PC turnover (in the *sec14^{ts} cki1* strain) and the other involving diminished PC biosynthesis (in the *cho1, cho2,* or *opi3* strain), both result in Opi$^-$ phenotypes and misregulation of *INO1*. We reason that all of these strains must have something in common and that this common metabolic condition must be related to the signal controlling *INO1* derepression/repression.

III. Model for the Regulation of Genes Containing UAS$_{INO}$

In the preceding discussion of the Opi$^-$ phenotypes of mutants defective in each of the reactions leading from PA to PC (Fig. 1) we argued that the signal could lie at the beginning or the end of the reaction sequence. In the *sec14^{ts} cki1* strain shifted to the restrictive temperature, two precursors, PA and choline, are produced in equal proportion via PC turnover. PA is also the precursor acted on by the gene product defective in the *cdg1* mutant (168, 216), which lies at the beginning of the series of mutants with Opi$^-$ phenotypes. Furthermore, we have demonstrated that free choline is not responsible for the signal (38). Thus, we are proposing a model (Fig. 2) in which the metabolic signal for *INO1* derepression is generated by PA or a closely related metabolite.

The features of the complex regulation controlling genes containing UAS$_{INO}$ that must be explained by our model are summarized in Table VI. We propose that a metabolic signal communicates via the *OPI1* gene product. The *OPI1* gene product transmits a signal controlling the *INO2* and *INO4* gene products, which, in turn, bind to UAS$_{INO}$, thus activating transcription. Based on the phenotypes of the *opi1, ino2,* and *ino4* mutants, it can be concluded that Opi1p exerts negative control and that Ino2p and Ino4p are required for transcriptional activation. The metabolite that acts as a signal is produced in an early step of phospholipid biosynthesis. The working hypothesis of our laboratory is that the signal is phosphatidic acid (PA), but it could, in fact, be a metabolite derived from PA, such as DGPP or DAG

TABLE VI
Major Features of the Regulation Controlling *INO1* and Coregulated Genes

Feature	Description	Ref.
The coregulated genes all contain one or more copies of a version of the repeated element; UAS_{INO}	The consensus sequence for the element is 5′ CATGTAAAT 3′ and contains within it the core binding site (CANNTG) for DNA-binding proteins of the bHLH class	10, 11, 62, 91, 93, 97, 213, 218
The *INO2* and *INO4* loci encode DNA-binding proteins of the bHLH class and have been demonstrated to form a heterodimer that binds to UAS_{INO}	When either *INO2* or *INO4* is deleted, the cell is unable to derepress any of the UAS_{INO}-containing genes	10, 11, 88, 98, 122, 231
The *OPI1* gene product is required for repression of all genes containing UAS_{INO}	The *OPI1* gene product contains a leucine zipper and polyglutamine stretches, motifs found in DNA-binding proteins, but has not been shown to bind DNA directly. Deletion of the *OPI1* gene leads to constitutive overexpression of all genes containing UAS_{INO}. The *opi1* mutants fail to repress UAS_{INO}-containing genes in response to phospholipid precursors and do not turn off these genes in stationary phase or in response to general nutrient depletion	10, 11, 97, 114, 210, 219
Expression of genes containing UAS_{INO} is responsive to growth phase and the overall metabolic state of the cell as well as to the availability of phospholipid precursors such as inositol and choline	Genes containing UAS_{INO} are maximally expressed during logarithmic phase in cells growing in the absence of inositol. If inositol is added under such conditions, genes containing UAS_{INO} are repressed. If choline is added in the presence of inositol, further repression occurs, but if it is added in the absence of inositol, it has little effect. Even in the absence of inositol, genes containing UAS_{INO} are repressed in stationary phase	10, 11, 38, 88, 97, 240
The expression of the structural genes containing UAS_{INO} is sensitive to the overall pattern of phospholipid biosynthesis and turnover	Mutants that have a lesion in the major biosynthetic pathway leading from PA to PC via methylation of PE exhibit inositol excretion phenotypes and fail to repress the coregulated genes in response to inositol. Repression in response to inositol can be restored by supplying an intermediate that enters the CDP-choline pathway and permits PC biosynthesis. Accelerated phospholipase D-mediated PC turnover can also lead to an inositol excretion phenotype. Under these conditions, the *INO1* gene is simultaneously derepressed, even in the presence of inositol, and it continues to be expressed in stationary phase	35, 37, 38, 129–131, 158, 223

(Figs. 1 and 2; Table VI), or lyso-PA, which is a precursor to PA and can also be derived from it by the action of a lipase. We propose that high rates of PA production and/or low rates of PA utilization lead to derepression (i.e., serving as a positive signal or inducer).

According to our model (Fig. 2), the Opi$^-$ phenotype in mutants having a block in the reaction series PA → CDP-DG → PS → PE → → → PC is the result of "damming-up" of upstream metabolites, including PA. When this occurs, the cellular response is depression of all of the coordinately regulated enzymes containing UAS$_{INO}$ including all of the enzymes in the pathway, PA → PC, which catalyze reactions that make use of metabolites downstream of PA. The derepression of these downstream enzymes tends to eliminate any damming-up of precursors such as PA in wild-type cells. However, in each of the mutants defective in PC biosynthesis via the PE methylation pathway, there is a metabolic bottleneck that cannot be relieved by derepressing the system. In contrast, in the $sec14^{ts}$ $cki1$ strain, the balance is shifted due to excess production of PA via turnover of PC rather than a bottleneck in PA utilization. Because PA is a very transient precursor that is rapidly used in downstream reactions, the sensing mechanism that produces the signal must be very sensitive to the flux of metabolites in the pathway.

In wild-type cells, our model predicts that the early precursors such as PA are made *de novo* somewhat in excess of their rate of utilization when inositol is absent from the medium. The build-up of these precursors generates a signal that results in derepression of the *INO1* gene, leading to inositol production, which is used in PI production, thereby drawing on the pool of early precursors such as PA. When exogenous inositol is present, the rate of PI production increases sufficiently to lower pools of PA and other precursors to the point of repressing the system. Thus, inositol availability affects the overall rate of phospholipid precursor output versus input in wild-type cells. This results in repression, however, only when the other major branches of the pathways that utilize PA, and other early precursors, are unimpeded and/ or when rates of lipid turnover contributing to the inflow of PA are normal. In wild-type cells, the presence of choline leads to a further degree of repression, but only if inositol is already present (*10, 88*). In *cho1*, *cho2*, and *opi3* mutants, the addition of choline to the growth medium restores the metabolic system to a state in which it can be repressed in response to inositol (*37, 129, 158*). When choline is supplied to a *cho1*, *cho2*, or *opi3* mutant, major net synthesis of PC occurs via the CDP-choline route using DAG as an intermediate in lipid biosynthesis (Fig. 1). DAG is derived from PA by dephosphorylation (*187*) (Fig. 1). Thus, choline supplementation of these mutants relieves the precursor "dam" that keeps early precursor levels high by opening a "sluice gate" via an alternative pathway branchpoint, downstream of PA.

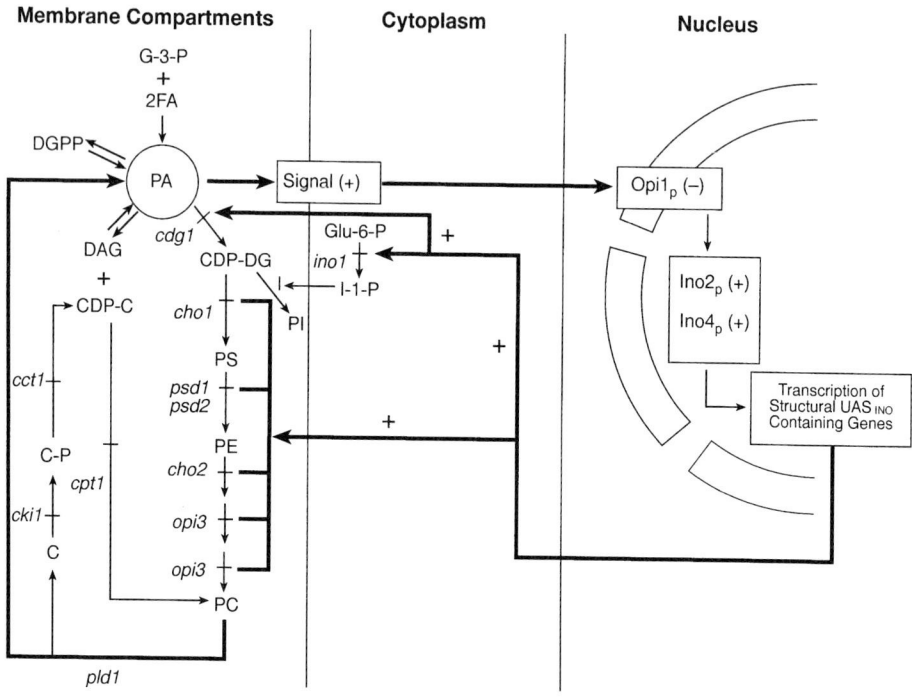

FIG. 2. Model for the transcriptional regulation of the inositol-sensitive genes containing upstream activating sequence (UAS$_{INO}$). This schematic depicts a subset of the phospholipid biosynthetic reactions of S. cerevisiae shown in Fig. 1. The model proposes that the relative balance of phosphatidic acid (PA) production versus utilization provides the signal for derepression of genes containing UAS$_{INO}$. However, closely related metabolites such as diacylglycerol pyrophosphate and sn-1,2-diacylglycerol (DGPP and DAG) and 1-acylglycerol-3-phosphate (lyso-PA) cannot be ruled out on the basis of current evidence. When PA is produced more rapidly than it is used, the system is in the "on" position; when PA is used more rapidly than it is produced, the system is "off." Thus, the signal produced is positive (+). The signal (+) is transmitted, possibly via an unknown signal transduction cascade, via the *OPI1* gene whose product plays a negative role (−). The *INO2* and *INO4* gene products, both of which play a positive role (+), are known to bind directly to UAS$_{INO}$. The balance of PA output versus utilization is influenced by at least four complex metabolic factors: (1) *General cellular metabolic conditions, including levels of basic metabolites such as phosphorus and carbon*. When general metabolic activity is low, as in stationary phase, PA production will be very low and the system will be in the "off" state. Any metabolic condition that leads to ongoing PA production in stationary phase, such as continuing phospholipid turnover, can lead to stationary phase expression of the coregulated genes. (2) *Phospholipid turnover*. PA produced by phospholipase D-mediated turnover can, under certain circumstances, constitute a major component of the total PA pool and can lead to unscheduled derepression of the genes containing UAS$_{INO}$. (3) *Availability of inositol and choline during active growth*. Addition of inositol to the growth medium of actively growing cells causes a shift in the pattern of phospholipid biosynthesis in which the rate of synthesis of phosphatidylinositol (PI) rises at the expense of precursors such as PA and cytidinediphosphate diacylglycerol (CDP-DG), causing the coregulated genes to be repressed. Choline enters

Our model also explains why genes containing UAS$_{INO}$ are repressed (do not derepress) in stationary phase whether inositol is present or not (38, 114, 240). If major cellular metabolites (i.e., carbon or phosphorous) are limiting, PA and other early precursors will not be made *de novo* and there will be no build-up of such precursors to be relieved by turning on the *INO1* gene and other enzymes of phospholipid metabolism. Thus, we observe that wild-type strains show leveling-off of the activity of the *INO1 lacZ* reporter construct (35) or comparable repression of the native *INO1* transcript (38, 114, 240) in stationary phase. However, at 37°C, the *sec14^{ts} cki1* strain continues to express the *INO1* reporter construct in stationary phase (35), consistent with the idea that there is ongoing production of PA via lipid turnover, which, according to our model, could cause derepression of *INO1*, even in stationary phase.

The *cho2 cki1* strain in stationary phase may present a special case for testing the model. In this strain, the *INO1* gene exhibits partial repression in response to inositol in logarithmic phase and then derepresses dramatically as cells enter stationary phase (38). In this strain, the build-up of phospholipid precursors caused by the *cho2* lesion may take some time to clear, as cells enter stationary phase. Indeed, in the *cho2 CKI1* strain, the stationary phase repression of *INO1* is much slower than in wild type (38), whereas in the *cho2 cki1*, such repression never occurs. It may well be that clearing of upstream precursors from the pathway during stationary phase is partially dependent on the CDP-choline pathway, especially in strains with reduced PE methylation capacity.

Finally, this model provides a framework for interpreting the finding that many classes of mutants, some seemingly unrelated to phospholipid metabolism, exhibit Ino$^-$ and Opi$^-$ phenotypes (Tables I–V). Some of these mutants may have the Ino$^-$ or Opi$^-$ regulatory phenotypes because they are global transcriptional regulators that interact at some level with the very sensitive transcriptional controls that mediate *INO1* expression. The transcriptional regulation of *INO1* may be more sensitive to perturbation of the overall transcriptional apparatus than the controls of many other less highly regulated genes. It could well be that very direct and/or sensitive communication occurs between the general transcription machinery and the specific regulatory factors such as Ino2p, Ino4p, and Opi1p that serve to activate/in-

---

at a different point in the pathway and draws initially on reserves of DAG, which, in turn, draws on PA, thus contributing to the degree of repression. (4) *Overall metabolic flow in the pathway leading from PA to phosphatidylcholine (PC) biosynthesis.* Any genetic lesion that institutes a significant block in the *de novo* pathway for PC biosynthesis (i.e., PA → CDP-DG → PS → PE → → → PC) causes a build-up of precursors, including PA. However, when a precursor such as choline that can draw on DG for the synthesis of PC via the Kennedy pathway, the cell becomes capable of repression in response to inositol.

activate the transcription of *INO1*. For example, transcription of the *INO1* gene requires the activity of the SWI/SNF complex (246–249). Mutations in histone H4 are capable of bypassing the requirement at the *INO1* locus for the SWI/SNF complex, which is believed to act by remodeling chromatin (249). However, histone H4 mutations, which are capable of alleviating the Ino$^-$ phenotype of *snf2* (*swi2*) mutants, did not suppress the Ino$^-$ phenotype of an *ino2* mutant (249). Thus, it appears that the *INO1* locus may require the function of the SWI/SNF complex to facilitate the binding of the Ino2p/Ino4p complex by remodeling the chromatin structure at its binding site (UAS$_{INO}$).

Mutants defective in processes not immediately involved in transcription, however, could conceivably have Ino$^-$ and Opi$^-$ phenotypes due to mechanisms that are much less direct, perhaps involving general effects on cellular metabolism. The regulatory model we have proposed in Fig. 2 predicts that a mutation that interferes with the flow of metabolites upstream from PA (including basic nutrients such as carbon, phosphorous, and nitrogen) could lead to *INO1* repression, under certain growth conditions, even in the absence of inositol. A mutation in an adjacent metabolic network, such as carbon or phosphorous metabolism, which restricts or slows the flow of precursors into PA biosynthesis without blocking growth altogether, could conceivably exhibit repressed, or nearly repressed, levels of *INO1* RNA even in the absence of inositol. Such a mutant might be an inositol auxotroph or at least show growth stimulation in the presence of inositol. On the other hand, mutations that influence the flow of any phospholipid metabolite downstream from PA, without fully blocking growth, could produce an Opi$^-$ phenotype. Mutations that slow the rate of membrane growth, without influencing formation of PA and other early precursors of phospholipid biosynthesis, could be in this category. This could be at least part of the explanation for Opi$^-$ phenotypes associated with some mutants in the secretory pathway.

These speculations remain to be tested. However, the model presented in Fig. 2 provides a framework for further exploration of this complex regulatory network. Its basic features will be tested in the course of the analysis of the many new mutants and combinations of mutants that will become available due to the construction of disruption mutations for all of the open reading frames identified in the completed *S. cerevisiae* genome project.

How applicable is our model to the analysis of other eukaryotic organisms? This is perhaps the most exciting aspect of the model. The model predicts a major role for PA in growth control in yeast as in mammals. The rapid induction of *INO1* when PLD-mediated turnover is elevated provides a convenient marker for this event in yeast. Thus, *INO1* reporter constructs and the Opi$^-$ phenotype can be used in devising new and powerful genetic approaches for the study of the role of PLD and PC turnover. Likewise, the production of choline that occurs during PLD-mediated turnover can be conveniently detected in yeast using the plate assay for the Opc⁻ phenotype.

Will yeast be a good model for analyzing the role of PA, PC turnover, and PLD in higher eukaryotes? The report by Patton-Vogt et al. (35) suggests a role for PLD activity and PA production in the secretory pathway of yeast, as in mammals. PA, lyso-PA, and DAG are all believed to act as mitogenic signals in mammals (250, 251). The hydrolysis of PC via PLD is a widespread phenomenon in many cell types. PLD activity in animal cells is dramatically increased in response to a variety of extracellular signals, including growth factors, hormones, cytokines, and neurotransmitters (251, 242). In fact, many mitogens that bind to the cell surface stimulate PC hydrolysis through the activation of PLD. Platelet-derived growth factor (PDGF) is a potent mitogen for fibroblasts and smooth muscle cells and is also one of the most powerful activators of PLD (253). Furthermore, phorbol esters, agents known to promote tumor formation and to stimulate protein kinase C (PKC), activate PLD activity in most cells tested, providing another link between PLD activity and growth control (253).

A wealth of recent evidence has also implicated PLD and subsequent PA formation in vesicular trafficking (48, 251, 253). The mode of action of PA in these events is unclear, as is the way that these findings correlate to the mitogenic stimulation of PLD. PA may act by changing the lipid bilayer properties such that vesicular budding and fusion events are facilitated. Alternatively, PA may directly regulate target proteins important for membrane trafficking events. Clearly, membrane trafficking is a fundamental cellular operation that must be maintained if cell division and mitogenesis are to occur properly. Although there have been numerous provocative publications on the biological roles of PLD and PA in growth control, signal transduction, and membrane trafficking, the exact molecular details remain to be determined.

Yeast should be an excellent system for exploring these questions because of its powerful genetics and the wealth of data available from its completed genome project. What role will the fission yeast, S. pombe, play in such studies? This organism is, by some accounts, almost as remotely related to S. cerevisiae as it is to higher eukaryotes (8). Yet, like S. cerevisiae, it is approachable using the powerful tools of yeast molecular genetics. In addition, the completion of its relatively small genome will undoubtedly precede that of any mammal, including humans. Thus, this organism will provide the first test bed for determining the antiquity of the various cellular regulatory networks discovered and analyzed in S. cerevisiae. With respect to the basic pathways of phospholipid biosynthesis, S. cerevisiae and S. pombe look similar (13, 39). Both classes of the phospholipid methyltransferases and most of the fundamental biochemical reactions of phospholipid metabolism must have been present in their last common ancestor. However, at first glance, the regulatory network controlling phospholipid metabolism does not look identical in the two organisms, and laboratory strains of S. *pombe* are inositol auxotrophs, apparently lacking a functional *INO1* homolog (*13*). The phospholipid methyltransferases are the only activities of phospholipid biosynthesis in S. *pombe* for which structural genes have been cloned, and these genes do not show clear transcriptional regulation in response to inositol and choline. Yet, the transcripts of the S. *pombe* PLMT genes are controlled in response to growth phase like their homologs in S. *cerevisiae* (*39, 164*). Thus, a great deal must yet be done to test the degree of similarity among eukaryotic organisms with respect to the regulation of membrane lipid metabolism and its relationship to growth control.

IV. Summary and Future Directions

We believe that our model for regulation of phospholipid biosynthesis in S. *cerevisiae* is consistent in its basic features with the available evidence. Many details remain to be worked out, however, including the steps involved in transmitting the signal to the general transcriptional apparatus, the precise role of the *OPI1* gene product, and the way in which the binding and/or activation and inactivation of the Ino2p/Ino4p complex are controlled. It may well be that the signal derived from PA metabolism is transmitted via one of the signal tansduction cascades already characterized in yeast. If this is true, mutants defining that pathway should have phenotypes related to some of those described here. For example, Cox *et al.* (*253a*) showed that the unfolded protein response pathway and regulation in response to inositol share common control elements. We are currently exploring this and other signal transduction pathways to determine how the signal for response to inositol is carried and connected to cellular regulatory networks. The sequences of the *OPI1*, *INO2*, and *INO4* gene products all contain potential phosphorylation sites (*210, 234, 238*) and there is evidence that Ino2p is phosphorylated (*234, 254*). Is phosphorylation involved in the modulation of the activity of Ino2p or the other regulatory factor? Why is expression of the *INO1* gene so acutely sensitive to perturbations in the general transcriptional apparatus of the cell? How are other levels of regulation, such as biochemical regulation of the enzyme activities, coordinated with the overall mechanism of the transcriptional control? What role does the intracellular trafficking of lipids play in the regulation? Will any of the features of this regulation, including its coupling with lipid turnover, be found in other eukaryotic cells?

Clearly, the questions remaining to be explored are fundamental. However, the molecular genetics of yeast has now developed to the point where the answers to such questions can be obtained, and a detailed understanding of complex genetic regulatory networks is feasible.

Acknowledgments

We are indebted to our colleagues Robert Lester, George Carman, Sepp D. Kohlwein, Fritz Paltauf, Dennis Voelker, Dennis Vance, Günther Daum, John Lopes, Miriam Greenberg, Pat McGraw, Peter Griac, Margaret Kanipes, Tony Graves, Kim Slekar, Vladimir Jiranek, Avula Sreenivas, Susan Dowd, Vincent Bruno, and Monica Ruiz-Noriega, all of whom have provided us with stimulating discussion, helpful advice, and access to their unpublished data during preparation of this review.

We are expecially grateful to Sue Haslett for her expert assistance (and patience) in preparing this manuscript, for maintaining our reference database, and for communicating with our colleagues on several continents.

This work was supported by NIH GM-19629 to SAH; JP-V was supported in part by a fellowship from the American Heart Association.

References

1. G. Ourisson and Y. Nakatani, *Chem. Biol.* **1,** 11 (1994).
2. W. Dowhan, *Annu. Rev. Biochem.* **66,** 199 (1997).
3. A. Goffeau *et al.*, *Science* **274,** 546 (1996).
4. E. Pennisi, *Science* **272,** 1736 (1996).
5. V. Smith, K. N. Chous, D. Lashkari, D. Botstein, and P. O. Brown, *Science* **274,** 2069 (1996).
6. K. Nasmyth, *Science* **274,** 1643 (1996).
7. S. G. Oliver, *Nature, (London)* **379,** 597 (1996).
8. M. Sipiczki, in "Molecular Biology of the Fission Yeast" (A. Nasim, P. Young, and B. F. Johnson, eds.), p. 431. Academic Press, San Diego, 1989.
9. P. Russell and P. Nurse, *Cell* **45,** 781 (1986).
10. F. Paltauf, S. Kohlwein, and S. A. Henry, in "The Molecular and Cellular Biology of the Yeast *Saccharomyces*" (J. Broach, E. Jones, and J. Pringle, eds), p. 415. Cold Spring Harbor Laboratory, Cold Spring Harbor, New York, 1992.
11. M. L. Greenberg and J. M. Lopes, *Microbiol. Rev.* **60,** (1996).
12. C. Kent, *Annu. Rev. Biochem.* **64,** 315 (1995).
13. S. Fernandez, M. J. Homann, S. A. Henry, and G. M. Carman, *J. Bacteriol.* **166,** 779 (1986).
14. A. L. Majumder, M. D. Johnson, and S. A. Henry, *Biochim. Biophys. Acta* **1348,** 245 (1997).
15. R. L. Lester and M. R. Steiner, *J. Biol. Chem.* **243,** 4889 (1968).
16. S. Steiner and R. L. Lester, *Biochim Biophys. Acta* **260,** 82 (1972).
17. R. T. Talwalkar and R. L. Lester, *Biochim. Biophys. Acta* **306,** 412 (1973).
18. J. L. Patton and R. L. Lester, *Arch. Biochem. Biophys.* **292,** 70 (1992).
19. K. R. Auger, C. L. Carpenter, L. C. Cantley, and L. Varitcovski, *J. Biol Chem.* **264,** 20181 (1989).
20. C. A. Flanagan *et al.*, *Science* **262,** 1444 (1993).
21. S. Yoshida, Y. Ohya, M. Goebl, A. Nakano, and Y. Anraku, *J. Biol. Chem.* **269,** 1166 (1994).
22. P. J. Trotter and D. R. Voelker, *J. Biol. Chem.* **270,** 6062 (1995).
23. J. H. Stack and S. D. Emr, *J. Biol. Chem.* **269,** 31552 (1994).

24. S. B. Helliwell *et al.*, *Mol. Biol. Cell* **5,** 105 (1994).
25. P. W. Greenwell *et al.*, *Cell* **82,** 823 (1995).
26. D. M. Morrow, D. A. Tagle, Y. Shiloh, F. S. Collins, and P. Hieter, *Cell* **82,** 831 (1995).
27. R. Kato and H. Ogawa, *Nucleic Acids Res.* **22,** 3104 (1994).
28. T. A. Weinert, G. L. Kiser, and L. H. Hartwell, *Genes Dev.* **8,** 652 (1994).
29. M. R. Steiner and R. L. Lester, *Biochim. Biophys. Acta* **260,** 222 (1972).
30. M. S. Bae-Lee and G. M. Carman, *J. Biol. Chem.* **259,** 10857 (1984).
31. P. M. Gaynor and G. M. Carman, *Biochim. Biophys. Acta* **1045,** 156 (1990).
32. C. J. Waechter and R. J. Lester, *J. Bacteriol.* **105,** 837 (1971).
33. C. Waechter and R. Lester, *Arch. Biochem. Biophys.* **158,** 401 (1973).
34. E. P. Kennedy and S. B. Weiss, *J. Biol. Chem.* **222,** 193 (1956).
35. J. L. Patton-Vogt *et al.*, *J. Biol. Chem.* **272,** 20873 (1997).
36. K. D. Atkinson *et al.*, *J. Bacteriol.* **141,** 558 (1980).
37. E. F. Summers, V. A. Letts, P. McGraw, and S. A. Henry, *Genetics* **120,** 909 (1988).
38. P. Griac, M. J. Swede, and S. A. Henry, *J. Biol. Chem.* **271,** 25692 (1996).
39. M. I. Kanipes and S. A. Henry, *Biochim. Biophys. Acta* **1348,** 134 (1997).
40. J. Bremer and D. M. Greenberg, *Biochim. Biophys. Acta* **35,** 287 (1959).
41. Z. Yao and D. E. Vance, *J. Biol. Chem.* **263,** 2998 (1988).
42. D. E. Vance, C. J. Walkey, and Z. Cui, *Biochim. Biophys. Acta* **1348,** 142 (1997).
43. K. S. Lee *et al.*, *J. Biol. Chem.* **269,** 19725 (1994).
44. J. S. Flick and J. Thorner, *Mol. Cell. Biol.* **13,** 5861 (1993).
45. K. Rose, S. A. Rudge, M. A. Frohman, A. J. Morris, and J. Engebrecht, *Proc. Natl. Acad. Sci. U.S.A.* **92,** 12151 (1995).
46. K. M. Ella, J. W. Dolan, C. Qi, and K. E. Meier, *Biochem. J.* **314,** 15 (1996).
47. M. Waksman, Y. Eli, M. Liscovitch, and J. E. Gerst, *J. Biol. Chem.* **271,** 2361 (1996).
48. M. Liscovitch, *J. Lipid Mediat. Cell Signal.* **14,** 215 (1996).
49. S. M. Honigberg, C. Conicella, and R. E. Esposito, *Genetics* **130,** 703 (1992).
50. S. M. Hammond, Y. M. Altshuller, T. C. Sung, S. A. Rudge, K. Rose, J. Engbrecht, A. J. Morris, and M. A. Frohman, *J. Biol. Chem.* **270,** 29640 (1995).
51. J. A. Mayr, S. D. Kohlwein, and F. Paltauf, *FEBS Lett.* **393,** 236 (1996).
52. M. Waksman, X. Tang, Y. Eli, J. E. Gerst, and M. Liscovitch, *J. Biol. Chem.* **272,** 36 (1997).
53. W. E. Payne and M. Fitzgerald-Hayes, *Mol. Cell. Biol.* **13,** 4351 (1993).
54. C. Schomerus and H. Küntzel, *FEBS Lett.* **307,** 249 (1992).
54a. N. J. Marini *et al.*, *EMBO J.* **15,** 3040 (1996).
55. W. W. Angus and R. L. Lester, *Arch. Biochem. Biophys.* **151,** 483 (1972).
56. P. T. Hawkins, L. R. Stephens, and J. R. Piggott, *J. Biol. Chem.* **268,** 3374 (1993).
57. W. W. Angus and R. L. Lester, *J. Biol. Chem.* **250,** 22 (1975).
58. F. Paltauf, E. Zinser, and G. Daum, *Biochim. Biophys. Acta* **835,** 322 (1985).
59. J. L. Patton, L. Pessoa-Brandao, and S. A. Henry, *J. Bacteriol.* **177,** 3379 (1995).
60. J.-I. Nikawa, Y. Tsukagoshi, and S. Yamashita, *J. Biol. Chem.* **266,** 11184 (1991).
60a. J. L. Patton-Vogt and S. A. Henry, *Genetics*, in press (1998).
61. S. A. Henry, in "Molecular Biology of the Yeast *Saccharomyces:* Metabolism and Gene Expression" (J. N. Strathern, E. W. Jones, and J. R. Broach, eds.), p. 101. Cold Spring Harbor Laboratory, Cold Spring Harbor, New York, 1982.
62. G. M. Carman and S. A. Henry, *Annu. Rev. Biochem.* **58,** 635 (1989).
63. D. M. Nikoloff and S. A. Henry, *Annu. Rev. Genet.* **25,** 559 (1992).
64. M. J. White, J. M. Lopes and S. A. Henry, *Adv. Microbial Physiol.* **32,** (1991).
65. G. M. Carman and G. M. Zeimetz, *J. Biol. Chem.* **271,** 13293 (1996).
66. S. D. Kohlwein, G. Daum, R. Schneiter, and F. Paltauf, *Trends Cell Biol.* **6,** 260 (1996).
67. M. J. Kelley, A. M. Bailis, S. A. Henry, and G. M. Carman, *J. Biol. Chem.* **263,** 18078 (1988).

68. W.-I. Wu, et. al., *J. Biol. Chem.* **270,** 13171 (1995).
69. W.-I. Wu, Y.-P. Lin, E. Wang, A. H. Merrill, Jr., and G. M. Carman, *J. Biol. Chem.* **268,** 13830 (1993).
70. V. M. McDonough et al., *J. Biol. Chem.* **270,** 18774 (1995).
71. J. J. Quinlan et al, *J. Biol. Chem.* **267,** 18013 (1992).
72. A. J. Kinney and G. M. Carman, *Proc. Natl. Acad. Sci. U.S.A.* **85,** 7962 (1988).
73. K. Lai and P. McGraw, *J. Biol. Chem.* **269,** 2245 (1994).
74. K. S. Robinson, K. Lai, T. A. Cannon, and P. McGraw, *Mol. Biol. Cell* **7,** 81–89 (1996).
74a. Skinner et al., *Proc. Natl. Acad. Sci. U.S.A.* **92,** 112 (1995).
75. E. Zinser and G. Daum, *Yeast* **11,** 493–536 (1995).
76. E. Zinser et al., *J. Bacteriol.* **173,** 2026–2034 (1991).
77. B. Gaigg, R. Simbeni, C. Hrastnik, F.Paltauf, and G. Daum, *Biochim. Biophys. Acta* **1234,** 214 (1995).
78. A. Leber, C. Hrastnik, and G. Daum, *FEBS Lett.* **377,** 271 (1995).
79. R. Schneiter and S. D. Kohlwein, *Cell* **88,** 431 (1997).
80. R. Schneiter et al., *Mol. Cell. Biol.* **16,** 7161 (1996).
81. S. Hamamatsu, I. Shibuya, M. Takagi, and A. Ohta, *FEBS Lett.* **348,** 33 (1994).
82. P. Griac and S. A. Henry, in "NATO ASI Series: Molecular Dynamics of Biological Membranes" (J. A. F. Op den Kamp, ed.), p. 339. Springer-Verlag, Berlin/Heidelberg, 1996.
83. V. A. Bankaitis, D. E. Malehorn, S. D. Emr and R. Greene, *J. Cell Biol.* **108,** 1271 (1989).
84. J. F. Aitken, G. P. vanHeusden, M. Temkin, and W. Dowhan, *J. Biol. Chem.* **265,** 4711 (1990).
85. V. Bankaitis, J. Aitkin, A. Cleves, and W. Dowhan, *Nature (London)* **347,** 561 (1990).
86. A. E. Cleves et al., *Cell* **64,** 789 (1991).
87. A. Cleves, T. McGee, and V. Bankaitis, *Trends Cell. Biol.* **1,** 30 (1991).
88. J. P. Hirsch and S. A. Henry, *Mol. Cell. Biol.* **6,** 3320 (1986).
89. L. S. Klig, M. J. Homann, G. M. Carman, and S. A. Henry, *J. Bacteriol.* **162,** 1135 (1985).
90. A. M. Bailis, M. A. Poole, G. M. Carman, and S. A. Henry, *Mol. Cell. Biol.* **7,** 167 (1987).
91. T. Kodaki, K.Hosaka, J.-I. Nikawa, and S. Yamashita, *J. Biochem.* **109,** 276 (1991).
92. P. M. Gaynor et al., *Biochim. Biophys. Acta* **1090,** 326 (1991).
93. J. M. Lopes, J. P. Hirsch, P. A. Chorgo, K. L. Schulze, and S. A. Henry, *Nucleic Acids Res.* **19,** 1687 (1991).
94. J. M. Lopes, K. L. Schulze, J. W. Yates, J. P. Hirsch, and S. A. Henry, *J. Bacteriol.* **175,** 4235 (1993).
95. K. A. Hudak, J. M. Lopes, and S. A. Henry, *Genetics* **136,** 475 (1994).
96. K. H. Slekar and S. A. Henry, *Nucleic Acids Res.* **23,** 1964 (1995).
97. N. Bachhawat, Q. Ouyang, and S. A. Henry, *J. Biol. Chem.* **270,** 25087 (1995).
98. D. M. Nikoloff and S. A. Henry, *J. Biol. Chem.* **269,** 7402 (1994).
99. J. C. Jackson and J. M. Lopes, *Nucleic Acids Res.* **24,** 1322 (1996).
100. M. R. Culbertson, T. F. Donahue, and S. A. Henry, *J. Bacteriol.* **126,** 232 (1976).
101. I. Chen and F. C. Charalompous, *Biochem. Biophys. Res. Commun.* **19,** 144 (1965).
102. O. Hoffmann-Ostenhof and F. Pittner, *Can. J. Chem.* **60,** 1863 (1982).
103. F. A. Loewus and M. W. Loewus, *Ann U. Rev. Plant Physiol.* **34,** 137 (1983).
104. F. A. Loewus, in "Inositol Metabolism in Plants" (D. J. Morre, W. F. Boss, and F. A. Loewus, eds.), p. 13. Wiley-Liss, New York, 1990.
105. T. Maeda and F. Eisenberg, Jr., *J. Biol. Chem.* **255,** 8458 (1980).
106. T. Donahue, Ph.D. Thesis, Albert Einstein College of Medicine, Bronx, New York, 1975.
107. T. F. Donahue, and S. A. Henry, *J. Biol. Chem.* **256,** 7077 (1981).
108. M.Murray and M. L. Greenberg, *Mol. Microbiol.* **26,** 481 (1997).
109. A. S. Fischl and G. M. Carman, *J. Bacteriol.* **154,** 304 (1983).

110. A. S. Fischl, M. J. Homann, M. A. Poole, and G. M. Carman, *J. Biol. Chem.* **261,** 3178 (1986).
111. J.-I. Nikawa and S. Yamashita, *Eur. J. Biochem.* **143,** 251 (1984).
112. J.-I. Nikawa, T. Kodaki, and S. Yamashita, *J. Biol. Chem.* **262,** 4876 (1987).
113. M. S. Anderson and J. M. Lopes, *J. Biol. Chem.* **271,** 26596 (1996).
114. V. Jiranek, J. A. Graves, and S. A. Henry, *Microbiology,* in press (1998).
115. M. R. Culbertson and S. A. Henry, *Genetics* **80,** 23 (1975).
116. T. F. Donahue and S. A. Henry, *Genetics.* **98,** 491 (1981).
117. M. R. Culbertson, T. F. Donahue, and S. A. Henry, *J. Bacteriol.* **126,** 243 (1976).
118. M. Culbertson, Ph.D. Thesis, Albert Einstein College of Medicine, Bronx, New York, 1975.
119. A. Majumder, S. Duttagupta, P. Goldwasser, T. Donahue, and S. Henry, *Mol. Gen. Genet.* **184,** 347 (1981).
120. L. S. Klig and S. S. Henry, *Proc. Natl. Acad. Sci. U.S.A.* **81,** 3816 (1984).
121. M. Dean-Johnson and S. A. Henry, *J. Biol. Chem.* **264,** 1274 (1989).
122. B. S. Loewy and S. A. Henry, *Mol. Cell. Biol.* **4,** 2479 (1984).
123. S. Henry, D. Hoshizaki, A. Bailis, M. Homann, and G. Carman, *CNRS-SISERM Int. Symp. NATO Workshop, 1986.*
124. M. Greenberg, Ph.D. Thesis, Albert Einstein College of Medicine, Bronx, New York, 1980.
125. M. L. Greenberg, B. Reiner, and S. A. Henry, *Genetics* **100,** 19–33 (1982).
126. M. J. Swede, K. A. Hudak, J. M. Lopes, and S. A. Henry, *Meth. Enzymol.* **209,** 21 (1992).
127. M. Greenberg, P. Goldwasser, and S. Henry, *Mol. Gen. Genet.* **186,** 157 (1982).
128. M. L. Greenberg, L. S. Klig, V. A. Letts, B. S. Loewy, and S. A. Henry, *J. Bacteriol.* **153,** 791 (1983).
129. P. McGraw and S. A. Henry, *Genetics* **122,** 317 (1989).
130. L. S. Klig *et al., J. Bacteriol.* **170,** 1878 (1988).
131. H. Shen and W. Dowhan, *J. Biol. Chem.* **271,** 29043 (1996).
132. C. Scafe *et al., Mol. Cell. Biol.* **10,** 1270 (1990).
133. C. Scafe, M. Nonet, and R. A. Young, *Mol. Cell. Biol* **10,** 1010 (1990).
134. C. Scafe *et al., Nature (London)* **347,** 491 (1990).
135. N. A. Woychik and R. A. Young, *Mol. Cell. Biol.* **9,** 2854 (1989).
136. L. J. Gansheroff, C. Dollard, P. Tan, and F. Winston, *Genetics* **139,** 523 (1995).
137. K. M. Arndt, S. Ricupero-Hovasse, and F. Winston, *EMBO J.* **14,** 1490 (1995).
138. S. Roberts and F. Winston, *Mol. Cell. Biol.* **16,** 3206 (1996).
139. C. L. Peterson, W. Kruger, and I. Herskowitz, *Cell* **64,** 1135 (1991).
140. A. E. Cleves, P. J. Novick, and V. A. Bankaitis, *J. Cell Biol.* **109,** 2939 (1989).
141. E. A. Whitters, A. E. Cleves, T. P. McGee, H. B. Skinner, and V. A. Bankaitis, *J. Cell Biol.* **122,** 79 (1993).
142. S. Kagiwada *et al., Genetics* **143,** 685 (1996).
143. J.-I. Nikawa, M. Akiyoshi, S. Hirata, and T. Fukuda, *Nucleic Acids Res.* **24,** 4222 (1996).
144. J. S. Cox and P. Walter, *Cell* **87,** 391 (1996).
145. V. A. Letts and I. W. Dawes, *J. Bacteriol.* **156,** 212 (1983).
146. M. Swede, Ph.D. Thesis, Carnegie Mellon University, Pittsburgh, Pennsylvania, 1994.
147. T. Kodaki and S. Yamashita, *J. Biol. Chem.* **262,** 15428 (1987).
148. V. Letts and I. Dawes, *Biochem. Soc. Trans.* **7,** 976 (1979).
149. L. Kovac, I. Gbelska, V. Poliachova, J. Subik, and V. Kovacova, *Eur. J. Biochem.* **111,** 491 (1980).
150. S. Yamashita, A. Oshima, J.-I. Nikawa, and K. Hosaka, *Eur. J. Biochem.* **128,** 589 (1982).
151. R. H. Hjelmstad and R. M. Bell, *J. Biol. Chem.* **263,** 19748 (1988).
152. R. H. Hjelmstad and R. M. Bell, *J. Biol. Chem.* **265,** 1755 (1990).
153. K. Hosaka, J. Nikawa, T. Kodaki, and S. Yamashita, *J. Biochem.* **111,** 352 (1992).
154. T. S. Tillman and R. S. Bell, *J. Biol. Chem.* **261,** 9144 (1986).

155. G. A. Scarborough and J. F. Nyc, *J. Biol. Chem.* **242,** 238 (1967).
156. C. C. Lindegren, G. Lindegren, E. Shult, and Y. L. Hwang, *Nature (London)* **194,** 260 (1962).
157. K. Atkinson, S. Fogel, and S. A. Henry, *J. Biol. Chem.* **255,** 6653 (1980).
158. V. A. Letts and S. A. Henry, *J. Bacteriol.* **163,** 560 (1985).
159. C. Sperka-Gottlief *et al., Yeast* **6,** 331 (1990).
160. T. Kodaki and S. Yamashita, *Eur. J. Biochem.* **185,** 243 (1989).
161. S. A. Henry *et al., Fifth Int. Yeast Symp., London, Ontario, Can., 1981.*
162. W. Preitschopf *et al., Curr. Genet.* **23,** 95 (1993).
163. J. E. Hill, C. Chung, P. McGraw, E. Summers, and S. A. Henry, in "Biochemistry of Cell Walls and Membranes in Fungi" (P. J. Kuhn, A. P. J. Trinci, M. J. Jung, M. W. Goosey, and L. G. Copping, eds.), p. 246. Springer-Verlag, Berlin and New York, 1990.
164. M. I. Kanipes, Ph.D. Thesis, Carnegie Mellon University, Pittsburgh, Pennsylvania, 1997.
164a. M. I. Kanipes, J. E. Hill, and S. A. Henry, *Genetics* In press (1998).
165. Z. Cui, J. Vance, M. Chen, D. Voelker, and D. Vance, *J. Biol. Chem.* **268,** 16655 (1993).
166. C. J. Clancey, S.-C. Chang, and W. Dowhan, *J. Biol. Chem.* **268,** 24580 (1993).
167. P. J. Trotter, J. Pedretti, and D. R. Voelker, *J. Biol. Chem.* **268,** 21416 (1993).
168. H. Shen, P. N. Heacock, C. J. Clancey, and W. Dowhan, *J. Biol. Chem.* **271,** 789 (1996).
169. Y. Tsukagoshi, J.-I. Nikawa, and S. Yamashita, *Eur. J. Biochem.* **169,** 477 (1987).
170. K. Hosaka, K. Tsumomu, and S. Yamashita, *J. Biol. Chem.* **264,** 2053 (1989).
171. R. H. Hjelmstad and R. M. Bell, *J. Biol. Chem.* **266,** 5094 (1991).
172. R. H. Hjelmstad and R. M. Bell, *J. Biol. Chem.* **266,** 4357 (1991).
173. C. R. McMaster and R. M. Bell, *J. Biol. Chem.* **269,** 28010 (1994).
174. C. R. McMaster and R. M. Bell, *J. Biol. Chem.* **269,** 14776 (1994).
175. S. A. Minskoff *et al., J. Bacteriol.* **174,** 5702 (1992).
176. P. V. Racenis *et al., J. Bacteriol.* **174,** 5702 (1992).
177. V. A. Letts, Ph.D. Thesis, University of Edinburgh, Edinburgh, Scotland, 1980.
178. R. L. Lester, G. B. Wells, G. Oxford, and R. C. Dickson, *J. Biol. Chem.* **268,** 845 (1993).
179. M. M. Nagiec, G. B. Wells, R. L. Lester, and R. C. Dickson, *J. Biol. Chem.* **268,** 22156 (1993).
180. G. B. Wells and R. C. Lester, *J. Biol. Chem.* **258,** 10200 (1983).
181. W. J. Pinto *et al., Fed. Proc.* **45,** 1826 (1986).
182. W. J. Pinto *et al., J. Bacteriol.* **174,** 2565 (1992).
183. W. J. Pinto, G. W. Wells, and R. L. Lester, *J. Bacteriol.* **174,** 2575 (1992).
184. K. Athenstaedt and G. Daum, *J. Bacteriol.* **179,** 7611 (1997).
185. W.-I. Wu *et al., J. Biol. Chem* **271,** 1868 (1996).
186. Y.-P. Lin and G. M. Carman, *J. Biol. Chem.* **264,** 8641 (1989).
187. K. R. Morlock, J. J. McLaughlin, Y.-P. Lin, and G. M. Carman, *J. Biol. Chem.* **267,** 3586 (1991).
188. D. A. Dillon *et al., J. Biol. Chem.* **272,** 10361 (1997).
188a. D. Toke, A., *et al., J. Biol. Chem.* **273,** 3278 (1998).
189. T. Munnik, T. de Vrije, R. F. Irvine, and A. Musgrave, *J. Biol. Chem.* **271,** 15708 (1996).
190. S. A. Henry, K. D. Atkinson, A. I. Kolat, and M. R. Culbertson, *J. Bacteriol.* **130,** 472 (1977).
191. B. A. Hanson and R. L. Lester, *J. Bacteriol.* **142,** 79 (1980).
192. G. Becker and R. L. Lester, *J. Biol. Chem.* **252,** 8684 (1977).
193. K. D. Atkinson, A. I. Kolat, and S. A. Henry, *J. Bacteriol.* **132,** 806 (1977).
194. P. Novick, C. Field, and R. Schekman, *Cell* **21,** 205 (1980).
195. S. A. Henry, T. F. Donahue, and M. R. Culbertson, *Mol. Gen. Genet.* **143,** 5 (1975).
196. A. J. Shatkin and E. L. Tatum, *Am J. Bot.* **48,** 760 (1961).
197. C. J. Waechter, M. R. Steiner, and R. L. Lester, *J. Biol. Chem.* **244,** 3419 (1969).

198. M. A. Carson, M. Emala, P. Hogsten, and C. J. Waechter, *J. Biol. Chem.* **259,** 6267 (1984).
199. S. Yamashita and A. Oshima, *Eur. J. Biochem.* **104,** 611 (1980).
200. M. A. Poole, M. J. Homann, M. S. Bae-Lee, and G. M. Carman, *J. Bacteriol.* **168,** 668 (1986).
201. M. J. Homann, S. A. Henry, and G. M. Carman, *J. Bacteriol.* **163,** 1265 (1985).
202. M. L. Greenberg, A. Hubbell, and C. Lam, *Mol. Cell. Biol.* **8,** 4773 (1988).
203. E. Lamping, S. D. Kohlwein, S. A. Henry, and F. Paltauf, *J. Bacteriol.* **173,** 6432 (1991).
204. J. H. Overmeyer and C. J. Waechter, *J. Biochem. Biophys.* **290,** 511 (1991).
205. M. J. Homann, M. A. Poole, P. M. Gaynor, C.-T. Ho, and G. M. Carman, *J. Bacteriol.* **169,** 533 (1987).
206. K. R. Morlock, Y.-P. Lin, and G. M. Carman, *J. Bacteriol.* **170,** 3561 (1988).
207. J. Ko, S. Cheah, and A. S. Fischl, *J. Bacteriol.* **176,** 5181 (1994).
208. B. Loewy, Ph.D. Thesis, Albert Einstein College of Medicine, Bronx, New York, 1985.
209. B. Loewy, J. Hirsch, M. Johnson, and S. Henry, *in* "Yeast Cell Biology" (J. Hicks, ed.), p. 551. Alan R. Liss, New York, 1986.
210. J. A. Graves, Ph.D., Carnegie Mellon University, Pittsburgh, Pennsylvania, 1996.
211. V. A. Letts, L. S. Klig, M. Bae-Lee, G. M. Carman, and S. A. Henry, *Proc. Natl. Acad. Sci. U.S.A.* **80,** 7279 (1983).
212. J.-I. Nikawa, Y. Tsukagoshi, T. Kodaki, and S. Yamashita, *Eur. J. Biochem.* **167,** (1987).
213. T. Kodaki, J. Nikawa, K. Hosaka, and S. Yamashita, *J. Biochem.* **173,** 7992 (1991).
214. A. M. Ballis, J. M. Lopes, S. D. Kohlwein, and S. A. Henry, *Nucleic Acids Res.* **20,** 1411 (1992).
215. K. Kiyono *et al.*, *J. Biochem.* **102,** 1089 (1987).
216. L. S. Klig, D. K. Hoshizaki, and S. A. Henry, *Curr. Genet.* **13,** 7 (1988).
217. D. K. Hoshizaki, J. E. Hill, and S. A. Henry, *J. Biol. Chem.* **265,** 4736 (1990).
218. J. P. Hirsch, Ph.D. Thesis, Albert Einstein College of Medicine, Bronx, New York, 1987.
219. M. J. White, J. P. Hirsch, and S. A. Henry, *J. Biol. Chem.* **266,** 863 (1991).
220. D. M. Nikoloff, P. McGraw, and S. A. Henry, *Nucleic Acids Res.* **20,** 3253 (1992).
221. C. I. Hammond, P. Romano, S. Roe, and P. Tontonoz, *Cell. Mol. Biol. Res.* **39,** 561 (1994).
222. K. Hosaka, J. Nikawa, T. Kodaki, and S. Yamashita, *J. Biochem. (Tokyo)* **115,** 131 (1994).
223. H. Shen and W. Dowhan, *J. Biol. Chem.* **272,** 11215 (1997).
224. K. Hosaka, T. Murakami, T. Kodaki, J.-I. Nikawa, and S. Yamashita, *J. Bacteriol.* **172,** 2005 (1990).
225. P. Griac, *J. Bacteriol.* **179,** 5843 (1997).
226. M. Hasslacher, A. S. Ivessa, F. Paltauf, and S. D. Kohlwein, *J. Biol. Chem.* **268,** 10946 (1993).
227. A. Li and M. Brendel, *Mol. Gen. Genet.* **241,** 680 (1993).
228. P. McGraw and K. Lai, *Yeast* **8,** S510 (1992).
229. J. Nikawa, K. Hosaka, Y. Tsukagoshi, and S. Yamashita, *J. Biol Chem.* **265,** 15996 (1990).
230. J. M. Lopes and S. A. Henry, *Nucleic Acids Res.* **19,** 3987 (1991).
231. J. Ambroziak and S. A. Henry, *J. Biol. Chem.* **269,** 15344 (1994).
232. T. K. Blackwell and H. Weintraub *Science,* **250,** 1104 (1990).
233. J. Koipally *et al.*, *Yeast* **12,** 653 (1995).
234. J. Ambroziak, Ph.D. Thesis, Carnegie Mellon University, Pittsburgh, Pennsylvania, 1994.
235. S. Schwank, R. Ebbert, K. Rautenstrauss, E. Schweizer, and H.-J Schuller, *Nucleic Acids Res.* **23,** 230 (1995).
236. C. Murre *et al.*, *Cell* **58,** 537 (1989).
237. B. P. Ashburner and J. M. Lopes, *Proc. Natl. Acad. Sci. U.S.A.* **92,** 9722 (1995).
238. M. Nikoloff, Ph.D. Thesis, Carnegie Mellon University, Pittsburgh, Pennsylvania, 1993.
239. H.-J. Schüller, K. Richter, B. Hoffmann, R. Ebbert, and E. Schweizer, *FEBS Lett.* **370,** 149 (1995).

240. E. Lamping, F. Paltauf, S. A. Henry, and S. D. Kohlwein, *Genetics* **137**, 55 (1995).
241. P. Griac and S. Henry, In preparation (1998).
242. E. Summers, Albert Einstein College of Medicine, Yeshiva University, Bronx, New York, 1992.
243. S. A. Henry *et al.*, *in* "Dynamics of Membrane Assembly" (J. A. F. Op den Kamp, ed.), p. 33. Springer-Verlag, Barcelona/Budapest, 1992.
243a. A. Sreenivas, J. L. Patton-Vogt, V. Bruno, P. Griac, and S. A. Henry, *J. Biol. Chem.*, in press (1998).
244. T. P. McGee, H. B. Skinner, E. A. Whitters, S. A. Henry, and V. A. Bankaitis, *J. Cell Biol.* **124**, 273 (1994).
245. B. G. Kearns *et al.*, *Nature (London)* **387**, 101 (1997).
246. C. L. Peterson and I. Herskowitz, *Cell* **68**, 573 (1992).
247. C. L. Peterson A. Dingwall, and M. P. Scott, *Proc. Natl. Acad. Sci. U.S.A.* **91**, 2905 (1994).
248. C. L. Peterson and J. W. Tamkun, *Trends Biochem. Sci.* **20**, 143 (1995).
249. M. S. Santisteban, G. Arents, E. N. Moudrianakis, and M. M. Smith, *EMBO J.* **16**, 2493 (1997).
250. M. R. Boarder, *Trends Pharmacol. Sci.* **15**, 57 (1994).
251. J. H. Exton, *J. Biol. Chem.* **272**, 15579 (1997).
252. J. H. Exton, *Biochim Biophys. Acta* **1212**, 26 (1994).
253. Z. Kiss, *in* "Chemistry and Physics of Lipids," p. 81. Elsevier Science Ireland, 1996.
253a. J. S. Cox *et al.*, *Mol. Cell. Biol.* **8**, 1805 (1997).
254. S. Roe, Ph.D. Thesis, Wesleyan University, Middletown, Connecticut, 1995.
255. M. A. Kuziora *et al.*, *J. Biol. Chem.* **258**, 11648 (1983).
256. M. Schweizer, C. Lebert, J. Holtke, L. M. Roberts and E. Schweizer, *Mol. Gen. Genet.* **194**, 457 (1984).
257. M. Schweizer *et al.*, *Mol. Gen. Genet.* **203**, 479 (1986).
258. W. Al-Feel, S. S. Chirala, and S. J. Wakil, *Proc. Natl. Acad. Sci. U.S.A.* **89**, 4534 (1992).
259. J. E. Stukey, V. M. McDonough, and C. E. Martin, *J. Biol. Chem.* **264**, 16537 (1989).
260. R. J. Duronio, L. J. Knoll, and J. I. Gordon, *J. Cell Biol.* **117**, 515 (1992).
261. D. R. Johnson, L. J. Knoll, N. Rowley, and J. I Gordon, *J. Biol. Chem.* **269**, 18037 (1994).
262. D. R. Johnson, L. J. Knoll, D. E. Levin, and J. I. Gordon, *J. Cell Biol.* **127**, 751 (1994).
263. R. Buede, C. Rinker-Schaffer, W. J. Pinto, R. L. Lester, and R. C. Dickson, *J. Bacteriol.* **173**, 4325 (1991).
264. M. M. Nagiec, G. B. Wells, and R. C. Dickson, *Proc. Natl.Acad. Sci. U.S.A.* **91**, 7899 (1994).
265. C. Zhao, T. Beeler, and T. Dunn, *J. Biol. Chem.* **269**, 21480 (1994).
266. M. M. Nagiec *et al.*, *J. Biol. Chem.* **272**, 9809 (1997).
267. M. A. Brostrom and E. T. Browning, *J. Biol. Chem.* **248**, 2364 (1973).
267a. F. Jiang, H. S. Rizavi, and M. L. Greenberg, *Mol. Microbiol.* **26**, 481 (1997).
268. G. M. Carman, C. J. Belunis, and J. T. Nickels, Jr., *Meth. Enzymol.* **209**, 183 (1992).
269. K. M. Ella, J. W. Dolan, and K. E. Meier, *Biochem. J.* **307**, (1995).
270. Z. Li, E. Haase and M. Brendel, *Curr. Gen.* **19**, 423 (1991).
271. M. Nonet and R. Young, *Genetics* **123**, 715 (1989).
272. G. S. Roeder, C. Beard, M. Smith, and S. Keranen, *Mol. Cell. Biol.* **5**, 1543 (1985).
273. H. Wang, L. Reynolds-Hager, and D. J. Stillman, *Mol. Gen Genet.* **245**, 675 (1994).
274. J.-I. Nikawa and S. Yamashita, *Mol. Microbiol* **6**, 1441 (1992).
275. J.-I. Nikawa, A. Murakami, E. Esumi, and K. Hosaka, *J. Biochem.* **118**, 39 (1995).
276. S. M. Swift, Ph.D. Thesis, University of Maryland—Baltimore County, Maryland, 1997.

Inosine-5'-Monophosphate Dehydrogenase: Regulation of Expression and Role in Cellular Proliferation and T Lymphocyte Activation

ALBERT G. ZIMMERMANN,*
JING-JIN GU,* JOSÉE LALIBERTÉ,*
AND BEVERLY S. MITCHELL*,†

Departments of *Pharmacology and
†Medicine
University of North Carolina
Chapel Hill, North Carolina 27599

I. Introduction	182
II. IMPDH Activity and Cellular Proliferation	183
III. De Novo Purine Biosynthesis and Salvage	184
IV. Cellular IMPDH Activity Comprises Activities of Two Isozymes	190
A. cDNA Cloning, Protein Structure, and Function	190
B. IMPDH mRNA Expression and Activity	191
C. Comparison of the IMPDH Type I and Type II Genes	196
V. IMPDH Activity and Cellular Signaling	200
VI. In Vitro Effects of IMPDH Inhibition on the Immune System	201
VII. Clinical Applications for IMPDH Inhibitors	202
VIII. Summary	203
References	204

Guanine nucleotide synthesis is essential for the maintenance of normal cell growth and function, as well as for cellular transformation and immune responses. The expression of two genes encoding human inosine-5'-monophosphate dehydrogenase (IMPDH) type I and type II results in the translation of catalytically indistinguishable enzymes that control the rate-limiting step in the *de novo* synthesis of guanine nucleotides. Cellular IMPDH activity is increased more than 10-fold in activated peripheral blood T lymphocytes and is attributable to the increased expression of both the type I and type II enzymes. In contrast, abrogation of cellular IMPDH activity by selective inhibitors prevents T lymphocyte activation and establishes a requirement for elevated IMPDH activity in T lymphocytic responses. In order to assess the molecular mechanisms governing the expression of the IMPDH type I and type II genes in resting and activated peripheral blood T lymphocytes, we have cloned the human IMPDH type I and type II genes and characterized their genomic organization and their respective 5'-flanking regions. Both genes contain 14 highly conserved exons that vary in size from 49 to 207 base pairs. However, the intron structures are completely divergent, resulting in disparities in gene length (18 kilobases for type I and 5.8 kilobases for type II). In addition, the 5′-regulatory sequences are highly divergent; expression of the IMPDH type I gene is controlled by three distinct promoters in a tissue specific manner while the type II gene is regulated by a single promoter and closely flanked in the 5′ region by a gene of unknown function. The conservation of the IMPDH type I and type II coding sequence in the presence of highly divergent 5′-regulatory sequences points to a multifactorial control of enzyme expression and suggests that tissue-specific and/or developmentally specific regulation of expression may be important. Delineation of these regulatory mechanisms will aid in the elucidation of the signaling events that ultimately lead to the synthesis of guanine nucleotides required for cellular entry into S phase and the initiation of DNA replication. © 1998 Academic Press

I. Introduction

Guanine nucleotides are involved as substrates, activators, and regulators in many important anabolic processes in the cell, including biosynthesis of RNA, DNA, and protein and transmembrane signaling. Inosine-5′-monophosphate dehydrogenase (IMPDH) is the essential rate-limiting enzyme in *de novo* synthesis of guanine nucleotides and is accounted for by the expression of two enzymes, termed IMPDH type I and type II, that are the products of two distinct genes. The products of these genes exhibit extensive homology at the nucleic acid and amino acid levels, suggesting strong evolutionary pressure for conservation of function. An important role for IMPDH in cell proliferation became evident in 1975 when Jackson and co-workers (76) observed that IMPDH activity is increased in hepatomas and regenerating liver as compared to normal liver. Subsequent investigations in a wide variety of cell lines complemented these initial studies and established a close correlation between total cellular IMPDH activity and cellular proliferation. In addition, transformed cells were shown to express high levels of IMPDH activity. The close correlation between increased cellular enzyme activity and elevated IMPDH type II mRNA levels has implicated increased expression of the IMPDH type II gene in this up-regulation. More recent studies on T lymphocytes demonstrate that the IMPDH type I gene is similarly up-regulated in activated cells, suggesting a role for both gene products in these cells. Clinical use of selective inhibitors of IMPDH reveals that the immune system is exquisitely sensitive to inhibition of this enzyme. IMPDH inhibitors prevent both B and T lymphocyte activation *in vitro* and are currently used in clinical practice as highly efficacious agents in the treatment of allograft rejection through immunosuppression.

In view of the important role of this metabolic pathway in normal and neoplastic cells and in the immune system, we will review the basic biochemistry relevant to IMPDH type I and type II activity, expression, and function.

II. IMPDH Activity and Cellular Proliferation

Activity of the enzyme inosine-5′-monophosphate dehydrogenase (EC 1.1.1.205) varies with many aspects of cellular function. The IMPDH enzyme is positioned at the branch point of the *de novo* adenine and guanine nucleotide biosynthetic pathways and is the first committed step in the synthesis of guanine nucleotides from IMP (29, 75, 136, 163). The importance of IMPDH in cellular metabolism is underscored by the observations that both mRNA transcript and activity levels are low in quiescent cells (83) and are dramatically up-regulated in replicating (75, 76, 99, 103) and malignant, transformed cells (25, 73, 98, 165). In contrast, cells that are induced to differentiate exhibit decreased enzyme activity (27, 82, 91, 101).

This biosynthetic pathway is important for both B and T lymphocytes, which appear to rely on *de novo* synthesis to provide the guanine nucleotides required for mitogen- and antigen-initiated proliferative responses (5). The critical requirement for IMPDH is evident from several studies that have demonstrated a close association between IMPDH activity and proliferation in a variety of cell types (28, 75, 76, 120, 131). In addition, the linear relationship between IMPDH activity and growth rates in tumors of diverse tissue origin, as compared to their normal cellular counterparts, has linked elevated expression of IMPDH with the malignant phenotype (25, 99) and established that the activity of the enzyme strongly correlates with the onset and rate of cellular division, as opposed to the presence or degree of neoplastic transformation (56).

Selective inhibition of cellular IMPDH activity with mycophenolic acid (MPA) or other drugs results in a cessation of DNA synthesis (23, 24, 38, 44, 158) and a cell cycle arrest at the G_1–S boundary (38, 149). This inhibition of cell replication is dose and time dependent and the direct consequence of a reduction in cellular guanine ribo- and deoxyribonucleotide pools, because exogenous guanosine is able to abrogate the inhibition by being converted to guanine and salvaged into the guanine nucleotide pool (1, 6, 27, 35, 79, 80, 90, 112). Inhibitors of IMPDH enzymatic activity have also been demonstrated to possess antineoplastic (92, 132, 156), antiviral (93, 145, 166), antiparasitic (160), and immunosuppressive (53, 97, 107) activities. In addition, IMPDH inhibitors have been shown to induce terminal differentiation *in vitro* in a variety of human tumor cell lines, including leukemic (27, 78, 79, 80, 82, 90, 112, 137, 173, 174), breast cancer (15, 132), and melanoma (80) cells.

The general observation that IMPDH enzymatic activity levels are elevated in proliferating versus quiescent cells and down-regulated in cells induced to differentiate suggests an important role for guanine nucleotides in the control of normal cell growth. The functional importance of IMPDH in maintaining nucleotide pools through the *de novo* biosynthetic pathway and the sensitivity of lymphocyte activation to inhibition of its activity have led to a number of investigations into the regulation of IMPDH type I and type II gene expression in these cells and into the effects of IMPDH inhibitors on lymphocyte function and proliferation. In order to understand these relationships, it is important to gain further insights into the specific role of IMPDH in guanine nucleotide biosynthesis.

III. *De Novo* Purine Biosynthesis and Salvage

Figure 1 presents a schematic outline of the *de novo* biosynthetic and salvage pathways for purine nucleotides, including positive- and negative-feedback regulation of enzyme activities. The ribose phosphate portion of the purine nucleotide is initiated from the 5-phospho-α-D-ribosyl-1-pyrophosphate (PRPP) precursor, which in turn is synthesized from ATP and ribose-5-phosphate. PRPP interacts with glutamine in a reaction catalyzed by phosphoribosylpyrophosphate amidotransferase (EC 2.4.2.14) to form 5-phosphoribosylamine, which is converted through a 10-step anabolic pathway to the purine compound inosine-5'-monophosphate, the precursor metabolite for both adenine and guanine nucleotides.

In the bifurcating *de novo* purine nucleotide biosynthetic pathway, the adenylosuccinate synthetase enzyme is the first committed step toward adenine nucleotide synthesis. The product of this reaction, adenosine-5'-monophosphate (AMP), is formed by the condensation of IMP with aspartic acid in the presence of GTP, resulting in the formation of adenylosuccinate. Adenylosuccinate is subsequently cleaved by adenylosuccinate lyase (EC 4.3.2.2) to form fumarate and AMP (Fig. 1). The complementary guanine nucleotide biosynthetic pathway is initiated by the conversion of IMP to xanthosine-5'-monophosphate (XMP) by the enzyme IMPDH. IMPDH cata-

FIG. 1. Summary of biosynthetic pathways involved in the *de novo* and salvage synthesis of purine nucleotides. Solid lines represent metabolic conversions and broken lines depict feedback regulatory pathways. Salvage pathways (S) are designated; (1) phosphoribosylpyrophosphate amidotransferase; (2) adenylosuccinate synthase; (3) inosine-5'-monophosphate dehydrogenase; (4) adenylosuccinate lyase; (5) guanosine-5'-monophosphate synthetase; (6)hypoxanthine–guanine phosphoribosyl transferase; (7) adenine phosphoribosyl transferase; (8) adenosine deaminase; (9) 5'-nucleotidase; (10) ribonucleotide reductase; (11) purine nucleoside phosphorylase; (12) deoxycytidine kinase; (13) adenosine kinase.

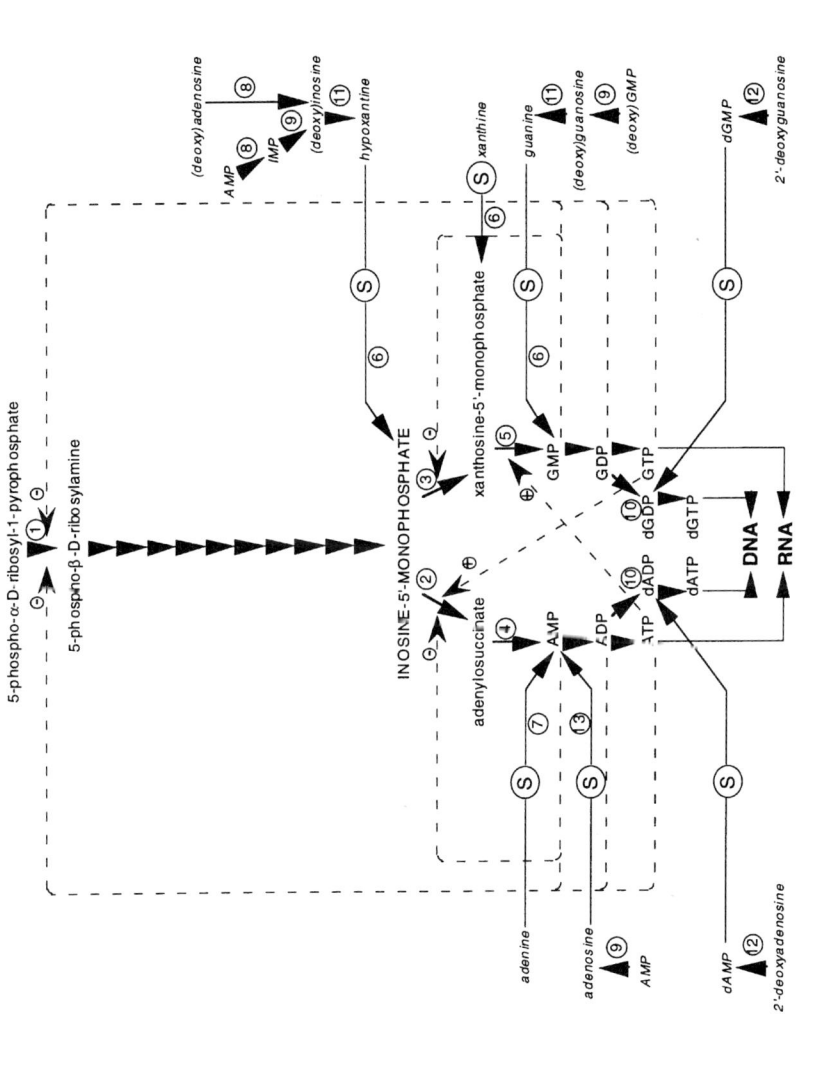

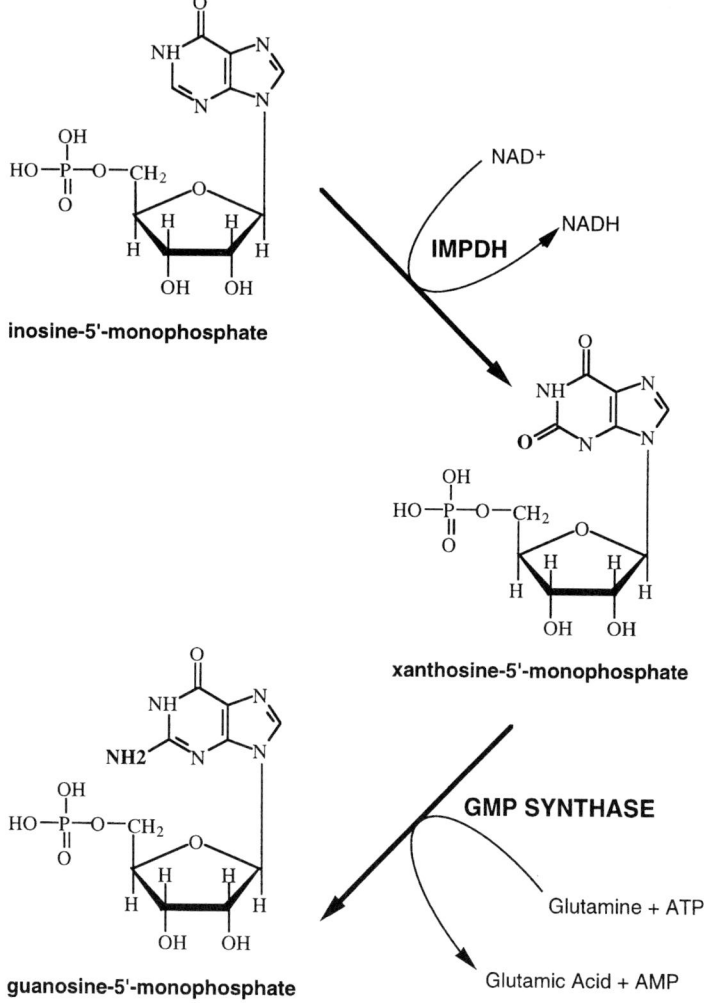

FIG. 2. The IMPDH and GMP synthetase reactions.

lyzes the nicotinamide adenine dinucleotide (NAD) cofactor-dependent irreversible oxidation of IMP. The product of this reaction, XMP, is rapidly aminated to guanosine-5'-monophosphate (GMP) by the enzyme GMP synthetase (EC 6.3.4.1) in the presence of ATP and glutamine (Figs. 1 and 2). The requirements for GTP in AMP synthesis and ATP in GMP synthesis result in a positive-feedback mechanism whereby GTP promotes AMP and ATP promotes GMP synthesis, respectively. The purine ribonucleoside monophosphate AMP is phosphorylated to ADP and ATP by adenylate kinase, while GMP is phosphorylated to GDP by GMP kinase and GTP by GDP kinase. Both nucleoside triphosphates serve as substrates for RNA synthesis, energy donors for metabolic processes, and signaling intermediates. The ribonucleotide diphosphates (ADP, GDP) are also converted by the enzyme ribonucleotide reductase (EC 1.17.4) to the corresponding deoxyribonucleotide diphosphates (dADP and dGDP), which are then phosphorylated to dATP and dGTP to serve as substrates for DNA synthesis (Fig. 1).

The enzymes involved in the interconversion of IMP to AMP and GMP are regulated by allosteric mechanisms. The enzymes catalyzing the initial commitment steps in the *de novo* synthesis of adenine and guanine nucleotides, adenylosuccinate synthase and IMPDH, are inhibited through negative feedback by their corresponding products, AMP and GMP (75, 143). In addition, adenine and guanine mono-, di-, and triphosphates inhibit the rate-limiting enzyme for the *de novo* synthesis of IMP, PRPP amidotransferase, thus regulating the formation of 5-phosphoribosyl-1-amine from PRPP and glutamine. Furthermore, guanine nucleotides are potent inhibitors of PRPP synthase and the formation of PRPP (177), the limiting and regulatory substrate for PRPP amidotransferase (64, 170). As a consequence, a decreased cellular content of guanine nucleotides results in activation of these enzymes, thereby enhancing net purine biosynthesis. Interestingly, PRPP amidotransferase in human lymphocytes is inhibited by adenine nucleotides (AMP and ADP) but activated by guanine nucleotides (GMP, GDP, and GTP) (45). The catalytic activity of ribonucleotide reductase is decreased by the binding of dATP, whereas binding of dGTP stimulates ADP reduction (146). Consequently, an excess of adenine nucleotides, and/or depletion of guanine nucleotides, can decrease the pool of PRPP, whereas an excess of dATP can inhibit ribonucleotide reductase activity.

Salvage of purines can occur by two general mechanisms. First, there are two cellular purine phosphoribosyltransferases that are responsible for the conversion of adenine (APRT) and hypoxanthine and guanine (HGPRT) to their corresponding ribonucleotides AMP, IMP, and GMP. In these reactions, PRPP serves as the ribose phosphate donor (Fig. 1). Second, salvage may occur through a cycle in which IMP and GMP, as well as their respective deoxyribonucleotides, are converted to the nucleosides inosine, deoxyinosine, guanosine, and deoxyguanosine by purine 5'-nucleotidase (5'-NT). These purine ribonucleosides and 2'-deoxyribonucleosides are converted to hypoxanthine or guanine by purine nucleoside phosphorylase (PNP), resulting in ribose-1-phosphate or 2'-deoxyribose-1-phosphate as phosphorolysis products. Hypoxanthine and guanine can then be phosphoribosylated by HGPRT in the presence of PRPP to form IMP and GMP, thereby completing the cycle. In purine nucleotide catabolism, AMP, adenosine, and deoxyadenosine are deaminated by adenosine deaminase to form IMP, inosine, and deoxyinosine, respectively (89). The nucleosides are subsequently converted to hypoxanthine by the activity of PNP and can be phosphoribosylated by HGPRT to form IMP. As illustrated in Fig. 1, the phosphorylation of adenosine, 2'-deoxyadenosine, and 2'-deoxyguanosine can be carried out by two kinases, adenosine and deoxycytidine kinase, resulting in salvage of these metabolic intermediates.

It has been suggested that the guanine nucleotide pool destined for incorporation into DNA is highly compartmentalized and that only the pathway involving dGTP formation by ribonucleotide reductase is capable of the efficient delivery of dGTP for DNA synthesis (108). The inability of exogenous deoxyribonucleosides to circumvent the inhibition of DNA synthesis occurring with inhibitors of ribonucleotide reductase is consistent with this hypothesis (123). The observation that the cellular GMP pool constitutes the smallest ribonucleotide pool of both purines and pyrimidines (22, 95) further suggests that this pool might be the most sensitive to modulation. Indeed, it has been demonstrated using radionucleotide tracer studies that the large adenine nucleotide pool provides a substantial delay in the incorporation of labeled precursors into DNA, whereas the guanine nucleotide precursor pool is much smaller and more readily incorporated into DNA (38). Natsumeda *et al.* (104) also observed that a higher percentage of labeled guanine, compared to labeled adenine, was incorporated into the acid-insoluble fraction of rat hepatoma cells after pulse labeling, suggesting a smaller functional pool and a higher turnover rate of guanine than adenine nucleotides for nucleic acid synthesis. This finding does not apply directly to precursors for RNA synthesis, because both adenine and guanine tracers are rapidly incorporated into RNA, implicating a distinct cellular compartment of ribonucleotide precursors for RNA synthesis (38). It has been observed that the ratio of guanine:adenine nucleotides in rat liver hepatomas increases from 17% in slow growing to 79% in rapidly growing cells (74). In addition, IMP was found to be preferentially channeled into guanine nucleotide synthesis during the transition from rest to exponential growth in rat hepatoma 3924A cells (103), leading to a more marked expansion of GTP than of ATP pools (164). This purine metabolic imbalance in proliferating cells has been attributed to increased IMPDH activity (76, 162). Szkeres *et al.* (149) demonstrated that IMPDH activity was significantly increased in S-phase-enriched fractions of HL-60 and K562 cells, whereas GTP concentrations were also significantly increased in S-phase-enriched HL-60 cells. In contrast, a greater than 50% decrease in ATP and GTP concentrations was observed in nondividing cells, consistent with the previous demonstration of greater than 95% decreases in *de novo* purine biosynthesis in lymphoblasts at high cell density (62). Lymphocytes responding to antigenic and mitogenic stimuli also exhibit a sustained increase in PRPP concentrations and incorporation of [^{14}C]glycine into purine nucleotides (66, 67). Increased activity of PRPP synthetase is also associated with elevated IMPDH activity, supporting the view that levels of IMPDH activity are strongly regulated by the availability of purine nucleotides (56). These observations suggest that proliferating cells preferentially expand the guanine nucleotide pool to supply limiting precursors for RNA and DNA synthesis.

Defects in purine salvage metabolism have been associated with several diseases, including gout, immunodeficiency, Lesch–Nyhan syndrome, myopathies, skin diseases, parasitic infections, and neoplasias (43, 124, 153, 162). The *de novo* purine nucleotide biosynthetic pathway seems to be especially important for the overall function and development of human T and B lymphocytes. The observation that children deficient in HGPRT (Lesch–Nyhan syndrome), the major enzyme in the purine salvage pathway, have essentially normal numbers and function of T and B lymphocytes indicates that the high rate of *de novo* synthesis in these cells can compensate for the absence of this salvage enzyme (5, 85, 126). Salvage pathways therefore seem to contribute little to guanine nucleotide pool maintenance in lymphocytes. Individuals with the Lesch–Nyhan syndrome do, however, have distinctive neurological deficits, including mental retardation and self-mutilation. The activity of HGPRT is higher in brain than in other tissues, whereas the activity of PRPP amidotransferase is low (146), suggesting a higher dependency on the salvage of exogenous purines in the brain. PNP deficiency, on the other hand, is associated with T lymphocyte depletion in the presence of little, if any, B cell dysfunction (46). The phenotype of PNP deficiency has been attributed to the intracellular accumulation of dGTP in thymocytes, resulting in allosteric inhibition of ribonucleotide reductase and DNA synthesis. Deficiencies of both PNP and the salvage enzyme HGPRT have been associated with increased IMPDH activity (65). Reconstitution of HGPRT into deficient cell lines results in normalization of IMPDH activity (56). Adenosine deaminase deficiency, an inherited immunodeficiency in children, is characterized by a selective decrease in functional T and B lymphocytes in the presence of normal numbers of neutrophils, erythrocytes, and platelets and abnormal brain function (47, 68), and is due at least in part to the selective accumulation of dATP in lymphoid precursor cells. ADA deficiency also leads to accumulation of adenine ribonucleotides, which inhibit the rate-limiting enzyme for the *de novo* purine biosynthesis, PRPP synthase, and leads to reduced levels of PRPP. This compound is a critical intermediate in purine biosynthesis by both the *de novo* and salvage pathways and reiterates the importance of *de novo* purine biosynthesis for lymphocytic cells.

IV. Cellular IMPDH Activity Comprises Activities of Two Isozymes

A. cDNA Cloning, Protein Structure, and Function

Human IMPDH exists as two isoforms, type I and type II, as was demonstrated by the cloning of two distinct cDNAs encoding active IMPDH enzymes. The cDNAs for both IMPDH type I and type II contain open reading frames corresponding to 514 amino acids and proteins of 56 kDa (26, 105). The two enzymes exhibit 84% identity at the amino acid level, diverging by a total of 84 amino acids; of these amino acids, 32 are nonconservative substitutions (105). Both native (49, 172) and recombinant (20) isoforms of the two proteins were demonstrated by density gradient centrifugation to consist of homotetramers of the 56-kDa subunit. This finding was recently confirmed by X-ray diffraction analysis of crystallized Chinese hamster IMPDH type II (135).

Kinetic characterizations of human type I and type II recombinant IMPDH enzymes have demonstrated a common mechanism for both enzymes in which catalysis proceeds via an ordered Bi–Bi kinetic mechanism whereby IMP binding precedes the binding of NAD to the free enzyme. Following substrate binding and hydride transfer, IMPDH subsequently releases reduced NADH, followed by XMP (20, 83). This catalytic mechanism is in agreement with that reported for native bacterial (19), protozoan (71, 159), and mammalian enzymes (10, 11, 20, 63, 65, 172). Further kinetic analysis has revealed that the human IMPDH type II enzyme requires binding of an essential monovalent potassium cation prior to the binding of IMP and NAD (171). In addition, Chinese hamster IMPDH type II was shown to be locally reorganized and stabilized by the binding of IMP or XMP, but not of NAD (109).

The type I and type II IMPDH enzymes have nearly identical k_{cat} values in the range of 1.8 and 1.4 sec^{-1}, K_m values for IMP of 14 and 9 μM, and K_m values for NAD of 42 and 32 μM, respectively (59). Sensitivities to inhibition by the reaction products, NADH and XMP, were also similar for the two isoforms: the K_i values of type I and type II IMPDH were 102 and 90 μM for NADH, and 80 and 94 μM for XMP, respectively (20, 65, 172). GMP, as the end product of the reaction, inhibits both IMPDH type I and type II with a K_i of 450 μM, whereas dGMP, GDP, GTP, and ATP do not inhibit either isoform at concentrations up to 1 mM (20). The two isoforms exhibit only slightly different pH optima of 7.7 and 7.9 for the type I and type II enzyme, respectively (102).

IMPDH enzymes have been isolated from a variety of eukaryotic and prokaryotic sources and exhibit a high level of amino acid conservation (16,

26, 49, 77, 105, 134, 155, 168). Molecular mass, kinetic parameters, and the hydrophobic nature of purified or partially purified IMPDH enzymes from different sources, including hamster V79 cells (69), rodent cells (21, 72, 110, 172), and human placenta (65), are all similar, demonstrating a high degree of conservation of structure and kinetic parameters among species.

Using a property pattern approach, Bork et al. (18) identified a nucleotide-binding region in the *Escherichia coli* IMPDH sequence. The sequence is highly conserved among known IMPDH sequences from *E. coli* (152, 154), *Bacillus subtilis* (77), *Leishmania donovani* (168), humans and Chinese hamsters (26, 105), and mice (155), and most likely represents the substrate binding site. There is also strong evidence for a thiol group in the IMP binding site in independent studies examining the inactivation of bacterial (13, 19, 48, 60) and mammalian (12) enzymes by 6-Cl-IMP. Indeed, recent more detailed analysis of the IMPDH type II IMP binding site demonstrated the covalent modification of a single cysteine residue, Cys-331, by 6-Cl-IMP and EICARMP, compounds that irreversibly inactivate IMPDH (13, 161). This Cys-331 has also been shown to form a covalent bond with [^{14}C]IMP in the presence of NAD (70, 86). These findings were recently confirmed by X-ray defraction analysis of Chinese hamster IMPDH type II in complex with IMP, which demonstrated the formation of a covalent bond between the C-2 carbon of IMP and the sulfur atom of Cys-331 to yield an oxidized IMP thioimidate intermediate (135).

B. IMPDH mRNA Expression and Activity

As discussed, IMPDH catalytic activity and the generation of guanine nucleotides have been closely linked to cell proliferation. The presence of two structurally and kinetically similar IMPDH enzymes in humans has raised the question of their relative contributions to intracellular enzyme activity and guanine nucleotide biosynthesis and has led to a detailed examination of mRNA expression of the IMPDH type I and type II isoforms and total cellular IMPDH activity in normal and neoplastic cells and tissues.

A survey of steady-state IMPDH type I and type II mRNA expression in human adult and fetal tissues demonstrated expression of both isoforms in all tissues examined, with quantitatively greater expression of the type II isoform in most tissues, with the major exception of peripheral blood leukocytes (127). A lower steady-state level of IMPDH type I compared to the type II transcript has been consistently observed and has led to the conclusion that the IMPDH type I gene is constitutively expressed, unrelated to either cell proliferation, malignant transformation, or differentiation (98, 99). Studies examining the expression of IMPDH in peripheral blood T lymphocytes and B and T lymphocytic cell lines have clearly established that the 1.9-kb IMPDH type II mRNA transcript is the predominant species in these cells

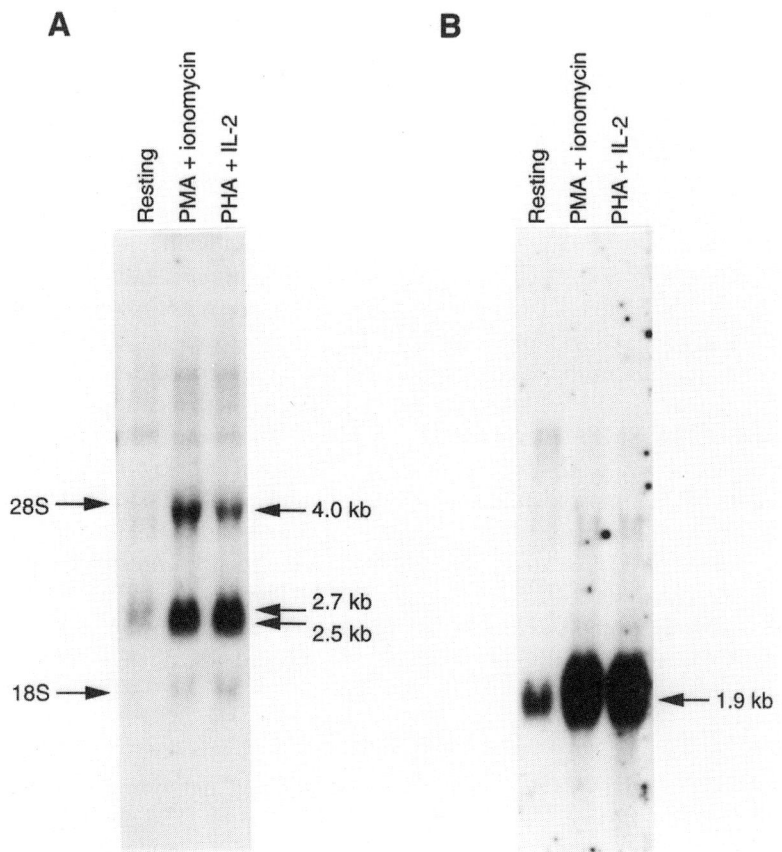

FIG. 3. Northern blot analysis of IMPDH type I and type II mRNA expression in resting and activated human peripheral blood T lymphocytes. Cells were maintained in the absence of stimulation or treated with phorbol myristate acetate (PMA, 10 ng/ml) and ionomycin (250 ng/ml) or phytohemagglutinin-L (PHA, 5 μg/ml) and IL-2 (10 U/ml) for 24 hr. RNA was isolated, 15 μg was electrophoresed on a denaturing formaldehyde–agarose gel, transferred to a Zetaprobe GT membrane, and sequentially probed with IMPDH type I (A) and type II (B) cDNAs. Replicated 18 and 28S ribosomal markers are shown to the left of both panels.

and is markedly up-regulated in activated T lymphocytes (32, 58, 83, 98, 99). In addition, Nagai et al. (99) demonstrated that the IMPDH type II mRNA transcript is increased by Epstein–Barr viral transformation of normal lymphocytes. IMPDH type I and type II gene expression is induced 10-fold at the mRNA level in activated peripheral blood T lymphocytes versus resting cells, and is associated with a 15-fold increase in total cellular IMPDH activity and 6-fold increase in GTP levels over a 72-hr period of activation (*32*). Recent Northern blot analyses have confirmed this increase and also revealed the induction of three distinct transcripts for the IMPDH type I gene versus one for the type II gene (Fig. 3).

Malignant transformation is associated with a shift toward anabolic metabolism, which, in the case of purines is characterized by a significant increase in the activity of IMPDH, ribonucleotide reductase, PRPP amidotransferse, and succinyl-AMP synthetase, whereas salvage enzyme activities are altered less markedly (*106, 118, 121, 162*). IMPDH activities in acute myelogenous leukemia, acute lymphocytic leukemia, and tonsillar B lymphocytes are significantly increased compared to that in normal bone marrow cells or peripheral blood lymphocytes, whereas IMPDH activity in chronic lymphocytic leukemia was found to be higher but not statistically different from that in normal peripheral blood (*17, 119*). Increases in enzyme activity of seven- to ninefold have been observed in the human leukemic cell lines K562 and HL-60, in agreement with the increase in the total amount of protein (*83*). The elevated activities observed in these tumors have been attributed to specific up-regulation of the IMPDH type II gene, because only the IMPDH type II transcript is increased in expression relative to nontransformed cells (*83, 98*).

Induction of terminal differentiation in the myeloid leukemia cell line HL-60 is associated with a selective down-regulation of guanine nucleotide synthesis and depletion of intracellular guanine nucleotide pools (*91*), as well as with a 55% decrease in IMPDH activity (*101*). The decrease in guanine nucleotide pools has been demonstrated to correlate closely with a reduction in the levels of the IMPDH type II transcript (*27, 82, 91*), with little or no change in the levels of the type I mRNA transcript (*99*). Of additional interest in relation to the differential expression of these two genes is the fact that inhibition of IMPDH catalytic activity leads to increased expression of the type II, but not the type I, gene at the mRNA level. Inhibition of IMPDH activity *in vitro* (*88, 92, 137, 167, 178*) and *in vivo* (*125*) results in increases in the IMP nucleotide pool, providing higher concentrations of substrate. In addition, increases in the steady-state levels of IMPDH type II mRNA, protein, and catalytic activity have been observed in HL-60 cells following IMPDH inhibition (*27, 38, 80, 82*). Increases in IMPDH activity have also been reported for HGPRT-deficient erythrocytes (*114, 148*), suggesting that the cells are attempting to compensate for the decrease in cellular guanine nucleotides by a feedback mechanism to increase the level of IMPDH gene expression.

It has been suggested that steady-state IMPDH type II but not type I mRNA levels in the promyelocytic cell line HL-205 and melanoma cell line SK-MEL-131 are inversely regulated by intracellular guanine nucleotide concentrations (51). A decrease in guanine nucleotide pools resulting from IMPDH inhibition caused an increase in expression. Furthermore, increases in guanine nucleotide levels due to enhanced salvage of guanine caused a decrease in type II mRNA and cellular IMPDH protein levels (51). Attempts to determine the mechanism underlying the regulation of IMPDH type II expression by guanine nuclcotide levels revealed no direct effect on transcription as measured by nuclear run-on assays nor any effect on mRNA stability. These data were interpreted as showing that the regulation of IMPDH type II occurs at the level of posttranscriptional nuclear processing (51). Although the definitive molecular mechanisms underlying these apparent feedback regulatory mechanisms have not been determined, the striking differences in the regulation of type I and type II IMPDH mRNA levels point to differential functions for the two isozymes in cells and tissues and/or distinct roles during development.

In comparison with the type II gene, the transcriptional regulation of IMPDH type I is quite complex. The expression of the IMPDH type I gene in lymphocytic cells is mediated by three distinct mRNA transcript of 4.0, 2.7, and 2.5 kb (58). These three transcripts have been associated with the presence of three distinct promoter regions, P1, P2, and P3, and are thus variable in the extent of the 5′ untranslated regions that are encoded by three exons in the 5′ flanking region of the gene. The 4.0-kb mRNA species is expressed preferentially in activated T lymphocytes and monocytes and is low to undetectable in most lymphoid and nonlymphoid tumor cell lines. The 2.7-kb transcript has to date been observed only in selected tumor cell lines, whereas the 2.5-kb transcript is universally expressed. The use of reporter gene constructs has provided evidence that the P1, P2, and P3 promoters are differentially utilized in T versus B lymphoid cell lines and additional evidence supports tissue-specific differences in the effects of upstream DNA elements in modulating IMPDH type I transcription (58). In summary, the regulation of the IMPDH type I gene is complex and may well play important roles in the tissue-specific and developmental expression of this isozyme.

The increase in IMPDH type II gene expression in activated T lymphocytes was demonstrated to occur primarily at the level of transcription, because type II mRNA stability was unaltered between resting and activated T lymphocytes (176). Expression of the IMPDH type II gene in T lymphocytic cells is mediated by a 5′ flanking region of 197 bp that includes a short 50-bp 5′ untranslated region. The 5′ flanking region consisting of 463 bp upstream of the translation initiation site was demonstrated to confer induced transcription and differential regulation on a chloramphenicol acetyltransferase reporter gene when transfected into Jurkat T cells and human peripheral blood T lymphocytes, respectively (175). DNase I-hypersensitive site analysis of the genomic DNA identified a region of transcriptional activity in

TABLE I
RELATIVE TRANSCRIPTIONAL ACTIVITY OF IMPDH TYPE I
AND TYPE II PROMOTER–CAT CONSTRUCTS[a]

Promoter	CAT activity (fold increase)	
	Jurkat	Raji
pCATBasic	1	1
IMPDH type I	19	46
IMPDH type II	1391	3128

[a] The T and B cell lines Jurkat E6-1 and Raji were transiently transfected with 10 μg of pCATBasic reporter gene constructs containing the proximal 791-bp IMPDH type I P3 promoter or the 461-bp IMPDH type II promoter. Cells were harvested at 48 hr and extracted for CAT activity assays. Values represent the mean of duplicate determinations and are expressed as fold increase over vector alone.

the 5′ flanking region of the gene. Fine mapping by *in vivo* footprinting demonstrated five transcription factor binding sites that are occupied in both resting and activated peripheral blood T lymphocytes; these are tandem CRE motifs, an Sp1 site, an overlapping Egr-1/Sp1 site, and a novel palindromic octamer sequence (POS) (*176*). The tandem CRE and POS sites are of major functional importance as judged by mutational and electrophoretic mobility shift analyses and provide evidence that expression of the human IMPDH type II gene is regulated by the nuclear factors ATF-1, ATF-2, and an as yet unidentified POS-binding protein. Additional major protein–DNA interactions do not occur within the promoter region following T lymphocyte activation, indicating a requirement for additional protein–protein interactions and/or posttranslational modifications of prebound transcription factors to account for the observed increase in IMPDH type II gene expression. Transfection studies using Jurkat E6-1 T cells and Raji B cells demonstrated that the type II promoter is 73- and 68-fold more active in these cells than the most active type I promoter P3 (Table I). This evidence for differential regulation of the IMPDH type I and type II genes at the transcriptional level argues strongly for significant but discrete roles for each gene product in cellular metabolism.

Western blot analysis of extracts obtained from resting and activated peripheral blood T lymphocytes mirrored the Northern blot results and demonstrated an increased level of both the type I and type II IMPDH enzymes in activated cells, although the type II protein levels are significantly higher than the type I levels under both resting and stimulated conditions, suggesting that the type II gene is responsible for the major portion of total cellular IMPDH activity in these cells (Fig. 4). The extent to which the type I and type II

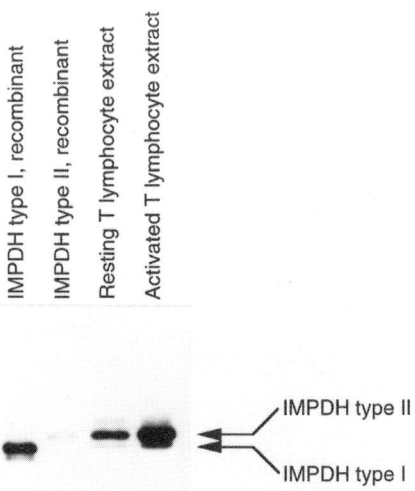

FIG. 4. Western blot analysis of IMPDH type I and type II protein expression in resting and activated human peripheral blood T lymphocytes. Four nanograms of recombinant IMPDH type I (lane 1), type II (lane 2), and 25 μg total cellular T lymphocyte extract obtained from resting (lane 3) and PMA + ionomycin-treated T lymphocytes (lane 4) were analyzed on a 12% SDS–PAGE gel, transferred to a PVDF membrane, and blotted with an antibody generated against IMPDH type I that also cross-reacts with IMPDH type II.

IMPDH enzymes contribute to overall cellular activity in these cells remains to be determined, but the observation that both genes are induced suggests that their gene products are necessary for T lymphocyte activation. Our increased knowledge of the elements regulating IMPDH type I and type II gene expression should lead to elucidation of their functional roles and potentially provide an approach to modulating their respective expression in a differential manner through selective interruption of signaling events leading to enhanced transcription.

C. Comparison of the IMPDH Type I and Type II Genes

The IMPDH type I and type II transcripts are the products of two distinct but closely related genes (57, *175*), located on chromosomes 7q31.3–q32 and 3p21.2 → p24.2, respectively (50, 57). The recent cloning of both genes by our group has allowed a structural comparison of their genomic organization. The genes are remarkably similar at the genomic level, containing coding exons of identical size and highly conserved exon–intron boundaries (Fig. 5 and Table II). The IMPDH type II gene is a relatively small

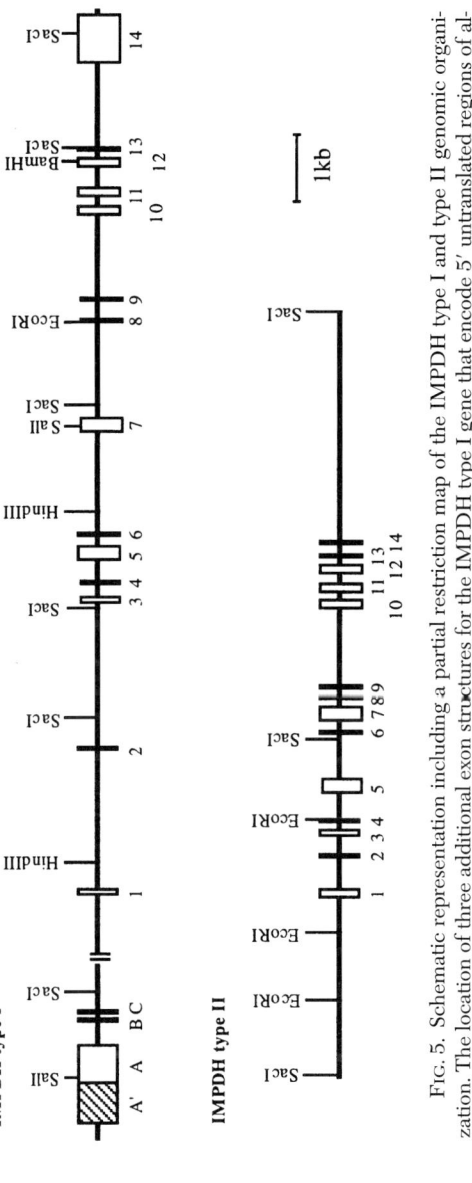

FIG. 5. Schematic representation including a partial restriction map of the IMPDH type I and type II genomic organization. The location of three additional exon structures for the IMPDH type I gene that encode 5′ untranslated regions of alternative transcripts are depicted (A′–C).

TABLE II
Conservation of Structure at Exon–Intron Boundaries of the IMPDH Type I and Type II Genes[a]

Exon (bp)	Intron (bp)		Exon
A (~1200)	AGCCCGAGAGgtgacgccca	(265)	aaaatcttagCCCTAGATTG	A′
B (44)	CCCCGTTCAGgtggctccaa	(88)	ttcctcctagAACTATCTTC	B′
C (64)	GCAGGGCTAGgtgaggacac	(~1700)	ctcgcggcagCATGGCGGAC GACGGCAGAG	1
1 (99)	TCACCTACAAgtaagcgccg	...(~2000)...	gtcttcacagCGACTTCCTG	2
(148)	TCACCTACAAgtgcgggcct	.....(445)...	tccctcgcagTGACTTTCTC	
	T Y N		D F L	
2 (49)	TGATGAGGTGgtgagtacct	..(~2300)...	ccatcctcagGACCTGACCT	3
	AGACCAGGTGgtgagtatga	.....(225)...	gtctcctcagGACCTGACTT	
	D E/Q V		D L T	
3 (102)	TGCCATGGCTgtgagttaca	......(137)...	tactctcgagCTGATGGGAG	4
	AGCAATGGCFgtgagcccat	......(107)...	tatcctgtagCTTACAGGCG	
	A M A		L M/T G	
4 (75)	GAAGGTCAAGgcaagtacca	.....(278)...	ctgcccgcagAAGTTTGAAC	5
	GAAAGTGAAGgtcagaaggg	.....(327)...	ccctttccagAAATATGAAC	
	K V K		K F/Y E	
5 (207)	CCTCAGTGAGgtacctgcag	......(149)...	ggccccacagGTGATGACGC	6
	CTTGGAAGAGgtgggtgcca	......(657)...	tcccacgtagATAATGACAA	
	L S/E E		V/I M T	
6 (88)	AGCAAGAAAGgtaccaggga	...(~1500)...	tccttgtcagGGAAGCTGCC	7
	AGCAAGAAGGgtaagtccta	........(73)...	ctgaccacagGAAAGTTGCC	
	S K K G		K L	
7 (200)	CATAGTCTTGgtaaggcccc	...(~1400)...	tttctcacagGACTCGTCCC	8
	AGTGGTTTTGgtgagctgct	.......(77)...	cttgtcctagGACTCTTCCC	
	I/V V L		D S S	
8 (91)	GGGGGGAACGgtgagtgcgg	....(236)...	ccaccccagTGGTGACAGC	9
	GGAGGCAATGgtaaggcaag	..... (99)...	ttcaccatagTGGTCACTGC	
	G G M V		V T	
9 (96)	ACCCAGGAAGgtgggtgtcg	..(~1300)...	cctcccccagTGATGGCCTG	10
	ACGCAGGAAGgtaagaatat	....(1065)...	atctcaacagTGCTGGCCTG	
	T Q E V		M/L A	
10 (144)	GCCTCCACAGgtgaggggag	......(94)...	gcccacgtagTGATGATGGG	11
	GCCTCCACAGgtgaggcagt	......(83)...	ctgtccgcagTCATGATGGG	
	A S T V		M M	
11 (145)	GATACTTCAGgttccctgac	.....(288)...	ctccacgcagCGAGGGGGAT	12
	GATATTTCAGgtgggacagg	.......(94)...	ctccctgcagTGAAGCTGAC	
	Y F S		E G/A D	
12 (144)	CTGTCCTTCGgtgagtgctg	.......(93)...	ttcctcctagGTCCATGATG	13
	CCCAAGTCCGgtgagcttgg	.......(80)...	ccttcttcagAGCCATGATG	
	V L/V R		S/A M M	
13 (84)	GCCTGCACTCgtaagtgtgg	..(~1200)...	gtccccacagTTACGAAAAG	14
	GCCTCCATTCgtaagtcacc	.......(89)...	ctgcctgcagGTATGAGAAG	
	L H S		Y E K	
14 (734)	TCAAGCAGGT			
(75)	TTTAGAAAGA			

[a]Sequences for the IMPDH type I gene are italicized.

gene of 5.8 kb consisting of 14 exons varying in size from 49 to 207 bp and introns from 73 to 1065 bp and is closely flanked by two genes with unknown function (52, 175, 176). In contrast, the IMPDH type I gene is a larger gene of more than 18 kb with intron sizes varying from 88 to greater than 3200 bp (58). The increased size of the type I gene is due primarily to larger intron sizes, excluding introns 4 and 5, and the presence of three additional noncoding exons with accompanying introns at the 5' end of the gene (Fig. 5). The observation that intron sizes and sequence of the two IMPDH genes are divergent suggests that they have evolved following an initial gene duplication event. In addition, evolutionary conservation of exon structure and amino acid sequence in these proteins that have kinetic properties that are largely indistinguishable (20, 59, 105) again argues strongly that the two genes have considerable and nonoverlapping functional importance for cells.

The presence of multiple, processed pseudogenes for IMPDH type I has previously been documented in the human genome (33, 37). This observation in conjunction with the increase in IMPDH type I intron size suggests that the gene duplication event originated with the IMPDH type II gene. In contrast, it is also possible, due to the expanded intron sizes of the IMPDH type I gene, that this gene represents an older version of the two genes that were independently derived by gene duplication from a no longer existing primordial IMPDH gene. It has also been demonstrated that the IMPDH type I and type II genes are widely distributed, from primate to avian species, but are absent in yeast (33, 127), suggesting that a gene duplication event might have occurred in early eukaryotes. IMPDH cDNAs sequenced from a variety of nonmammalian sources, including *Drosophila melanogaster* (GenBank L14847), *Trypanosoma brucei* (GenBank M97794), *E. coli* (154), and *B. subtilis* (77), appear on phylogenetic analysis to be equally divergent from the mammalian IMPDH type I and type II coding regions, precluding definitive conclusions on the evolutionary events that led to the existence of the two highly conserved IMPDH genes. These observations do, however, suggest that there is a strong evolutionary selection for preservation of genomic structure and ultimately protein function for both IMPDH genes.

Although the reasons for conservation of both genes remain speculative, two major hypotheses should be explored. First, it is possible that differences in tissue-specific or stage of development-specific regulation of expression may be required. Although the relative expression of IMPDH type I and type II has not been examined at different stages of embryogenesis, there is little evidence for strict tissue-specific differences in expression of the type II gene in human tissues (33, 127). In contrast, expression of IMPDH type I from upstream promoters P1 and P2 appears to be highly tissue specific (58). It will be important to explore developmental requirements for the expression of each gene and attempts are in progress to engineer functional IMPDH type

I and type II gene knock-outs in mice. Secondly, it has been suggested that peripheral blood T lymphocytes rely more specifically on the *de novo* synthesis of guanine nucleotides. Because these cells appear to be the most sensitive to inhibition of IMPDH activity, it is conceivable that the increased expression of IMPDH type II during cellular proliferation is a critical prerequisite for cellular division in these and selected other cells. Thus, the ability to increase rapidly IMPDH activity in cells dependent on *de novo* guanine nucleotide biosynthesis would require the expression of a second gene that can be growth regulated.

V. IMPDH Activity and Cellular Signaling

The cellular protein p53 serves as a tumor suppressor and functions by initiating G_1 arrest in cells that are subjected to DNA damage. A proposed role for p53 in the regulation of guanine nucleotide biosynthesis resulted from the demonstration that down-regulation of IMPDH enzymatic activity occurred with increased p53 expression (129). Addition of the guanine nucleotide precursor, xanthosine, overcame the growth inhibitory effect of p53 expression, further supporting a role for guanine nucleotides in growth inhibition mediated by p53. A more recent study demonstrated that increased p53 expression induced a switch from exponential growth kinetics to the linear division kinetics observed in renewing stem cells *in vivo* (130). The addition of xanthosine prevented the switch to linear cell division kinetics, suggesting the presence of a functional link between guanine nucleotide metabolism, renewal stem cell growth kinetics, and the expression of p53. These results, if reproducible, would implicate the purine nucleotide biosynthetic pathway as a cellular target for the antiproliferative action of p53. More recent studies on the effects of ribonucleotide depletion demonstrated that inhibition of IMPDH with MPA results in a reversible G_0/G_1 cell cycle arrest in normal human fibroblasts. The arrest coincided with prolonged induction of p53 and the Cdk inhibitor p21 and with dephosphorylation of pRb. This G_0/G_1 arrest is p53 dependent because cells lacking this protein progress through to S phase in the presence of MPA, suggesting that p53 could serve as a metabolite sensor that is activated by the depletion of ribonucleotides and induce a quiescent state in cells that is reversible by the addition of exogenous guanine (87). In contrast, studies on the effect of MPA on T lymphocyte activation demonstrate cell cycle inhibition in early to mid-G_1 that is associated with inhibition of pRb phosphorylation, but little p53 or p21 expression (179). Instead, cell cycle arrest is associated with a lack of reduction in the Cdk inhibitor p27 and with a lack of induction of cyclin D3/Cdk6 kinase activity. These results, in conjunction with those of Linke *et al.* (87), suggest that there may be cell type-specific mechanisms for monitoring guanine nucleotide pools, but that depletion of these pools may lead to inhibition of cell cycle progression in many cell types.

VI. *In Vitro* Effects of IMPDH Inhibition on the Immune System

IMPDH inhibitors have been demonstrated to have a more selective *in vitro* and *in vivo* antiproliferative effect on T and B lymphocyte response to a variety of stimuli (*41, 42*). Therapeutically attainable levels of the potent, uncompetitive and selective inhibitor, MPA, readily inhibit the proliferative response of human peripheral blood lymphocytes with an IC_{50} of less than 100 nM, without having antiproliferative activity against fibroblasts or endothelial cells (*6, 41*). The IMPDH type II enzyme, which predominates in proliferating B and T lymphocytes as well as in neoplastic cells (*98, 99*), has been reported to exhibit a 4.8-fold increase over the type I enzyme in sensitivity to inhibition by MPA (*20*), although no significant difference in K_i values was found in a second study by Hager *et al.* (*59*). The potential for differential sensitivity of the type I and type II enzymes to inhibition is an appealing concept that has been used to explain the relative lack of serious systemic side effects of IMPDH inhibitors in clinical trials (*36, 39, 140*), although a careful comparison of the kinetic determinants of a series of inhibitors has demonstrated no major differences between the two enzymes (*59*). It has also been suggested that lymphocytes preferentially rely on *de novo* purine synthesis for proliferation, whereas other cell types such as rat hepatoma cells clearly circumvent inhibition by utilizing the salvage pathways (*2, 3, 5, 40, 44*). Clearly, given the *in vitro* effects of IMPDH inhibition on T cell activation, including inhibition of cell cycle progression in G_1 and its reversibility with guanine (*179*), the expansion of the guanine ribo- and deoxyribonucleotide pools through *de novo* synthesis is an essential requirement for lymphocyte proliferative responses. In support of this hypothesis are the observations that two of the IMPDH type I mRNA transcripts (4.0 and 2.5 kb) as well as the 1.9-kb type II mRNA are up-regulated by more than 10-fold following stimulation of human peripheral blood T lymphocytes with phorbol ester and the calcium ionophore, ionomycin (*32, 58*). These increases in mRNA levels are transcriptionally mediated and correspond to a 15-fold increase in total cellular IMPDH activity and a 6-fold increase in GTP levels over a 72-hr period of stimulation (*32*), strongly supporting the view that both gene products contribute to the observed increase in guanine nucleotide biosynthesis in activated T lymphocytes.

Several studies have addressed whether inhibitors of IMPDH might affect other pathways involved in lymphocyte activation. In contrast to cyclosporine, which inhibits IL-2 production, inhibitors of IMPDH have not been demonstrated to interfere with cytokine synthesis (100, 158), and the effects of IMPDH inhibitors are additive to those of cyclosporine in inhibiting activation *in vitro* (158). In addition, there is a complete lack of specificity of these inhibitors for the pathway of T cell activation, be it initiated through anti-CD3 in the presence or absence of anti-CD28, a response to irradiated allogeneic cells, or activation with phorbol ester and ionomycin (158), suggesting that the initial signal transduction pathway is not of importance in their mechanism of action. Finally, there have been reported to be effects on the glycosylation of membrane proteins by several investigators, the *in vivo* significance of which have not been determined. (7, 137).

VII. Clinical Applications for IMPDH Inhibitors

Several IMPDH inhibitors have been used clinically and two—mycophenolate mofetil (MMF), the prodrug morpholinoethyl ester of MPA, and mizoribine or bredninin—have been found to be very effective immunosuppressive drugs in the setting of renal transplantation and, to a more limited extent, in autoimmune disease. MMF has been demonstrated to prolong the survival of allografts and to reverse acute rejection in animal and human studies (36, 96, 117, 140) and is effective both in the prevention of rejection and in the treatment of ongoing rejection (30, 61, 96, 116). This drug also exerts a preventive effect on the development and progression of proliferative vasculopathy, which is a critical pathologic lesion in chronic rejection (4, 96). In phase I clinical trials (31), cadaveric renal transplant recipients receiving standard doses of cyclosporine and prednisone combined with increasing doses of MMF (100–3500 mg/day) exhibited an inverse correlation between the incidence of acute rejection episodes and the dose of MMF (141), as well as stabilization or improvement in renal function in nearly 70% of patients suffering refractory rejection following treatment with high-dose steroids and OKT3 (36). Phase III clinical trials using MMF for the prevention of renal allograft rejection have demonstrated a significant decrease in allograft rejection in regimens containing MMF (138, 139). Broader clinical use of MMF may be complicated by drug toxicity (31, 115, 128, 139), and has resulted in efforts to increase drug efficacy and decrease toxic side effects.

Mizoribine is the most potent of the competitive inhibitors of IMPDH identified to date. This drug is activated to its corresponding monophosphate by the enzyme adenosine kinase. Although an effective inhibitor of T cell activation *in vitro*, its effects at high concentration are not completely reversible with guanine, implying a second mechanism of action in this concentration range (34, 35). This drug is in widespread use in Japan as a primary immunosuppressive agent in human renal transplantation (9, 14, 84, 94, 150, 151) and has also been found to be effective in certain autoimmune disorders, such as lupus nephritis. As suggested by *in vitro* studies (158), it has a strong synergistic effect when used in combination with cyclosporine to prolong the survival of canine renal (8, 54, 55, 111, 113), and rat cardiac and lung allografts (147).

IMPDH inhibitors also have potential in the treatment of malignant and viral diseases. Tiazofurin, a synthetic nucleoside analog of the antiviral agent ribavirin, has proved to be an effective agent in the treatment of refractory granulocytic leukemias in clinical trials and has resulted in decreased numbers of blasts and increased differentiation of leukemic cells in patients with chronic granulocytic leukemia in blast crisis (156, 157). Ribavirin has potent antiviral activity (133, 169), whereas tiazofurin has more widespread antineoplastic activity (122). Selenazofurin, a seleno analog of tiazofurin, has demonstrated effectiveness against both viral infections and animal tumors (81, 142, 144). Finally, the IMPDH inhibitor 5-ethynyl-1-β-D-ribofuranosylimidazole-4-carboxamide 5′-monophosphate, the active metabolite of EICAR, has potent antiviral and antileukemic activity (161). The wide spectrum of clinical applications of IMPDH inhibitors and in particular their potent activity against the immune system holds tremendous promise for future drug development and for exploration of the molecular mechanisms underlying these effects.

VIII. Summary

In conclusion, work from our laboratory and from others has identified IMPDH as an important pharmacologic target. The provision of guanine nucleotides through *de novo* purine biosynthesis is clearly critical in the biology of the immune system and in the regulation of cellular differentiation. The two enzymes that comprise total cellular IMPDH activity are regulated very differently and the complexities that underlie the transcriptional regulation of each gene will have potential relevance for differentially manipulating their expression. Although the type I enzyme has not been crystallized, the kinetic data obtained with purified recombinant enzymes indicate that it will be very difficult to obtain competitive or uncompetitive inhibitors that will be truly selective for either of the isoenzymes. Nevertheless, the lack of widespread serious side effects of IMPDH inhibitors used clinically and in animal models indicate that the biology of certain elements of the immune system in general, and possibly of T lymphocytes in particular, renders them highly susceptible to guanine nucleotide depletion. Specific elucidation of the nature of this cell type-specific response, possibly a sensing mechanism involving GTP as a phosphate donor or important cofactor, could lead to important insights into mechanisms of immunomodulation and the regulation of cell differentiation.

Acknowledgments

This work was supported by NIH Grants CA64192 and CA34085.

References

1. A. C. Allison, S. J. Almquist, C. D. Muller, and E. M. Eugui, *Transplant. Proc.* **23** (Suppl. 2), 10 (1991).
2. A. C. Allison and E. M. Eugui, *Springer Sem. Immunopathol.* **14**, 353 (1993).
3. A. C. Allison and E. M. Eugui, *Clin. Trnasplant.* **7**, 96 (1993).
4. A. C. Allison and E. M. Eugui, *Transplant. Proc.* **26**, 3205 (1994).
5. A. C. Allison, T. Hovi, R. W. E. Watts, and A. D. B. Webster, *Lancet* **2**, 1179 (1975).
6. A. C. Allison, W. J. Kowalski, C. D. Muller, and E. M. Eugui, *Ann. N.Y. Acad. Sci.* **696**, 63 (1993).
7. A. C. Allison, W. J. Kowalski, C. J. Muller, R. V. Waters, and E. M. Eugui, *Transplant. Proc.* **25** (Suppl. 2), 67 (1993).
8. H. Amemiya, S. Suzuki, S. Niiya, H. Watanabe, and T. Kotake, *Transplantation*, **46**, 768 (1988).
9. H. Amemiya, S. Suzuki, H. Watanabe, R. Hayashi, and S. Niiya, *Transplant. Proc.* **21**, 956 (1989).
10. J. H. Anderson and A. C. Sartorelli, *Fed. Proc.* **26**, 730 (1967).
11. J. H. Anderson and A. C. Sartorelli, *J. Biol. Chem.* **243**, 4762 (1968).
12. J. H. Anderson and A. C. Sartorelli, *Biochem. Pharm.* **18**, 2737 (1969).
13. L. C. Antonio, K. Straub, and J. C. Wu, *Biochemistry* **33**, 1760 (1994).
14. K. Aso, H. Uchida, K. Sato, K. Yokota, T. Osakabe, Y. Nakayama, M. Ohkubo, K. Kumano, T. Endo, K. Koshiba, K. Watanabe, and N. Kashiwagi, *Transl. Proc.* **19**, 1955–1958.
15. S. Bacus, K. Kiguchi, D. Chin, C. R. King, and E. Huberman, *Mol. Carcinogen.* **3**, 350 (1990).
16. J. T. Beck, S. Zhao, and C. C. Wang, *Exp. Parasitol.* **78**, 101 (1994).
17. H. J. Becker and G. W. Löhr, *Klin. Wochenschr.* **57**, 1109 (1979).
18. P. Bork and C. Grunwald, *Eur. J. Biochem.* **191**, 347 (1990).
19. L. W. Brox and A. Hampton, *Biochemistry* **7**, 2589 (1968).
20. S. F. Carr, E. Papp, J. C. Wu, and Y. Natsumeda, *J. Biol. Chem.* **268**, 27286 (1993).
21. M. B. Cohen, *Somat. Cell Mol. Genet.* **13**, 627 (1987).
22. M. B. Cohen, J. Maybaum, and W. Sadée, *J. Chromatog.* **198**, 435 (1980).
23. M. B. Cohen, J. Maybaum, and W. Sadée, *J. Biol. Chem.* **256**, 8713 (1981).
24. M. B. Cohen and W. Sadée, *Cancer Res.* **43**, 1587 (1983).
25. F. R. Collart, C. B. Chubb, B. L. Mirkin, and E. Huberman, *Cancer Res.* **52**, 5826 (1992).
26. F. R. Collart and E. Huberman, *J. Biol. Chem.* **263**, 15769 (1988).
27. F. R. Collart and E. Huberman, *Blood* **75**, 570 (1990).

28. D. A. Cooney, Y. Wilson, and E. McGee, *Anal. Biochem.* **130**, 339 (1983).
29. G. W. Crabtree and J. F. Henderson, *Cancer Res.* **31**, 985 (1971).
30. A. M. D'Alessandro, M. Rankin, J. McVey, G. R. Hafez, H. W. Sollinger, M. Kalayoglu, and F. O. Belzer, *Transplant. Proc.* **55**, 695 (1993).
31. G. M. Danovitch, *Kidney Int.* **48** (Suppl. 52), S93 (1995).
32. J. S. Dayton, T. Lindsten, C. B. Thompson, and B. S. Mitchell, *J. Immunol.* **152**, 984 (1994).
33. J. S. Dayton and B. S. Mitchell, *Biochem. Biophys. Res. Commun.* **195**, 897 (1993).
34. J. S. Dayton and B. S. Mitchell, in "Recent Developments in Transplantation Medicine" (H. Sollinger, ed.), p. 129. Physicians & Scientists Publishing Co., Inc., Glenview, Illinois, 1994.
35. J. S. Dayton, L. A. Turka, C. B. Thompson, and B. S. Mitchell, *Mol. Pharmacol.* **41**, 671 (1992).
36. M. H. Deierhoi, H. W. Sollinger, A. G. Diethelm, F. O. Belzer, and R. S. Kauffman, *Transplant. Proc.* **25**, 693 (1993).
37. N. A. Doggett, D. F. Callen, Z. L. Chen, S. Moore, J. G. Tesmer, L. A. Duesing, and R. L. Stallings, *Genomics* **18**, 687 (1993).
38. D.-S. Duan and W. Sadee, *Cancer Res.* **47**, 4047 (1987).
39. W. W. Epinette, C. M. Parker, E. L. Jones, and M. C. Greist, *J. Am. Acad. Dermatol.* **17**, 962 (1987).
40. E. M. Eugui and A. C. Allison, *Ann. N.Y. Acad. Sci.* **685**, 309 (1993).
41. E. M. Eugui, S. Almquist, C. D. Muller, and A. C. Allison, *Scand. J. Immunol.* **33**, 161 (1991).
42. E. M. Eugui, A. Mirkovich, and A. C. Allison, *Scand. J. Immunol.* **33**, 175 (1991).
43. W. N. Fishbein, V. W. Armbrustmacher, and J. L. Griffen, *Science* **200**, 545 (1978).
44. T. J. Franklin and J. M. Cook, *Biochem. J.* **113**, 515 (1969).
45. R. C. Garcia, P. Leoni, and A. C. Allison, *Biochem. Biophys. Res. Commun.* **77**, 1067 (1977).
46. E. R. Giblett, A. J. Ammann, D. W. Wara, R. Sandman, and L. K. Diamond, *Lancet* **1**, 1010 (1975).
47. E. R. Giblett, J. E. Anderson, F. Cohen, B. Pollara, and H. J. Meurwissen, *Lancet* **2**, 1067 (1972).
48. H. J. Gilbert and W. T. Drabble, *Biochem. J.* **191**, 533 (1980).
49. H. J. Gilbert, C. R. Lowe, and W. T. Drabble, *Biochem. J.* **183**, 481 (1979).
50. D. Glesne, F. Collart, T. Varkony, H. Drabkin, and E. Huberman, *Genomics* **16**, 274 (1993).
51. D. A. Glesne, F. R. Collart, and E. Huberman, *Mol. Cell. Biol.* **11**, 5417 (1991).
52. D. A. Glesne and E. Huberman, *Biochem. Biophys. Res. Commun.* **205**, 537 (1994).
53. M. F. Goldsmith, *J. Am. Med. Assoc.* **263**, 1184 (1990).
54. C. R. Gregory, I. M. Gourley, G. R. Cain, T. W. Broaddus, L. D. Cowgill, N. H. Willits, J. D. Patz, and G. Ishizaki, *Transplantation* **45**, 856 (1988).
55. C. R. Gregory, I. M. Gourley, S. C. Haskins, G. R. Cain, G. Ishizaki, L. D. Cowgill, N. H. Willits, and J. D. Patz, *Am. J. Vet. Res.* **49**, 305 (1988).
56. H. E. Gruber, I. Jansen, R. C. Willis, and J. E. Seegmiller, *Biochim. Biophys. Acta* **846**, 135 (1985).
57. J. J. Gu, K. Kaiser-Rogers, K. Rao, and B. S. Mitchell, *Genomics* **24**, 179 (1994).
58. J. J. Gu, J. Spychala, and B. S. Mitchell, *J. Biol. Chem.* **272**, 4458 (1997).
59. P. W. Hager, F. R. Collart, E. Huberman, and B. S. Mitchell, *Biochem. Pharm.* **49**, 1323 (1995).
60. A. Hampton and A. Nomura, *Biochemistry* **6**, 679 (1967).
61. L. Hao, K. J. Lafferty, A. C. Allison, and E. M. Eugui, *Transplant. Proc.* **22**, 1659 (1990).

62. M. S. Hershfield and J. E. Seegmiller, *J. Biol. Chem.* **251,** 7348 (1976).
63. S. D. Hodges, E. Fung, D. J. McKay, B. S. Renaux, and F. F. Snyder, *J. Biol. Chem.* **264,** 18137 (1989).
64. E. W. Holmes, J. A. McDonald, J. M. McCord, J. B. Wyngaarden, and W. N. Kelley, *J. Biol. Chem.* **248,** 144 (1973).
65. E. W. Holmes, D. M. Pehlke, and W. N. Kelly, *Biochim. Biophys. Acta* **364,** 209 (1974).
66. T. Hovi, A. C. Allison, and J. Allsop, *FEBS Lett.* **55,** 291 (1975).
67. T. Hovi, A. C. Allison, O. Raivio, and A. Vaheri, *in* "Purine/Pyrimidine Metabolism," p. 225. Ciba Found. Symp., 1977.
68. T. Hovi, J. F. Smyth, A. C. Allison, and S. C. Williams, *Clin. Exp. Immunol.* **23,** 395 (1976).
69. E. Huberman, C. K. McKeown, and J. Friedman, *Proc. Natl. Acad. Sci. U.S.A.* **78,** 3151 (1981).
70. J. A. Huete-Perez, J. C. Wu, F. G. Whitby, and C. C. Wang, *Biochemistry* **34,** 13889 (1995).
71. D. J. Hupe, B. A. Azzolina, and N. D. Behrens, *J. Biol. Chem.* **261,** 8363 (1986).
72. T. Ikegami, Y. Natsumeda, and G. Weber, *Life Sci.* **40,** 2277 (1987).
73. O. Itoh, S. Kuroiwa, S. Atsumi, K. Umezawa, T. Takeuchi, and M. Hori, *Cancer Res.* **49,** 996 (1989).
74. R. C. Jackson, M. S. Lui, T. J. Boritzki, H. P. Morris, and G. Weber, *Cancer Res.* **40,** 1286 (1980).
75. R. C. Jackson, H. P. Morris, and G. Weber, *Biochem. J.* **166,** 1 (1977).
76. R. C. Jackson, G. Weber, and H. P. Morris, *Nature (London)* **256,** 331 (1975).
77. N. Kanzaki and K. Miyagawa, *Nucleic Acids Res.* **18,** 6710 (1990).
78. S. M. Kharbanda, M. L. Sherman, D. R. Spriggs, and D. W. Kufe, *Cancer Res.* **48,** 5965 (1988).
79. K. Kiguchi, F. R. Collart, C. Henning-Chubb, and E. Huberman, *Exp. Cell Res.* **187,** 47 (1990).
80. K. Kiguchi, F. R. Collart, C. Henning-Chubb, and E. Huberman, *Cell Growth Differ.* **1,** 259 (1990).
81. J. J. Kirsi, J. A. North, P. A. McKernan, B. N. Murray, P. G. Canonico, J. W. Huggins, P. C. Srivastava, and R. K. Robins, *Antimicrob. Agents Chemother.* **24,** 353 (1983).
82. R. D. Knight, J. Mangum, D. L. Lucas, D. A. Cooney, E. C. Khan, and D. G. Wright, *Blood* **69,** 634 (1987).
83. Y. Konno, Y. Natsumeda, M. Nagai, Y. Yamaji, S. Ohno, K. Suzuki, and G. Weber, *J. Biol. Chem.* **266,** 506 (1991).
84. R. Kusaba, O. Otubo, H. Sugimoto, I. Takahashi, Y. Yamada, J. Yamauchi, N. Akiyama, and T. Inou, *Proc. Eur. Dial. Transplant. Assoc.* **18,** 420 (1981).
85. M. Lesch and W. L. Nyhan, *Am. J. Med.* **36,** 561 (1964).
86. J. O. Link and K. Staub, *J. Am. Chem. Soc.* **118,** 2091 (1996).
87. S. P. Linke, K. C. Clarkin, A. Di Leonardo, A. Tsou, and G. M. Wahl, *Genes Dev.* **10,** 934 (1996).
88. J. K. Lowe, L. Brox, and J. F. Henderson, *Cancer Res.* **37,** 736 (1977).
89. J. M. Lowenstein, *Physiol. Rev.* **52,** 383 (1972).
90. D. L. Lucas, R. K. Robins, R. D. Knight, and D. G. Wright, *Biochem. Biophys. Res. Commun.* **115,** 971 (1983).
91. D. L. Lucas, H. K. Webster, and D. G. Wright, *J. Clin. Invest.* **72,** 1889 (1983).
92. M. S. Lui, M. A. Faderan, J. J. Liepnieks, Y. Natsumeda, E. Olah, H. N. Jayaram, and G. Weber, *J. Biochem.* **259,** 5078 (1984).
93. F. Malinoski and V. Stollar, *Virology* **110,** 281 (1981).
94. F. Marumo, M. Okubo, K. Yokota, H. Uchida, K. Kumano, T. Endo, K. Watanabe, and N. Kashiwagi, *Transplant. Proc.* **20,** 406 (1988).

95. J. Maybaum, F. K. Klein, and W. Sadée, *J. Chromatog.* **188,** 149 (1980).
96. R. E. Morris, E. G. Hoyt, M. P. Murphy, E. M. Eugui, and A. C. Allison, *Transplant. Proc.* **22,** 1659 (1990).
97. R. E. Morris, J. Wang, J. R. Blum, T. Flavin, M. P. Murphy, S. J. Almquist, N. Chu, Y. L. Tam, M. Kaloostian, A. C. Allison, and E. M. Eugui, *Transplant. Proc.* **23** (Suppl. 2), 19 (1991).
98. M. Nagai, Y. Natsumeda, Y. Konno, R. Hoffman, S. Irino, and G. Weber, *Cancer Res.* **51,** 3886 (1991).
99. M. Nagai, Y. Natsumeda, and G. Weber, *Cancer Res.* **52,** 258 (1992).
100. S. E. Nagy, J. P. Andersson, and U. G. Anderrson, *Immunopharmacology* **26,** 11 (1993).
101. H. Nakamura, Y. Natsumeda, N. Nagai, J. Takahara, S. Irino, and G. Weber, *Leuk. Res.* **16,** 561 (1992).
102. Y. Natsumeda and S. F. Carr, *Ann. N.Y. Acad. Sci.* **696,** 88 (1993).
103. Y. Natsumeda, T. Ikegami, K. Murayama, and G. Weber, *Cancer Res.* **48,** 507 (1988).
104. Y. Natsumeda, T. Ikegami, E. Olah, and G. Weber, *Cancer Res.* **49,** 88 (1989).
105. Y. Natsumeda, S. Ohno, H. Kawasaki, Y. Konno, G. Weber, and K. Suzuki, *J. Biol. Chem.* **265,** 5292 (1990).
106. Y. Natsumeda, N. Prajda, J. P. Donohue, J. L. Glober, and G. Weber, *Cancer Res.* **44,** 2475 (1984).
107. P. H. Nelson, E. Eugui, C. C. Wang, and A. C. Allison, *J. Med. Chem.* **33,** 833 (1990).
108. B. T. Nguyen and W. Sadée, *Biochem. J.* **234,** 263 (1986).
109. E. Nimmesgern, T. Fox, M. A. Fleming, and J. A. Thomson, *J. Biol. Chem.* **271,** 19421 (1996).
110. M. Okada, K. Shimura, H. Shiraki, and H. Nakagawa, *J. Biochem. (Tokyo)* **94,** 1065 (1983).
111. M. Okubo, Y. Masaki, K. Kamata, N. Sato, K. Inoue, and N. Umetani, *Transplant. Proc.* **21,** 1085 (1989).
112. E. Olah, Y. Natsumeda, T. Ikegami, Z. Kote, M. Horanyi, J. Szelenyi, E. Paulik, T. Kremmer, S. R. Hollan, J. Sugar, and G. Weber, *Proc. Natl. Acad. Sci. U.S.A.* **85,** 6533 (1988).
113. T. Osakabe, H. Uchida, Y. Masaki, K. Sato, Y. Nakayama, M. Ohkubo, K. Kumano, T. Endo, K. Watanabe, and K. Aso, *Transplant. Proc.* **21,** 1598 (1989).
114. D. M. Pehlke, J. A. McDonald, E. W. Holmes, and W. N. Kelley, *J. Clin. Invest.* **51,** 1398 (1972).
115. R. Pichlmayr, *Lancet* **345,** 1321 (1995).
116. K. P. Platz, W. O. Bechstein, D. Eckhoff, Y. Suzuki, and H. W. Sollinger, *Surgery* **110,** 736 (1991).
117. K. P. Platz, H. W. Sollinger, D. A. Hullett, D. E. Eckhoff, E. M. Eugui, and A. C. Allison, *Transplant. Proc.* **51,** 27 (1991).
118. N. Pradja, H. P. Morris, and G. Weber, *Cancer Res.* **39,** 3909 (1979).
119. G. M. Price, A. V. Hoffbrand, M. R. Taheri, and J. P. M. Evans, *Leuk. Res.* **11,** 525 (1987).
120. R. T. Proffitt, V. K. Pathak, D. G. Villacorte, and C. A. Presant, *Cancer Res.* **43,** 1620 (1983).
121. G. H. Reem and C. Friend, *Science* **157,** 1203 (1967).
122. R. K. Robins, P. C. Srivastava, V. L. Narayanan, L. Plowman, and K. D. Pauli, *J. Med. Chem.* **25,** 107 (1982).
123. F. W. Scott and D. R. Forsdyke, *Biochem. J.* **190,** 721 (1980).
124. J. E. Seegmiller, *in* "Metabolic Control and Disease" (P. K. Bondy and L. E. Rosenberg, eds.). Saunders, Philadelphia, Pennsylvania, 1980.
125. J. E. Seegmiller, A. I. Grayzel, and L. Liddle, *Nature (London)* **183,** 1463 (1959).
126. J. E. Seegmiller, F. M. Rosenbloom, and W. N. Kelly, *Science* **155,** 1682 (1967).
127. M. Senda and Y. Natsumeda, *Life Sci.* **54,** 1917 (1994).
128. L. M. Shaw, H. W. Sollinger, P. Halloran, R. E. Morris, R. W. Yatscoff, J. Ransom, I. Tsina,

P. Keown, D. W. Holt, R. Lieberman, A. Jaklitsch, and J. Potter, *Therapeut. Drug Monitor.* **17**, 690 (1995).
129. J. L. Sherley, *J. Biol. Chem.* **266**, 24815 (1991).
130. J. L. Sherley, P. B. Stadler, and D. R. Johnson, *Proc. Natl. Acad. Sci. U.S.A.* **92**, 136 (1995).
131. K. Shimura, M. Okada, H. Shiraki, and H. Nakagawa, *J. Biochem. (Tokyo)* **94**, 1595 (1983).
132. Y. Sidi, C. Panet, L. Wasserman, A. Cyjon, A. Novogrodsky, and J. Nordenberg, *Br. J. Cancer* **58**, 61 (1988).
133. R. W. Sidwell, J. H. Huffman, G. P. Khare, J. T. Witkowski, L. B. Allen, and R. K. Robins, *Science* **177**, 705 (1972).
134. C. D. Sifri, K. Wilson, S. Smolik, M. Forte, and B. Ullman, *Biochim. Biophys. Acta* **1217**, 103 (1994).
135. M. D. Sintchak, M. A. Fleming, O. Futer, S. A. Raybuck, S. P. Chambers, P. R. Caron, M. A. Murcko, and K. P. Wilson, *Cell* **85**, 921 (1996).
136. F. F. Snyder, J. F. Henderson, and D. A. Cook, *Biochem. Pharm.* **21**, 2351 (1972).
137. J. A. Sokoloski, O. C. Blair, and A. C. Sartorelli, *Cancer Res.* **46**, 2314 (1986).
138. H. W. Sollinger, *Kidney Int.* **48**, S14 (1995).
139. H. W. Sollinger, *Transplantation* **60**, 225 (1995).
140. H. W. Sollinger, F. O. Belzer, M. H. Deierhoi, A. G. Diethelm, T. A. Gonwa, R. S. Kauffman, G. B. Klintmalm, S. V. McDiarmid, J. Roberts, J. T. Rosenthal, and S. J. Tomlanovich, *Transplant. Proc.* **25**, 698 (1993).
141. H. W. Sollinger, M. H. Deierhoi, F. O. Belzer, A. Diethelm, and R. S. Kauffman, *Transplantation* **53**, 428 (1992).
142. P. C. Srivastava and R. K. Robins, *J. Med. Chem.* **26**, 445 (1983).
143. M. M. Stayton, *Curr. Top. Cell. Regul.* **22**, 103 (1983).
144. D. G. Streeter and R. K. Robins, *Biochem. Biophys. Res. Commun.* **115**, 544 (1983).
145. D. G. Streeter, J. T. Witkowski, G. P. Khare, R. W. Sidwell, R. J. Bauer, R. K. Robins, and L. N. Simon, *Proc. Natl. Acad. Sci. U.S.A.* **70**, 1174 (1973).
146. L. Stryer, in "Biosynthesis of Nucleotides" (W. H. Freeman, ed.), p. 602. Biochemistry, New York, 1988.
147. S. Suzuki, T. Hijioka, I. Sakakibara, and H. Amemiya, *Transplantation* **43**, 743 (1987).
148. L. Sweetman and W. L. Nyhan, *Arch. Intern. Med.* **130**, 214 (1972).
149. T. Szekeres, M. Fritzer, K. Pillwein, T. Felzmann, and P. Chiba, *Life Sci.* **51**, 1309 (1992).
150. A. Tajima, M. Hata, N. Ohta, Y. Ohtawara, K. Suzuki, and Y. Aso, *Transplantation* **38**, 116 (1984).
151. S. Takahara, Y. Fukunishi, Y. Kokado, Y. Ichikawa, M. Ishibashi, S. Nagano, and T. Sonoda, *Transplant. Proc.* **20**, 147 (1988).
152. M. S. Thomas and W. T. Drabble, *Gene* **36**, 45 (1985).
153. W. H. S. Thompson and I. Smith, *Metabolism* **27**, 151 (1978).
154. A. A. Tiedeman and J. M. Smith, *Nucleic Acids Res.* **13**, 1303 (1985).
155. A. A. Tiedeman and J. M. Smith, *Gene* **97**, 289 (1991).
156. G. J. Tricot, H. N. Jayaram, E. Lapis, Y. Natsumeda, C. R. Nichols, P. Kneebone, N. Heerema, and G. Weber, *Cancer Res.* **49**, 3696 (1989).
157. G. J. Tricot, H. N. Jayaram, C. R. Nichols, K. Pennington, E. Lapis, G. Weber, and R. Hoffman, *Cancer Res.* **47**, 4988 (1987).
158. L. A. Turka, J. Dayton, S. Sinclair, C. B. Thompson, and B. S. Mitchell, *J. Clin. Invest.* **87**, 940 (1991).
159. R. Verham, T. D. Meek, L. Hedstrom, and C. C. Wang, *Mol. Biochem. Parasitol.* **24**, 1 (1987).
160. C. C. Wang, R. Verham, H.-W. Chen, A. Rice, and A. L. Wang, *Biochem. Pharm.* **33**, 1323 (1984).

161. W. Wang, V. V. Papov, N. Minakawa, A. Matsuda, K. Biemann, and L. Hedstrom, *Biochemistry* **35,** 95 (1996).
162. G. Weber, *Cancer Res.* **43,** 3466 (1983).
163. G. Weber, M. S. Lui, Y. Natsumeda, and M. A. Faderan, *Adv. Enz. Regul.* **21,** 53 (1983).
164. G. Weber, E. Olah, J. E. Denton, M. S. Lui, E. Takeda, D. Y. Tzeng, and J. Ban, *Adv. Enz. Regul.* **19,** 87 (1981).
165. G. Weber, N. Prajda, and R. C. Jackson, *Adv. Enz. Regul.* **14,** 3 (1976).
166. R. H. Williams, D. H. Lively, D. C. DeLong, J. C. Cline, M. J. Sweeney, G. A. Poore, and S. H. Larsen, *J. Antibiot. (Tokyo)* **21,** 463 (1968).
167. R. C. Willis and J. E. Seegmiller, in "Purine Metabolism in Man. III" (A. Rapada, R. W. E. Watts, and C. H. M. M. De Bruyn, eds.), p. 237. Plenum, New York, 1980.
168. K. Wilson, F. R. Collart, E. Huberman, J. R. Stringer, and B. Ullman, *J. Biol. Chem.* **266,** 1665 (1991).
169. J. T. Witkowski, R. K. Robins, R. W. Sidwell, and L. N. Simon, *J. Med. Chem.* **14,** 1150 (1972).
170. A. W. Wood and J. E. Seegmiller, *J. Biol. Chem.* **248,** 138 (1973).
171. B. Xiang, J. C. Taylor, and G. D. Markham, *J. Biol. Chem.* **271,** 1435 (1996).
172. Y. Yamada, Y. Natsumeda, and G. Weber, *Biochemistry* **27,** 2193 (1988).
173. Y. Yamaji, Y. Natsumeda, Y. Yamada, S. Irino, and G. Weber, *Life Sci.* **46,** 435 (1989).
174. J. Yu, V. Lemas, T. Page, J. D. Connor, and A. L. Yu, *Cancer Res.* **49,** 5555 (1989).
175. A. G. Zimmermann, J. Spychala, and B. S. Mitchell, *J. Biol. Chem.* **270,** 6808 (1995).
176. A. G. Zimmermann, K. L. Wright, J. P.-Y. Ting, and B. S. Mitchell, *J. Biol. Chem.* **272,** 22913 (1997).
177. E. Zoref, A. De Vries, and O. Sperling, *J. Clin. Invest.* **56,** 1093 (1975).
178. E. Zoref-Shani, R. Lavie, Y. Bromberg, E. Beery, Y. Sidi, O. Sperling, and J. Nordenberg, *J. Cancer Res. Clin. Oncol.* **120,** 717 (1994).
179. J. Laliberté, A. Yee, Y. Xiong, and B. S. Mitchell, *Blood* **91,** 2896 (1998).

Structure and Function Analysis of *Pseudomonas* Plant Cell Wall Hydrolases

GEOFFREY P. HAZLEWOOD*
AND HARRY J. GILBERT[†]

*Laboratory of Molecular Enzymology
The Babraham Institute
Babraham, Cambridge CB2 4AT
United Kingdom
[†]Department of Biological & Nutritional
  Sciences
University of Newcastle upon Tyne
Newcastle upon Tyne NE1 7RU
United Kingdom

I. Introduction	212
II. Composition and Structure of Plant Cell Walls	213
A. Cellulose	213
B. Hemicellulose	214
C. Other Polysaccharides	215
III. Plant Cell Wall Hydrolases	215
A. Cellulases	216
B. Xylan Degrading Enzymes	216
C. Mannan-Degrading Enzymes	217
D. Arabinan- and Galactan-Degrading Enzymes	217
IV. *Pseudomonas fluorescens* subsp. *cellulosa*	217
V. Architecture of *Pseudomonas* Plant Cell Wall Hydrolases	218
A. Cellulases	218
B. Xylan-Degrading Enzymes	219
C. Other Hydrolases	221
VI. Function of Noncatalytic Domains	222
A. Cellulose-Binding Domains	223
B. NodB Domain	229
VII. Structures and Catalytic Mechanisms of *Pseudomonas* Plant Cell Wall Hydrolases	231
A. Xylanase A	233
B. Mannanase A	237
C. Galactanase A	238
References	239

Hydrolysis of the major structural polysaccharides of plant cell walls by the aerobic soil bacterium *Pseudomonas fluorescens* subsp. *cellulosa* is attributable to the production of multiple extracellular cellulase and hemicellulase enzymes, which are the products of distinct genes belonging to multigene families. Cloning and sequencing of individual genes, coupled with gene sectioning and functional analysis of the encoded proteins have provided a detailed picture of structure/function relationships and have established the cellulase-hemicellulase system of *P. fluorescens* subsp. *cellulosa* as a model for the plant cell wall degrading enzyme systems of aerobic cellulolytic bacteria. Cellulose- and xylan-degrading enzymes produced by the pseudomonad are typically modular in structure and contain catalytic and noncatalytic domains joined together by serine-rich linker sequences. The cellulases include a cellodextrinase; a β-glucan glucohydrolase and multiple endoglucanases, containing catalytic domains belonging to glycosyl hydrolase families 5, 9, and 45; and cellulose-binding domains of families II and X, both of which are present in each enzyme. Endo-acting xylanases, with catalytic domains belonging to families 10 and 11, and accessory xylan-degrading enzymes produced by *P. fluorescens* subsp. *cellulosa* contain cellulose-binding domains of families II, X, and XI, which act by promoting close contact between the catalytic domain of the enzyme and its target substrate. A domain homologous with NodB from rhizobia, present in one xylanase, functions as a deacetylase. Mannanase, arabinanase, and galactanase produced by the pseudomonad are single domain enzymes. Crystallographic studies, coupled with detailed kinetic analysis of mutant forms of the enzyme in which key residues have been altered by site-directed mutagenesis, have shown that xylanase A (family 10) has 8-fold α/β barrel architecture, an extended substrate-binding cleft containing at least six xylose-binding pockets and a calcium-binding site that protects the enzyme from thermal inactivation, thermal unfolding, and attack by proteinases. Kinetic studies of mutant and wild-type forms of a mannanase and a galactanase from *P. fluorescens* subsp. *cellulosa* have enabled the catalytic mechanisms and key catalytic residues of these enzymes to be identified. © 1998 Academic Press

I. Introduction

Plant biomass constitutes the largest reservoir of organic carbon in the biosphere; more than half of that carbon is contained in the cell wall, a structurally complex insoluble matrix composed predominantly of the polysaccharides cellulose and hemicellulose, with variable amounts of lignin. It has been estimated that 10^{11} tonnes of plant biomass are generated annually as a result of photosynthesis (1). Maintenance of the balance of the carbon cycle involves the continuous degradation of plant biomass and the eventual return of carbon to the atmosphere in the form of carbon dioxide. Microorganisms that break down the structural polysaccharides of plant cell walls have a key role in this important biological process. They include aerobic and anaerobic bacteria and fungi and can be found in most habitats. As a major natural depository for plant biomass, the soil is one of the main sites of degradation of plant tissue and is a source of many different plant cell wall-degrading bacteria and fungi (2). Within this ecosystem, microorganisms have adapted to an environment in which their main source of fermentable carbon is an abundant supply of plant cell walls, by evolving enzyme systems that efficiently degrade the major structural polysaccharides cellulose and hemicellulose, thus ensuring for themselves a supply of sugars that can be utilized to provide energy and carbon precursors for microbial growth. As well as playing a crucial role in the natural world, these microorganisms have more recently become a recognized source of enzymes that have novel and valuable uses in biotechnology (3, 4).

II. Composition and Structure of Plant Cell Walls

Much has been written about the composition and structure of plant cell walls (see, for example Refs. 5 and 6), but for the purpose of this review a somewhat simplified overview should be sufficient to highlight those aspects most relevant to the enzymology of cell wall breakdown. Chemical analysis indicates that plant cell walls contain mainly cellulose and hemicellulose, with a variable amount of lignin. In reality, the native cell wall is not a simple mixture of these different components, but is a structurally complex framework of cellulose microfibrils embedded in a matrix of other polymers, including xylans, mannans, pectins, and glycoproteins (6). The physical nature of the cell wall owes much to the intermolecular interactions and extensive cross-linking that occurs between the different components. Examples include hydrogen bonding within and between the cellulose and hemicellulose moieties (7, 8), the formation of covalent linkages between lignin and hemicelluloses (9, 10) and the covalent cross-linking of wall glycoproteins (11). A coextensive network of pectins that is independent of the cellulose–hemicellulose network also contributes to the complexity of the cell wall but apparently does not affect its structural integrity (5). Notwithstanding the fact that plant cell walls are chemically and structurally complex, much of our knowledge of the biochemistry of plant cell wall hydrolases is based on studies carried out using relatively pure cellulose and hemicellulose substrates extracted from cell wall-rich plant material.

A. Cellulose

Cellulose may be simply described as an unbranched homopolymer of β-1,4-linked glucose residues; individual chains or microfibrils containing from 100 to 1400 glucose units form bundles or fibrils in which the cellulose molecules are aligned parallel to the long axis and are maintained in this state through interchain hydrogen bonding and van der Waal's interactions, resulting in highly ordered regions interspersed with less ordered amorphous regions. Fibrils of native cellulose (cellulose I) exhibiting a high degree of parallelism contain a predominance of crystalline regions, as, for example, in cotton and bacterial microcrystalline cellulose (BMCC).

In contrast, processed cellulose such as Avicel, which is derived from bleached wood pulp, contains both crystalline and amorphous forms, and phosphoric acid-swollen cellulose (PASC) is regarded as largely amorphous. This relatively simplistic view of cellulose structure has proved useful in advancing understanding of the mode of action of cellulases, but is has become increasingly apparent that future mechanistic studies will need to take greater account of the fact that cellulose structure is highly variable and rarely uniform. For example, it is well documented that variations in intra- and intermolecular hydrogen bonding can have a marked effect on secondary and tertiary structure, respectively, with the result that two stable secondary structures and at least six different crystal lattice types are possible for cellulose. For a more detailed treatment of cellulose structure the reader is referred to Tomme *et al.* (*12*).

B. Hemicellulose

The hemicellulose fraction of plant cell walls comprises a group of heteropolymers that includes xylans, mannans, and noncellullosic glucans such as xyloglucan. For most species of land plants, xylan is the main component of the hemicellulose fraction (*13*, *14*). It comprises a backbone of 1,4-linked β-D-xylopyranose units substituted with mainly acetyl, arabinofuranosyl, and glucuronosyl residues. The nature of the side groups and the frequency of substitution are dependent on source. Thus hardwood xylan is typically O-acetyl-4-O-methylglucuronoxylan with, on average, some 10% of xylose units α-1,2 linked to a 4-O-methylglucuronic acid side chain, and 70% of xylose residues acetylated at the C-2 or C-3 positions. Softwood xylans have a shorter average chain length than do hardwood xylans and are commonly arabino-4-O-methylglucuronoxylans in which more than 10% of backbone xylose residues are substituted with α-1,3-linked arabinofuranosyl side chains. Although less branched than hardwood xylans, they are more heavily substituted with 4-O-methylglucuronic acid side chains.

Grass xylans are also 4-O-methylglucuroxylans, but contain less glucuronosyl side chains and have a larger content of arabinofuranosyl side chains at C-2 and C-3 of the main chain xylose residues, many of which are substituted at position 5 with feruloyl or *p*-coumaroyl residues. Such xylans also contain up to 5% by weight of O-acetyl groups linked to C-2 or C-3 of the main chain xyloses. Other forms of xylan, ranging from those with highly branched structures involving combinations of arabinose, galactose, and glucuronic acid, to homoxylans devoid of side chains, have been described, but the acetylated xylan of hardwoods and the arabinoxylan of softwoods and grasses are the most abundant forms. Within the plant cell wall, cross-linking of xylan molecules and their covalent association with lignin are mediated by the phenolic acid substituents of arabinofuranosyl side chains (15, 16).

Polysaccharides containing mannose alone or mannose in conjunction with galactose and glucose occur in significant quantities in the hemicellulose fraction from softwoods, the seeds of leguminous plants, and a variety of nuts and beans (17, 18). The simplest mannan is a linear homopolymer composed of β-1,4-linked mannopyranosyl residues. Galactomannan has a similar backbone structure, but is substituted at varying frequency with α-1,6-linked galactopyranosyl side chains. Glucomannans comprise linear chains of randomly distributed β-1,4-linked mannopyranosyl and glucopyranosyl residues, sometimes substituted with α-1,6-linked galactopyranosyl side chains (galactoglucomannan). In addition to galactopyranosyl side chains the backbone mannose residues of the different mannans are frequently substituted with O-acetyl groups at the C-2 and C-3 positions (18).

C. Other Polysaccharides

Other polysaccharides that should be considered within the context of plant cell wall hydrolysis include arabinans and galactans, which have been loosely termed hemicelluloses but are now believed to be derived either from cell wall glycoproteins or from the "hairy" regions of galacturonan (17). Arabinan is a branched homopolymer comprising a backbone of α-1,5-linked L-arabinofuranosyl residues, substituted with α-1,2- and α-1,3-linked L-arabinofuranosyl side chains, and is often joined to rhamnopyranosyl units derived from rhamnogalacturonan, via an α-1,2 linkage (17, 19). Several different forms of galactan are recognized. The simplest is a linear polymer composed of β-1,4-linked galactopyranosyl residues (17, 20). Other forms include the more complex branched arabinogalactans that typically occur as side chains attached to a rhamnogalacturonan backbone (17, 21). Type I arabinogalactan comprises galactan substituted either with single arabinosyl residues or with arabinan side chains via α-1,3 linkages. Type II arabinogalactan has a backbone of β-1,3-linked galactopyranosyl residues, some of which are substituted with either β-1,6-linked galactopyranosyl units, with α-1,3-linked arabinofuranosyl residues, or with α-1,3-linked arabinan side chains.

III. Plant Cell Wall Hydrolases

Enzymatic dissolution of plant cell walls is a unique and complex process; because it involves a highly heterogeneous substrate and occurs at the interface between liquid and solid phases, it cannot be adequately described in terms of classical enzymology, which relates more to soluble enzymes and substrates. From the brief description of the plant cell wall provided previously it will be apparent that hydrolysis of this chemically and structurally complex polysaccharide matrix poses a significant challenge. Microorganisms that are able to digest the cell wall meet the challenge by producing extracellular enzyme systems composed of large numbers of polysaccharide hydrolases with complementary specificities, which act cooperatively to break down cellulose, hemicellulose, and other cell wall polysaccharides, liberating soluble products that may be taken up and used to provide carbon and energy. It is perhaps not surprising that although distinct subsets of enzymes are required to hydrolyze the different cell wall polysaccharides, organisms that are able to digest cellulose also generally produce the enzymes to enable them to break down other plant cell wall polysaccharides.

Enzymes responsible for degrading the major cell wall polysaccharides have provided the main focus for numerous reviews (see, for example, Refs *1, 12,* and *22–27*) and much of the detail contained in these articles will not be repeated here. For the purpose of this account, a brief simplified summary of the types of enzymes involved in breaking down the different cell wall polysaccharides will be provided.

A. Cellulases

Efficient conversion of cellulose to glucose requires the concerted action of several enzymes; the first, endo-β-1,4-glucanase (EC 3.2.1.4), cleaves β-1,4-glycosidic bonds randomly within the cellulose chain. Multiple isoforms are normally produced and are particularly active against soluble glucans, cellulose derivatives such as carboxymethylcellulose (CMC), and highly disordered cellulose. The second enzyme, cellobiohydrolase (EC 3.2.1.91), is an exo-acting enzyme that releases cellobiose from the nonreducing end of cellulose microfibrils, and acts cooperatively with endoglucanase on highly ordered cellulose. The third enzyme, β-glucosidase (EC 3.2.1.21), hydrolyzes cellobiose and short-chain cellooligosaccharides, producing glucose and relieving end-product inhibition.

B. Xylan-Degrading Enzymes

Xylans from different sources vary considerably in composition and structure and so require the combined action of a large consortium of enzymes to effect complete hydrolysis. Endo-β-1,4-xylanase (xylanase) (EC 3.2.1.8), the counterpart of endoglucanase in cellulase systems, cleaves β-1,4-linkages randomly within the xylan backbone and, like endoglucanase, is typically produced in multiple isoforms. Exo-acting β-xylosidase (EC 3.2.1.37) hydrolyzes xylooligosaccharides and xylobiose to xylose. To alleviate the inhibitory effects of side chains and acetyl substituents on xylanase activity, the actions of a number of accessory enzymes are required to promote complete breakdown. These include α-L-arabinofuranosidase (EC 3.2.1.55), α-D-glucuronidase, acetyl esterase (EC 3.1.1.6), and phenolic acid esterase. The first of these enzymes releases α-L-arabinofuranosyl groups from arabinoxylan, whereas α-D-glucuronidase is required for cleavage of the α-1,2-glycosidic linkage between xylose and 4-O-methylglucuronic acid substituents. Removal of O-acetyl groups from the C-2 and C-3 positions of backbone xylose residues and the liberation of esterified ferulic and coumaric acids are mediated by acetylxylan esterase and phenolic acid esterases, respectively.

C. Mannan-Degrading Enzymes

Enzymatic hydrolysis of mannans also requires the cooperative actions of several enzymes. The backbone of the polysaccharide is cleaved randomly by endo-β-1,4-mannanase (EC 3.2.1.78) to produce mannobiose and longer mannooligosaccharides from which mannose is removed by exo-acting β-1,4-mannosidase (EC 3.2.1.25). Two other enzymes, α-galactosidase (EC 3.2.1.22) and β-glucosidase (EC 3.2.1.21), are also required for complete breakdown of the heteromannans glucomannan and galactoglucomannan.

D. Arabinan- and Galactan-Degrading Enzymes

Random cleavage of galactans by endo-acting galactanases results in the production of galactose and galactooligosaccharides. Two types of endo-galactanase have been described (22); the first attacks the linear β-1,4-linked galactans, the second is specific for the β-1,3-linkages of some arabinogalactans. The two main enzymes that attack arabinan are α-L-arabinofuranosidase (EC 3.2.1.55) and endo-α-1,5-arabinanase (EC 3.2.1.99). The former enzyme removes arabinose side chains, permitting the latter enzyme to attack the glycosidic linkages of the arabinan backbone, producing a mixture of arabinooligosaccharides.

IV. *Pseudomonas fluorescens* subsp. *cellulosa*

Included in the large number of cellulolytic aerobic bacteria isolated from soil are a number of strains that have been assigned to the genus *Pseudomonas*. The best studied of these is *Pseudomonas fluorescens* subsp. *cellulosa* (NCIMB 10462). This strain was first described in 1952 (28) and presumably was given a name that reflects its considerable capacity for hydrolyzing cellulose. Subsequent taxonomic studies have cast doubt over the status of strain NCIMB 10462 as a valid member of the genus, but, although it undoubtedly lacks many of the traits characteristic of fluorescent pseudomonads, a recent study (29) concluded that the genus *Pseudomonas* is appropriate and proposed the name of *P. cellulosa.*

Early attempts to characterize the plant cell wall hydrolases of *P. fluorescens* subsp. *cellulosa* using biochemical techniques provided evidence for the existence of multiple extracellular and periplasmic proteins with cellulase activity (30). Subsequent work has confirmed that the bacterium can hydrolyze cellulose, xylan, arabinan, galactan, and mannan, and can use all but the last polysaccharide as sole source of carbon and energy. Conversion of these plant cell wall polysaccharides to metabolizable sugars by strain NCIMB 10462 is mediated by extracellular polysaccharide hydrolases, which are induced by culturing with cellulose, CMC, and xylan, repressed during growth with cellobiose, glucose or xylose and secreted without apparently forming aggregates or associating with the cell surface (31). However, as with other microbial polysaccharide hydrolases, purification and characterization of multiple proteins with similar activities has proved to be difficult and the true complexity of the *P. fluorescens* subsp. *cellulosa* system has only become apparent through studies using recombinant DNA techniques, which have enabled single genes to be cloned and expressed in *Escherichia coli*, a bacterium devoid of plant cell wall hydrolases. These studies have revealed that the capacity of *Pseudomonas* strain NCIMB 10462 to digest plant cell wall polysaccharides can be attributed to the production of extracellular cellulase and hemicellulase enzymes, all of which are the products of distinct genes belonging to multigene families (32, 33). Knowledge of the primary structures of each gene, coupled with gene sectioning and functional analysis of the encoded protein, have provided a detailed picture of structure/function relationships and have established the cellulase–hemicellulase system of *P. fluorescens* subsp. *cellulosa* as a paradigm for the plant cell wall-degrading enzyme systems of aerobic cellulolytic bacteria (31).

V. Architecture of *Pseudomonas* Plant Cell Wall Hydrolases

A. Cellulases

The cellulase system of *P. fluorescens* subsp. *cellulosa* is composed of three different activities: endoglucanase, cellodextrinase, and β-glucan glucohydrolase. As with several other bacterial systems, endoglucanase is produced in multiple isoforms (32). Three different genes encoding endoglucanase have been isolated, *celA* (34), *celB* (35), and *celE* (36); in each case the encoded enzyme is modular in structure and contains a distinct catalytic domain and two other domains, joined together by serine-rich linker sequences (see Fig. 1). Mature forms of all three endoglucanases are preceded by signal peptides characteristic of secreted prokaryotic proteins. Based on hydrophobic cluster analysis (HCA), the catalytic domains of endoglucanases CELA, CELB, and CELE belong to glycosyl hydrolase families 9, 45, and 5, respectively (37). Functional analysis of the polypeptides encoded by gene fragments has shown that the highly conserved 100-residue domain located at the C terminus of CELE and CELA and at the N terminus of CELB is a high-affinity cellulose-binding domain (CBD), homologous with family II CBDs described in other cellulases, notably those produced by species of *Cellulomonas, Streptomyces,* and *Thermomonospora* (12).

A similar family II CBD is present in the N-terminal 100 residues of the fourth member of the *P. fluorescens* extracellular cellulase repertoire, namely, cellodextrinase C (CELC), which is encoded by the *celC* gene and has a catalytic domain belonging to glycosyl hydrolase family 5 (38). The third or middle domain of each of these cellulase enzymes contains a novel CBD of about 30 residues, first described in xylanase E (XYLE) from *P. fluorescens* subsp. *cellulosa* (39); thus each enzyme is composed of a catalytic domain to which two CBDs are attached. The fifth component of the cellulase system, 1,4-β-D-glucan glucohydrolase D (CELD), a family 3 glycosyl hydrolase, is encoded by the *celD* gene (40). Unlike other members of the enzyme system CELD is not modular and is not secreted. Association of CELD with the cell envelope of *P. fluorescens* subsp. *cellulosa* is apparently mediated by a hydrophobic N-terminal region linked to the catalytic domain by a short sequence rich in hydroxyamino acids (Fig. 1).

B. Xylan-Degrading Enzymes

Xylan-degrading activity of *P. fluorescens* subsp. *cellulosa* is attributable to a separate subset of at least six proteins encoded by multiple genes (33). Four genes (*xynA, B, E,* and *F*) coding for endoxylanases have been isolated (39, 41, 42). All four enzymes are modular in structure (see Fig. 1) and contain three or more distinct domains joined together by linkers rich in hydroxyamino acids. Based on sequence comparisons and hydrophobic cluster analysis, the catalytic domains of XYLA, B, and F, belong to glycosyl hydrolase family 10, whereas that of XYLE is in family 11. The N-terminal 100 residues of XYLA and XYLB comprise a family II CBD closely similar to that seen in CELA, B, C, and E; the middle domain of XYLA and the C-terminal domain of XYLE contain the same small novel CBD previously identified in the middle region of the above four *Pseudomonas* cellulases. A second novel CBD, with no homology to sequences contained in the protein sequence databases, has been demonstrated immediately downstream of the signal peptide and first linker sequence of XYLF (39). Interestingly, XYLE contained a fourth domain, homologous with the NodB protein of nitrogen-fix-

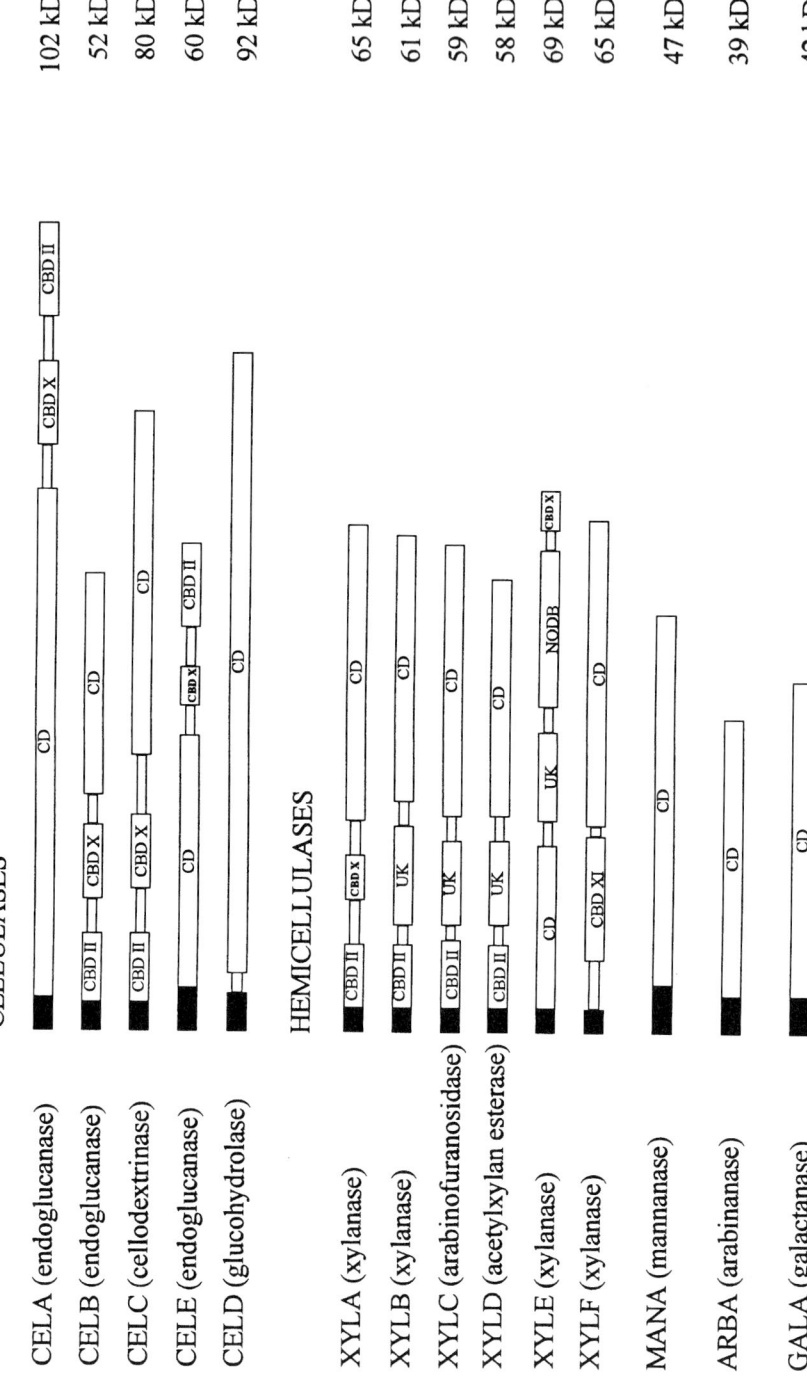

ing bacteria, which is present also in modular xylanases from *Cellulomonas fimi* (*43*) and *Cellvibrio mixtus* (*39*). This domain has recently been shown to deacetylate chemically acetylated xylan and, under the appropriate conditions, would be predicted to potentiate the activity of the xylanase catalytic domain by removing acetyl substituents, which can inhibit the activity of xylanases against naturally acetylated xylans (*44*).

Apart from endo-acting xylanases, a number of accessory enzymes, active against side chains and other substituents, are necessary for the complete breakdown of xylan, and in *P. fluorescens* subsp. *cellulosa*, genes coding for two such activities have been isolated and sequenced; the first, *xynC*, maps to within 150 nucleotides of the 3' end of *xynB* (see above) and encodes an arabinofuranosidase (XYLC) active against polymeric arabinoxylan but not against small synthetic substrates (*42*). The second gene, *xynD*, encodes an esterase that combines synergistically with xylanase to release acetic and phenolic acids from plant cell wall material (*45–47*). In both enzymes, the signal peptide was followed by a 100-residue family II CBD also found in other cellulases and hemicellulases from this bacterium, and a second domain closely similar to the domain of unknown function found previously in XYLB and XYLC (Fig. 1). Interestingly, nucleotides 114–931 of *xynB*, *xynC*, and *xynD* and residues 39–311 of XYLB, XYLC, and XYLD were identical, indicating precise replication in the N-terminal regions of the three enzymes (*42*, *45*).

C. Other Hydrolases

In addition to the cellulose- and xylan-degrading enzymes necessary for enzymatic dissolution of the major structural components of plant cell walls, *P. fluorescens* subsp. *cellulosa* produces a subset of enzymes that attack the smaller quantities of more accessible polysaccharides found associated with the cell wall matrix; these include mannans, arabinans, and galactans. Genes coding for enzymes active against each of these polysaccharides have been isolated and sequenced. Mannanase A (MANA) encoded by the *manA* gene is a family 26 glycosyl hydrolase active against galactomannan and manno-oligosaccharides, but with little activity against other cell wall polysaccharides (*48*). Arabinanase activity of *Pseudomonas* was attributed to an extracellular enzyme, ArbA, encoded by the *arbA* gene and induced by culturing with arabinan or arabinose, but repressed during culture with glucose (*49*). ArbA be-

---

←

FIG. 1. Architecture of *Pseudomonas fluorescens* subsp. *cellulosa* plant cell wall hydrolases. The different functional domains of the cellulases and hemicellulases are shown as follows: ■, signal peptide; CD, catalytic domain; CBD II, family II CBD; CBD X, novel CBD (family X proposed); CBD XI, novel CBD (family XI proposed); NODB, NodB homolog; UK, unknown function; ▭, linker. The numbers on the right are the sizes of the enzymes in kilodaltons.

longed to glycosyl hydrolase family 43, was homologous with arabinanase from *Aspergillus niger*, and hydrolyzed linear, but not branched, arabinan and arabinooligosaccharides. Interestingly, initial attack of arabinan by ArbA was random but subsequent release of exclusively arabinotriose indicated a processive mechanism and suggested that ArbA combines endo and exo modes of action. Galactanase A (GalA), an endo-β-1,4-galactanase encoded by the *ganA* gene from *P. fluorescens* subsp. *cellulosa*, had a similarly narrow substrate range, attacking only galactan and galactooligosaccharides (*50*). It exhibited significant sequence identity with a galactanase from *Aspergillus aculeatus* and, on that basis, belongs to the small glycosyl hydrolase family 53. Each of the three enzymes described above differed significantly from the other glycosyl hydrolases of *P. fluorescens* subsp. *cellulosa*. First, each occurred as a single copy, unlike the cellulases and xylanases of this organism, which are produced as multiple isoforms and are encoded by multigene families. Second, none displayed the modular architecture typical of the cellulases and xylanases (Fig. 1). Consequently, even small deletions from the 5′ or 3′ ends of the respective genes were sufficient to inactivate these single domain enzymes. Finally, CBDs, which are prevalent in cellulases and hemicellulases from *P. fluorescens* subsp. *cellulosa* and other aerobic bacteria, are absent from MANA, ArbA, and GalA, perhaps reflecting the greater accessibility of their target substrates in plant cell walls and a reduced requirement for polysaccharide-binding domains to enhance catalytic efficiency.

VI. Function of Noncatalytic Domains

The nature, distribution, and biological roles of the noncatalytic domains of modular cellulases and hemicellulases have been a major focus of recent research. In general, these noncatalytic domains tend to be of four main types: polysaccharide-binding domains (mainly CBDs), which mediate specific binding to cell wall polysaccharides, mainly cellulose (*12, 51, 52*); thermostabilizing domains, which are characteristic of certain xylanases from thermophilic (*53*) and mesophilic (*43*) bacteria and appear to confer stability in a broad sense on the enzymes that contain them; NodB domains homologous with a nodulation factor from rhizobia and present in a number of xylanases from aerobic bacteria (*39, 43, 52*); and docking domains, which are prevalent in the cellulases and hemicellulases from anaerobic bacteria and fungi, where they mediate formation of multiprotein complexes by binding to a receptor domain borne by a scaffolding protein (*27, 54*). A fifth type of noncatalytic domain particularly common in the cellulases of *C. fimi* is the fibronectin type III domain (*12*). Noncatalytic domains of the cellulases and hemicellulases from *P. fluorescens* subsp. *cellulosa* are of two types: cellulose-binding domains and NodB domains.

A. Cellulose-Binding Domains

Based on a combination of functional analysis and sequence comparisons, more than 100 CBDs have been identified and assigned to five main families (I, II, III, IV, and VI) and four minor families (V, VII, VIII and IX) (*12*). Available evidence suggests that CBDs from different families differ with respect to their biochemical and biophysical properties, but it is also becoming increasingly evident that even within families, CBDs display both structural and functional differences (*55, 56*). Structures have been determined for CBDs belonging to four of the nine families. The family I CBD, just over 30 residues in size, found in cellobiohydrolase I (CBHI) from *Trichoderma reesei* (*57*), and the larger family II CBD (110 residues), found in a xylanase/β-glucanase (Cex) from *C. fimi* (*58*), are both composed of antiparallel β-sheets that form a flat binding surface, featuring three exposed tyrosine or tryptophan residues that have been shown by mutational analysis to play a crucial role in binding the domains to cellulose (*55, 56, 59–61*), probably by stacking against the pyranose rings of crystalline cellulose. The family IV CBD, CBD_{N1} (152 residues), found in endoglucanase CenC from *C. fimi*, is composed of 10 β-strands folded into two antiparallel β-sheets, forming a jelly-roll β-sandwich containing a binding cleft rather than a flat binding surface (*62*). This fold is markedly different from those of CBD_{CBH1} and CBD_{Cex}, and the presence of a binding cleft or groove, containing a central strip of hydrophobic residues flanked on both sides by polar hydrogen-bonding groups, may explain why CBD_{N1}, unlike CBD_{CBH1} and CBD_{Cex}, binds selectively to amorphous cellulose and soluble cellooligosaccharides, but not to the flat surface of crystalline cellulose (*63, 64*). The fourth family for which a structural paradigm exists is family III. The crystal structure of the CBD (CBD_{Cip}; 155 residues) found in the cellulosome integrating protein (Cip) from *Clostridium thermocellum* has been elucidated at 1.75 Å resolution (*65*). The domain has the structure of a nine-stranded β-sandwich with jelly-roll topology and, though different from CBD_{Cex} in detail, bears some similarity to the family II CBD in that both consist of two β-sheets packed face to face in a β-sandwich. A planar linear strip of aromatic and polar residues exposed on the surface of CBD_{Cip} facilitates binding along a cellulose chain via a mechanism that involves hydrophobic interaction and hydrogen bonding.

Family II CBDs occur in each of the cellulases and four of the hemicellulases from *P. fluorescens* subsp. *cellulosa* (Fig. 2). The recent discovery of a second discrete CBD in each of the cellulases and XYLA has confirmed that the modular architecture of these enzymes is even more complex than

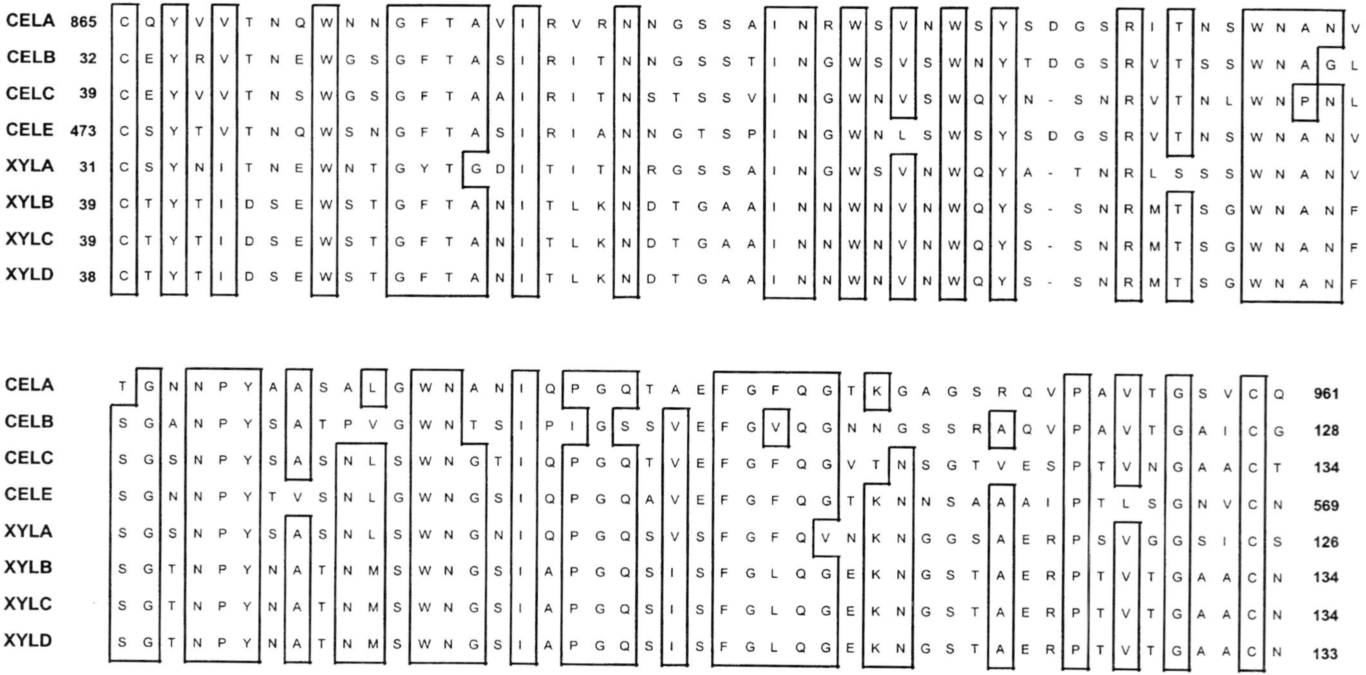

FIG. 2. Comparison of the amino acid sequences of family II CBDs in cellulases and hemicellulases from *P. fluorescens* subsp. *cellulosa*. Regions containing identical residues or conservative substitutions are boxed. The numbers denote the positions of the residues within their respective full-length sequences.

at first realized. In each case, the second binding domain is a small novel CBD distinct from any contained in the nine families already defined, and first characterized in XYLE from the pseudomonad (39) (Fig. 3). It has been shown that two CBDs, contained within a single polypeptide, can interact synergistically, resulting in higher affinity binding than observed for a single CBD (66). Whether this is true for the paired CBDs of *P. fluorescens* subsp. *cellulosa* cellulases and hemicellulases remains to be seen. A third novel type of CBD, also not previously described, has been demonstrated immediately upstream of the catalytic domain of XYLF (39). The high frequency with which CBDs occur in the cellulases and hemicellulases from *P. fluorescens* subsp. *cellulosa* suggests that the selective advantage conferred by these domains is so great that it has led to evolutionary subduction of less well-adapted enzymes.

A unifying role for CBDs has yet to be established and, in the absence of definitive experimental evidence, the biological function of CBDs continues to be a matter for conjecture. Similarly, although there is evidence indicating an important role for linker sequences in some enzymes, a wider biological function for these regions of modular polysaccharide hydrolases has not been unequivocally established. Research conducted on the CBD and linker sequences of the modular cellulases and hemicellulases from *P. fluorescens* subsp. *cellulosa* has contributed significantly to the debate.

Although the removal of CBDs from endo-β-1,4-glucanases and cellobiohydrolases does not affect the activity of the respective enzymes against soluble substrates, there is evidence that CBDs play a major role in potentiating the activity of cellulases against the more resistant forms of insoluble cellulose (36, 67, 68). For example, Din et al. (69) demonstrated that the isolated family II CBD from *C. fimi* endoglucanase A (CenA) could open up the structure of highly crystalline cellulose, making it more accessible to enzyme attack. Similarly, removal of the CBD from CenA diminished activity against BMCC, but not amorphous cellulose (68). A similar effect on crystalline cellulose has not been demonstrated for the isolated family II CBD from *P. fluorescens* XYLA. However, full-length CELE from the pseudomonad, containing a C-terminal family II CBD, was inactive against BMCC but displayed four times higher activity against Avicel than did a truncated form lacking the CBD; activities of the two forms of CELE against PASC did not differ (36). These results imply that the family II CBD from CELE does not enhance activity against all forms of cellulose. Thus it seems that family II CBDs can behave differently in different enzymes.

The original observation that CBDs occur with high frequency in *P. fluorescens* subsp. *cellulosa* hemicellulases, which have no activity against cellulosic substrates, provoked the suggestion that these domains potentiate the catalytic activity of hemicellulases by promoting prolonged close contact be-

XYLE	616	Q	C	N	W	G	T	F	Y	P	L	C	Q	T	S	G	W	G	W	E	N	S	R	S	C	I	S	T	S	T	C	N	S	Q	T	G	G	G	G	V	C	N	661
XYLA	183	Q	C	N	W	Y	G	T	L	Y	P	L	C	V	T	T	T	N	G	W	G	W	E	D	Q	R	S	C	I	A	R	S	T	C	A	A	Q	P	A	P	F	G	I	V	G	S	G	228
CELA	669	N	C	N	W	Y	G	T	L	Y	P	L	C	V	T	T	Q	S	G	W	G	W	E	N	S	Q	S	C	I	S	A	S	T	C	S	A	Q	P	A	P	Y	G	I	V	G	A	A	714
CELC	183	Q	C	N	W	Y	G	T	L	Y	P	L	C	V	S	T	S	G	W	G	W	E	N	N	R	S	C	I	S	P	S	T	C	S	A	Q	P	A	P	Y	G	I	V	G	G	S	228
CELB	180	A	C	N	W	Y	G	T	L	T	P	L	C	N	N	T	S	N	G	W	G	Y	E	D	G	R	S	C	V	A	R	T	C	S	A	Q	P	A	P	Y	G	I	V	S	T	S	225
CELE	380	Q	C	N	W	Y	G	T	L	Y	P	L	C	S	T	T	T	N	G	W	G	W	E	N	N	A	S	C	I	A	R	A	T	C	S	G	Q	P	A	P	W	G	I	V	G	G	S	425

FIG. 3. Comparison of the amino acid sequences of novel CBDs (family X) in xylanases (XYLA, XYLE) and cellulases (CELA, CELB, CELC, CELE) from *P. fluorescens* subsp. *cellulosa*. Regions containing identical residues or conservative substitutions are boxed. The numbers denote the positions of the residues within their respective full-length sequences.

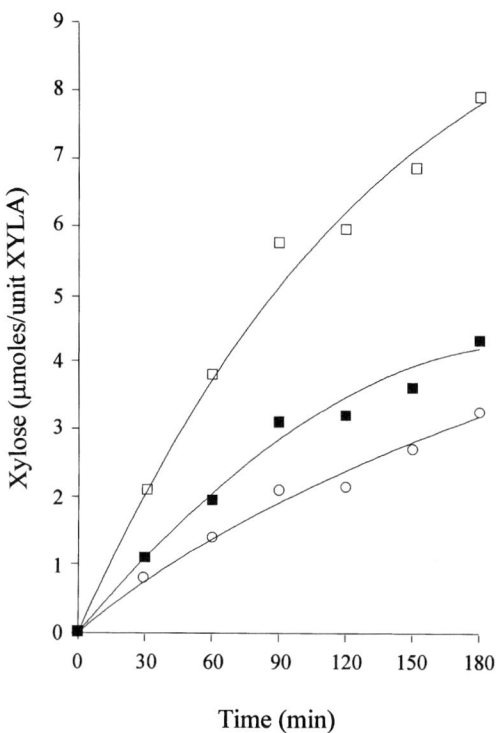

FIG. 1. Release of xylose from the cellulose-xylan complex by full length and truncated forms of *P. fluorescens* subsp. *cellulosa* xylanase A (XYLA). Full-length XYLA (□) and XYLA lacking the linker sequence (■) or both the linker sequence and CBD (○) were incubated with hardwood (birch) kraft pulp at 37°C and the release of xylose was measured. From Ref. 71.

tween these enzymes and their native substrates in plant cell walls (70). Studies focusing on engineered and full-length forms of a xylanase (XYLA) and arabinofuranosidase (XYLC) from *Pseudomonas*, both of which contain a family II CBD, have provided evidence for the biological roles of both CBDs and linkers, and further support for the overall view that tight binding of these hemicellulases to native plant cell wall material, and the inherent flexibility conferred by linkers, lead to enhanced catalytic activity (71). Removal of the linker sequences from XYLC did not affect catalytic activity against soluble arabinoxylan or its ability to bind to cellulose; this is consistent with the previous observation that linkerless derivatives of XYLA retained their full activity against soluble substrate and their capacity to bind to cellulose (70). However, the activities of both enzymes against their target substrates contained in cellulose–xylan complexes derived from plant cell walls were

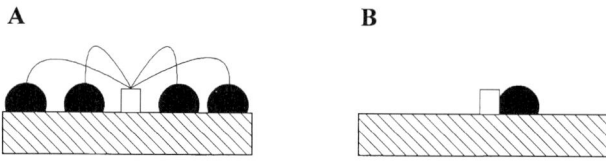

FIG. 5. Proposed model for the hydrolysis of plant cell wall xylan by *P. fluorescens* subsp. *cellulosa* xylanase A (XYLA). XYLA, which consists of a CBD (□), linker sequence (—), and catalytic domain (●), attaches to cellulose contained in the plant cell wall ( ) via the CBD. (A) The linker sequence confers flexibility on XYLA, enabling the catalytic domain to attack xylan at multiple sites. (B) The absence of linker sequences restricts the amount of substrate available to the catalytic domain when the enzyme is attached via the CBD to cellulose. From Ref. 71.

enhanced by the presence of the CBDs and linkers, respectively (Fig. 4). A possible explanation for this observation is that during hydrolysis of the xylan component by, for example, XYLA, the CBD restricts substrate availability by anchoring the enzyme at a fixed location. Structural flexibility conferred on the enzyme by the linker may increase the number of glycosidic bonds accessible to the catalytic domain (Fig. 5). The fact that removal of the linker sequences from *C. fimi* CenA reduced activity against soluble and crystalline substrates emphasizes the apparent lack of a unifying role for linker sequences.

The mechanism by which CBDs potentiate the catalytic activity of cellulases against highly crystalline substrates has provoked much interest. Several reports have suggested that these domains disrupt the ordered structure of polysaccharides, making them more accessible to enzymatic attack. However, whether CBDs appended to plant cell wall hydrolases other than cellulases exhibit the same biochemical properties and mechanisms of action as CBDs that are an integral part of naturally occurring cellulases is an important question. To address this issue, we have characterized, in detail, the biochemical properties of CBDs from a *Pseudomonas* cellulase (CELE) and xylanase (XYLA), and have investigated the mechanism by which they enhance catalytic activity. The data (Table I) demonstrate that the affinities of the two domains for crystalline and amorphous cellulose are similar. Binding of the xylanase CBD to soluble oligosaccharides was demonstrated by nuclear magnetic resonance (NMR) spectroscopy; pronounced spectral shifts were observed for signals corresponding to conserved aromatic residues in the presence of cellulooligosaccharides (Fig. 6). The affinity for soluble oligosaccharides was orders of magnitude less than for insoluble cellulose (Table II), suggesting that the xylanase CBD, in common with the cellulase CBD, has a strong preference for cellulose chains, and probably binds pref-

TABLE I
BINDING CHARACTERISTICS OF THE FAMILY II CBDs FROM *P. fluorescens*
XYLANASE A (XYLA) AND ENDOGLUCANASE E (CELE)

Binding domain	Cellulose	Relative affinity K_r (liters g^{-1})	Saturation (μmol g^{-1})
$CELE_{CBD}$	$BMCC^a$	17.7	15.2
$CELE_{CBD}$	$PASC^b$	15.5	9.2
$CELE_{CBD}$	Avicel	3.5	4.1
$XYLA_{CBD}$	$BMCC^a$	20.4	16.7
$XYLA_{CBD}$	$PASC^b$	14.4	9.0
$XYLA_{CBD}$	Avicel	3.0	4.5

[a]Bacterial microcrystalline cellulose.
[b]Phosphoric acid-swollen cellulose.

erentially to several cellulose chains held in a fixed conformation rather than to single flexible polysaccharide molecules.

In protein engineering experiments (G. P. Hazlewood and H. J. Gilbert, unpublished results), fusion of either the cellulase or xylanase CBD to the catalytic domain of a second cellulase enhanced the activity of the enzyme, providing evidence of a common role for CBDs regardless of their origins. Other experiments showed that the xylanase CBD enhanced the activity of cellulase or xylanase enzymes only when it was covalently attached to the respective catalytic domains. The CBD alone elicited no detectable change in the structural integrity of either crystalline cellulose or plant cell walls when assessed using the constant-load extension assay (creep assay) originally developed for evaluating the biological properties of expansins. These latter plant proteins are active in catalytic amounts and loosen the structure of plant cell walls by physically disrupting the close-packed structure of microfibrils at the cellulose/hemicellulose interface. Taken together, these data indicate that the xylanase CBD enhances catalytic activity not by disrupting the complex interactions between plant structural polysaccharides in the manner of expansins, but simply by promoting close contact between the catalytic domain of the enzyme and its target substrate.

B. NodB Domain

The second type of noncatalytic domain found in polysaccharide hydrolases from *P. fluorescens* subsp. *cellulosa*, the so-called NodB domain, occurs in XYLE only (39). It is, however, present in xylanases from other aerobic cellulolytic bacteria, notably xylanase A from *C. mixtus* (39) and xylanase D (XYLD) from *C. fimi* (52, 72), in an acetylxylan esterase from *Streptomyces*

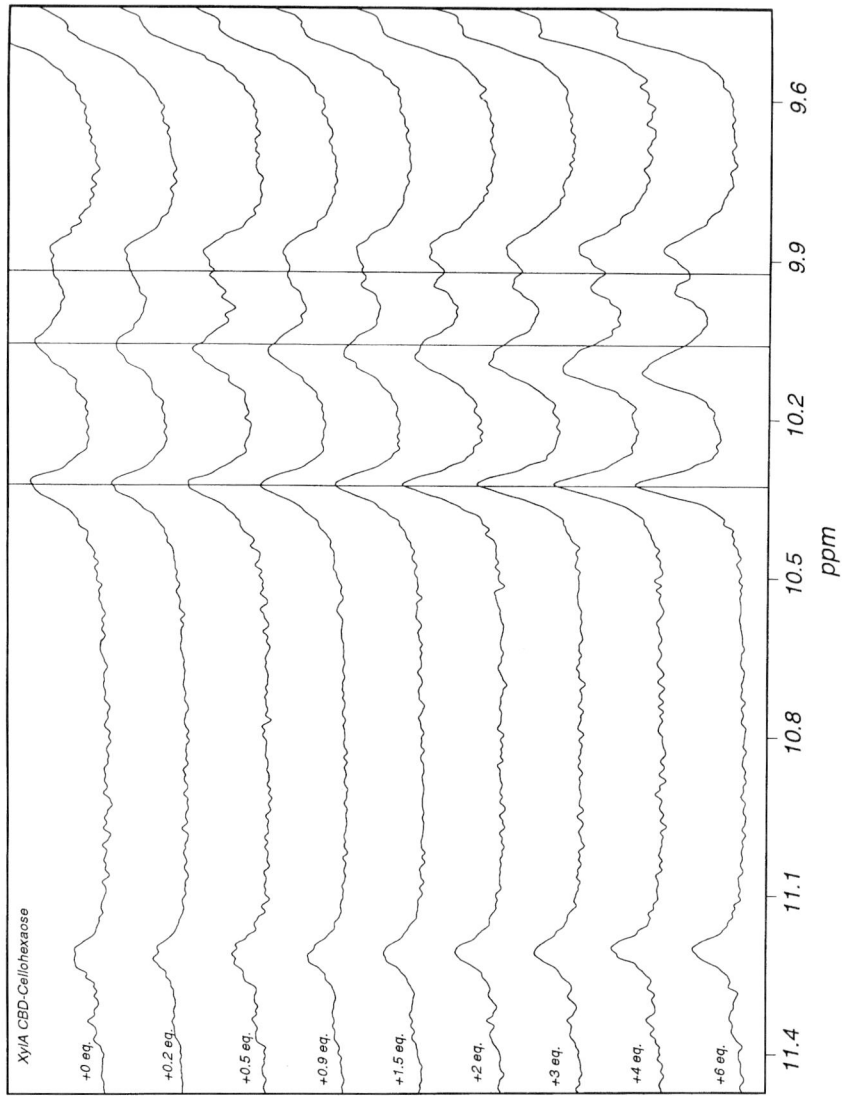

TABLE II
Affinity of the Family II CBDs from *P. fluorescens* Xylanase A (XYLA) and Endoglucanase E (CELE) for Cellulooligosaccharides and Cellulose

Substrate	Affinity constant K_a (M^{-1})
Cellohexaose	3.3×10^2
Cellopentaose	1.4×10^2
Cellotetraose	4.0×10^1
Cellotriose	1.6×10^1
BMCC[a]	5.8×10^6

[a]Bacterial microcrystalline cellulose.

lividans (73), and in a chitin deacetylase from *Mucor rouxii* (74) (Fig. 7). Sectioning of the *C. fimi xynD* gene has established that the NodB domain in XYLD has the capacity to release acetyl groups from chemically acetylated xylan (44). Thus a likely role for the NodB domains of xylan-degrading enzymes in general is to relieve the inhibition of xylanase activity that is known to result from acetyl substitution of the xylan backbone.

VII. Structures and Catalytic Mechanisms of *Pseudomonas* Plant Cell Wall Hydrolases

Enzymatic hydrolysis of the glycosidic bonds contained in plant structural polysaccharides proceeds via general acid catalysis and has a critical requirement for amino acid residues to act as proton donor and nucleophile/base, respectively (74). The reaction is stereospecific and proceeds by mechanisms that lead either to retention or to inversion of configuration at the anomeric center. Retaining enzymes cleave their substrates via a double-displacement mechanism; in the first phase, the glycosidic bond is cleaved by a combination of nucleophilic attack at the anomeric carbon and protonation of the glycosidic oxygen, generating a glycosyl–enzyme intermediate. In the second stage, the enzyme–substrate complex is hydrolyzed by water attacking the anomeric carbon of the glycosyl group. Inverting enzymes pro-

---

←

FIG. 6. One-dimensional NMR spectrum showing the shifting of signals corresponding to tryptophan residues during titration of the family II CBD (0.5 mM) from *P. fluorescens* xylanase A (XYLA) with increasing concentrations of cellohexaose (from 0 to 6 molar equivalents). Vertical lines denote the positions of signals corresponding to three tryptophan residues, two of which experience pronounced shifts on addition of increasing concentrations of cellohexaose.

C. fimi XYLD	V	G	L	T	F	D	D	D	G	P	N	T	G	T	T	N	Q	I	L	S	T	L	T	Q	Y	G	A	T	A	-	T	V	F	P	T	G	Q	N	A	Q	G	N	P	S	L	M	Q	A	Y	K	N	407
C. mixtus XYLA	V	G	L	T	F	D	D	D	G	P	A	G	A	N	T	T	-	T	L	V	N	L	L	L	K	Q	N	L	T	P	-	V	T	W	F	V	Q	G	N	Y	V	A	A	N	S	N	L	M	S	Q	L	L	S	451
P. fluorescens XYLE	V	G	I	T	F	D	D	D	G	P	N	S	N	T	A	-	T	L	L	V	N	L	L	R	Q	N	L	T	P	V	T	W	F	N	Q	G	Q	N	V	V	A	S	N	A	H	L	M	S	Q	Q	L	S	448
S. lividans axeA	V	G	L	T	F	D	D	D	G	P	-	S	G	G	S	T	Q	S	L	L	N	A	L	R	Q	N	G	L	R	A	-	T	M	F	N	Q	G	Q	Y	V	A	A	Q	N	P	S	L	V	R	A	Q	V	D	94
M. rouxii chitin deacetylase	W	G	L	T	Y	D	D	D	G	P	N	C	-	S	H	N	A	F	Y	D	Y	L	Q	E	Q	K	L	K	A	S	-	M	F	Y	I	G	S	N	V	V	D	W	P	Y	G	A	M	R	G	V	V	D	207
R. meliloti NodB	-	I	Y	L	T	F	D	D	D	G	P	N	P	H	C	T	P	E	I	L	D	V	L	A	E	Y	G	V	P	A	-	T	F	F	V	I	G	T	Y	A	K	S	Q	P	E	L	I	R	R	I	V	A	71

C. fimi XYLD	A	G	V	Q	I	-	G	N	H	S	W	D	H	P	H	L	V	N	M	S	Q	S	D	M	Q	S	Q	L	T	R	T	Q	Q	A	I	Q	Q	T	A	G	-	V	T	P	T	L	F	R	P	P	Y	456
C. mixtus XYLA	V	G	-	E	V	Q	N	N	H	S	Y	T	H	P	H	L	I	-	N	L	G	Y	Q	Q	I	-	V	Y	D	E	L	N	R	T	N	Q	A	-	P	K	P	T	L	F	R	P	P	Y	499
P. fluorescens XYLE	V	G	-	E	V	H	N	N	H	S	Y	T	H	P	H	M	T	S	W	T	Y	Q	A	Q	V	V	Y	D	E	L	N	R	T	N	Q	A	-	P	K	P	T	L	F	R	P	P	Y	496
S. lividans axeA	A	G	M	W	V	A	N	H	S	Y	T	H	P	H	M	T	Q	L	G	Q	A	Q	M	D	S	E	I	-	S	R	T	Q	Q	A	I	-	G	R	Q	-	A	V	P	A	A	V	142
M. rouxii chitin deacetylase	-	G	H	H	I	-	A	S	H	T	W	S	H	P	Q	M	T	T	K	T	N	Q	E	V	L	A	E	F	Y	Y	T	Q	K	A	I	-	K	L	A	T	G	-	L	T	P	R	Y	W	R	P	P	Y	255
R. meliloti NodB	E	G	H	E	V	A	N	H	T	M	T	H	P	D	L	S	T	C	G	P	H	E	V	E	R	E	I	-	V	E	A	S	E	A	I	-	A	L	V	L	R	P	R	S	D	T	Y	E	A	P	Y	121

C. fimi XYLD	G	E	S	N	A	T	L	R	Q	V	E	S	S	L	G	L	R	E	I	-	I	W	D	V	D	S	Q	D	W	N	N	A	S	A	S	Q	-	-	-	-	-	-	-	R	Q	A	495
C. mixtus XYLA	G	E	V	N	A	N	V	N	Q	A	A	A	Q	A	L	G	L	R	V	-	I	T	W	N	V	D	S	Q	D	W	N	G	A	S	A	T	A	-	-	-	-	-	-	-	A	N	A	538
P. fluorescens XYLE	G	E	L	N	S	T	I	Q	Q	A	A	A	Q	A	L	G	L	R	V	V	T	W	D	V	D	S	Q	D	W	N	G	A	S	A	T	A	-	-	-	-	-	-	-	A	N	A	535
S. lividans axeA	R	Q	T	N	A	T	L	R	S	V	E	A	K	Y	G	L	T	E	V	-	V	W	D	V	D	S	Q	D	W	N	A	S	T	D	A	-	-	-	-	-	-	-	V	Q	A	181
M. rouxii chitin deacetylase	G	D	I	D	D	R	V	R	W	I	A	S	Q	L	G	L	T	A	V	-	I	W	N	L	D	T	D	D	W	S	A	G	V	T	T	T	V	E	A	V	E	Q	S	Y	S	D	Y	I	-	A	M	G	305
R. meliloti NodB	G	V	W	S	E	E	A	L	T	R	S	A	S	A	G	L	T	A	I	-	H	W	S	A	D	P	R	D	W	S	R	P	G	A	N	A	-	-	-	-	-	-	-	V	D	A	160

C. fimi XYLD	A	S	R	-	L	T	N	G	Q	I	I	L	M	H	D	509
C. mixtus XYLA	A	-	N	Q	L	Q	N	G	Q	V	I	L	M	H	D	552
P. fluorescens XYLE	A	-	N	Q	L	G	N	G	Q	V	I	L	M	H	D	549
S. lividans axeA	V	S	R	-	L	G	N	G	Q	V	I	L	M	H	D	195
M. rouxii chitin deacetylase	T	N	G	T	F	A	N	S	G	N	I	V	L	T	H	320
R. meliloti NodB	V	L	D	S	V	R	P	G	A	L	V	L	L	H	D	175

FIG. 7. Comparison of the amino acid sequences of the NodB-like domains of *Pseudomonas fluorescens* xylanase E (XYLE), *Cellulomonas fimi* xylanase D (XYLD), and *Cellvibrio mixtus* xylanase A (XYLA) with the sequences of the acetylxylan esterase (Axe A) from *Streptomyces lividans*, a chitin deacetylase (Cda) from *Mucor rouxii*, and NodB from *Rhizobium meliloti*. Regions containing identical residues or conservative substitutions are boxed. The numbers denote the positions of the residues within their respective full-length sequences.

ceed via a mechanism in which protonation of the glycosidic oxygen and bond cleavage are accompanied by simultaneous attack of a water molecule that has been activated by the catalytic base residue; the resulting product has stereochemistry opposite to that of the substrate. The catalytic nucleophile is typically within 5.5 Å of the proton donor in retaining enzymes, but is more distant (10 Å) in inverting enzymes because of the need to accommodate a water molecule between the catalytic base residue and the sugar.

Based on similarities in their amino acid sequences, the current total of almost 1000 known glycosyl hydrolases can be fitted into a classification comprising some 60 different families (37). This sequence-based classification has proved to be enormously beneficial in reflecting common structural features and evolutionary relationships and as a tool for deducing mechanistic information. Crystal structures have been determined for 20 or so glycosyl hydrolases (75), but there are no structural paradigms for more than half of the 60 families.

A. Xylanase A

1. STRUCTURE OF CATALYTIC DOMAIN

Pseudomonas fluorescens subsp. *cellulosa* produces cellulases that belong to glycosyl hydrolase families 3 (CELD), 5 (CELC and CELE), 9 (CELA), and 45 (CELB) and xylanases belonging to families 10 (XYLA, XYLB, XYLF) and 11 (XYLE). The crystal structure of one of the family 10 xylanases (XYLA) has been determined at 1.8 Å resolution and a crystallographic R factor of 0.166 (76, 77). The catalytic domain of XYLA (M_r 39,000) was hyperexpressed in *E. coli*, purified, and crystallized. A glutamate residue, with the potential to act as the catalytic nucleophile, was identified within the catalytic domain of XYLA on the basis of its conservation among members of family 10. The structure of XYLA was solved using an engineered catalytic core in which the nucleophilic glutamate residue was replaced by a cysteine, enabling production of high-quality mercurial derivatives and the preparation of an inactive enzyme–substrate complex in the crystal. These studies showed that the catalytic core of XYLA (and, by inference, those of other family 10 xylanases) has the architecture of the 8-fold α/β barrel or TIM barrel, with the nucleophilic glutamate (E246) located at the carboxy end of β-strand 7 of the barrel and the catalytic acid–base (E127) within 5.5 Å at the end of β-strand 4 (Fig. 8). Similar 8-fold α/β architecture has been demonstrated for representatives of glycosyl hydrolase families 1, 2, 5, 17, 30, 35, 39, and 42; for each of the nine members of this superfamily or clan (GH-A), the catalytic residues and catalytic mechanism are strictly conserved, indicating that all have diverged from a common ancestor (75). Compared with other 8-fold α/β barrels within the superfamily, XYLA has one additional α-

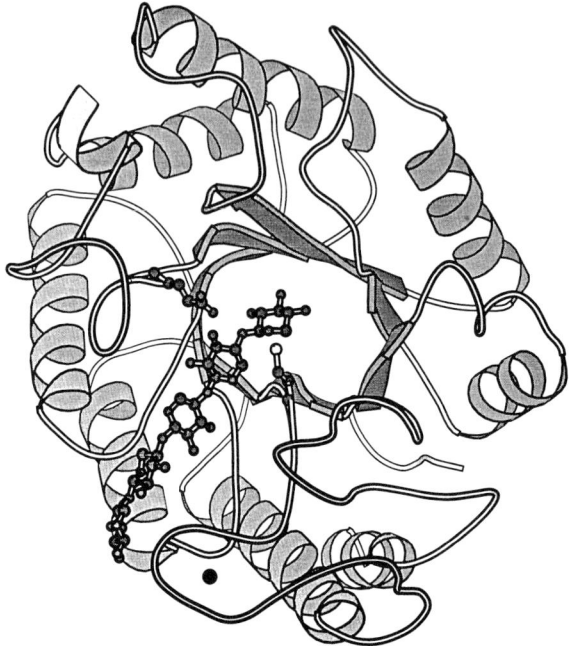

FIG. 8. Eight-fold α/β-barrel architecture of the catalytic domain of xylanase A (XYLA) from *P. fluorescens* drawn using MOLSCRIPT. The eight β-strands, shown as arrows, form a barrel surrounded by eight α-helices. Ca^{2+} binding (●) stabilizes the loop after strand 7 and the substrate-binding subsites are formed by loops after strands 4 to 7. The active site nucleophile (E246) is located on strand 7 and the acid–base catalyst (E127) is on strand 4. The substrate, xylopentaose, and active site glutamates are shown using solid bonds. From Ref. 76.

helix and an atypically long loop after strand 7, which appeared by X-ray crystallography to be stabilized by calcium. Xylopentaose soaked into crystals of the inactive enzyme occupied an active-site cleft formed by the longer loops that follow β-strands 4 and 7; five substrate-binding subsites (A to E) were identified within the cleft, with the cleaved bond located between subsites D and E, indicative of an exo mode of action. A cluster of conserved residues in the substrate-binding cleft adjacent to subsite E was indicative of an additional subsite (subsite F), which was obscured from xylopentaose by contacts within the crystal (Fig. 9). The presence of subsite F was confirmed by biochemical analysis that showed that (1) X_2 and X_3 were the main products formed from xylan and xylooligosaccharides (X_3 to X_6) and (2) activity against X_6 was higher than against X_5 or any other xylooligosaccharides. These results indicated that XYLA has an endo mode of action and at least six substrate-binding subsites, with at least two xylose-binding sites on each side of the catalytic residues (78).

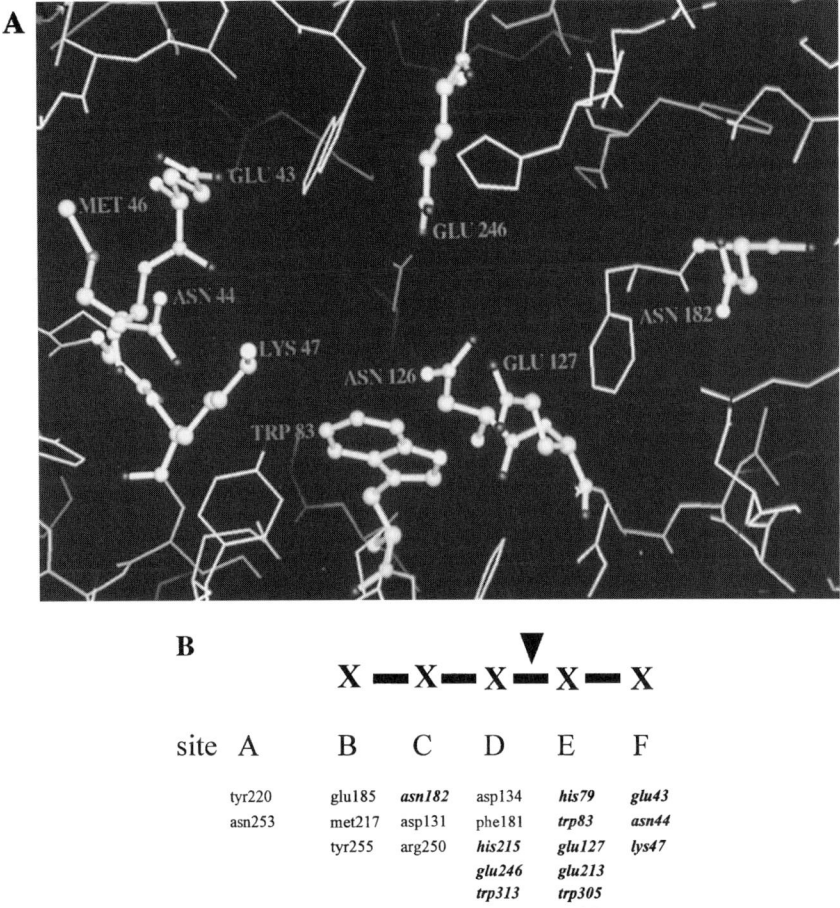

FIG. 9. Substrate-binding subsites of xylanase A (XYLA) from *P. fluorescens*. (A) Residues contributing to subsites C, D, E, and F are represented in ball and stick configurations. Glutamate 43, asparagine 44, and lysine 47 are adjacent to each other on the surface of the active site cleft and have the potential to form a xylose binding site. Methionine 46 is located close to these residues but is not on the surface of the active site. Asparagine 126 is in close proximity to the acid–base catalyst, glutamate 127, and thus could play an important role in the function of the catalytic carboxylic acid residue. Tryptophan 83 is positioned between lysine 47 and asparagine 126, and thus these two residues could influence the fluorescence of the aromatic amino acid. Asparagine 182 is also on the surface of the active site cleft and is positioned the appropriate distance away from the two key catalytic residues, glutamate 127 and glutamate 246, to occupy xylose binding site C of XYLA. (B) The disposition of amino acids along the substrate-binding cleft is shown in simplified form. The position occupied by xylopentaose relative to substrate-binding subsites is shown; ▼, the bond cleaved during hydrolysis. Amino acids conserved in family 10 glycosyl hydrolases are shown in italics. From Ref. 78.

The functional importance of subsite F and the roles of other conserved residues located on the surface of the active-site cleft of family 10 xylanases were investigated by site-directed mutagenesis, involving the substitution of alanine (A) for glutamate 43 (E43A), asparagine 44 (N44A), lysine 47 (K47A), and methionine 46 (M46A) (78) (Fig. 9). The results showed that binding to the F site is essential for efficient hydrolysis of xylooligosaccharides. Mutants N44A and E43A were much less active than is the wild-type enzyme against X_3, X_4, and X_5, but their activity against xylan was unaltered. K47A was less active against xylan and had little activity against xylooligosaccharides. These data were consistent with a reduction in the capacity of the F site to bind substrate, leading to random binding of oligosaccharides along the cleft at sites A to D, and the production of dead-end complexes; activity against polymeric xylan remained unchanged or only slightly reduced because binding of xylan at sites A to D would place adjacent xylose residues in sites E and F. Thus the primary role of the F subsite of XYLA is to prevent small oligosaccharides from forming nonproductive enzyme–substrate complexes.

We have also investigated the role of amino acids at other xylose-binding pockets within the extended substrate-binding cleft of *Pseudomonas* XYLA. Of particular interest was residue N182, which is located at the boundary between subsites B and C (Fig. 9). In *Streptomyces lividans* xylanase A, an enzyme that also belongs to glycosyl hydrolase family 10, the equivalent residue, N173, is clearly important for the subsite B xylose-binding pocket. However, N182A, N182R, and N182D mutants of *Pseudomonas* XYLA generated xylotriose and xylose from xylotetraose and not xylobiose as produced by the wild-type enzyme. These data suggest that in XYLA, N182 plays an important role in the function of subsite C, but not subsite B, and provide evidence that residues that are highly conserved in xylanases belonging to the same glycosyl hydrolase family do not necessarily play equivalent roles in enzyme function.

2. Calcium-Binding Site

Other mutagenesis studies conducted with XYLA have shown that the calcium-binding site, which is unique among family 10 xylanases, stabilizes the structure of the extended loop 7 and protects XYLA from thermal inactivation, thermal unfolding, and attack by proteinases (79) (Fig. 8). Binding of calcium to XYLA occurred with 1:1 stoichiometry and a K_a of 4.9×10^4 M^{-1}, and did not affect catalytic activity. However, a general increase in the stability of XYLA on ligand binding was reflected in an increase in melting temperature of the enzyme from 60.8 to 66.5°C and acquired resistance against degradation by chymotrypsin. Replacement of the calcium-binding domain within loop 7 with the corresponding shorter loop from another fam-

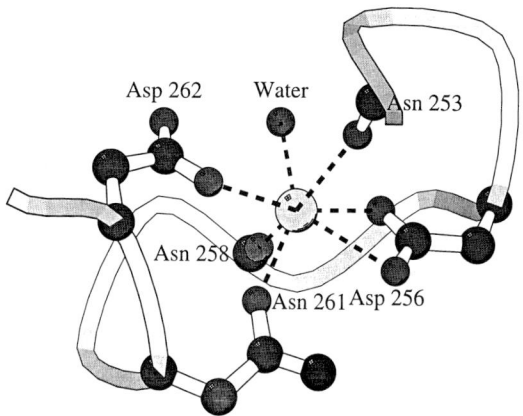

FIG. 10. Calcium-binding loop of xylanase A (XYLA) from *P. fluorescens*. Amino acid residues comprising the calcium-binding loop are represented in ball and stick configurations. Ca^{2+} is shown in the center (○). Dotted lines depict the electrostatic interactions between residues within the calcium-binding loop of XYLA and a water molecule. From Ref. 79.

ily 10 xylanase (Cex from *C. fimi*) (*80*) did not alter the biochemical properties of XYLA significantly. Substitution of alanine for aspartate 256, asparagine 261, and aspartate 262 resulted in nonbinding mutants and demonstrated the pivotal role of these residues in calcium binding (Fig. 10). Comparison of the sequences of 28 family 10 xylanases showed that only 4 others contained an extended loop 7. Phylogenetic analysis revealed that 3 of these, together with XYLA from *P. fluorescens* subsp. *cellulosa*, constitute a subgroup of family 10 xylanases that may have evolved from a common ancestor containing a DNA insertion in the region encoding loop 7.

B. Mannanase A

Apart from X-ray crystallographic determination of the tertiary structures of enzyme–substrate or enzyme–inhibitor complexes, a number of other approaches have been used to facilitate identification of the key catalytic residues of glycosyl hydrolases. These include carrying out detailed kinetic analysis on mutant enzymes in which the putative catalytic carboxylate residues have been altered by site-directed mutagenesis. This latter approach has been used for two *P. fluorescens* subsp. *cellulosa* enzymes belonging to glycosyl hydrolase families for which no structural paradigms exist.

The first enzyme, mannanase A (MANA), belongs to family 26. Analysis of the stereochemical course of mannotetraose hydrolysis by purified recombinant MANA showed that the configuration of the anomeric carbon was retained on cleavage of the middle glycosidic bond, indicating a double-displacement mechanism (*81*). Hydrophobic cluster analysis (HCA) revealed that two glutamate and two aspartate residues were conserved in all members of family 26, and identified sequence motifs that indicated that MANA was related to glycosyl hydrolases of the GH-A superfamily, all of which have 8-fold α/β barrel architecture. Site-directed mutagenesis was carried out to test the prediction that E320 and E212 are the catalytic nucleophile and acid–base, respectively. Replacement of the conserved aspartates with alanine and glutamate had no dramatic effect on catalytic activity. In contrast, substituting alanine or aspartate for either of the conserved glutamates substantially decreased activity against galactomannan, mannotetraose, and 2,4-dinitrophenyl-β-mannobioside (2,4-DNPM). The apparent K_m of E320A was similar to that of wild-type MANA, but a large reduction in K_m was seen for E212A. Analysis of the pre-steady-state kinetics of 2,4-DNPM hydrolysis by E212A revealed a rapid initial rate of 2,4-dinitrophenol (2,4-DNP) release, as would be predicted for the hydrolysis by MANA of a substrate with a good leaving group, followed by a rapid decay leading to a slow steady-state release of 2,4-DNP as deglycosylation became rate limiting. During hydrolysis of 2,4-DNPM by E320A, glycosylation was apparently rate limiting, and there was no significant pre-steady-state burst of 2,4-DNP release. These results were consistent with the view that E212 and E320, respectively, are the catalytic acid–base and nucleophile residues of the retaining glycosyl hydrolase MANA.

C. Galactanase A

A second *Pseudomonas* enzyme for which the catalytic mechanism and key catalytic residues have been identified in the absence of detailed structural information is the endo-β-1,4-galactanase, GalA, which belongs to glycosyl hydrolase family 53 (50). Analysis of the stereochemical course of 2,4-dinitrophenyl-β-galactobiose (2,4-DNPG$_2$) hydrolysis by GalA in the presence of D$_2$O revealed that the β-conformation of the anomeric carbon was retained on cleavage of the aglycone bond, suggesting that the hydrolysis of glycosidic bonds by GalA proceeds by a double-displacement mechanism. HCA indicated that GalA and, by inference, other members of family 53 are related to the GH-A superfamily, members of which have an 8-fold α/β barrel structure. Furthermore, it predicted that E161 and E270 were the key catalytic acid–base and nucleophile residues, respectively. GalA mutants in which E161 and E270 had been replaced by alanine or aspartate were unaltered in conformation, as evidenced by circular dichroism spectroscopy, but were virtually inactive against galactan. E161A exhibited a much lower K_m for 2,4-DNPG$_2$ than did wild-type GalA and elicited a rapid initial pre-steady-state burst of 2,4-DNP release; no such pre-steady-state burst was seen during hydrolysis of 2,4-DNPG$_2$ by mutant E270A. These data are consistent with the view that E161 and E270 are the catalytic acid–base and nucleophile residues of the retaining enzyme GalA.

ACKNOWLEDGMENTS

The authors gratefully acknowledge the continuing financial support of the Biotechnology and Biological Sciences Research Council, and the substantial contributions made to this research by their colleagues and co-workers.

REFERENCES

1. G. R. Stephens and G. H. Heichel, *Biotechnol. Bioeng. Symp.* **5,** 27 (1975).
2. L. G. Ljungdahl and K.-E. Eriksson, *Adv. Microbial Ecol.* **8,** 237 (1985).
3. H. J. Gilbert and G. P. Hazlewood, *J. Gen. Microbiol.* **139,** 187 (1993).
4. K. K. Y. Wong and J. N. Saddler, *in* "Hemicellulose and Hemicellulases" (M. P. Coughlan and G. P. Hazlewood, eds.), p. 127. Portland Press, London and Chapel Hill, 1993.
5. M. C. McCann and K. Roberts, *in* "The Cytoskeletal Basis of Plant Growth and Form" (C. W. Lloyd, ed.), p. 109. Academic Press, New York, 1992.
6. N. C. Carpita and D. M. Gibeaut, *Plant J.* **3,** 1 (1993).
7. B. S. Valent and P. Albersheim, *Plant Physiol.* **54,** 105 (1974).
8. M. McNeil, A. G. Darvill, S. C. Fry, and P. Albersheim, *Annu. Rev. Biochem.* **53,** 625 (1984).
9. T. K. Kirk, *in* "The Filamentous Fungi" (J. E. Smith, D. R. Berry, and B. Kristiansen, eds.), Vol. 4, p. 266. Edward Arnold Ltd., London, 1983.
10. T. W. Jeffries, *Biodegradation* **1,** 163 (1990).
11. M. T. Esquerre-Tugaye and D. Mazau, *Physiol. Vegetale* **19,** 415 (1981).
12. P. Tomme, R. A. J. Warren, and N. R. Gilkes, *Adv. Microbial Physiol.* **37,** 1 (1995).
13. T. E. Timell, *Wood Sci. Technol.* **1,** 45 (1967).
14. R. L. Whistler and E. L. Richards, *in* "The Carbohydrates–Chemistry and Biochemistry" (W. Pigman and D. Horton, eds.), Vol. 2A, p. 447. Academic Press, New York, 1970.
15. A. Scalbert, B. Monties, J.-Y. Lallemand, E. Guittet, and C. Rolando, *Phytochemistry* **24,** 1359 (1985).
16. I. Mueller-Harvey, R. D. Hartley, P. J. Harris, and E. H. Curzon, *Carbohydr. Res.* **148,** 71 (1986).
17. A. M. Stephen, *in* "The Polysaccharides" (G. O. Aspinall, ed.), Vol. 2, p. 98. Academic Press, London, 1983.
18. J. Puls and J. Schuseil, *in* "Hemicellulose and Hemicellulases" (M. P. Coughlan and G. P. Hazlewood, eds.), p. 1. Portland Press, London and Chapel Hill, 1993.
19. M. Tanaka, A. Abe, and T. Uchida, *Biochim. Biophys. Acta* **658,** 377 (1981).
20. R. R. Selvendran, *in* "Dietary Fibre" (G. G. Birch and K. J. Parker, eds.), p. 95. Applied Science Publishers, London and New York, 1983.
21. G. O. Aspinall, *in* "The Biochemistry of Plants" (J. Preis, ed.), Vol. 3, p. 473. Academic Press, New York, 1980.
22. R. F. H. Dekker and G. N. Richards, *Adv. Carbohydr. Chem. Biochem.* **32,** 277 (1976).
23. M. P. Coughlan, *in* "Microbial Enzymes and Biotechnology" (W. M. Fogarty and C. T. Kelly, eds.), p. 1. Elsevier Applied Science, London and New York, 1990.
24. G. P. Hazlewood and H. J. Gilbert, *in* "Hemicellulose and Hemicellulases" (M. P. Coughlan and G. P. Hazlewood, eds.), p. 103. Portland Press, London and Chapel Hill, 1993.

25. M. P. Coughlan and G. P. Hazlewood, *Biotechnol. Appl. Biochem.* **17,** 259 (1993).
26. P. Béguin and J.-P. Aubert, *FEMS Microbiol. Rev.* **13,** 25 (1994).
27. P. Béguin and M. Lemaire, *Crit. Rev. Biochem. Mol. Biol.* **31,** 201 (1996).
28. K. Ueda, S. Ishikawa, T. Itami, and T. Asai, *J. Agric. Chem. Soc. Jpn.* **26,** 35 (1952).
29. C. Dees, D. Ringelberg, T. C. Scott, and T. J. Phelps, *Appl. Biochem. Biotechnol.* **51/52,** 263 (1995).
30. K. Yamane and H. Suzuki, *Meth. Enzymol.* **160,** 200 (1988).
31. G. P. Hazlewood, J. I. Laurie, L. M. A. Ferreira, and H. J. Gilbert, *J. Appl. Bacteriol.* **72,** 244 (1992).
32. H. J. Gilbert, G. Jenkins, D. A. Sullivan, and J. Hall, *Mol. Gen. Genet.* **210,** 551 (1987).
33. H. J. Gilbert, D. A. Sullivan, G. Jenkins, L. E. Kellett, N. P. Minton, and J. Hall, *J. Gen. Microbiol.* **134,** 3239 (1988).
34. J. Hall and H. J. Gilbert, *Mol. Gen. Genet.* **213,** 112 (1988).
35. H. J. Gilbert, J. Hall, G. P. Hazlewood, and L. M. A. Ferreira, *Mol. Microbiol.* **4,** 759 (1990).
36. J. Hall, G. W. Black, L. M. A. Ferreira, S. J. Millward-Sadler, B. R. S. Ali, G. P. Hazlewood, and H. J. Gilbert, *Biochem. J.* **309,** 749 (1995).
37. B. Henrissat and A. Bairoch, *Biochem. J.* **316,** 695 (1996).
38. L. M. A. Ferreira, G. P. Hazlewood, P. J. Barker, and H. J. Gilbert, *Biochem. J.* **279,** 793 (1991).
39. S. J. Millward-Sadler, K. Davidson, G. P. Hazlewood, G. W. Black, H. J. Gilbert, and J. H. Clarke, *Biochem. J.* **312,** 39 (1995).
40. J. E. Rixon, L. M. A. Ferreira, A. J. Durrant, J. I. Laurie, G. P. Hazlewood, and H. J. Gilbert, *Biochem. J.* **285,** 947 (1992).
41. J. Hall, G. P. Hazlewood, N. S. Huskisson, A. J. Durrant, and H. J. Gilbert, *Mol. Microbiol.* **3,** 1211 (1989).
42. L. E. Kellett, D. M. Poole, L. M. A. Ferreira, A. J. Durrant, G. P. Hazlewood, and H. J. Gilbert, *Biochem. J.* **272,** 369 (1990).
43. J. H. Clarke, K. Davidson, H. J. Gilbert, C. M. G. A. Fontes, and G. P. Hazlewood, *FEMS Microbiol. Lett.* **139,** 27 (1996).
44. J. I. Laurie, J. H. Clarke, A. Ciruela, C. B. Faulds, G. Williamson, H. J. Gilbert, J. E. Rixon, J. Millward-Sadler, and G. P. Hazlewood, *FEMS Microbiol. Lett.* **148,** 261 (1997).
45. L. M. A. Ferreira, T. M. Wood, G. Williamson, C. Faulds, G. P. Hazlewood, G. W. Black, and H. J. Gilbert, *Biochem. J.* **294,** 349 (1993).
46. C. B. Faulds, M.-C. Ralet, G. Williamson, G. P. Hazlewood, and H. J. Gilbert, *Biochim. Biophys. Acta* **1243,** 265 (1995).
47. B. Bartolomé, C. B. Faulds, P. A. Kroon, K. Waldron, H. J. Gilbert, G. P. Hazlewood, and G. Williamson, *Appl. Environ. Microbiol.* **63,** 208 (1997).
48. K. L. Braithwaite, G. W. Black, G. P. Hazlewood, B. R. S. Ali, and H. J. Gilbert, *Biochem. J.* **305,** 1005 (1995).
49. V. A. McKie, G. W. Black, S. J. Millward-Sadler, G. P. Hazlewood, J. I. Laurie, and H. J. Gilbert, *Biochem. J.* **323,** 547 (1997).
50. K. L. Braithwaite, T. Barna, T. D. Spurway, S. J. Charnock, G. W. Black, N. Hughes, J. H. Lakey, R. Virden, G. P. Hazlewood, B. Henrissat, and H. J. Gilbert, *Biochemistry* **36,** 15489 (1997).
51. N. R. Gilkes, B. Henrissat, D. G. Kilburn, R. C. Miller, and R. A. J. Warren, *Microbiol. Rev.* **55,** 303 (1991).
52. S. J. Millward-Sadler, D. M. Poole, B. Henrissat, G. P. Hazlewood, J. H. Clarke, and H. J. Gilbert, *Mol. Microbiol.* **11,** 375 (1994).
53. C. M. G. A. Fontes, G. P., Hazlewood, E. Morag, J. Hall, B. Hirst, and H. J. Gilbert, *Biochem. J.* **307,** 151 (1995).

54. C. Fanutti, T. Ponyi, G. W. Black, G. P. Hazlewood, and H. J. Gilbert, *J. Biol. Chem.* **270,** 29314 (1995).
55. D. M. Poole, G. P. Hazlewood, N. S. Huskisson, R. Virden, and H. J. Gilbert, *FEMS Microbiol. Lett.* **106,** 77 (1993).
56. N. Din, I. J. Forsythe, L. D. Burtnick, N. R. Gilkes, R. C. Miller, Jr., R. A. J. Warren, and D. G. Kilburn, *Mol. Microbiol.* **11,** 747 (1994).
57. P. J. Kraulis, G. M. Clore, M. Nilges, T. A. Jones, G. Petterson, J. Knowles, and A. M. Gronenborn, *Biochemistry* **28,** 7241 (1989).
58. G.-Y. Xu, E. Ong, N. R. Gilkes, D. G. Kilburn, D. R. Muhandiram, M. Harris-Brandts, J. P. Carver, L. E. Kay, and T. S. Harvey, *Biochemistry* **34,** 6993 (1995).
59. T. Reinikainen, L. Ruohonen, T. Nevanen, L. Laaksonen, P. Kraulis, T. A. Jones, J. K. C. Knowles, and T. T. Teeri, *Proteins: Struct. Funct. Genet.* **14,** 475 (1992).
60. T. Reinikainen, O. Teleman, and T. T. Teeri, *Proteins: Struct. Funct. Genet.* **22,** 392 (1995).
61. M. Linder, M.-L. Mattinen, M. Kontelli, G. Lindberg, J. Ståhlberg, T. Drakenberg, T. Reinikainen, G. Petterson, and A. Annila, *Protein Sci.*, **4,** 1056 (1995).
62. P. E. Johnson, M. D. Joshi, P. Tomme, D. G. Kilburn, and L. P. McIntosh, *Biochemistry* **35,** 14381 (1996).
63. P. Tomme, A. L. Creagh, D. G. Kilburn, and C. A. Haynes, *Biochemistry* **35,** 13885 (1996).
64. P. E. Johnson, P. Tomme, M. D. Joshi, and L. P. McIntosh, *Biochemistry* **35,** 13895 (1996).
65. J. Tormo, R. Lamed, A. J. Chirino, E. Morag, E. A. Bayer, Y. Shoham, and T. A. Steitz, *EMBO J.* **15,** 5739 (1996).
66. M. Linder, I. Salovuori, L. Ruohonen, and T. T. Teeri, *J. Biol. Chem.* **271,** 21268 (1996).
67. P. Tomme, H. van Tilbeurgh, G. Petersson, J. van Damme, J. Vandekerckhove, J. Knowles, T. Teeri, and M. Claeyssens, *Eur. J. Biochem.* **170,** 575 (1988).
68. J. B. Coutinho, N. R. Gilkes, R. A. J. Warren, D. G. Kilburn, and R. C. Miller, Jr., *FEMS Microbiol. Lett.* **113,** 211 (1993).
69. N. Din, N. R. Gilkes, B. Tekant, R. C. Miller, Jr., R. A. J. Warren, and D. G. Kilburn, *Bio/technology* **9,** 1096 (1991).
70. L. M. A. Ferreira, A. J. Durrant, J. Hall, G. P. Hazlewood, and H. J. Gilbert, *Biochem. J.* **261** (1990).
71. G. W. Black, J. E. Rixon, J. H. Clarke, G. P. Hazlewood, M. K. Theodorou, P. Morris, and H. J. Gilbert, *Biochem. J.* **319,** 515 (1996).
72. G. W. Black, G. P. Hazlewood, S. J. Millward-Sadler, J. I. Laurie, and H. J. Gilbert, *Biochem. J.* **307,** 191 (1995).
73. F. Shareck, P. Biely, R. Morosoli, and D. Kluepfel, *Gene* **153,** 105 (1995).
74. D. Kafetzopoulos, G. Thireos, J. N. Vournakis, and V. Bouriotis, *Proc. Natl. Acad. Sci. U.S.A.* **90,** 8005 (1993).
75. G. Davies and B. Henrissat, *Structure* **3,** 853 (1995).
76. G. W. Harris, J. A. Jenkins, I. Connerton, N. Cummings, L. LoLeggio, M. Scott, G. P. Hazlewood, J. I. Laurie, H. J. Gilbert, and R. W. Pickersgill, *Structure* **2,** 1107 (1994).
77. G. W. Harris, J. A. Jenkins, I. Connerton, and R. W. Pickersgill, *Acta Crystallogr.* **D52,** 393 (1996).
78. S. J. Charnock, J. H. Lakey, R. Virden, N. Hughes, M. L. Sinnott, G. P. Hazlewood, R. Pickersgill, and H. J. Gilbert, *J. Biol. Chem.* **272,** 2942 (1997).
79. T. D. Spurway, C. Morland, A. Cooper, I. Sumner, G. P. Hazlewood, A. G. O'Donnell, R. W. Pickersgill, and H. J. Gilbert, *J. Biol. Chem.* **272,** 17523 (1997).
80. D. Tull, S. G. Withers, N. R. Gilkes, D. G. Kilburn, R. A. J. Warren, and R. Aebersold, *J. Biol. Chem.* **266,** 15621 (1991).
81. D. N. Bolam, N. Hughes, R. Virden, J. H. Lakey, G. P. Hazlewood, B. Henrissat, K. L. Braithwaite, and H. J. Gilbert, *Biochemistry* **35,** 16195 (1996).

Regulation of the Spatiotemporal Pattern of Expression of the Glutamine Synthetase Gene

HELEEN LIE-VENEMA,[1]
THEODORUS B. M. HAKVOORT,
FORMIJN J. VAN HEMERT,
ANTOON F. M. MOORMAN, AND
WOUTER H. LAMERS[2]

Department of Anatomy and Embryology
University of Amsterdam
Academic Medical Center
1105 AZ Amsterdam, The Netherlands

I. Introduction	244
II. Spatiotemporal Aspects of Glutamine Synthetase Expression	246
A. Tissue-Specific Expression of GS and Its Functional Implications	246
B. Development of GS Expression	263
III. Levels of Regulation of Glutamine Synthetase Expression	266
A. Posttranscriptional Regulation	266
B. Pretranslational Regulation	270
IV. Glutamine Synthetase Gene: Evolution, Structure, and Regulatory Regions	273
A. Avian and Mammalian Gene Structure	275
B. Regulatory Regions of Mammalian GS Genes	276
C. Regulation of the Chicken GS Gene	288
V. Concluding Remarks	293
References	295

Glutamine synthetase, the enzyme that catalyzes the ATP-dependent conversion of glutamate and ammonia into glutamine, is expressed in a tissue-specific and developmentally controlled manner. The first part of this review focuses on its spatiotemporal pattern of expression, the factors that regulate its levels under (patho)physiological conditions, and its role in glutamine, glutamate, and ammonia metabolism in mammals. Glutamine synthetase protein stability is more than 10-fold reduced by its product glutamine and by covalent modifications. During

---

[1] Present address: Department of Clinical Viro-Immunology, Central Laboratory of the Netherlands Red Cross Blood Transfusion Service, Plesmanlaan 125, 1066 CX Amsterdam, The Netherlands.

[2] To whom correspondence may be addressed.

late fetal development, translational efficiency increases more than 10-fold. Glutamine synthetase mRNA stability is negatively affected by cAMP, whereas glucocorticoids, growth hormone, insulin (all positive), and cAMP (negative) regulate its rate of transcription. The signal transduction pathways by which these factors may regulate the expression of glutamine synthetase are briefly discussed.

The second part of the review focuses on the evolution, structure, and transcriptional regulation of the glutamine synthetase gene in rat and chicken. Two enhancers (at −6.5 and −2.5 kb) were identified in the upstream region and two enhancers (between +156 and +857 bp) in the first intron of the rat glutamine synthetase gene. In addition, sequence analysis suggests a regulatory role for regions in the 3' untranslated region of the gene. The immediate-upstream region of the chicken glutamine synthetase gene is responsible for its cell-specific expression, whereas the glucocorticoid-induced developmental appearance in the neural retina is governed by its far-upstream region. © 1998 Academic Press

I. Introduction

The ATP-dependent conversion of glutamate and ammonia into glutamine, catalyzed by glutamine synthetase (GS; EC 6.3.1.2), was first demonstrated over 60 years ago by Krebs (1). In animal tissues, it is the only enzyme capable of *de novo* biosynthesis of glutamine. GS functions both to remove ammonia and/or glutamate and to produce glutamine. GS is the high-affinity system for removal of ammonia, a highly toxic compound that is formed in all tissues. At rest, humans produce approximately 90 mmol of ammonia per day, 90% of which derives from the portal-drained viscera and 10% from the kidney (2). The ammonia produced by the intestines is removed on passage of the blood through the liver. Following maximum exercise, the production of ammonia in muscle exceeds that of the intestines and kidney combined, due to an increased flux through the purine nucleotide cycle (3). In the central nervous system, high extracellular levels of glutamate are toxic, because glutamate is an excitatory neurotransmitter. Equally important, cellular glutamine (and alanine) biosynthesis permits the transport of nitrogen in a nontoxic form between the tissues in the body (4, 5). In this way, glutamine is an important mediator of amino acid catabolism by serving as a substrate for both hepatic urea synthesis and renal ammoniagenesis to maintain acid–base homeostasis. Furthermore, this glutamine serves, together with glucose, as primary respiratory fuel for enterocytes, macrophages, lymphocytes, and other rapidly dividing cells. The amide group of glutamine is used as the nitrogen source for the synthesis of amino acids, nucleotides, and amino sugars (6–9).

Because net absorption of dietary glutamine from the gut is small (10–12), glutamine homeostasis is primarily regulated by the rate of its synthesis and utilization within the body. Primary sites of glutamine synthesis are skeletal muscle (13), lung (14–16), and adipose tissue (17, 18), but in hyperammonemia, kidney (19), brain (20, 21), and liver (22) are also net producers. The gut is the major site of glutamine utilization (11, 23), whereas the kidney assumes this role in acidotic conditions (24, 25) and the liver in sepsis (26). Perhaps to subserve these interorgan fluxes, glutamine is the amino acid present in the highest concentration in mammalian blood (0.5 to 0.8 mM) and has one of the highest turnover rates (27, 28). Nevertheless, even though inhibition of the activity of GS in rats for 4 days by methionine sulfoximine, an irreversible inhibitor of GS, causes a decline in circulating arterial as well as in liver, kidney, and muscle glutamine levels by 30–50%, organ physiology is only mildly impaired (29–31). This finding shows that fairly wide changes in glutamine levels are tolerated fairly well and that acute changes in glutamine biosynthetic capacity are not required to preserve organ function.

A highly characteristic and functionally important feature of glutamine metabolism is the topographic distribution of cells engaged in glutamine biosynthesis and degradation. Prime examples are nervous system, liver, and kidney. In the nervous system, glutamine is synthesized by the astrocytes to be used by the neurons as a precursor for the synthesis of the neurotransmitters glutamate and γ-aminobutyric acid (GABA) (see Section II,A,2). In kidney, but particularly in liver, glutaminase is confined to the upstream parenchymal cells to facilitate removal of glutamine from the circulation, whereas GS is confined to the downstream parenchymal cells to add back glutamine to the circulation (see Sections II,A,1 and II,A,3). Cellular GS levels in brain, liver, and kidney are high. On the other hand, GS is homogeneously expressed at a relatively low level in the organs that are primarily involved in glutamine production, such as muscle and adipose tissue (see Sections II,A,4. and II,A,7).

The cloning of the GS gene of several mammalian and avian species (32–36) has permitted the beginning of the dissection of the molecular mechanisms underlying the spatiotemporal and quantitative aspects of its expression. On the one hand, these studies are facilitated by the fact that GS is present only as a single copy in the haploid genome. On the other hand, its dual function, that is to remove ammonia and/or glutamate and to produce glutamine, and the fact that it is expressed at very high levels in subsets of cells in some tissues, and at moderate to low levels in all cells of other tissues, suggest that the regulatory elements of the gene may be complex. Mapping and characterizing these regulatory regions requires comparison of the expression patterns of reporter genes driven by portions of the regulatory regions with endogenous expression of the gene. In this study we therefore first review the tissue distribution of glutamine synthetase in mammals and summarize what is known about its local regulatory factors and function. In the second part, we review what is known about the gene and its regulatory regions in the liver of mammals, but also that in the neural retina of the chick, because it is a well-established and well-studied model in developmental biology.

II. Spatiotemporal Aspects of Glutamine Synthetase Expression

This section describes the main features of the expression of the GS gene in mammals: its tissue distribution and cell specificity, the factors that affect its cellular level under (patho)physiological conditions and during development, and the functional implications of its expression pattern and level.

A. Tissue-Specific Expression of GS and Its Functional Implications

GS is believed to be expressed at low levels in virtually all cells, but some cells contain more than 500×10^6 GS molecules per cell, which amounts to 3.5% of cellular protein. Table I is an inventory of the tissue distribution of GS activity in rats and mice, showing that tissue distribution in both species is comparable. Because the cellular distribution of GS is directly related to its function, data on the cellular distribution of GS expression are also included in Table I. Although accurate quantitative data on the volume percentage of GS-positive cells in tissues or organs are not yet available, such data can be generated by morphometric analysis of immunohistochemically stained sections.

1. LIVER

   a. Topography. Whereas GS is homogeneously distributed in the liver of uricotelic animals (birds and reptiles) (*37*), the enzyme is confined to a two- to three-cell-thick layer of hepatocytes surrounding the efferent central veins in mammals, that is, to less than 7% of the hepatocytes (*33, 37–44*). The specific activity of GS in liver is approximately 6 ± 0.5 μmol g tissue^{-1} min^{-1} (see Table I), or approximately 85 μmol g tissue^{-1} min^{-1} in the pericentral cells. Because purified mammalian GS has a specific activity of about 12 μmol product mg protein^{-1} min^{-1} (see Section IV), the pericentral cells contain approximately 7 mg enzyme per gram, that is, 160 μM or 3.5% of cellular protein. The strictly pericentral expression pattern of GS develops in the late fetal period (*39, 45*) (see Section II,B,2) and is very stable thereafter. In adult rodent liver, the pattern of expression of GS is identical at the mRNA and protein level (*40, 41*), indicating a pretranslational regulation of this position-specific expression.

TABLE I
Compilation of GS Activities[a]

Organ	GS activity RLU	Ref.	Localization of expression	Ref.	Approximate cellular concentration of GS protein
Liver					
Rat	1.0 ± 0.1 (14)	69, 83, 212, 213, 238, 239, 252, 307, 320, 410, 450, 496–498	Pericentral hepatocytes (rat, mouse, pig, monkey, man)	37, 38, 40–43	160 μM
Mouse	1.0 (2)	210, 328			
Caput epididymis					
Rat	1.6 (1)	307	Principal cells in epithelium of ductus epididymis, more abundant in caput than in cauda (rat, mouse)	210, 307	60 μM
Total epididymis					
Mouse	0.8 (1)	210			
Stomach					
Rat	0.83 (1)	252	Gastric mucosa, except for mucous cells (mouse)	210	15 μM
Mouse	0.5 (1)	210			
Brain					
Rat	0.7 ± 0.1 (10)	69, 83, 238, 239, 252, 307, 320, 410, 496	Astrocyte, oligodendrocytes (rat, mouse, cow, man) ependymal cells	121–123, 126, 128, 129, 131, 329	40 μM
Mouse	1.4 (2)	210, 328			
Kidney					
Rat	0.5 ± 0.1 (12)	69, 212, 213, 237–239, 252, 307, 320, 410, 496, 499	Proximal straight tubule (rat); whole nephron, most abundant in proximal tubule (rabbit); absent (pig, man)	207, 209, 210	30 μM
Mouse	0.3 (2)	210, 328			

continues

TABLE I (Continued)

Organ	GS activity RLU	Ref.	Localization of expression	Ref.	Approximate cellular concentration of GS protein
Testis					
Rat	0.5 ± 0.1 (4)	239, 307, 410, 496	Maturing spermatozoa, interstitial cells of Leydig (rat, mouse)	210, 310	n.d.
Mouse	0.5 (1)	210			
Brown adipose tissue					
Rat	0.5 (2)	497	Adipocytes (mouse)	210	8 μM
Mouse	0.6 (1)	210			
Skin					
Rat	0.4 (3)	212, 252, 325	Hair follicles[b]	—	n.d.
Retina					
Rat	0.4 ± 0.24 (3)	69, 410, 450	Müller glial cells (rat)	132, 133	n.d.
Mouse	1.5 (1)	328			
Submaxillary gland					
Rat	0.3 (1)	498	Acini[b]	—	n.d.
Spleen					
Rat	0.2 ± 0.08 (4)	69, 239, 320, 410	n.d.[c]	—	
Mouse	0.07 (1)	210			
Small intestine					
Rat	0.16 ± 0.06 (3)	212, 239, 252	Crypt cells (rat, mouse)	210, 282	n.d.
Mouse	0.04 (1)	210			

Tissue	Activity	References	Localization	Km
White adipose tissue				
Rat	0.15 ± 0.06 (4)	212, 213, 252, 413, 497	Adipocytes (mouse)	6 μM
Mouse	0.03 (1)	210		
Skeletal muscle				
Rat	0.09 ± 0.02 (13)	69, 83, 212, 213, 237–239, 252, 307, 325, 413, 497, 499	Perinuclear (mouse), highest level in fast-twitch white fibers, lowest in slow-twitch red fibers (rat), absent from smooth muscle (mouse)	1 μM
Mouse	0.05 (2)	210, 328		
Lung				
Rat	0.09 ± 0.03 (4)	69, 213, 239, 320	Epithelium of respiratory bronchioles (mouse and rat[b])	6 μM
Mouse	0.4 (1)	210		
Heart				
Rat	0.06 ± 0.03 (3)	69, 238, 239	n.d.	—
Thymus				
Rat	0.02 (2)	69, 239	n.d.	—

[a]Activities were measured by several laboratories in tissue homogenates of organs of the rat and mouse; data include cellular localization of GS expression. The cellular concentration of GS protein in the cells expressing GS was obtained by relating activity and localization. Because the absolute amounts of GS activity varied with the method by which GS activity was determined (radioactively, colorimetric, or with HPLC technology) and the laboratory in which the activity was determined, GS activities in each study were normalized according to the procedure described below. For recalculation of GS activities to the same tissue base, see ref. 495. Systematic differences between each pair of studies were eliminated by calculating the ratio of the enzyme activity levels in every organ analyzed in both studies. The average ratio of enzyme levels of all organs in both studies generates a conversion factor between the two studies. By repeating this comparison of each study with all other studies available, a general conversion factor can be obtained. All available activity data of each study are then normalized by dividing them with the corresponding conversion factor. Enzyme activities of all studies thus normalized are used to calculate a normalized average enzyme activity for each organ. Normalized enzyme activity in the liver is used as reference activity (relative liver units; RLU). The average GS activity in rat and mouse liver was determined to be 6 μmol γ-glutamylhydroxamate g fresh tissue^{-1} min^{-1} at 37°C [due to an error in a conversion factor, the reported GS activities were 10-fold too high in our previous studies (74, 210, 280)]. Data are shown ± SEM; the number of analyses is given in parentheses. GS protein content as measured by Western blot (239) showed the same tendency as the GS activity measurements and was therefore included in this compilation.

[b]Our own unpublished observations.

[c]n.d., Not determined.

b. Regulation of Expression. Blood-borne factors (*46–48*), the innervation pattern (*49*), cell–cell interactions between hepatocytes and endothelial cells (*50–52*), the position of the hepatocyte on the portocentral axis (*53–55*), and the cell lineage (*56*) do not appear to be important regulators of position-specific expression of hepatic GS. Instead, all experimental approaches point to the (vascular) architecture of the liver as a major determinant of its expression pattern. Perhaps the most powerful arguments stem from experiments that showed that the normal spatiotemporal appearance of GS in hepatocytes *in situ* (*39, 41, 57*) can be reproduced outside the liver. Transplantation of embryonic hepatocytes to the spleen revealed that only hepatocytes in lobular structures acquire the capacity to express high levels of GS protein, whereas outside such an architectural context, expression of GS expression becomes prematurely arrested (*58, 59*). The initial induction of GS expression appears to depend solely on a critical size of the cellular agglomerate (*60*), whereas the subsequent accumulation of GS to high cellular levels is confined to hepatocytes surrounding the efferent veins of the newly formed lobules (*58, 59*). These observations, as well as the total absence of an effect of the source of afferent blood (portal, systemic arterial, or systemic venous) on GS expression in the liver (*46, 48*), suggested to us that (secreted) products of the upstream periportal hepatocytes are necessary to permit the accumulation of GS in downstream, pericentral hepatocytes. Indeed, a secreted product from FL-ET-14 cells is able to induce GS expression in periportal hepatocytes (*51*). Unfortunately, neither the identity of the inducing factor nor the parentage of this cell line is known.

GS activity levels in hepatocytes are mildly decreased (approximately 20%) in adrenalectomized rats (*61*), whereas hypophysectomy causes a more severe 50% decline (*61–64*). These effects are more pronounced at the mRNA level (*64, 65*). Because glucocorticosteroid supplementation increases GS in adrenalectomized animals (*66*) and growth hormone supplementation increases GS in hypophysectomized animals (*61–64*), these effects can be ascribed to glucocorticosteroids and growth hormone, respectively. *In vivo*, a single dose of glucocorticosteroids induces a rapid, transient increase in GS mRNA with maximum levels being reached at 6 hr (*67*). In cultured hepatocytes, only the combination of glucocorticosteroids and growth hormone induces GS expression (*68*), although in certain preparations, dexamethasone alone suffices (*56*). Even though gender does not affect GS activity levels in rat liver (*47, 69*), the size of the GS-positive zone in male liver is larger than that in female liver, and is dependent on testosterone levels (*47*). Conditions that are associated with increased hepatocellular cyclic AMP levels (diabetes, fasting, high-protein diet) (*70*) decrease GS mRNA and protein levels (*65*), typically by decreasing the number of GS-positive cells (*65, 71*). Insulin, on the other hand, may stimulate GS activity (*72*). Like growth hormone, it can exert its effect *in vitro* only in conjunction with glucocorticosteroids (56). Finally, it has been observed that thyroid hormones induce GS expression in hypothyroid rats (73).

After partial hepatectomy, the GS-positive zone shrinks to approximately two-thirds of its size, even though the diameter of the lobule returns to its original value (74–76). Conversely, after selective destruction of the pericentral hepatocytes with carbon tetrachloride or bromobenzene, GS-positive cells develop from the remaining GS-negative hepatocytes, but only in the most downstream zone (50, 52, 77, 78). A difference between the two models of liver damage is that, in the first case, mitotic activity is concentrated in the periportal cells, whereas in the latter model, it is concentrated in the pericentral cells (78), implying that the life history of the upstream cells also plays a role. In fact, when, as a result of continued exposure to hepatotoxic agents (79) or cardiac failure (R. H. Lekanne Deprez and W. H. Lamers, unpublished observations), cirrhosis develops and mitoses expand to the periportal area, GS expression is virtually lost. However, in cirrhotic liver, the architecture of the liver has changed from lobular to acinar, that is, the downstream zone no longer occupies the center of the perfusion unit but, instead, has come to lie at the periphery of it (80). As a result, the exposure of the downstream cells to secreted products of the upstream cells also declines.

Liver atrophy resulting from chronic calory (46, 81–84) or protein restriction (85, 86) also causes a reduction in hepatic GS activity and the size of the GS-positive pericentral zone. In fact, differences in the degree of liver atrophy in the pair-fed control rats explain why some find decreased GS levels after portocaval shunt (83) and others do not (82), and why some find decreased GS levels in the liver of tumor-bearing rats (84, 86) and others do not (85). Interestingly, in animals carrying a subcutaneous hepatoma, decreased hepatic levels of GS were associated with increased GS levels in the tumor due to an increased half-life of the protein in these tumors (85).

Hepatocellular carcinoma is often characterized by a very high expression of GS (87–90). All available evidence suggests that GS-positive carcinomas develop from GS-positive foci in the pericentral zone (87, 89, 90), and that the development of these foci is due to the enhanced transformation in the pericentral cells (48). It is not yet known how these neoplastic cells can maintain such a high level of GS expression.

c. Function. One of the major functions of the liver is the maintenance of ammonia and bicarbonate homeostasis in the body. Normally, glutamine flows from the periphery to the splanchnic bed to serve as fuel for the intestinal mucosa and, together with alanine, as substrate for hepatic urea synthesis (25). Glutamine metabolism in the liver is heterogeneously distributed, with catabolism via the ammonia-activated, hepatic form of glutaminase strictly confined to the upstream periportal hepatocytes (*91–93*) and biosynthesis via glutamine synthetase confined to the downstream pericentral hepatocytes.

Glutamine catabolism in the liver is controlled by the activity of hepatic glutaminase and by glutamine transport (*94*). Hepatic glutaminase is activated by its substrate, ammonia ($K_{0.5}$ = 0.3 mM), that is, at physiological portal concentrations. This "feed-forward activation" of hepatic glutaminase (*95*) allows the liver to increase its glutamine catabolism when the organism is confronted with a glutamine load (*5*). Channeling of ammonia between hepatic glutaminase and carbamoylphosphate synthetase, the first enzyme of the ornithine cycle, in the mitochondrion partially compensates for the low affinity of the latter enzyme for ammonia. Furthermore, the hormonal regulation of the expression of both enzymes (glucocorticoids, cyclic AMP) is similar (for reviews, see Refs. *95–97*). At physiological concentrations of circulating glutamine, the Na^+- and pH-dependent system N is responsible for 80–90% of glutamine transport across the plasma membrane of livers cells from fed rats (*98–102*). Starvation, proinflammatory cytokines, eicosanoids, and glucocorticoids, that is, factors associated with catabolic illness (*99, 103–105*), as well as extra dietary glutamine (*106, 107*), stimulate Na^+-dependent glutamine transport and utilization. Because glutamine utilization in the gut declines under these conditions, the liver becomes the major organ of glutamine consumption (*26, 108*). In metabolic acidosis, on the other hand, glutamine flow is directed away from the splanchnic bed toward the kidney (*25*). Under these conditions, urea synthesis is inhibited (*109–111*), because, among other reasons, amino acid transport and the flux through glutaminase and carbonic anhydrase are inhibited (*95, 96*). The enhanced glutamine synthesis by the liver that is often reported to accompany metabolic acidosis can at least partly be ascribed to the NH_4Cl loading that is used to induce the acidosis (*112, 113*).

About one-third of the physiological portal ammonia load reaches the pericentral hepatocytes (*114, 115*). In the absence of downstream glutamine synthetase, e.g., in cirrhosis, circulating ammonia is fivefold elevated to 0.2–0.3 mM (*79, 116*; our own unpublished observations in rats with right-sided heart failure), clearly demonstrating that the ornithine cycle functions as the low-affinity system for ammonia detoxification and pericentral glutamine synthetase functions as the high-affinity scavenging system. Perhaps as a compensatory mechanism, hepatic glutaminase is up-regulated under these conditions to maximize the flux through the ornithine cycle (*116*).

Glutamate transport is appropriately concentrated in the pericentral zone of the liver (*99, 117–119*). Whereas hepatic glutamate transport is quantitatively largely Na^+ independent, the Na^+-dependent form only is strongly up-regulated by glucocorticoids and diabetes (*99, 117*). Because of the pericentral localization of ornithine aminotransferase, transamination of ornithine has been suggested as another source of glutamate for glutamine synthesis (*120*).

2. NERVOUS SYSTEM

a. Topography. GS is predominantly expressed in the protoplasmic (type 1) astrocytes of gray matter (*121–124*). This distribution is more pronounced in humans than in rats, where lower levels of GS are also found in the fibrillary astrocytes of white matter (*121, 122*). Within the astrocytes, GS is concentrated in mitochondria-rich subdomains near glutamatergic synapses (*123, 125*), reflecting its role in the intercellular glutamate metabolism. Oligodendrocytes, the glial cells that produce myelin, also express GS, but at four- to fivefold lower levels than astrocytes (*126–129*). Finally, ependymal cells, the ciliated cells lining the ventricles of the brain, express trace amounts of GS (*122*), but their expression can be induced *in vitro* by hormones (*130, 131*). In rat and chicken retina, Müller cells, the glial cells of the retina, also express high levels of GS (*132–134*). In the gut, GS is expressed in the enteric glia, that is, in the supporting cells of the myenteric and submucosal plexuses (*135–137*). GS activity is fivefold higher in purified astrocytes of the forebrain than in a homogenate (*126*), corresponding with the earlier finding that ammonia fixation into glutamine occurs in a compartment that comprises less than 20% of all brain ammonia (*138, 139*). Moreover, astroglia are believed to occupy approximately 20–25% of brain volume (*126, 140*). Together, these data indicate that GS concentrations in astrocytes amounts to fivefold the overall concentration in brain, or approximately 40 μM (Table I).

b. Regulation of Expression. Most data on the hormonal regulation of GS expression in astrocytes are derived from cultures of neonatal cerebrum or cerebellum after 1–3 weeks in culture. In these studies, GS expression activity is often used as a marker to follow differentiation of glial cell populations, high levels of GS indicating a higher degree of astrocyte-like differentiation. Under these *in vitro* conditions, glucocorticosteroids are always stimulatory agents (*66, 130, 141–149*), but the effects of thyroid hormones on GS expression are limited (*66, 130, 131, 141, 149–152*).

When interpreting the available information on peptide hormones and their intracellular second messengers, it has to be kept in mind that the growth- and differentiation-inducing activities of these factors have negative and positive effects, respectively, on GS expression. Furthermore, the level of GS expression appears to have a maximum, beyond which induction of expression is not possible (*142, 144, 153*), so that additive effects are not always found. Exposure of astrocytes to acidic or basic fibroblast growth factor (aFGF, bFGF) (*130, 154–157*), epidermal growth factor (EGF) (*130, 158,*

159), platelet-derived growth factor (PDGF) (*130, 157*), cyclic AMP (*66, 130, 144, 147, 160–162*), staurosporine (*163*), or phorbol ester (*164*) enhances differentiation more than proliferation (*154, 160*), and consequently, enhances GS expression. Cyclic AMP induces GS only in early passage, that is, relatively undifferentiated, but not in late, more differentiated C-6 glioma cells (*165*). Similarly, exposure of freshly isolated neonatal astrocytes to cyclic AMP for less than a week induces GS expression (*66, 130, 144, 147, 161*), whereas exposure of astrocytes that are already morphologically differentiated (*146*) or gliomas already containing high levels of GS (*145*) is associated with down-regulation of GS levels. On the other hand, thrombin stimulates astrocyte growth without inducing differentiation and, hence, does not stimulate GS expression in astrocyte cultures (*166, 167*), whereas interleukin-4 at low concentration stimulates growth and at high concentration stimulates differentiation and stimulation of GS activity (*168*). Other cytokines (growth hormone, tumour necrosis factor-α, γ-interferon, and interleukin-1α, -1β, or -6) do not affect GS activity in astrocytes (*61, 62, 169, 170*). Transforming growth factor-β (TGF-β) suppresses stimulation of GS expression (*169, 171*), possibly by sensitizing astrocytes to the mitogenic effects of bFGF (*171*). However, it does not have a comparable effect on other mitogenic factors, such as EGF, PDGF, or thrombin (*171*). Endothelin-1 stimulates growth of differentiated astrocytes and induces dedifferentiation, which is associated with decreased GS levels (*162*). Exposure of the cultures to insulin stimulates GS expression (*141*). Platelet-activating factor (PAF) stimulates GS in both early-passage (undifferentiated) and late-passage (differentiated) C-6 glioma cells (*165*). In view of the differential effects of growth and differentiation on GS expression in glial cells (*172*), it is premature at present to link any of these growth factors directly to the regulation of expression of GS. Moreover, even though the regulation of expression of GS in astrocytes and oligodendrocytes shares factors such as glucocorticosteroids and cyclic AMP, both cell types respond differentially to growth factors (*130*).

c. Function. Glutamine plays an important role as a precursor for neuronal glutamate and γ-aminobutyric acid (GABA) biosynthesis. Glutamate is the major excitatory transmitter in the brain and GABA the major inhibitory transmitter. Glutamine is synthesized from glutamate and ammonia in the astrocytes and is transported to the neurons. After having served as neurotransmitter, glutamate is taken up from the synaptic cleft by astrocytes. The metabolic compartmentation of GABA and glutamate in neurons, and glutamine in the astrocytes, is known as the intracellular "glutamate–glutamine shuttle" (*173, 174*). The rates of both uptake and release of glutamate by astrocytes and neurons far exceed those of glutamine (*175*), suggesting that the availability of glutamate for intracellular metabolism can be rapidly modulated (cf. *176*). In astrocytes, the rate of uptake of glutamate exceeds that of release, whereas the reverse is the case in neurons, resulting in a net transport of glutamate from the neurons to the astrocytes. Conversely, there is a net transport of glutamine from astrocyte to neuron (*177, 178*). The observation that the loss of glutamate from glutamatergic neurons exceeds their uptake of glutamine led to the finding that α-ketoglutarate and alanine also serve as precursors for the neuronal biosynthesis of glutamate (*175, 179, 180*). These substrates are, like glutamine, provided to the neurons by the astrocytes, because only the latter cells express pyruvate carboxylase to enable anaplerosis of tricarboxylic acid cycle intermediates.

Brain NH_4^+ metabolism *in vivo* has been studied with ^{13}C and ^{15}N nuclear magnetic resonance (NMR) (*181–186*). Approximately 30% of glutamate is derived from glutamine and a similar percentage comes from glutamate biosynthesis via glutamate dehydrogenase. The remaining glutamate may be produced by transamination (*175, 187*). *In vivo*, brain GS functions with apparent K_M for arterial NH_4^+ of approximately 2 mM (*183, 185*). The low flux through the enzyme [calculated to be approximately 2–10% of its capacity (*183–185, 188*)] can be ascribed to the low circulating NH_4^+ levels (50–100 μM (*189*), a relatively low astrocyte glutamate concentration [approximately 1–3 mM (*184, 190–192*), i.e., below K_M], and an ATP level of approximately 2.5 mM (*193*). The role of low circulating NH_4^+ levels is underscored by the finding that the rate of glutamine synthesis increases twofold in hyperammonemic rats (*188*). Phosphate-activated glutaminase (*181*) and glutamate dehydrogenase (*182*) enzymes function at only 1 and 0.1% of their capacity, respectively. These striking kinetic data underscore the importance of transport for neurotransmitter homeostasis.

Hyperammonemia contributes to the development of encephalopathy in at least two ways. Signs of coma appear at blood ammonia levels of 0.5–1.0 mM and brain ammonia levels of 3–4 mM (*194*). Hyperammonemia causes an enhanced flux through GS, which, in turn, induces swelling of the astrocyte and the development of cerebral edema (*177, 195–197*). Astrocyte swelling and brain edema originate from glutamine biosynthesis, because inhibition of GS activity by methionine sulfoximine prevents the symptoms (*198–200*). The degenerative changes in the astrocytes cause a loss of AMPA- and kainate-type of glutamate transporters (*201, 202*) and lead to accumulation of extracellular glutamate. Excess extracellular glutamate is neurotoxic, because it increases intracellular [Ca^{2+}] via activation of the *N*-methyl-D-aspartic acid (NMDA)-type glutamate receptor (*203, 204*). Thus, the final steps in the development of hyperammonemic encephalopathy resemble those in ischemia, hypoxia, and hypoglycemia (*180*). These, as well as experimentally induced metabolic encephalopathies, are accompanied by elevated brain ammonia (*189*), possibly reflecting astrocyte dysfunction.

3. Kidney

a. Topography. GS is expressed in rat, mouse, hamster, guinea pig, rabbit, sheep, and pigeon kidney but not in pig, dog, cat, monkey, human, or chicken kidney (*19, 205, 206*). The reason for these species differences is not known, but the excretion of more acidic urine, allowing more NH_4^+ excretion, has been associated with the absence of GS (*207*). Highest levels of GS are present in the distal portion of the proximal convoluted tubules and the proximal portion of the straight tubules of the rat (*207–209*) and mouse (*210*). In the mouse, but not in the rat, GS is also expressed in the proximal portion of the proximal convoluted tubules (H. Lie-Venema and W. H. Lamers, unpublished results), whereas in the rabbit, highest levels are reported in the proximal portion of the proximal convoluted tubules (*207*). In the tubular epithelial cells of the rat kidney, GS concentration amounts to approximately 30 μM (Table I).

b. Regulation of Expression. GS activity levels in kidney are, as in liver, decreased to approximately 70% in hypophysectomized rats (*61, 62*). These effects can be ascribed to growth hormone, because its administration to hypophysectomized animals restores GS levels (*61, 62*). In diabetic rats, GS levels are unaltered (*211*). Starvation causes a twofold increase in GS level in the kidney (*212*), but acidosis does not (*205, 207, 209, 213*).

c. Function. Kidneys of animals that do not express GS in their nephrons are net glutamine consumers, whereas the kidneys of animals that do express GS in their nephrons can either consume or produce glutamine depending on their acid/base balance (*5, 19*). The capacity for renal glutamine biosynthesis is thought to be important for species that constitutively (herbivores) or conditionally produce alkaline urine, because ammonia cannot be efficiently excreted into the urine under these conditions (*5*). The kidneys of rats take up or release little glutamine under normal conditions, with both glutamine synthesis and glutaminolysis taking place simultaneously (*27*). In this species, renal glutamine consumption is strongly stimulated in acidosis, diabetes, and high-protein diet (*5*), whereas glutamine release prevails after an ammonia load (*19*) and endotoxin treatment (*214*). Similar observations were made with rabbit kidney tubules (*215*).

Both in species with and without renal GS, acidosis, in particular metabolic acidosis, is combatted by redirecting interorgan glutamine flow away from the splanchnic bed to the kidneys. Glutamine was recognized as the major substrate for renal ammoniagenesis during chronic metabolic acidosis more than 50 years ago (*216*). A greatly enhanced extraction of glutamine by the kidneys is responsible for the 40% decline in circulating glutamine levels under these conditions, which in turn causes the decrease in glutamine utilization in the viscera. Stimulation of renal glutamine uptake and utilization is dependent on the stimulation of glutamine oxidation by glucocorticoids and growth hormone, which is necessary for net bicarbonate generation (25, 217–219), and on the stimulation of NH_4^+ excretion by growth hormone (220). The minimally effective circulating concentration of glucocorticoids is approximately 0.7 μM cortisol or 0.01 μM dexamethasone (218). Even though phosphate-activated glutaminase is expressed in the mitochondria of all cells of the nephron, up-regulation of glutaminase expression and ammonia excretion in chronic metabolic acidosis occurs only in the proximal convoluted tubule (208, 221–223). Nevertheless, the adaptive changes in glutaminase expression trail those in ammoniagenesis (208, 224, 225). Because a decrease in pH lowers the K_M of α-ketoglutarate dehydrogenase for α-ketoglutarate (226), the flux through this enzyme is thought to be the pace-generating step in acute acidosis (223). On the other hand, kidneys from animals with a high GS content are glucose consumers (227), and glutamine synthesis from ammonia and glucose is associated with a low flux through α-ketoglutarate dehydrogenase (227, 228).

4. MUSCLE

a. Topography. Striated muscle fibers contain only very low concentrations of GS (Table I). Nevertheless, because of its total mass, muscle GS represents a sizable portion of total body GS and muscle serves as a major source of glutamine to maintain glutamine homeostasis in the circulation (229–231). GS activity in muscle is approximately 0.5 μM g tissue^{-1} min^{-1}, so that its cellular concentration is about 1 μM (Table I). Slow-twitch red muscle fibers contain the lowest levels of GS protein, followed by the fast-twitch red muscle fibers and fast-twitch white muscle fibers, the concentration difference being twofold between the first and the last fiber type. (232–236). Muscle mRNA content increases in the same order, but in such a way that the protein:mRNA concentration ratio in slow- and fast-twitch red fibers is twice as high as in fast-twitch white fibers (232, 233), indicating that posttranscriptional control mechanisms play a role in muscle GS expression. GS expression in heart muscle is very low (237–239).

b. Regulation of Expression. Muscle GS levels are enhanced by glucocorticosteroids (233, 239–248), by conditions that are accompanied by a rise in circulating glucocorticoids, such as inflammation, sepsis, and endotoxemia (8, 234, 235, 244, 249–251), and fasting (242, 244, 252). The minimally effective circulating concentration of glucocorticoids to induce GS expression is approximately 1 µM dexamethasone or 50 µM cortisol. Denervation increases GS activity levels very rapidly (<1 hr) in slow-twitch red and fast-twitch white muscle (see below) (232, 253).

c. Function. It has been known for more than 25 years that muscle is a major site of glutamine synthesis in the body, with branched-chain amino acids as major donors of the amino group (254, 255). Furthermore, muscle functions as a sink to remove excess blood ammonia by increasing glutamine synthesis, if a "cafeteria" diet is fed (256, 257), or if liver function is moderately impaired (21, 83, 258–260), but not if it is severely impaired (261). Under these conditions, muscle GS levels are increased (83, 257). The physiological role of glucocorticoids is to mobilize glutamine from muscle to be utilized by the kidney for ammoniagenesis when metabolic acidosis develops (217, 225). Because the uptake of glutamine in the kidney trails its mobilization in muscle, blood glutamine levels temporarily rise before declining to approximately 60% of control levels (25, 219, 225).

Conditions that are associated with chronically increased levels of circulating glucocorticoids [sepsis (13, 262), burn injury (263), fasting (264), metabolic acidosis (217)] or decreased levels of insulin (diabetes) are accompanied by a net release of glutamine from muscle (234–236, 265) and muscle wasting. Mobilization and export of glutamine from muscle is very rapid and sensitive to tetradotoxin. For this reason, the muscular depletion of glutamine that is caused by both glucocorticoids and denervation is thought to result from a decrease in the Na^+ electrochemical gradient to drive the Na^+-dependent (system N^M) glutamine transporter (266–268; reviewed in Refs. 229, 269, and 270). This line of reasoning is underscored by the observation that slow-twitch red muscle, in which the Na^+ electrochemical gradient is largely dependent on its innervation, is more resistant to glucocorticoid-induced atrophy than is fast-twitch white muscle (271, 272). The increases in GS levels in atrophying muscles are regulated both at the pretranslational and posttranscriptional levels (232, 253). The latter can be ascribed to the lowered intracellular glutamine concentration (see Section III,A,1), because the increase in GS level can be prevented by glutamine infusion (233, 243) or exercise (240, 241, 273). The wasting that characterizes chronic mobilization of muscle glutamine is ascribed to the sensitivity of the balance of muscle protein synthesis and breakdown to intracellular glutamine levels (274, 275). Furthermore, glutamate uptake into muscle to enable replenishment of glutamine stores appears to become down-regulated as well in severe catabolic conditions (276), possibly by the same mechanism as glutamine transport.

5. Gastrointestinal Tract

a. Topography. The presence of GS in the intestines was first shown functionally, because glutamine did not decrease below 0.2 mM in the perfusate of isolated preparations unless methionine sulfoximine was added (23). Although all parts of the gastrointestinal tract synthesize GS, expression is not homogeneously distributed. In rodents, ruminants, and chickens, cellular GS mRNA and protein levels are highest in the stomach, followed by colon, with lowest levels in the small intestines (210, 277–279). In both rodents (280) and sheep (279), GS activity is highest in the distal, acid-producing part of the stomach (210, 280). Assuming that GS-positive mucosal cells occupy about 60% of antrum pyloricum, cellular GS concentration is approximately 15 μM in the stomach (Table I). In the small intestine of rat and mouse, GS mRNA and protein are located in the crypt cells of the intestinal mucosa (210, 282), whereas in mouse stomach and colon GS mRNA is also found closer to the luminal surface (210).

b. Regulation of Expression. Even though GS levels are stimulated by glucocorticosteroids in Caco cells (283), GS levels in the small and large intestine are not affected by these hormones (277, 284). Lipopolysaccharide, cytokines, and prostaglandins do not affect GS levels in Caco cells (283).

c. Function. The mucosa of the intestine is the major site for catabolism of the glutamine that is released into the circulation by other tissues (10, 11, 285), although glutamine synthesis in the gut amounts to 30–35% of break down (10). Because its extraction from the circulation is linearly related to blood glutamine concentration, intestinal uptake is decreased in conditions associated with decreased blood glutamine levels, such as excess glucocorticoids, starvation, metabolic acidosis, and diabetes (5, 6). The switch from transamination and alanine release to oxidation and NH_4^+ release further shows that glutamine becomes an important metabolic fuel under these conditions. In the gut and in the liver, the decline and increase, respectively, of glutamine utilization [because of the activation of hepatic glutaminase by the increased portal ammonia load (5); see Section II,A,1,c] make the liver the major organ of glutamine consumption under these conditions (26, 108). The extremely low intracellular level of glutamine in enterocytes (approximately 0.2 mM) (10) points to the importance of glutamine transport for flux, but precise data on its regulation probably have to await the molecular cloning of the transporters.

6. Lung

a. Topography. In mouse and rat lung GS protein and mRNA can be visualized in the epithelium of the bronchioli (210); P. J. Blommaart, M. J. B.

van den Hoff, and W. H. Lamers, unpublished observations). GS activity is approximately 0.5 μmol g^{-1} min^{-1} (Table I). Assuming that the epithelium of the respiratory bronchioli represents approximately 15% of the lung mass, the GS concentration in these cells amounts to approximately 6 μM. Although GS mRNA and protein are also present in long-term cultures of pulmonary endothelial cells (*286*), which account for approximately 50% of lung mass in rat lung (*287*), it is difficult to relate data from these findings to the *in vivo* situation because data on cellular concentrations in freshly isolated endothelial cells are not available.

b. Regulation of Expression. GS mRNA and protein levels in lung are stimulated 20–700% by treatment with glucocorticosteroids (*16, 239, 244*), endotoxins (sepsis) (*244, 250, 288, 289*), and starvation (*244*). The effects of glucocorticosteroid treatment on mRNA levels and enzyme activity are maximal at 4 hr and 8 hr, respectively (*290*). GS mRNA synthesis is also induced by glucocorticosteroids in lung-derived established cell lines of epithelial (L2 cells) (*291*) and endothelial (MPEC cells) (*286*) origin, but lipopolysaccharide, cytokines, or prostaglandins do not affect GS levels in these cells (*286, 291*).

c. Function. Like muscle, lung is a quantitatively important site for glutamine biosynthesis (*14, 292*). As in muscle, glucocorticosteroids and endotoxins quickly but temporarily stimulate glutamine release in both rats (*16, 288, 289, 293*) and humans (*289, 294, 295*).

7. ADIPOSE TISSUE

a. Topography. GS activity in brown adipose tissue amounts to approximately 50% of that in liver, whereas that in white adipose tissue is only approximately 15% (Table I). Assuming a homogeneous distribution and taking into account that about 35% of cell volume in brown adipose tissue and 70% of cell volume in white adipose tissue will be occupied by fat stores (*296, 297*), cytosolic GS concentrations in both adipose tissues amount to approximately 7 μM. In agreement with this finding, GS activity per milligram of cellular protein is similar for white and brown adipose tissue (*212*).

b. Regulation of Expression. GS activities in both types of adipose tissue increase transiently (twofold) on exposure to low temperatures (*296, 298*), but only GS in white adipose tissue appears to be sensitive to fasting (*296, 297*).

The regulation of GS expression is best investigated in 3T3-L1 adipocytes (*299–301*). The rate of GS transcription increases more than 100-fold in differentiating preadipocytes, so that GS can be used as a marker for adipocyte differentiation. Retinoic acid at high concentration increases steady-state GS mRNA levels 2-fold (302). Glucocorticosteroids enhance GS expression in 3T3 fibroblasts (303) and in fully differentiated 3T3-L1 adipocytes (299–301, 304), whereas insulin and cyclic AMP suppress its expression (299–301, 304). However, as in astrocytes, cyclic AMP also stimulates differentiation of 3T3-L1 cells and, hence, temporarily GS activity (305).

c. Function. Adipose tissue serves, together with muscle and lung, as a glutamine-producing tissue (18, 306).

8. REPRODUCTIVE ORGANS

a. Epididymis. i. TOPOGRAPHY. GS expression is highest in the caput epididymis and declines toward the cauda in both rats (307) and mice (210). We have identified the epithelial cells that line the epididymal duct as the GS-containing cells (210). Approximately 30% of epididymal mass is occupied by tubular epithelium (210), so that its concentration in the tubular cells of the caput amounts to 60 μM.

ii. FUNCTION. Glutamate levels in the seminal fluid decline from approximately 60 mM in the caput to less than 5 mM in the cauda (308, 309), suggesting that GS is involved in its removal.

b. Testis. i. TOPOGRAPHY. In testis, GS is present in the intertubular cells of Leydig that surround the microvessels (210, 310). Because Leydig cells share the expression of glial fibrillary protein and the glucose transporter GLUT-1 with astroycytes, they may play an important role in the maintenance of the blood–testis barrier, much like the astrocytes do in the maintenance of the blood–brain barrier (310). Within the seminiferous tubules, GS is expressed only in spermatids during the final phase of spermiogenesis (H. Lie-Venema and W. H. Lamers, unpublished observation), that is, it is transcribed from the postmeiotic haploid genome.

ii. FUNCTION. Restriction of GS expression to the second half of spermiogenesis implies participation in the cycle of the seminiferous epithelium and explains the marked regional heterogeneity in expression.

c. Ovary. i. TOPOGRAPHY. The first mammalian glutamine synthetase gene was cloned from Chinese hamster ovary cells, in which the gene was amplified as a result of inhibition of GS activity by methionine sulfoximine (32). Expression of GS in the theca cells, the female homolog of the Leydig cells, remains to be shown.

ii. REGULATION OF EXPRESSION. In these fibroblast-like cells, GS transcription is stimulated by glucocorticosteroids and insulin, but cyclic AMP has no effect (311).

9. Lymphoid and Other Rapidly Dividing Tissues

a. Topography Although lymphocytes are avid consumers of glutamine, low levels of GS are found in spleen, thymus, and lymphocytes (Table I). Whether GS is expressed in a specific subset of the cells of the immune system is not known.

b. Regulation of Expression. GS levels in thymus (*69*) and leukemic cells (*312–314*) are up-regulated by glucocorticosteroids.

c. Function. Rapidly dividing cells (lymphocytes, macrophages, reticulocytes, endothelial cells, enterocytes, hair follicles, established cell lines, and tumor cells) are all characterized by a high rate of glutamine and glucose utilization (*7, 8, 185, 315–317*). Glucose and glutamine feed into the oxidative and biosynthetic branches of the glycolytic and glutaminolytic pathways, respectively, with the flux through the oxidative branches being in vast excess ($>$100-fold) over that through the biosynthetic branches. However, only a small fraction of glutamine and glucose is fully oxidized, with glutamate, aspartate, alanine, and lactate being the major end-products. The high rates of glutaminolysis and glycolysis are thought to provide the biosynthetic branch with a "dynamic" substrate buffer, that is, intermediates can be tapped off when required for biosynthesis of nucleic acids without appreciably affecting the concentration of these intermediates. The inherently inefficient utilization of glutamine and glucose as fuels in terms of energy provision would then be the cost that has to be paid for this precision in control (so called branchpoint sensitivity) (*176, 230, 255, 316*). The glutamine required to feed the rapidly dividing cells is provided via the circulation by glutamine-producing organs. Particularly in prolonged (surgery, trauma, sepsis, burns) and chronically catabolic conditions (cancer), the rapidly proliferating immune or tumor cells obtain a large fraction their glutamine from muscle, explaining the muscle wasting frequently observed (*255, 318*).

10. Summary

In the previous sections we discussed the tissue distribution and cell specificity of GS expression, the factors that affect its cellular level under (patho)physiological conditions, and the functional implications of the GS expression pattern and level. In all tissues GS expression is cell specific, but the contribution of GS-positive cells to the organ differs, ranging from less than 10% of the cells in liver to more than 95% of the cells in muscle and adipose tissue.

Of the humoral factors that have been investigated, only glucocorticoids are well-established inducers of GS levels. In liver and kidney, growth hormone enhances GS expression, whereas cyclic AMP suppresses it. In brain, it is difficult to distinguish the direct and indirect (differentiation-promoting) effects of mitogens and cytokines on GS expression in astroyctes.

The major function of GS in muscle, lung, and adipose tissue is the production of glutamine. Ammonia and glutamate detoxification are its main functions in the liver and brain, and probably also in the epididymis. In the kidney of animals that produce alkaline urine, ammonia detoxification is its most important role. The function of GS in the gastrointestinal tract and in lymphoid tissues awaits further elucidation, although a role in the production of substrates for nucleic acid synthesis is often mentioned.

B. Development of GS Expression

In this section, we will discuss the interorgan similarities in the temporal pattern of the developmental appearance of GS expression, as well as developmental changes in the cellular distribution of GS-positive cells in liver and central nervous tissue.

1. Patterns in the Developmental Appearance of GS

The developmental appearance of GS expression in the various organs follows similar patterns. In rat liver, brain, and kidney, GS activity increases biphasically (Fig. 1A), with a neonatal upsurge to 40–70% of adult levels and a second, stronger upsurge that leads to adult levels of expression in the third week of life (*319–322*). In between both surges, enzyme levels decrease to 20–50% of the adult level. The developmental pattern in mouse liver closely resembles that in rat liver (*323*). In muscle, white adipose tissue, and skin (Fig. 1C), the neonatal peak in GS activity exceeds the adult level by approximately threefold, to decline to this level gradually in the late suckling and weaning period (*278, 322, 324, 325*). The course of the developmental appearance in stomach and small intestine (Fig. 1B) is intermediate between those in liver and adipose tissue, the neonatal peaks reaching 100–150% of the adult levels (*278, 322, 324*). The high levels of GS activity during the first upsurge probably obscure the preweaning upsurge as a separate event in muscle, white adipose tissue, and skin. The main difference between the respective patterns is, therefore, in the relative amplitude of the neonatal increase. However, the cellular concentration of GS in the respective neonatal organs differs less than that in the adult organs (cf. Table I). In the preweaning period, cellular GS becomes further up-regulated in organs of the first group, but is apparently down-regulated in the last group.

Glucocorticoids are probably among the most important of the factors that govern the developmental appearance of GS. Circulating free corticosterone levels peak at 1 day before birth and in the third week of postnatal life in the rat (*326, 327*), that is, coincident with the rising phases of GS levels.

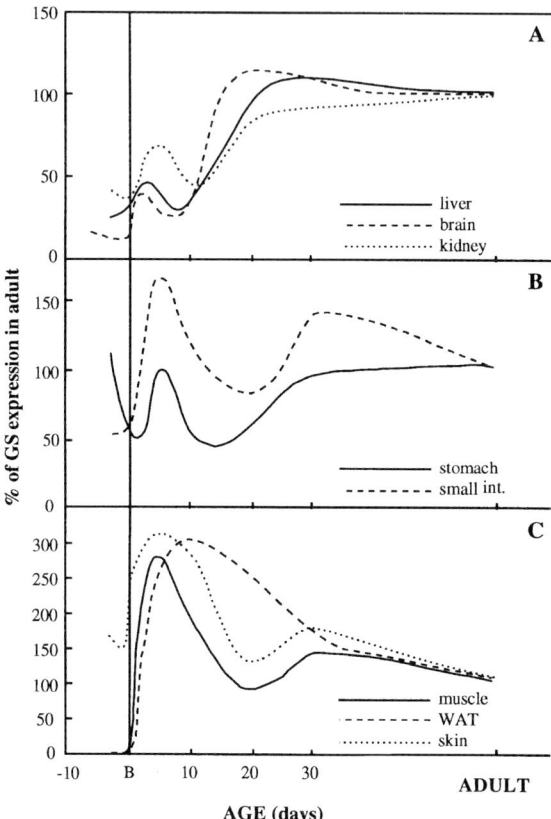

FIG. 1. Developmental appearance of GS activity in organs of the rat. (A) In liver, brain, and kidney, the perinatal peak in activity amounts to 40–70% of the adult level and is followed by an increase to the adult level in the third week of life. (B) In stomach and small intestine, the neonatal peak in GS activity amounts to 100–150% of the adult level. (C) In muscle, white adipose tissue (WAT), and skin, the neonatal peak in GS activity exceeds the adult level by approximately threefold, to decline gradually in the late suckling and weaning period (B, day of birth). For a comparison of adult levels, see Table I. (Data from Refs. 39, 278, 319–322, 324, 325, and 333).

Furthermore, GS expression is precociously inducible by cortisol in brain of suckling rats (142, 153) and in brain, retina, and kidney of suckling mice (328). Although glucocorticoids could not be demonstrated to have an effect on GS expression in suckling mouse liver (328), the similarity of the developmental profile of GS in the liver with that in brain, as well as with other glucocorticoid-inducible genes in liver, strongly points to a prominent role for glucocorticoids (65). The capacity of glucocorticoids to induce GS expression in brain diminishes in the course of development (153), that is, with increasing levels in GS and endogenous corticosterone. The molecular mechanism underlying the glucocorticoid-dependent induction of GS expression in the astroglia of the chicken neural retina will be discussed in Section IV,C.

The ratio between GS protein (activity) and GS mRNA in the liver gradually increases in the late fetal period and becomes stable after the first neonatal day (65). Postnatal changes in GS activity closely resemble those in GS mRNA, both in liver and brain (280, 328–330). These findings indicate that posttranscriptional control mechanisms of GS expression predominate before birth and pretranslational control mechanisms predominate after birth. This switch in the control of GS expression is discussed more extensively in Section III,A,3.

2. DEVELOPMENTAL CHANGES IN THE CELLULAR DISTRIBUTION OF GS-POSITIVE CELLS

Topographical aspects of developmental changes in GS expression were investigated in the liver and the central nervous system. GS mRNA is still homogeneously expressed in the liver at the end of organogenesis in the rat (57, 60) and mouse (41, 280). In line with the low GS protein:mRNA ratios (see Section II,B,1), only a faint, homogeneous immunohistochemical staining of GS protein can be detected at 4–5 days before birth (29, 60, 331). The strictly pericentral expression pattern of GS develops during the late fetal period and is closely linked to the establishment of the locular architecture of the liver (45, 58, 59). In the fetal period, expression of the GS mRNA is specifically enhanced in pericentral hepatocytes, whereas expression in the periportal hepatocytes ceases, leading to the establishment of the steep portocentral gradient in expression that is characteristic for the definitive distribution of GS mRNA. At 2 days before birth, rat (39, 57) and mouse (41, 280, 331) liver first show the mature heterogeneous staining pattern for GS, albeit that the GS-positive pericentral region is still wider than after birth (41, 45, 280). Birth is not a determinant in the establishment of the lobular architecture and the zonal pattern of GS expression, as is demonstrated by the strictly pericentral expression pattern of GS in precocial rodents, such as the spiny mouse, at 4 days before birth (45). In human livers, GS protein is not present at 6 months after birth (A. F. M. Moorman and J. L. M. Vermeulen, unpublished observation), but has acquired a pericentral expression pattern at 2 years of age (42).

The developmental increase of the GS levels in brain is tightly associated with astrocyte differentiation, the timing of which differs slightly for different brain regions (142, 332, 333). The acquisition of GS expression in glial cells also depends on the interaction with neurons (334). In the rat fetus, GS immunoreactivity can first be detected in the ependymal lining of the central canal of the neural tube at ED14 and subsequently in the fibers that radiate from the ependym ("radial glia") and along which nerve cells migrate (*329, 335*). Thereafter, GS-positive glia cells increase in number with time (*335*).

III. Levels of Regulation of Glutamine Synthetase Expression

A. Posttranscriptional Regulation

Studies in a number of organs, in which both GS protein or activity and mRNA levels were measured, revealed that the ratio between protein and mRNA can differ considerably in various tissues, indicating that posttranscriptional regulatory mechanisms are involved. Thus, GS protein:mRNA ratios are similar in brain, epididymis, kidney, and liver, but are up to 10-fold lower in lymphoid organs, small intestine, lung, and muscle (*210, 239*). In muscle, white (phasic/glycolytic) fibers have a twofold lower GS protein:mRNA ratio compared to red (tonic/oxidative) fibers (*232*). Glucocorticoid treatment causes a further twofold decrease in the GS protein:mRNA ratio in lung (*290*), and thyroid hormone treatment of hypothyroid animals causes a 1.3-fold decrease in liver (*73*). Such findings point to differences in translational efficiency or protein stability.

1. REGULATION OF GS PROTEIN STABILITY

a. In Vivo. In liver and in brain, GS protein has a half-life of 4–5 days (*64, 85*). In the absence of growth hormone, GS half-life in liver increases more than twofold, whereas excess growth hormone causes a decline to 60% in stability (*64*). Growth hormone has no effect on GS protein in brain (*64*), possibly because of the blood–brain barrier. The mechanism via which growth hormone affects GS protein stability is not known. Protein starvation also causes a decline to 60% in GS protein stability, but increases GS stability threefold in subcutaneously implanted Morris hepatoma cells (*85*).

b. In Vitro. In cultured cells, GS protein has a much shorter half-life. GS half-life in differentiated 3T3-L1 adipocytes is approximately 24 hr. In these cells, exposure to insulin increases GS protein stability 1.3-fold, whereas exposure to cyclic AMP decreases GS stability to 70% (*299*); GS half-life in 3T3-L1 cells is not affected by exposure to glucocorticoids (*300*). GS half-life in long-term primary cultures of astrocytes is approximately 18 hr and decreases to 12.5 hr following glucose deprivation (*336*).

In rat hepatoma cells, the half-life of GS protein is reported to be approximately 32 hr in the absence of glutamine in the medium and 10- to 15-fold less in the presence of 5 mM glutamine. The half-maximal effect is obtained at 0.2–0.4 mM glutamine. Cellular enucleation as well as inhibitors of transcription and translation prevent glutamine-induced degradation, whereas glucocorticoids do not (*337–341*). Quantitatively similar observations were made in L6 (*342, 343*) and neonatal (*344*) skeletal muscle cells, in HL-60, CEM-C7, and MOLT 4 leukemia cells (*312, 345, 346*), in mouse GF1 fibroblasts (*35*), in CHO cells (*347*), in Neuro-2A neuroblastoma cells (*35, 348*), in C-6 glioma cells, and in neonatal glial cells (*348*), the notable exception being 3T3-L1 adipocytes (*349*). The available data suggest the existence of a labile intracellular glutamine-sensor protein that, when glutamine is bound, associates with GS protein and targets it for degradation (*340, 350*). However, even though this hypothesis is over 15 years old, the hypothesized protein has not yet been identified.

As stated above, the half-maximal effect of glutamine on GS stability is reached at 0.2–0.4 mM glutamine in the medium. The Na^+-dependent glutamine transporter (system N^M) has, in the presence of a physiological concentration of insulin, the corresponding properties in muscle *in vivo* (*266, 274*) and is responsible for the high distribution ratio of free glutamine in muscle (*229, 270, 274*). Muscle glutamine concentrations have been reported to vary between 5 and 30 mM. Apparently, intracellular glutamine concentrations in tissues, in which GS protein typically has a short half-life (low protein:mRNA ratio), are relatively high and regulated by a Na^+-dependent transporter, whereas tissues in which GS protein has a relatively long half-life have a relatively low intracellular glutamine concentration.

2. COVALENT MODIFICATION OF GS

Both bacterial and mammalian GS proteins are very sensitive to functional inactivation by oxidation (*351–353*). Metal-catalyzed oxidation of GS occurs when a cation capable of redox cycling, such as Fe^{2+} or Cu^{2+}, binds to the Mg^{2+}-binding site of GS. Subsequent reaction with H_2O_2 or O_2 generates an active-oxygen species that oxidizes a histidine residue in the Mg^{2+}-binding region (*354–356*). Oxidatively modified GS is degraded rapidly (*357*). Free radical peptides that are formed on fragmentation of β-amyloid, the protein that deposits in the plaques that characterize Alzheimer's disease, also facilitate oxidative inactivation of GS (*358*). This explains the correlation between the density of β-amyloid deposits and locally decreased GS levels in Alzheimer disease (*359, 360*). Inactivation of GS, which is also found in other neurodegenerative diseases (*360*), may be at the base of the memory problems associated with these conditions (*361*). The selective stimulation of GS expression in cultured astrocytes that is seen after prolonged exposure (48–96 hr) to the most potent generator of peptide radicals, the β-amyloid peptide β25–35 (*362*), may well be caused by a similar mechanism of gene amplification that causes overexpression of GS on enzyme inhibition by methionine-sulfoximine in CHO (*32*) and 3T6 cells (*363*).

GS can also be inactivated by administration of hepatotoxic doses of the analgesic paracetamol (acetaminophen). In fact, in pericentral liver damage due to paracetamol, GS is the first protein that becomes inactivated by forming an arylated adduct with an activated metabolite of paracetamol (*364*).

3. Translation of GS mRNA in Fetal Liver

As was noted in Section II,B,1, the ratio of GS protein:mRNA in liver tissue increases approximately 17-fold in the late fetal period, coincident with rapid growth of the fetus, whereas the ratio does not change significantly after birth (*65*) (Fig. 2). The same phenomenon is found for the protein:mRNA ratios of two other hepatic ammonia-metabolizing enzymes, carbamoyl-phosphate synthetase (CPS; EC 6.3.4.16) and glutamate dehydrogenase (GDH; EC 1.4.1.3) (*365, 366*). Apart from altered protein stability, an in-

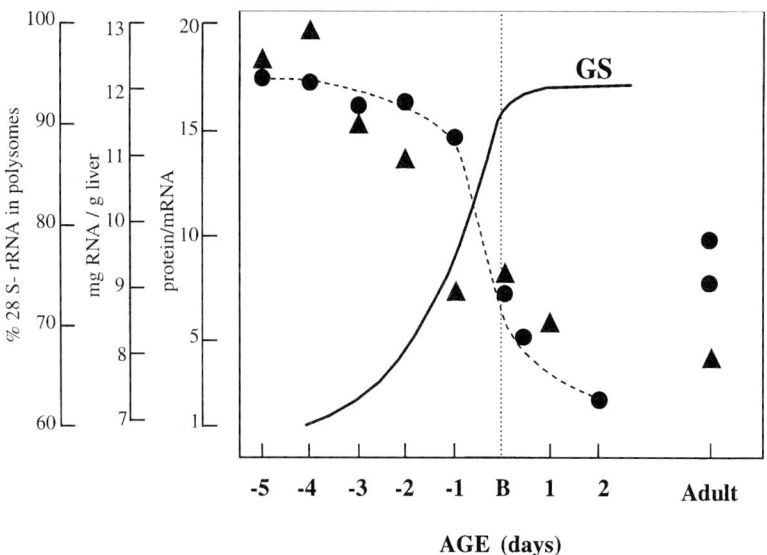

FIG. 2. Perinatal changes in the GS protein:mRNA ratio, RNA content, and the proportion of 28S rRNA in polysomes in rat liver. During late gestation, the protein:mRNA ratio of GS (solid line) in the fetal rat liver increases, whereas the concentration of total RNA (▲) and the fraction of the 28S rRNA that is present in polysomes (●--●) decrease (B, day of birth). (Data on the GS protein/mRNA ratio modified from Ref. 65.)

crease in the protein:mRNA ratio can be caused by changes in the translocation of the transcript from the nucleus to the cytosol, or by an altered rate of translation. There is no indication that control is exerted at the level of mRNA translocation across the nuclear membrane, because *in situ* hybridization experiments do not indicate a prenatal shift in the localization of GS mRNA from the nucleus to the cytosol of the hepatocyte (41, 57, 280).

Control at the level of initiation of protein synthesis is often marked by unmasking of the transcript from stored messenger ribonucleoprotein particles (mRNPs) (367). We therefore analyzed the polysome profiles of cytoplasmic extracts of rat liver at various ages (F. J. van Hemert, L. Boon, J. Salvadó, A. F. M. Moorman, and W. H. Lamers, unpublished results). However, no mobilization of the GS mRNA from mRNPs were observed between 5 days before birth (ED17) and the day of birth (Fig. 3). Instead, GS mRNA was found to be occupied by approximately 12 ribosomes at all ages. Because an actively translating ribosome covers about 90 nucleotides, and the reading frame of the GS transcript is 1122 nt in length (34, 36, 368), synthesis of the GS polypeptide chain proceeds with a maximal load of ribosomes on the GS mRNA. Thus, the increase in protein:mRNA ratios during late gestation cannot be attributed to enhanced mobilization of GS mRNA into polysomes. Similar results were obtained for GDH and CPS mRNA.

Analysis of the polysome profiles revealed two additional features that point to control of translation at the level of polypeptide elongation in the fetal liver. First, the proportion of polysomes relative to monosomes decreases during fetal development (Fig. 3), and simultaneously the percentage of the 60S ribosomal subunit (28S rRNA) present in polysomal fractions declines from 95% at ED17 to 70% at the day of birth (Fig. 2). At the same time the total RNA content in the liver decreases to 70% of prenatal values (65) (Fig. 2). Together, these observations show that the ribosomal content, that is, the protein synthetic capacity, of the hepatocyte decreases to 50% of prenatal values in the same time period that the protein:mRNA ratios increase. One explanation for this paradox can be that a considerable but declining fraction of the polysomes are half-mers, which are detectable as biphasic peaks in the polysomal profile of rat liver at ED18 (Fig. 4). The presence of half-mer polysomes indicates impaired binding of the 60S ribosomal subunit to the mRNA or 60S ribosomal subunit deficiency (369, 370), and implies a low rate of elongation. The late fetal increase in translational efficiency of GS, CPS, and GDH mRNAs is therefore probably caused by a more efficient elongation of the polypeptide chain, which in turn leads to a higher concentration of free ribosomal subunits. Indeed, half-mers disappear as birth approaches (not shown) and are absent in the adult polysome profile (Fig. 4).

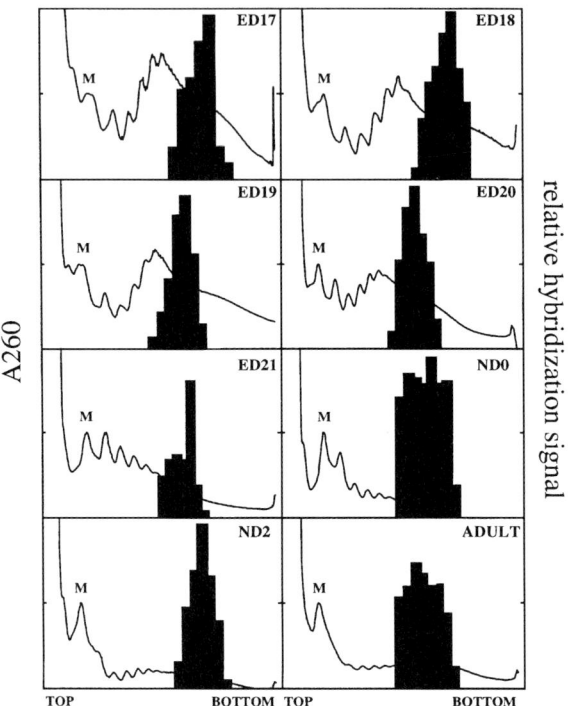

FIG. 3. Ribosomal occupancy of GS mRNA during development. Cytoplasmic extracts were prepared from rat liver (5 days before birth to adult) in a buffer containing cycloheximide (0.2 mg/ml) to prevent peptide elongation and ribosome sliding (491). Coagulation of mRNPs was prevented by addition of 1% deoxycholate. Thirty fractions were obtained after centrifugation (90 min, 200,00 g at 5°C) of the extract (0.5 ml or 20 A_{260} units) in an isokinetic sucrose gradient (0.5–1.25 M) (492). The presence of mono- and polysomes in the fractions was determined by spectrophotometry (A_{260}) and is indicated by a solid line. The polysomal profiles are depicted with the top of the monosomal peak (near the top of the gradient) positioned half of the height of the absorbance axis of each panel. The relative abundance of GS mRNA in each fraction was determined by dot–blot hybridization (493) and PhosphorImager analysis. The amount of GS mRNA signal in each fraction is indicated by black bars. Irrespective of the age of the animal (ED17 to adult), approximately 12 ribosomes occupy the GS mRNA. No mobilization of GS mRNA from mRNPs to polysomes was observed in the course of development. Note that from ED17 to 2 days after birth, the proportion of polysomes relative to monosomes decreases.

B. Pretranslational Regulation

We have described the effects of endocrine and paracrine factors on the expression of GS in tissues in Section II,A, primarily because often it has not yet been sorted out which of these effects is directly mediated by the regulatory regions of the GS gene, and which is indirect, that is, resulting from

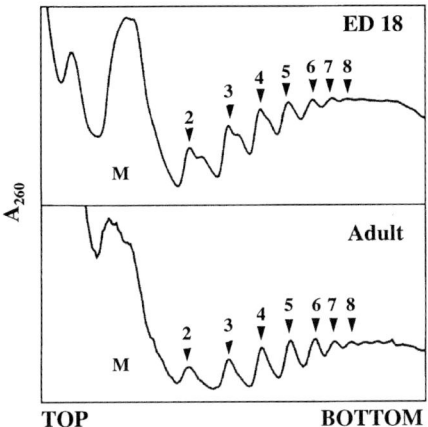

FIG. 4. Polysome profiles of cytoplasmic extracts of fetal (ED18) and adult rat liver. High-resolution polysome profiles of cytoplasmic extracts of fetal (ED18) and adult rat liver were obtained by centrifugation through a linear sucrose gradient (0.5–1.5 M, 4 hr, 110,000 g at 5°C). The number of ribosomes in the polysomal peaks is indicated; the monosomal peak (M) is near the top of the gradient. Polysome half-mers, revealed by biphasic peaks, are present in the fetal (upper panel), but not in the adult (bottom panel) polysomes.

the expression of genes upstream in the signal transduction pathway. Furthermore, the observation that the effects of endocrine or paracrine factors on GS expression apparently have an upper limit is pertinent. Thus, it has been shown in developing brain that GS expression in astrocytes is more sensitive to hormones in perinatal animals with low levels of GS expression than is the case in weaned animals, in which GS expression has reached an upper level (*142, 153*). Similar observations have been made *in vitro* (*144*). These findings show that as gene expression begins to approach its maximum rate, the effect of addition of most regulatory factors declines, making it more difficult to delineate their role. This phenomenon may explain why modulation of GS expression is most clearcut in tissues and cells that contain relatively low levels of GS. In this section, we will confine ourselves to the pretranslational effects of hormones, that is, on GS mRNA stability and GS transcription. (Regulatory regions in the GS gene are discussed in Section IV,B.)

1. FACTORS AFFECTING GS mRNA STABILITY

The half-life of GS mRNA is approximately 2 hr in 3T3-L1 adipocytes (*371*) and approximately 6 hr in L6 muscle cells (*245*). GS mRNA stability is not affected by glucocorticoids (*245, 371*) or insulin (*371*), but declines to approximately 50% in the presence of cyclic AMP (*371*). If hormone treatment induces gene transcription, but does not change mRNA stability, the time required to reach half-maximal concentration after hormonal induction reflects mRNA half-life (372). Accordingly, half-maximum GS mRNA levels after glucocorticoid treatment are observed after 2–3 hr in lung (290) and liver (67). Therefore, the half-life of GS mRNA in most tissues appears to be 2–6 hr.

2. FACTORS AFFECTING GS TRANSCRIPTION

The regulation of the rate of transcription of the GS gene has been determined in 3T3-L1 adipocytes (304), in CHO cells (311), and in L6 muscle cells (245). In all cell types, glucocorticoids stimulate gene transcription. Insulin decreases it in 3T3-L1 cells, but increases GS transcription in CHO cells. Cyclic AMP also decreases GS transcription in 3T3-L1 cells and is without effect in CHO cells.

3. SIGNAL-TRANSDUCTION PATHWAYS

Although considerable progress has been made in mapping the regions of the DNA that link signal transduction pathways with GS gene expression in the chicken (see Section IV,C), such data are not yet available for mammalian GS. Here, we will therefore very briefly summarize the signal transduction pathways of the factors that have been implicated in GS expression in mammals (see Section II,A,10).

The biological effects of glucocorticoids are mediated by the glucocorticoid receptor (GR). On ligand binding, this receptor dissociates from its chaperone HSP90, enters the nucleus, dimerizes, and transactivates expression by binding to glucocorticoid response elements in the regulatory regions of genes. Transactivation of expression by glucocorticoid receptors is mediated by steroid receptor coactivator (SRC) and CREB–binding protein (CBP), so-called coactivator proteins that do not bind DNA, but enhance its accessibility to transcription factors by modifying the nucleosomal histones (373) (for a brief review, see Ref. 374). Cyclic AMP modulates transcription by activating the catalytic subunit of protein kinase A (PKA), thereby inducing its nuclear translocation. Within the nucleus, PKA-dependent phosphorylation of cyclic AMP response element-binding (CREB) protein at Ser-133 allows binding of CBP and transcriptional activation. CREB, being a basic leucine zipper (bZIP) protein, can interact with other bZIP transcription factors, such as c-Fos, c-Jun, and nuclear factor-kappa B (NF-κB). Furthermore, it can interact with activated members of the steroid/thyroid hormone receptor superfamily, including GRs. These latter interactions inhibit gene transcription by forming complexes that cannot bind to the response elements or transactivate transcription, possibly by competing for a limiting pool of CBP ("squelching") (375). The association of GRs with CREB may be implicated in the down-regulation of GS by cyclic AMP.

Insulin utilizes multiple signal transduction pathways. It initiates its biological effects by binding to the insulin receptor, resulting in sequential tyrosine phosphorylation of the insulin receptor and intracellular substrates, such as insulin receptor substrate-1 (IRS-1) and IRS-2. These, in turn, activate the phosphatidylinositol 3-kinase (PI3-kinase) and/or the ras/mitogen-activated protein kinase (MAPK) cascade (376).

The signal transduction pathway, by which growth hormone (GH) affects GS expression, is not yet known, but GHR signaling is often mediated via members of the Jak family of tyrosine kinases that act via the STAT family of transcription factors (377, 378).

IV. Glutamine Synthetase Gene: Evolution, Structure, and Regulatory Regions

GS genes or cDNAs have been cloned from a large number of species, ranging from Archaebacteria to Eubacteria, plants, and invertebrates to vertebrates, including humans. Two main types of GS enzymes are known: GSI and GSII (379–381). GSI has thus far been found only in prokaryotes and consists of 12 identical subunits arranged in two rings of six. GSII is present in both eukaryotes and prokaryotes and consists of 8 subunits (2 rings of 4) that lack the C-terminal portion of GSI. GSIII, a 6-subunit type of GS, with subunits that are larger than those of GSI, has thus far been detected in only three bacterial species (381). Analysis of GS sequences has been used to support the division of the extant living organisms into Eubacteria, Archaebacteria, and eukaryotes (380, 382), and to trace evolutionary lineages in vertebrates (383–385). Comparison of GSI and GSII sequences further revealed that the two types arose by a gene duplication that preceded the divergence of pro- and eukaryotes by perhaps as much as a billion years, qualifying it as a genuinely ancient enzyme (379) and as a good molecular clock (386). As expected, conservation at the active site is most pronounced (387). A vertebrate phylogenetic tree that is well in accordance with accepted taxonomy was produced by alignment of amino acid sequences of GS enzymes (Fig. 5).

In several eukaryotes, GSII exists as different isoforms, each having its specific tissue and developmental expression pattern. For example, distinct GS genes are expressed in chloroplasts and cytosol of higher plants. Moreover, different GS genes are expressed in the cytosol of nodules, roots, and leaves (388). *Drosophila melanogaster* also has separate genes for cytosolic and mitochondrial GS (388, 389). The spiny lobster is reported to have only one GS gene (390), but whether other arthropods have separate GS genes for their cytosol and mitochondria is not yet clear. Furthermore, echinoderms (391) and vertebrates [elasmobranchs (382, 392), amphibians (393), reptiles (394), birds (384, 395), and mammals (33, 34, 36)] have only a single GS

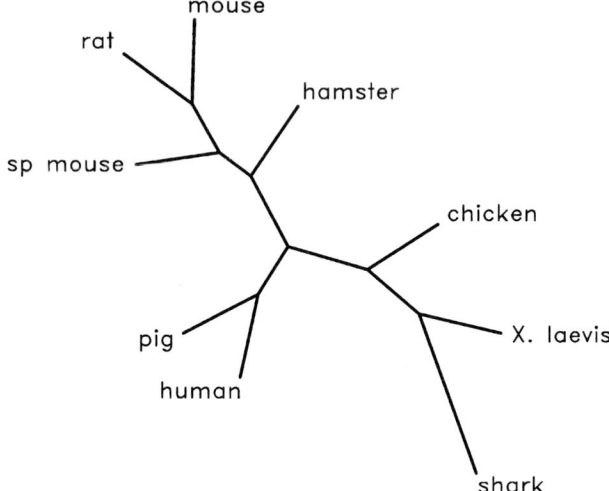

FIG. 5. Unrooted phylogenetic tree of vertebrates GS. Amino acid sequences were deduced from the cDNA sequences of nine vertebrate species (34, 368, 383, 393, 395, 414, 417); the porcine cDNA sequence was retrieved from GeneBank (locus SSGLUSYN; B. E. Loveland, unpublished result); the Chinese hamster cDNA sequence was determined by R. H. Wilson and co-workers (personal communication); the spiny mouse cDNA was recently cloned in our laboratory (W. T. Labruyère and W. H. Lamers, unpublished result). Trees were inferred by means of the PROTPARS module of PHYLIP (100 bootstraps) (494), using randomized input order of the sequences. The CONSENSE module was used to calculate the consensus tree. The Xenopus sequence may contain errors: position 1190 of the cDNA is, for instance, a G, instead of the T that creates the TGA stop codon in all other vertebrates GS sequences.

gene. In elasmobranchs, the mitochondrial GS isoform in liver and kidney and the cytosolic GS isoform in brain and spleen arise as a result of differential usage of two in-frame initiation codons at the N terminus of the same mRNA (383). In birds, mitochondrial GS in liver and cytosolic GS in brain have identical N termini (384), showing that differential targeting to the mitochondrial or the cytosolic compartment is possible with the same primary translation product. In mammals, GS is a purely cytosolic enzyme (37, 384).

The kinetic properties of mammalian and avian GS are similar. Purified GS has a specific activity of ~12 μmol product mg protein^{-1}m min^{-1} at 37°C (γ-glutamyltransferase assay)^3 (396–409). The K_M for glutamate is approximately 5 mM (85, 207, 238, 242, 244, 336, 343, 344, 397, 399, 410–413); for ammonia, 0.2 mM (238, 244, 397, 410, 413) (the K_M for hydroxylamine, the ammonia analog used in the γ-glutamyltransferase assay, is also ~0.2 mM) (238); and for ATP, approximately 1.5 mM (207, 237, 238, 397, 413).

A. Avian and Mammalian Gene Structure

In this section, the structure of the avian and mammalian GS genes will be discussed together, whereas their regulation will be discussed separately (in Sections IV,B and IV,C, respectively). This is because the regulatory regions of avian and mammalian genes show very little similarity (Fig. 6), and because the studies of the regulation of the chicken GS gene have focused on the hormone inducibility and cell specificity of its expression in retinal glia cells, whereas studies in mammals have mainly focused on GS expression in liver.

The first higher vertebrate GS sequence to be isolated was that of the Chinese hamster (*32, 414*), followed by the human (*88, 415–417*), chicken (*384, 395, 418*), cow (*37*), mouse (*33, 328*), and rat (*34, 36, 368*) GS cDNAs and genes. GS pseudogenes have been reported to be present in the mouse (*33, 330, 419*), rat (L. P. W. G. van de Znade and W. H. Lamers, unpublished results), and human (*420, 421*) genomes. The functional human GS gene was mapped to the long arm of chromosome 1, between bands q23 and q31 (*421–423*), whereas the pseudogene was mapped to the chromosomal region 9p13 (*421*). The rat GS gene was mapped to RNO13q22 (*423*).

The structural GS gene in mammals is approximately 10 kb in length and has its mRNA sequence contained in seven exons (*33, 34, 36, 424*). As a typical example, the structure of the rat GS gene and its corresponding mRNA is depicted in Fig. 7. The transcription unit of the chicken GS gene, also containing seven exons, spans 7kb, the difference with mammals being accounted for by differences in the length of the introns. The exon–intron boundaries are conserved between mammals and birds except for those in the 5′ and 3′ untranslated regions (UTRs) (Table II), with a similarity of the nucleotide sequence in the coding region of 87%. The UTRs of mammals share a better than 80% similarity, whereas the similarity between mammals and chicken is very poor (Fig. 8; see Section IV,B,3).

Mammalian GS genes can give rise to a major mRNA transcript of 2.8 kb and a minor mRNA of 1.4 kb in most tissues (see Fig. 7). Vertebrate GS genes use only the canonical AATAAA sequence (*425*) to enforce polyadenylation of the 2.8-bp mRNA, whereas that of the 1.4-kb species is less defined. The 2.8- and 1.4-kb GS mRNA species differ only in the length of their 3′ UTR, due to the use of different polyadenylation sites. An additional mRNA of ap-

---

[3] The biosynthetic assay, utilizing labeled glutamate as substrate ([^{14}C]glutamate + ammonia + ATP → glutamine + ADP), is chosen for its specificity and sensitivity (*410*). However, because of its convenience the γ-glutamyltransferase assay (glutamine + hydroxylamine → glutamyl-γ-hydroxamate + ammonia; transamidation) is most frequently used (*410a*). In contrast to the synthetic transfer assay (glutamate + hydroxylamine + ATP → γ-glutamyl hydroxamate + ADP), this assay does correspond with GS activity (*69*).

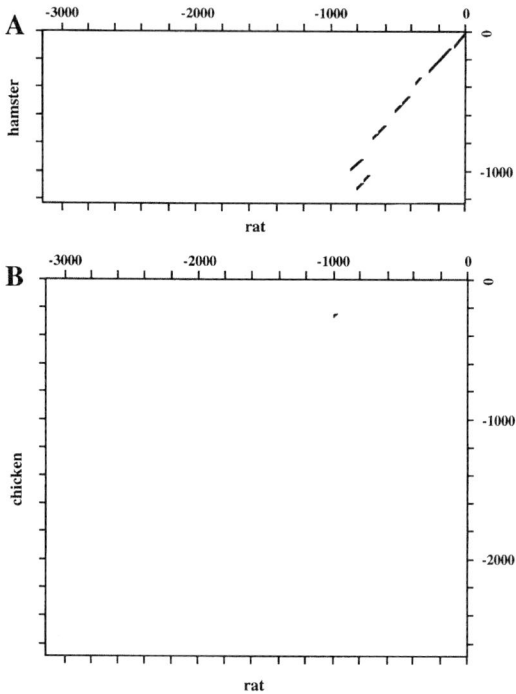

FIG. 6. Comparison of the upstream region of avian and mammalian GS genes. The upstream region of the rat GS gene [from −3150 bp to the transcription start site (36, 426); W. T. Labruyère and W. H. Lamers, unpublished results] was compared with (A) the upstream region of the Chinese hamster GS gene [from −1212 bp to the transcription start site (R. H. Wilson, personal communication] and (B) the upstream region of the chicken GS gene [from −2667 bp to the transcription start site (395)] by means of dotplot analysis (Genetics Computing Group, Wisconsin; package 9.0) employing a window of 40 nucleotides and a stringency of 27. It should be noted that by direct comparison of the rat and chicken upstream regions, several additional, shorter regions with sequence similarity were found (see Table III).

proximately 2.0 kb is transcribed in testes and adipocytes (33, 210, 239). Thus far, no functional differences in the respective transcripts have been described.

B. Regulatory Regions of Mammalian GS Genes

In addition to the exons, the sequences of the upstream region of the rat GS gene up to position −2520 relative to the transcription start site (tss), the first 1938 nucleotides (nt) of the first intron, and the 295 nt of the region immediately downstream of the last exon are publicly available (34, 36, 426, 427). Although little direct proof of the involvement of specific transactiva-

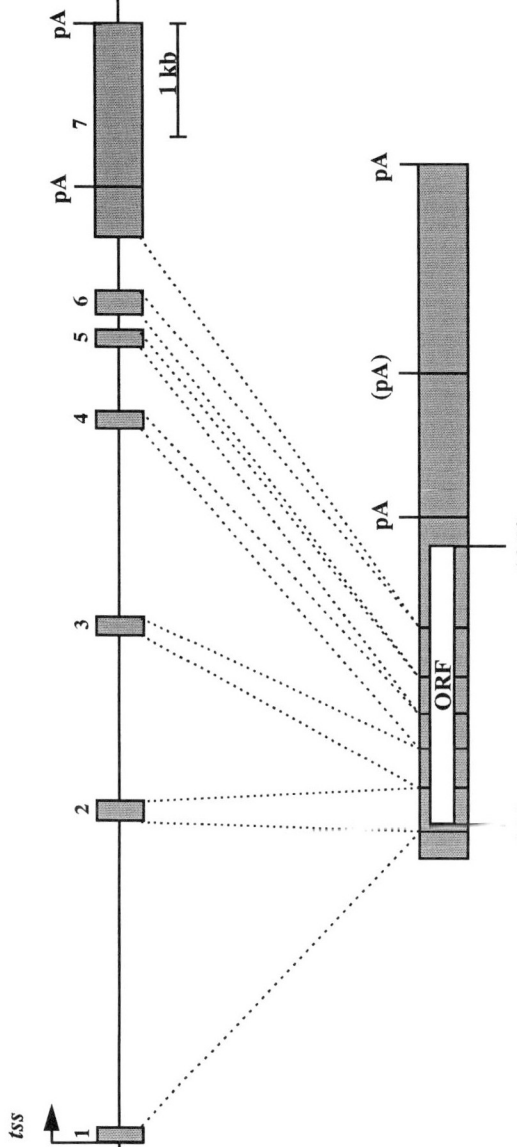

FIG. 7. Schematic drawing of the rat GS gene and its corresponding mRNA. The transcribed region of the GS gene spans 10 kb, comprising 7 exons (tss, transcription start site). Its corresponding mRNA contains an open reading frame (ORF) between nucleotides 132 and 1,253. The poly(A) addition sites (pA) relate to the 1.4- and 2.8-kb mRNAs that are detected in most tissues. The pA between brackets indicates the putative polyadenylation site of the 2.0-kb mRNA in testis and adipocytes. (Data modified from Ref. 34).

TABLE II
Comparison of Exon–Exon Junctions and Exon Lengths[a]

Exon		Exon length (bp)[b]	Exon sequence		
			Begin		End
Exon 1	C	66	+1	cacagccgag .... cccagcccag	+66
	M	118	+1	gggcgcggcc .... caccgctctg	+118
	R	118	+1	ctgcagagcg .... caccgctctg	+118
Exon 2	C	191	+67	gacagccctc .... AGCCTGGAAG	+257
	M	179	+119	aacaccttcc .... TGTGTGGAAG	+297
	R	179	+119	aacaccttcc .... TGTGTAGAAG	+297
Exon 3	C	162	+258	ATCTCCCCGA .... CAGTCTGCAG	+419
	M	162	+280	AGTTACCTGA .... AAACCTGCAG	+459
	R	162	+280	AGTTACCCGA .... AAGCCCGCAG	+459
Exon 4	C	147	+420	ACACAAATCT .... GGACCCCAAG	+566
	M	147	+460	AGACCAACTT .... GGACCCCAAG	+606
	R	147	+460	AGACCAACCT .... GGACCCCAAG	+606
Exon 5	C	128	+567	GTCCGTACTA .... GCCAGCCCAG	+694
	M	128	+607	GCCCGTATTA .... GCCTGCCCAG	+734
	R	128	+607	GACCCTATTA .... GCCTGCCCAG	+734
Exon 6	C	200	+695	TGGGAGTTCC .... GAGGTCTCAA	+894
	M	200	+735	TGGGAATTCC .... ATGGTCTGAA	+934
	R	200	+735	TGGGAATTCC .... ATGGTCTGAG	+934
Exon 7	C	1837	+895	GCACATCGAG .... tatcctgaaa	+2371
	M	1837	+935	GTGCATTGAG .... aaacctcaaa	+2771
	R	1856	+935	GTGCATTGAG .... aaacctcaaa	+2790

[a]Data are from Refs. 33, 34, 36, and 395. The nucleic acid sequence at the exon–exon junction is written in capital letters when it is in the coding region, and localization in the mRNA sequence is indicated relative to the transcription start site (+1).
[b]Data are for chicken (C), mouse (M), and rat (R) GS cDNA.

tor or suppressor proteins in mammalian GS expression is available, potential regulatory elements both upstream and downstream of the tss of the rat GS gene have been identified by DNaseI hypersensitivity mapping and transient transfection assays (36, 426–429). Furthermore, studies have been started to delineate the role of these regions in the regulation of GS expression *in vivo* (210, 323). The results of the aforementioned approaches are discussed in this section. Apart from the basal promoter, at least three regions are involved in the regulation of the spatiotemporal expression pattern of GS, viz. the far-upstream region, the first intron, and the 3′ UTR.

1. Upstream Regulatory Regions

Comparison of the upstream regions of the rat gene (36, 426) and the first 1200 bp of the Chinese hamster gene (R. H. Wilson, personal communica-

ly as a homodimer (*432*), by forming ternary complexes with members of the bZIP family of transcription factors and DNA (*433–435*).

More upstream, the 372-bp region between positions −2520 and −2148 was found to contain a strong enhancer that stimulates reporter gene expression more than 15-fold in conjunction with the heterologous TK81 promoter and 6-fold in conjunction with the GS promoter. Furthermore, this enhancer functions almost 7-fold more efficiently in HepG2 hepatoma cells than in mouse embryonic fibroblast (MEF) cells (*426*). An additional enhancer element has been reported to be present −6.5 kb (*428*).

b. Functional Delineation of Regulatory Regions in Vivo. Even though transfection experiments *in vitro* can identify the presence of regulatory elements that confer hormonal responsiveness and/or cell specificity on expression of a gene, stable transfections *in vivo*, that is, in transgenic mice, are necessary to establish definitively the role of a particular element in the regulation of the spatiotemporal pattern of expression of a gene. This is particularly true for a gene such as GS, which is expressed in a highly specific manner, but at variable levels in many different tissues (see Table I).

To assess the regulatory capacities of the upstream enhancer, Lie-Venema *et al.* (*323*) generated transgenic mice carrying the bacterial chloramphenicol acetyltransferase (CAT) reporter gene under the control of either 3150 bp (GSL mice) or 495 bp (GSK mice) of the upstream region of the rat GS gene. CAT expression is hardly detectable in GS-positive organs of GSK mice, showing that, in contrast to observations *in vitro*, the promoter is insufficient to cause transcriptional activation *in vivo*. A similar observation was made in transgenic mice carrying as much as 2100 bp of the 5′ upstream region of the rat carbamoylphosphate synthetase gene (*436*), but not in transgenic mice carrying only 450 bp of the 5′ region of the rat phospho*enol*pyruvate carboxykinase (PEPCK) gene (*437*). A presumably important difference between the three genes is that only in the PEPCK gene the promoter and enhancer are integrated into a single unit. These findings therefore indicate that *in vivo* the promoter does not suffice for transcriptional activity of a gene.

The 3150-bp upstream regulatory region, including the upstream enhancer that was identified *in vitro*, transactivates reporter gene expression *in vivo* and is a major determinant for the spatiotemporal pattern of expression in liver (*210, 280, 323*). In particular, we observed that the CAT reporter gene is expressed only in the pericentral hepatocytes of the liver. In addition, the perinatal and preweaning increases in GS activity in the liver correspond with similar increases in CAT expression.

To determine the role of the upstream regulatory element in GS expression in other organs in which the gene is expressed, CAT and GS mRNA ex-

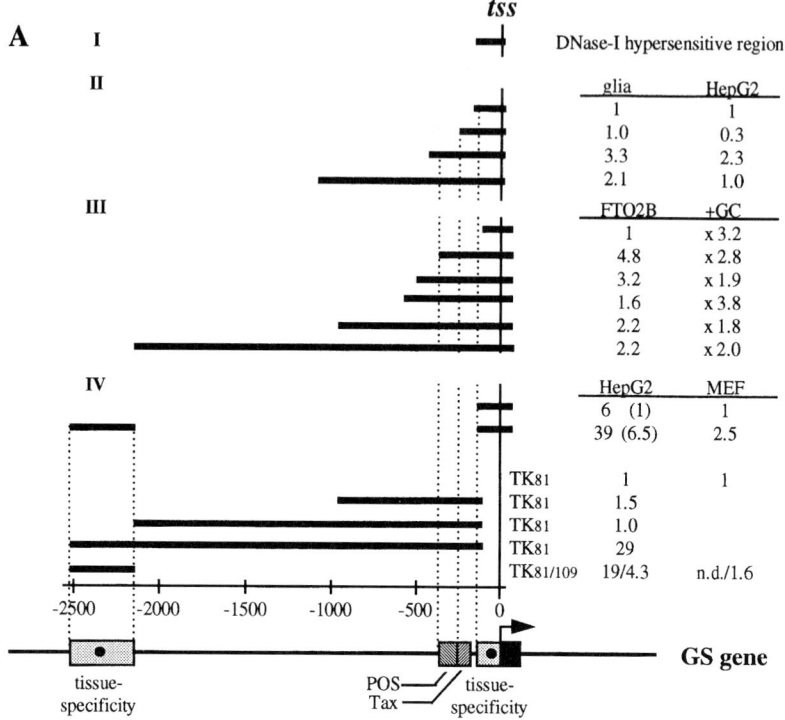
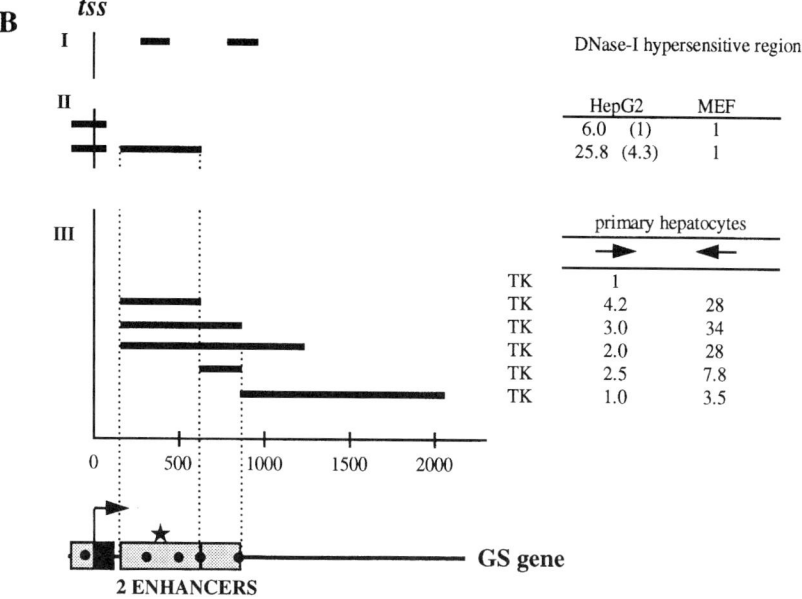

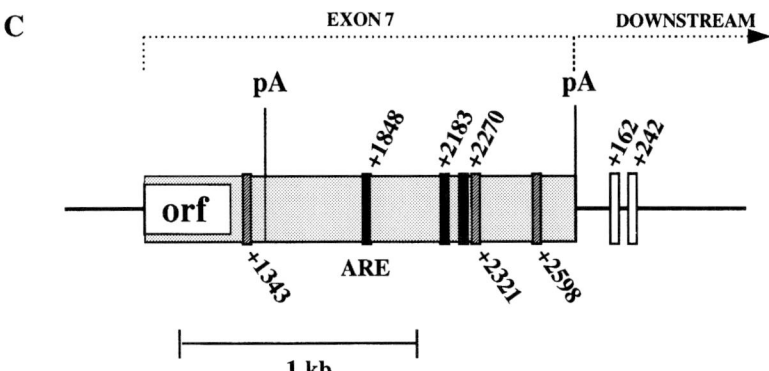

FIG. 9. Identification of regulatory regions in (A) the upstream region, (B) the first intron, and (C) the 3′ UTR of the mammalian GS gene. (A) Regulatory regions in the upstream region up to −2520 bp were identified by DNase hypersensitivity assays [I (426)] and transient transfections [II (36); III, IV (26, 428)]. The sequences identified as regulatory regions are depicted at the bottom of the panel. Consensus sequences for Sp1 sites are indicated (●) within the regulatory elements. A DNaseI-hypersensitive region was found in the basal promoter region (I). The first 153 bp of the basal promoter confer tissue specificity on reporter gene expression in hepatoma cells (HepG2, FTO2B) and glial cells when compared to mouse embryonic fibroblasts (MEF; IV), HeLa, and PC12 (not shown) cells. The sequence between −180 and −251 bp in the immediate upstream region can enhance expression of a reporter gene in both glial and HeLa cells, but only if the HTLV-1 viral protein Tax is cotransfected (429). Expression in hepatoma and glial cells is enhanced by the sequence (POS) between −251 and −368 bp (cf. II and III). Although a weak sequence similarity with a GRE is present at −406 bp, no glucocorticoid (GC) inducibility via this element could be shown (III). The upstream enhancer between −2148 and −2520 bp functions better in hepatoma cells than in fibroblasts, both in the context of the GS promoter and in the context of the heterologous viral thymidine kinase (TK) promoter (IV). (B) DNaseI-hypersensitive sites in the first intron (I) colocalize with two enhancer sequences that were identified by transient transfections of HepG2 and MEF cells (II) and primary hepatocytes (III) (tss, transcription start site) (426, 427). The position of the enhancer elements in the GS is depicted at the bottom of the panel. Consensus sequences for Sp1 sites are indicated (●) within the regulatory elements. The most 5′ enhancer confers tissue specificity on the expression of the reporter gene (II). The position of the active center of this enhancer was deduced from the strong dependence of enhancer activity on the orientation of the DNA element (III) and is indicated with a star. (C) Sequence similarity in the 3′ UTR of mammalian and chicken GS genes is high in three distinct regions (black boxes). At nt +1848 in the rat GS cDNA an AU-rich element (ARE) is present, which may be involved in the regulation of GS mRNA stability. The significance of the other two regions showing a high degree of conservation in human, rat, mouse, spiny mouse, Chinese hamster, and chicken genes is not known. Additional regions with a high degree of similarity in human, rat, mouse, spiny mouse, and Chinese hamster genes are indicated by cross-hatched boxes, whereas regions with sequence similarity in the region downstream of the rat and mouse genes are depicted as open boxes. The nucleotide sequences of these conserved regions are shown in Table IV.

pression was compared by ribonuclease-protein analysis and by *in situ* hybridization (210). In addition to the liver, the upstream regulatory region is also active with respect to the level and the topography of expression in the gastrointestinal tract (stomach, small intestine, and colon), epididymis, and skeletal muscle. On the other hand, GS gene expression is not or is only partly regulated by the upstream regulatory region up to position −3150 in the brain, kidney, adipocytes, spleen, lung, and testis, showing that additional regulatory elements must be active in these organs.

Comparison of CAT and GS mRNA expression in developing GSL mice revealed that the 5′ regulatory region is active and involved in the regulation of GS expression throughout development in pericentral hepatocytes, intestines, and epididymis, and that the 5′ regulatory region is not active at any stage during the development of the periportal hepatocytes and white adipocytes. In fetal stomach, muscle, brown adipose tissue, kidney, lung, and testis, the 5′ regulatory region is very active but, unexpectedly, becomes repressed during subsequent development (280).

When the CAT/GS mRNA and enzyme activity ratios were compared, considerable differences were found, particularly in spleen, jejunum, and muscle (210, 323). This finding probably reflects the pronounced differences in GS protein stability, although differences in mRNA stability cannot be excluded (see Sections III,A and II,B).

2. REGULATORY REGIONS IN THE FIRST INTRON

The presence of regulatory sequences within the first intron of the rat GS gene was suggested by the identification of DNaseI-hypersensitive sites from position +275 to +425 and from position +780 to +950 relative to the tss (426). The region between +156 and +857[4] enhances reporter gene expression in HepG2 hepatoma cells 15-fold in conjunction with the TK81 promoter and 4-fold in conjunction with the GS promoter, whereas the enhancer is inactive in MEF fibroblast cells (426). More recently, it was shown by the same group that the active element is localized in the region between +156 and +632, if tested in cultures of primary hepatocytes, and that this fragment, as well as the original +156 to +857 fragment are much less active in HepG2 cells (427). In conjunction with the TK81 promoter, the enhancer shows strong dependence on the distance from the promoter. We have observed a similar effect when the enhancer of the carbamoylphosphate synthetase gene was tested in front of the TK81 promoter (438). Using the results of the latter experiments for calibration, we estimate that the active center of the intron enhancer is positioned at approximately +350 bp from the tss. The similar activity of this enhancer in cultures of primary hepatocytes that are enriched in either periportal or pericentral cells indicates that it has a quantitative effect on GS expression

---

[4] Note that the nucleotide numbering is different in Refs 426 and 427: the first SmaI site in the intron is located at +254 in Ref. 426 and at +156 relative to the tss in Ref. 427.

in the liver, but does not play a role in determining the pattern of GS expression within the liver.

In addition to the proximal intronic enhancer at position +350, the region corresponding to the second DNaseI-hypersensitive site shows enhancer activity, albeit much weaker. As with the more 5' intron enhancer, the enhancer activity of the fragments was stronger in primary cultures of hepatocytes than in HepG2 cells and showed a strong orientation dependence, suggesting the presence of an active center at approximately +700 and perhaps a third one at +1150 (*427*).

3. Downstream Regulatory Regions

It is becoming increasingly clear that the 3' UTR of mRNAs can influence the localization, stability, and translation of eukaryotic mRNAs (*439, 440*), via interaction with the 5' UTR of the mRNA. In this way, the 3' UTR can control the decapping rate of mRNAs and influence the initiation of translation (*441*). Three regions with a very high degree of similarity in human, rat, mouse, Chinese hamster, and chicken mRNA are present in the 3' UTR of the GS mRNA (positions +1851 to +1873, +2197 to +2215 and +2270 to +2296 of the rat cDNA) (*33, 34, 36, 395, 417*; R. H. Wilson, personal communication) (see also Fig. 8 and Table IV). Of these regions, the one starting at position +1851 in the rat forms the completely conserved center of an AU-rich element (ARE), containing a canonical AUUUA pentamer. Two less conserved AREs, each containing an AUUUA pentamer, are found in the shark 3' UTR (*383*). AREs have been identified in the 3' UTRs of several mRNAs, mainly of those encoding protooncogenes, cytokines, and adhesion molecules (*441–443*), and are known to be among the sequences that control mRNA stability (*441, 442*). However, such a role for the AU-rich element in the GS mRNA remains to be demonstrated. The palindromic AU-rich region at position +2605 to +2823 in the 3' UTR of the chicken GS mRNA that is capable of forming a hairpin *in vitro* (*444*) is not found in the mammalian 3' UTRs. The conserved regions starting at positions +2197 and 2270 form the center of UG-rich elements. The second element contains two completely conserved stretches separated by a less conserved triplet. The function of this element is unknown.

The high sequence similarity of rat and mouse DNA immediately downstream of the GS gene (Table IV) suggests the presence of additional regions that may regulate GS expression at the level of transcription.

4. Summary

The available data indicate that at least three major regulatory elements are involved in the regulation of the spatiotemporal pattern of expression of the GS gene (see Fig. 9). First, the 5' region up to −3150 bp determines the

TABLE IV
Sequence Similarity in the 3′ UTR and Downstream Region of GS Genes[a]

Sequence	Species	Position in cDNA
(A)		
I		
GAAUAGUAUUUUUAUAUUUAAAUGUAAAGA.CAAAAA	Human	1875–1910
GAGUAGUAUUUUUAUAUUUAAAUGUAAAAA.CAAAAG	Hamster	1825–1857
AAGUAGUAUUUUUAUAUUUAAAUGUUAAAAACAAAAA	Spiny mouse	1893–1827
GAGUAGUAUUUUUAUAUUUAAAUGUUAAAAACAAAAA	Rat	1848–1884
GAGUAGUAUUUUUAUAUUUAAAUGUUAAAAACAAAAA	Mouse	1967–2003
AAGUAGUAUUUUUCUAUUUAAAUGUAAAAA.CGAACA	Chicken	1853–1888
No ARE found	Xenopus	–
CAAAACUGUUCUCUUAUUUAGAACAGUUAAUAACUCUU	Shark	1490–1527
GGACACUAACCUGUAAUUUAAUGCAGUCAACAUUGAUA	–	1552–1589
II		
AGGUUUAGAGAUAAGAGUUGGCUGGUCAACUUGAGCAUGUU	Human	2210–2250
AGAGUUAGCAGGAUGAGUUGGCUGGUCAACUUGAACAUUGU	Hamster	2150–2193
CAAGUUCCAAGUAUGAGUUGGCUGGUCAACUUGAACAUUGU	Spiny mouse	2220–2234
CGAGUUAGAAGUAUGAGUUGGCUGGUCAACUUGAACAUUGU	Rat	2183–2223
CGAGUUAGAAGUAUGAGUUGGCUGGUCAACUUGAACAUUGA	Mouse	2174–2204
GCGUUUGGAACUAGGAGUUGGCUGGUCAACUUCANCACGUU	Chicken	2209–2249
III		
GCAUGUCACUAAAGCAGGC...CUUUUGAU	Human	2292–2318
GCAUGUCACUAAAGCAGGC...CUUUUGAU	Hamster	2232–2258
ACAUGUAACUAAAGCAGGUCACUUUUGAU	Spiny mouse	2321–2348
GCAUGUCACUAAAGCGGGC...CUUUUGAU	Rat	2270–2296
GCAUGUCACUAAAAGGGC...CUUUUGAU	Mouse	2267–2293
GCACGUCACUAAAGCAGGG...CUUUUGAU	Chicken	2292–2318
(B)		
I		
GGGUGGAAUAUCAAGGUCGUUUUUUUCAUUC	Human	1327–1357
GGAUGGAAUAUCAAGGUC.UUUUU...AUUC	Hamster	1336–1366
GGAUGGAAUAUCAAGGUC.UUUUU...AUUC	Spiny mouse	1341–1368
GGAUGGAAUAUCAAGGUC.UUUUU...AUUC	Rat	1343–1369
GGAUGGAAUA.CAAGGUC.UUUUU...AUUC	Mouse	1320–1345

(continues)

TABLE IV (Continued)

Sequence	Species	Position in cDNA
II		
AAGUUUAGAUUUUAAUCAAAUUUGUAGGGUUUCU	Human	2347–2380
AAGUUCAGCUUUUAAUCAACUUUGUAGGGUUUCU	Hamster	2283–2316
AAGUUUAGGUUUUAAUCAACUCCGUAGGGUUUCU	Spiny mouse	2373–2406
AAGUUUAGAUUUUAAUCAAGUUCGUAGGGUUUCU	Rat	2321–2354
AAGUUUAGAUUUUAAUCAGAUUUGUAGGGUUUCU	Mouse	2319–2352
III		
GCAGGCCAGCUGUGG . . . UUUUCUUUUGCCAUGA	Human	2582–2612
GCAGGCUGCCUGUGG . . . UUUUCUCUUGCCAUGA	Hamster	2568–2608
GCAAGCCAGCUGUGG . . . UUUUCUCUUGCCAUGA	Spiny mouse	2642–2674
GCAGGCCAGCUGUGGUUUUUUUCUCUUGCCACGA	Rat	2598–2631
GCGGGCCAACUGUGG . . . UUUCUCUCUUGCCAUGA	Mouse	2584–2615
(C)		
I		
GAGACGCGGTCCTAGTGTCTGGCCT	Rat	+162
GAGACGGGGTCCTAGTGTCCGGCCT	Mouse	+147
II		
TCCCAATTTCATATTTGGTCTTATATTTTAAAAGCAGTGTTGTCTA	Rat	+162
TCCCAGTTTCATGTTAGGTCTTACATTTTGAAAGCAGTGTTGTCTA	Mouse	+147

[a]Core regions with perfect similarity are shown in boxes. (A) Regions in the 3′ UTR of mammalian and chicken GS genes. The minimal (pentameric) consensus sequence of the AU-rich element (ARE) us shown in a box, in boldface. (B) Human, Chinese hamster, spiny mouse, rat, and mouse GS genes have additional conserved regions in their 3′ UTRs. (C) In the rat and mouse GS genes, two regions immediately downstream of the 3′ UTR share the same sequence. The nucleotide immediately following the last nucleotide of the last exon is numbered +1. (Data from Refs. 33, 34, 36, 383, 395, and 417; Chinese hamster, R. H. Wilson, personal communication; spiny mouse, W. T. Labruyère and W. H. Lamers, unpublished results.)

spatiotemporal pattern of expression of GS in liver, intestine, and epididymis, but not in kidney, adipose tissue, testis, and brain. Second, the enhancer elements in the first intron may contribute predominantly to the level of GS expression in pericentral hepatocytes. Nevertheless, the intron region, or for that matter the putative enhancer at −6.5 kb of the tss or the downstream region, may also play a role in the expression of GS in the tissues in which the upstream regulatory region to −3.1 kb is insufficient for proper expression. Last, in the organs of transgenic GSL mice, in which the expression of CAT mRNA is relatively high compared to that of GS mRNA, for example, in the epididymis, spleen, small intestine, liver, and muscle, the stability of the GS mRNA may well be determined by RNA-binding proteins that recognize specific regions in the 3′ UTR of the transcript.

The presence of several enhancer elements at some distance from each other raises the question of how they communicate. It is known that DNA-bound Sp1 can associate two or more DNA sites in such a way that the intervening DNA is looped out (445). Therefore, the presence of several Sp1 sites within or near to the up- and downstream regulatory regions of the GS gene and in the basal promoter region tempted us to speculate that Sp1 bound to its consensus sites at positions −2343, −49, and +189 bp from the tss, and to one or more of the Sp1 sites more downstream in the GS gene, functions to position the upstream and intronic enhancer elements closely together (Fig. 10).

C. Regulation of the Chicken GS Gene

Here the focus is on the spatiotemporal regulation of GS expression in the developing chicken retina, the most extensively studied model system. As in mammals, expression is confined to glial (Müller) cells (446–448). Both developmental appearance and cell specificity of GS expression in the neural retina are regulated by the 5′ upstream region of the gene.

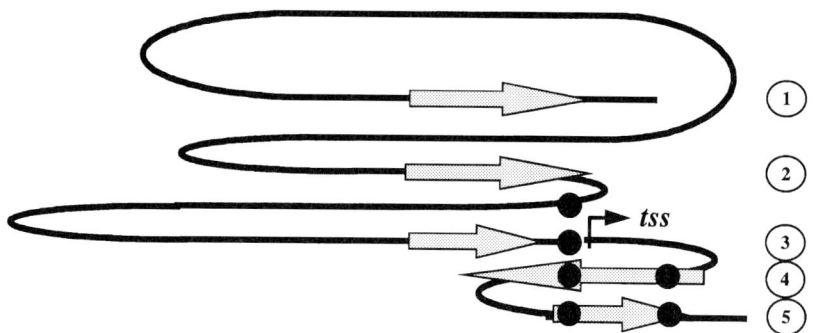

FIG. 10. Model to accommodate the interactions between the enhancer elements and the basal transcription machinery. DNA-bound Sp1 (●) can associate two or more DNA sites in such a way that the intervening DNA is looped out. The transcription start site (tss) is indicated by a black arrow. Enhancer elements are depicted as large grey arrows: (1) far upstream enhancer at −6.5 kb; (2) upstream enhancer between −2520 and −2148 bp from the tss; (3) stimulatory region between positions −368 and −251; (4) enhancer element in the first intron between +156 and +632 bp, and (5) enhancer element between +632 and +856 bp from the tss.

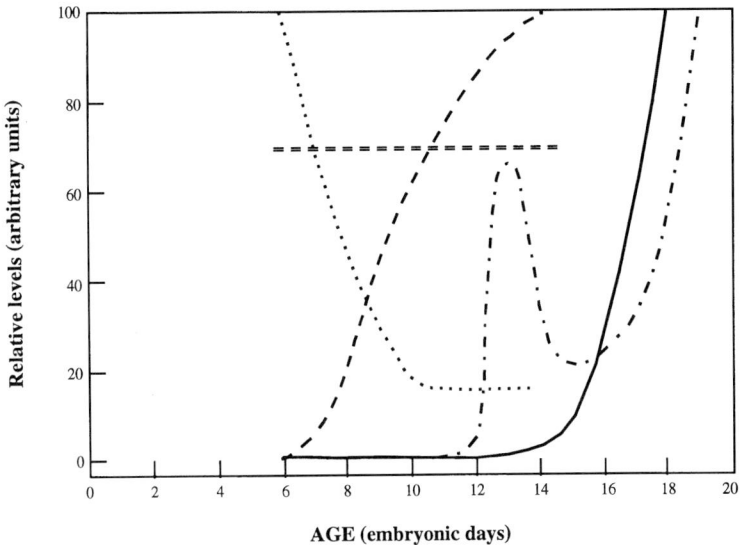

FIG. 11. The developmental appearance of factors that modulate GS expression in chicken neural retina, and their effects on GS expression. The age of the embryo in days of incubation of the egg is on the horizontal axis; cellular or plasma levels of the parameter investigated are on the vertical axis. All values are given as relative units, normalized against the highest value found in the period depicted, except those for glucocorticoid receptors (double dashes), which do not change between ED6 and ED12 and are arbitrarily set at 75%. The data were adapted from references given in the text. GS mRNA increases from ED15 onward (solid line), a few days after a steep transient rise in total plasma glucocorticoids (dashed–dotted line). GS mRNA can be induced precociously by glucocorticoid hormones (dashed line), but responsiveness to glucocorticoid hormones did not develop before ED8. Levels of c-Jun (dotted line) decrease between ED6 and ED8. Note the reciprocal change in the c-Jun level and glucocorticoid responsiveness.

1. Developmental Appearance

GS levels in chicken neural retina are very low until ED15 (449), but shortly before hatching (ED21) transcription of the GS gene increases more than 100-fold (418, 449–454). During subsequent development, the high levels of transcription are maintained (455). The developmental appearance of GS has been ascribed to the increase in circulating cortisol after ED15 (456), because glucocorticoids (cortisol) are able to enhance transcription of GS mRNA (446, 455). Interestingly, even though endogenous glucocorticoid receptor (GR) levels are high from ED5 onward (457–460), GS expression can be prematurely induced only from ED8 onward (418, 446, 448, 450–452, 455, 461). (Fig. 11). This points to posttranslational modification of the GRs and/or the presence of an inhibitory activity during the glucocorticoid-nonresponsive period. The development of hormone responsiveness of GS expression coincides with cessation of proliferation and the start of differentiation of retinal cells (*462–464*).

GS expression also increases in long-term cultures of embryonic retina in the absence of glucocorticoids. This increase, which is only 10% of that *in ovo*, reveals an additional, constitutive component involved in the regulation of GS expression (*454, 465, 466*). Similarly, the twofold prehatching increase in hepatic GS levels is barely dependent on the prehatching increase in circulating glucocorticoids (*418, 450*).

2. DEVELOPMENT OF HORMONAL INDUCIBILITY

Functional dissection of the 5′ upstream region of the chicken GS gene by transient transfections revealed a single glucocorticoid-response element (GRE) between -2120 and -2079 bp upstream of the tss (Fig. 12) (*467–469*). Structural dissection by footprint analysis with the DNA-binding domain of the GR delineated the boundaries of the GR-binding sequence to -2106 and -2079 bp (*467*). Within this sequence, the segment from -2102 to -2090 bp contains 8 of the 12 nucleotides of the canonical GRE (*470*).

Even though the retinal concentration of GR receptors is the same at ED6 and ED10, two important differences exist with respect to these time points. First, the GRs are still homogeneously distributed over the relatively undifferentiated ED6 retinal cells, but gradually accumulate in the Müller glial cells during subsequent development, so that by ED13, GR levels are fivefold higher in glial than in neural cells (*448*). Second, the ratio between the 95-kDa hyperphosphorylated GRs and 90-kDa GRs increases between ED6 and ED10 (*466, 471*). Phosphorylation of steroid receptors has been implicated in nuclear translocation, modulation of binding to response elements in target genes, and transcriptional activation (see, for example, Ref. *472*).

Using footprint analysis and transient transfection assays, a cis-acting element that is necessary to confer full glucocorticoid inducibility on the GS promoter in ED10 retinal cells (*467*) was identified 10 bp upstream of the GRE between -2120 and -2112 bp (Fig. 12). This element (5′CAGCGT-CA3′) bears resemblance to the binding sites for the bZIP transcription factors AP1 and ATF/CREB (*461, 467*). However, transient expression of ATF in ED6 retinal cells does not make GS expression inducible by glucocorticoids in these cells (*461*). Subsequently, retinal-activating protein (RAP), a member of the CCAAT/enhancer-binding protein (C/EBP) family of transcription factors, was identified as a candidate binding factor for the element juxtaposed to the GRE (*457*). RAP is expressed only in trace amounts at ED7 (glucocorticoid nonresponsive) and at elevated levels in ED12 retina (glucocorticoid responsive) (*457*). Furthermore, overexpression of the related pro-

REGULATION OF GS GENE EXPRESSION

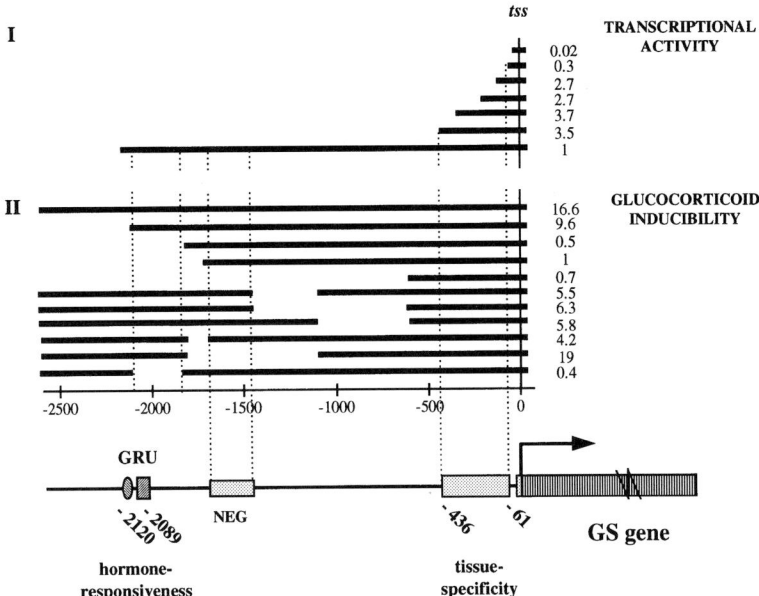

FIG. 12. Identification of regulatory regions in the upstream region of the chicken GS gene by transient transfections. (I) Transcriptional activity of upstream sequences of the chicken GS gene driving reporter gene expression in transfected ED12 neural retina cultures in the absence of hormones. Activity is expressed as fold stimulation, normalized to that of the longest upstream region that was used in the assay. The sequence between −436 and −61 bp relative to the transcription start site (tss) is able to drive reporter gene expression specifically to glia cells, and is not responsive to glucocorticoid hormones. (Data from Ref. 447.) (II) Hormonal inducibility of reporter gene expression in constructs where gene expression is driven by segments of the upstream region of the chicken GS gene. The fold induction of expression by glucocorticoid hormone is normalized to the hormonal inducibility of the sequence between −1849 and +34 bp relative to the tss. Glucocorticoid responsiveness is conferred by an element between −2120 and −2079 bp. In this region, an AP-1/ATF/CRE-like site and a single GRE were identified, together forming a glucocorticoid response unit (GRU). A region that negatively influences glucocorticoid hormone induction was identified between positions −1726 and −1494. (Data modified from Ref. 467.) The regions conferring glia specificity and glucocorticoid responsiveness to the expression of the reporter are indicated in the schematic representation of the GS gene at the bottom of the figure.

tein C/EBPα in ED7 retina cells causes a 1.5-fold induction in the hormone inducibility of GS transcription (457).

Because of the resemblance of the binding site to an AP1-binding site, the possibility that the glucocorticoid-nonresponsive state of the early neural retina is controlled by the Jun/Fos family of transcription factors (comprising

AP1) was investigated (*461, 473*). Expression of members of the c-Jun and c-Fos family of transcription factors has been associated with cell proliferation. Indeed, expression of c-Jun is high in early embryonic retina and declines with development (*474*) when retinal growth declines and glucocorticoid inducibility of endogenous GS rises (*462*) (Fig. 11). Overexpression of c-Jun in middevelopmental stages (ED10–13), that is, after endogenous GS expression has become glucocorticoid responsive, causes a pronounced decline in the glucocorticoid inducibility of the endogenous GS gene, suggesting the formation of inactive c-Jun–GR heterodimers (*474*). Interference of c-Jun with the binding of GRs to the GRE also explains why overexpression of GR in early-stage retinal cells can abrogate the inhibited glucocorticoid responsiveness of GS expression at this age (*461, 469*).

The regulation of GS expression in chick neural retina has been a paradigm for developmental processes that require, in order to proceed, complex cell–cell interactions between neurons and astrocytes in the context of an organ-specific architecture (*446, 467, 475–479*). It has now become apparent that monolayer cultures lose the capacity for glucocorticoid hormone-dependent activation of GS as a result of a dramatic decrease in the level of glucocorticoid receptor molecules that is accompanied by a reciprocal increase in the level of c-Jun (*480*). Increasing the level of expressed glucocorticoid receptor molecules by transfection (*461, 469*), or activation of protein kinase A by cyclic AMP, results in a recovery of the response (*480*), whereas disruption of the retinal architecture by oncogenes (v-*scr*) entails a reduced inducibility of GS by glucocorticoids (*462, 478*).

3. Cell Specificity of Expression

Although the expression in GR in neural retina becomes predominantly confined to Müller glia during development (*448*), steroid hormones are not essential to achieve Müller cell-specific GS expression (*447*). Just upstream of the GS promoter, a cis-acting element with -436 and -61 bp as provisional boundaries confers glial cell specificity on the expression of GS (*447*) (Fig. 12). Although the minimal sequence requirements and the responsible transcription factor(s) to achieve this cell specificity in expression still need to be determined, sequence similarity has been noted between the GC-rich sequences (nt -94 to -63) of the chicken GS gene (*395*) and sequences in the proximal promoters of genes encoding the chicken and human carbonic anhydrase II (*452, 481*), another glial marker enzyme. Although the rat gene also contains similar sequences in its promoter region (between -365 and -387 bp and between -97 and -128 bp in the tss) (*36, 426*), rat GS promoter–CAT constructs encompassing this region were not able to drive CAT expression to the astroglia in the central nervous system of transgenic mice (*210*).

4. Summary

Glucocorticoids increase GS levels in the chicken embryonic retina. However, the effectiveness of hormonal induction of GS expression depends on the balance between differentiation and proliferation in the (experimental) system. At the time that the circulating cortisol level increases, the 90kDa isoform is transformed into its hyperphosphorylated form (466, 471). When GR and transcription factors that facilitate its transcriptional activity (e.g., RAP) are present, expression of the GS gene is readily induced by glucocorticoids via its GRU (glucocorticoid response unit: a GRE and an ancillary site necessary for full activity) at −2.1 kb. On the other hand, when factors are present that interfere with GR transcriptional activity (e.g., c-Jun), glucocorticoid induction of GS expression is inhibited. c-Jun levels are elevated during the growth phase of the retina in early development and when neuron–glia interactions have been disrupted. On the other hand, when c-Jun levels are suppressed and RAP levels are increased during the prehatching phase of glia cell differentiation, or when cell–cell contacts are restored, glucocorticoid inducibility of GS expression is established.

The transcription factors that are involved in Müller cell specificity of retinal GS expression act via a region near the promoter region of the GS gene. However, their nature remains to be elucidated entirely.

V. Concluding Remarks

Figure 13 provides an overview of our present insight into the regulation of the expression of the GS gene. The molecular mechanisms underlying this regulation are only beginning to be understood. In this review, we aimed to provide data that will facilitate studies that strive for the identification and characterization of the DNA sequences that are necessary and sufficient to account for the spatiotemporal pattern of expression of GS. The first step in such a study is often the coarse delineation of sequences with regulatory properties with functional assays in cultured cells. Because the established cell lines that are often used combine convenience with peculiarities that result from their life *ex vivo*, the next step should, in our view, be the confirmation of the properties of these sequences *in vivo*, in transgenic mice. This approach offers the additional advantage that all other organs can be included in the analysis at the cost of little extra effort. A convenient way to assess the regulatory properties of a sequence with respect to tissue distribution, hormonal regulation, and developmental appearance is to compare the level of expression of the reporter gene used with the level of expression of the endogenous GS gene.

The study of glutamine metabolism reveals that, in addition to GS and

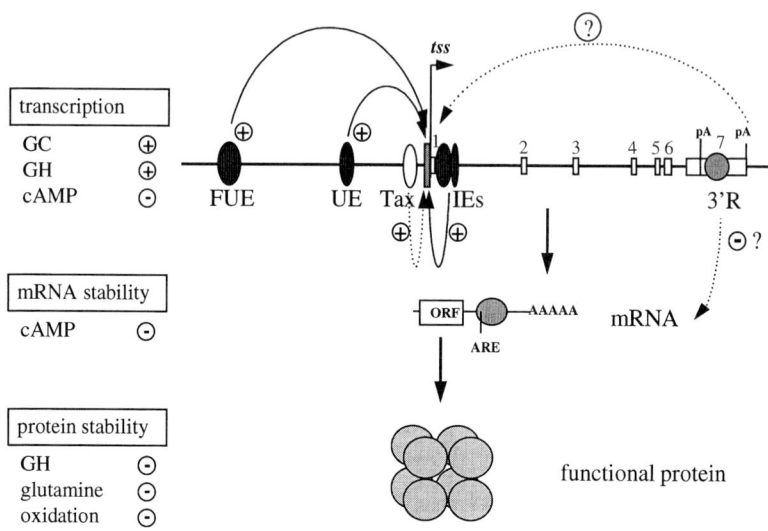

FIG. 13. Overview of the regulation of the GS gene. The enhancer elements found in the upstream region and in the first intron of the rat GS gene positively control its transcription from the basal promoter. Whereas the 5' region to −3150, including the upstream enhancer (UE), determines the spatiotemporal pattern of expression of GS in liver, intestine, and epididymis, the two enhancer (IEs) elements in the first intron appear to contribute predominantly to the level of GS expression, at least in liver. Sequences in the 3' UTR (3'R), one of which contains an AU-rich element (ARE), may be involved in the control of GS mRNA stability and translation. The question mark indicates the possible involvement of the conserved 3' end of the gene in transcription. The role of the far-upstream enhancer (FUE) at −6.5 kb from the tss in regulation of the spatiotemporal pattern of expression of the GS gene awaits further elucidation. Glucocorticoid (GC) and growth hormone (GH) increase transcription of the GS gene, whereas cAMP decreases it. The exact sites in the GS gene by which these effects are mediated are not yet known. The HTLV-1 protein Tax transactivates GS transcription (Tax). cAMP also modulates GS levels by affecting its mRNA stability. Growth hormone, glutamine, and oxidative modifications accelerate degradation of the functional protein.

glutaminases, glutamine and glutamate transporters play a major role. Thus far, these activities were mainly defined as operational parameters rather than as gene products (482–485). However, these functional properties appear to correspond only partially with single molecular identities (486, 487). A major task of future research will therefore be to determine the relative control of enzymes and transporters in glutamine metabolism.

Glutamine is now considered to be a semiessential amino acid, implying that its biosynthesis can become insufficient under catabolic conditions. Under these conditions, especially the tissues that are avid consumers of glutamine, such as lymphocytes, macrophages, reticulocytes, endothelial cells, enterocytes, and hair follicles (see Section II,A,9), become vulnerable. For this reason, glutamine is attracting considerable attention as a nutrient (*106, 488–490*). Because of the high catabolic capacity of the gut (*5*), such glutamine will only partially benefit further downstream glutamine consumers. Future studies should therefore address the role of GS in maintaining the glutamine balance by creating a loss of function ("knockout") mutant. This is of particular relevance because thus far, no syndrome has been described in which the gene encoding GS was affected, suggesting that loss of function is incompatible with life or that there is no phenotype. Furthermore, the multiple regulatory elements of the GS gene that are responsible for the more than 100-fold difference in cellular concentration of GS protein in different tissues (Table I) make GS a gene of choice in the search for cassettes that can be used to target cell- and tissue-specific expression of therapeutic genes.

Acknowledgment

We thank Prof. R. Charles for critical reading of the manuscript.

References

1. H. A. Krebs, *Biochem. J.* **29**, 1951 (1935).
2. J. R. Huizenga, C. H. Gips, and A. Tangerman, *Ann. Clin. Biochem.* **33**, 23 (1996).
3. G. van den Berghe, F. Bontemps, M. F. Vincent, and F. van den Bergh, *Prog. Neurobiol.* **39**, 547 (1992).
4. W. W. Souba, *J. Parent. Enteral Nutr.* **11**, 569 (1987).
5. T. C. Welbourne and S. Joshi, *J. Parent. Enteral Nutr.* **14**, 77S (1990).
6. F. Hartmann and M. Plauth, *Metabolism* **38**, 18 (1989).
7. K. Brand, W. Fekl, J. von Hintzenstern, K. Langer, P. Luppa, and C. Schoerner, *Metabolism* **38**, 29 (1989).
8. M. D. Caldwell, *Metabolism* **38**, 34 (1989).
9. S. S. Tate and A. Meister, in "The Enzymes of Glutamine Metabolism" (S. Prusiner and E. R. Staatman, eds.), p. 77. Oxford Univ. Press, London, 1973.
10. H. G. Windmüller and A. E. Spaeth, *J. Biol. Chem.* **249**, 5070 (1974).
11. G. H. Windmüller, *Adv. Enzymol.* **53**, 210 (1982).
12. H. G. Windmüller and A. E. Spaeth, *Am. J. Physiol.* **241**, E473 (1981).
13. J. Arnold, I. T. Campbell, T. A. Samuels, J. C. Devlin, C. J. Green, I. J. Hipkin, J. A. MacDonald, C. M. Scrimgeour, K. Smith, and M. J. Rennie, *Clin. Sci.* **84**, 655 (1993).
14. T. C. Welbourne, *Contrib. Nephrol.* **63**, 178 (1988).
15. W. W. Souba, K. Herskowitz, and D. A. Plumley, *J. Parent. Enteral Nutr.* **14**, 68S (1990).
16. M. S. M. Ardawi, *Clin. Sci.* **81**, 37 (1991).
17. K. N. Frayn, K. Khan, S. W. Coppack, and M. Elia, *Clin. Sci.* **80**, 471 (1991).
18. T. Kowalski and M. Watford, *Am. J. Physiol.* **266**, E151 (1994).
19. G. Baverel, C. Michoudet, and G. Martin, (D. Häussinger and H. Sies, eds.) *in* "Glutamine Metabolism in Mammalian Tissues" p. 187. Springer-Verlag, Berlin and New York, 1984.

20. V. Grill, O. Björkman, M. Gutniak, and M. Lindqvist, *Metabolism* **41,** 28 (1992).
21. A. J. L. Cooper and S. E. Grisolia, ed., in "Cirrhosis, Hepatic Encephalopathy, and Ammonia Toxicity," Vol. 1, p. 23. Plenum, 1990.
22. N. A. Farrow, K. Kanamori, B. D. Ross, and F. Parivar, *Biochem. J.* **270,** 473 (1990).
23. H. G. Windmüeller and A. E. Spaeth, *J. Biol. Chem.* **249,** 5070 (1974).
24. V. Phromphetcharat, A. Jackson, P. D. Dass, and T. C. Welbourne, *Kidney Int.* **20,** 598 (1981).
25. T. C. Welbourne, *Am. J. Physiol.* **253,** F1069 (1987).
26. W. W. Souba and T. R. Austgen, *J. Parent. Enteral Nutr.* **14,** 90S (1990).
27. A. C. Damian and R. F. Pitts, *Am. J. Physiol.* **218,** 1249 (1970).
28. E. J. Squires and J. T. Brosnan, *Biochem. J.* **210,** 277 (1983).
29. S. Heeneman and N. E. P. Deutz, *Clin. Sci.* **85,** 437 (1993).
30. S. Heeneman and N. E. P. Deutz, *Clin. Nutr.* **12,** 182 (1993).
31. S. Heeneman, C. H. C. Dejong, and N. E. P. Deutz, *Pflug. Arch. Eur. J. Phys.* **427,** 524 (1994).
32. P. G. Sanders and H. Wilson, *EMBO J.* **3,** 65 (1984).
33. C. F. Kuo and J. E. Darnell, Jr., *J. Mol. Biol.* **208,** 45 (1989).
34. L. P. W. G. van de Zande, W. T. Labruyère, A. C. Arnberg, R. H. Wilson, A. J. W. van den Bogaert, A. T. Das, D. A. J. van Oorschot, C. Fritjers, R. Charles, A. F. M. Moorman, and W. H. Lamers, *Gene,* **87,** 225 (1990).
35. A. Sandrasagra, G. Patejunas, and A. P. Young, *Arch. Biochem. Biophys.* **266,** 522 (1988).
36. J. F. Mill, K. M. Mearow, H. J. Purohit, H. Haleem-Smith, R. King, and E. Freese, *Mol. Brain Res.* **9,** 197 (1991).
37. D. D. Smith and J. W. Campbell, *Proc. Natl. Acad. Sci. U.S.A.* **85,** 160 (1988).
38. R. Gebhardt and D. Mecke, *EMBO J.* **2,** 567 (1983).
39. J. W. Gaasbeek Janzen, R. Gebhardt, C. H. J. ten Voorde, W. H. Lamers, R. Charles, and A. F. M. Moorman, *J. Histochem. Cytochem.* **35,** 49 (1987).
40. A. F. M. Moorman, P. A. J. de Boer, W. J. C. Geerts, L. P. W. G. van de Zande, R. Charles, and W. H. Lamers, *J. Histochem. Cytochem.* **36,** 751 (1988).
41. F. C. Kuo, K. E. Paulson, and J. E. Darnell, Jr., *Mol. Cell. Biol.* **8,** 4966 (1988).
42. A. F. M. Moorman, J. L. M. Vermeulen, R. Charles, and W. H. Lamers, *Hepatology* **9,** 367 (1989).
43. W. H. Lamers, A. Hilberts, E. Furt, J. Smith, C. N. Jones, C. J. F. van Noorden, J. W. Gaasbeek Janzen, R. Charles and A. F. M. Moorman, *Hepatology* **10,** 72 (1989).
44. G. T. M. Wagenaar, A. F. M. Moorman, R. A. F. M. Chamuleau, N. E. P. Deutz, C. de Gier-de Vries, P. A. J. de Boer, F. J. Verbeek, and W. H. Lamers, *Anat. Rec.* **239,** 441 (1994).
45. W. H. Lamers, J. W. Gaasbeek Janzen, A. te Kortschot, R. Charles, and A. F. M. Moorman, *Differentiation* **35,** 228 (1987).
46. G. T. M. Wagenaar, R. A. F. M. Chamuleau, J. G. de Haan, M. A. W. Maas, P. A. J. de Boer, F. Marx, A. F. M. Moorman, W. M. Frederiks, and W. H. Lamers, *Hepatology* **18,** 1144 (1993).
47. H. Sirma, G. M. Williams, and R. Gebhardt, *Liver* **16,** 166 (1996).
48. R. Gebhardt and G. M. Williams, *Carcinogenesis* **16,** 1673 (1995).
49. W. H. Lamers, K. E. Hoynes, D. Zonneveld, A. F. M. Moorman, and R. Charles, *Anat. Embryol.* **178,** 175 (1988).
50. F. C. Kuo and J. E. Darnell, Jr., *Mol. Cell. Biol.* **11,** 6050 (1991).
51. W. Schrode, D. Mecke, and R. Gebhardt, *Eur. J. Cell Biol.* **53,** 35 (1990).
52. L. Schöls, D. Mecke, and R. Gebhardt, *Histochemistry* **94,** 49 (1990).
53. G. Zajicek, R. Oren, and M. Weinreb, *Liver* **5,** 293 (1985).
54. N. Arber, G. Zajicek, and I. Ariel, *Liver* **8,** 80 (1988).

55. W. H. Lamers, *Hepatology* **12,** 372 (1990).
56. R. Gebhardt, J. Cruise, K. A. Houck, N. C. Luetteke, A. Novotny, F. Thaler, and G. Michalopoulos, *Differentiation* **33,** 45 (1986).
57. A. F. M. Moorman, P. A. J. de Boer, A. T. Das, W. T. Labruyère, R. Charles, and W. H. Lamers, *Histochem. J.* **22,** 457 (1990).
58. W. H. Lamers, W. Been, R. Charles, and A. F. M. Moorman, *Hepatology* **12,** 701 (1990).
59. R. G. E. Notenboom, P. A. J. de Boer, A. F. M. Moorman, and W. H. Lamers, *Development* **122,** 321 (1996).
60. A. L. Bennett, K. E. Paulson, R. E. Miller, and J. E. Darnell, Jr., *J. Cell Biol.* **105,** 1073 (1987).
61. B. S. Wong and A. Dunn, *Biochem. Biophys. Res. Commun.* **79,** 876 (1977).
62. B. S. Wong, M. E. Chenoweth, and A. Dunn, *Endocrinology* **106,** 268 (1980).
63. E. M. Nolan, J. N. Masters, and A. Dunn, *Mol. Cell. Endocrinol.* **69,** 101 (1990).
64. C. K. Lin and A. Dunn, *Life Sci.* **45,** 2443 (1989).
65. C. J. de Groot, C. H. J. ten Voorde, R. E. van Andel, A. te Kortschot, J. W. Gaasbeek Janzen, R. H. Wilson, A. F. M. Moorman, R. Charles, and W. H. Lamers, *Biochim. Biophys. Acta* **908,** 231 (1987).
66. D. T. Loo, M. C. Althoen, and C. W. Cotman, *J. Neurosci. Res.* **42,** 184 (1995).
67. M. S. Miller, G. S. Buzard, and A. E. McDowell, *Biochem. Pharmacol.* **45,** 1465 (1993).
68. R. Gebhardt and D. Mecke, *Eur. J. Biochem.* **100,** 519 (1979).
69. A. Herzfeld and N. A. I. Estes, *Biochem. J.* **133,** 59 (1973).
70. H. J. Seitz, M. J. Müller, P. Nordmeyer, W. Krone, and W. Tarnowski, *Endocrinology* **99,** 1313 (1976).
71. A. F. M. Moorman, P. A. J. de Boer, R. Charles, and W. H. Lamers, *FEBS Lett.* **276,** 9 (1990).
72. A. L. Kennan, *Endocrinology* **74,** 518 (1964).
73. R. Hartong, W. M. Wiersinga, W. H. Lamers, T. A. Plomp, M. Broenink, and M. H. van Beeren, *Horm. Metab. Res. Suppl.* **17,** 34 (1987).
74. G. T. M. Wagenaar, R. A. F. M. Chamuleau, C. W. Pool, J. G. de Haan, M. A. W. Maas, J. A. M. Korfage, and W. H. Lamers, *J. Hepatol.* **17,** 397 (1993).
75. R. Gebhardt, *Cancer Res.* **50,** 4407 (1990).
76. C. Wu, *Biochem. Int.* **8,** 679 (1984).
77. R. Gebhardt, H. Burger, H. Heini, K. Schreiber, and D. Mecke, *Hepatology* **8,** 822 (1988).
78. R. Gebhardt, F. Gaunitz, and D. Mecke, *Adv. Enzyme Regul.* **34,** 27 (1994).
79. R. Gebhardt and J. Reichen, *Hepatology* **20,** 684 (1994).
80. W. H. Lamers, F. J. Verbeek, A. F. M. Moorman, and R. Charles, in "RBC: Hepatocyte heterogeneity and liver function" (J. J. Gumucio, ed.), Vol. 19, p. 5. Springer Publ., New York, 1989.
81. G. T. M. Wagenaar, R. A. F. M. Chamuleau, M. A. W. Maas, K. de Bruin, J. A. M. Korfage, and W. H. Lamers, *Hepatology* **20,** 1532 (1994).
82. J. P. Colombo, C. Bachmann, E. Peheim, and J. Berüter, *Enzyme* **22,** 399 (1977).
83. G. Girard and R. F. Butterworth, *Dig. Dis. Sci.* **37,** 1121 (1992).
84. A. R. Quesada, M. A. Medina, J. Márquez, F. M. Sánchez-Jiménez, and I. Núnez de Castro, *Cancer Res.* **48,** 1551 (1988).
85. M. Louie, T. F. Deuel, and H. P. Morris, *J. Biol. Chem.* **253,** 6119 (1977).
86. C. Wu, E. H. Roberts, and J. M. Bauer, *Cancer Res.* **25,** 677 (1965).
87. T. H. Mauad, C. M. J. van Nieuwkerk, K. P. Dingemans, J. J. M. Smit, A. H. Schinkel, R. G. E. Notenboom, M. A. van den Bergh Weerman, R. P. Verkruisen, B. K. Groen, R. P. J. Oude Elferink, P. Borst, and G. J. A. Offerhaus, *Am. J. Pathol.* **145,** 1237 (1994).
88. L. Christa, M.-T. Simon, J.-P. Flinois, R. Gebhardt, C. Brechot, and C. Lasserre, *Gastroenterology* **106,** 1312 (1994).
89. R. Gebhardt, T. Tanaka, and G. M. Williams, *Carcinogenesis* **10,** 1917 (1989).

90. G. M. Williams, R. Gebhardt, H. Sirma, and F. Stenbäck, *Carcinogenesis* **14,** 2149 (1993).
91. D. Häussinger, W. Gerok, and H. Sies, *Trends Biochem. Sci.* **300** (1984).
92. M. Watford and E. M. Smith, *Biochem. J.* **267,** (1990).
93. A. F. M. Moorman, P. A. J. de Boer, M. Watford, M. A. Dingemanse, and W. H. Lamers, *FEBS Lett.* **356,** 76 (1994).
94. S. Y. Low, M. Salter, R. G. Knowles, C. I. Pogson, and M. J. Rennie, *Biochem. J.* **295,** 617 (1993).
95. D. Häussinger, D. *Biochem. J.* **267,** 281 (1990).
96. A. J. Meijer, W. H. Lamers, and R. A. F. M. Chamuleau, *Physiol. Rev.* **70,** 701 (1990).
97. R. Gebhardt, *Pharmacol. Ther.* **53,** 275 (1992).
98. S. Y. Low, P. M. Taylor, A. Ahmed, C. I. Pogson, and M. J. Rennie, *Biochem. J.* **278,** 105 (1991).
99. S. Y. Low, P. M. Taylor, H. S. Hundal, C. I. Pogson, and M. J. Rennie, *Biochem. J.* **284,** 333 (1992).
100. P. Fafournoux, C. Demigné, C. Rémésy, and A. le Cam, *Biochem. J.* **216,** 401 (1983).
101. H. M. Said, D. Hollander, and S. Khorchid, *Gastroenterology* **101,** 1094 (1991).
102. B. P. Bode, D. L. Kaminski, W. W. Souba, and A. P. Li, *Hepatology* **21,** 511 (1995).
103. C. P. Fischer, B. P. Bode, and W. W. Souba, *J. Trauma* **40,** 688 (1996).
104. B. K. Tamarappoo, M. Nam, M. S. Kilberg, T. C. Welbourne, *Am. J. Physiol.* **264,** E526 (1993).
105. C. P. Fisher, B. P. Bode, S. F. Abcouwer, G. C. Lukaszewicz, and W. W. Souba, *Shock* **3,** 315 (1995).
106. W. W. Souba, *N. Engl. J. Med.* **336,** 41 (1997).
107. N. J. Espat, K. T. Watkins, D. S. Lind, J. K. Weis, E. M. Copeland, and W. W. Souba, *J. Surg. Res.* **63,** 263 (1996).
108. T. R. Austgen, M. K. Chen, T. C. Flynn, and W. W. Souba, *J. Trauma* **31,** 742 (1991).
109. L. Boon and A. J. Meijer, *Eur. J. Biochem.* **172,** 465 (1988).
110. L. Boon, P. J. E. Blommaart, A. J. Meijer, W. H. Lamers, and A. C. Schoolwerth, *Contrib. Nephrol.* **110,** 133 (1994).
111. L. Boon, P. J. E. Blommaart, A. J. Meijer, W. H. Lamers, and A. C. Schoolwerth, *Contrib. Nephrol.* **138** (1994).
112. L. Boon, P. J. E. Blommaart, A. J. Meijer, W. H. Lamers, and A. C. Schoolwerth, *Am. J. Physiol.* **267,** F1015 (1994).
113. L. Boon, P. J. E. Blommaart, A. J. Meijer, W. H. Lamers, and A. C. Schoolwerth, *Am. J. Physiol.* **271,** F198 (1996).
114. D. Häussinger, *Eur. J. Biochem.* **133,** 269 (1983).
115. A. J. L. Cooper, E. Nieves, A. E. Coleman, S. Filc-DeRico, and A. S. Gelbard, *J. Biol. Chem.* **262,** 1073 (1987).
116. S. Kaiser, W. Gerok, and D. Häussinger, *Eur. J. Clin. Invest.* **18,** 535 (1988).
117. H. J. Burger, R. Gebhardt, C. Mayer, and D. Mecke, *Hepatology* **9,** 22 (1989).
118. B. Stoll, S. McNelly, H. P. Buscher, and D. Häussinger, *Hepatology* **13,** 247 (1991).
119. P. M. Taylor and M. J. Rennie, *FEBS Lett.* **221,** 370 (1987).
120. F. C. Kuo, W. L. Hwu, D. Valle, and J. E. Darnell, *Proc. Natl. Acad. Sci. U.S.A.* **88,** 9468 (1991).
121. A. Martinez-Hernandez, K. P. Bell, and M. D. Norenberg, *Science* **195,** 1356 (1977).
122. M. D. Norenberg, *J. Histochem. Cytochem.* **3,** 756 (1979).
123. W. Y. Ong, L. J. Garey, and R. Reynolds, *J. Neurocytol.* **22,** 893 (1993).
124. W. Y. Ong, L. J. Garey, S. K. Leong, and R. Reynolds, *J. Neurocytol.* **24,** 602 (1995).
125. A. Derouiche and M. Frotcher, *Brain Res.* **552,** 346 (1991).
126. F. A. Tansey, M. Farooq, and W. Cammer, *J. Neurochem.* **56,** 266 (1991).

127. F. D'Amelio, L. F. Eng, and M. A. Gibbs, *GLIA* **3,** 335 (1990).
128. W. Cammer, *J. Neuroimmunol.* **26,** 173 (1990).
129. R. A. Warringa, M. F. van Berlo, W. Klein, and M. J. Lopes-Cardozo, *Neurochemistry* **50,** 1461 (1988).
130. C. Fressinaud, H. Weinrauder, J. P. Delaunoy, G. Tholey, G. Labourdette, and L. L. Sarlieve, *J. Cell. Physiol.* **149,** 459 (1991).
131. M. N. Graff, D. Baas, J. Puymirat, L. L. Sarlieve, and J. P. Delaunoy, *Neurosci. Lett.* **150,** 174 (1993).
132. R. E. Riepe and M. D. Norenburg, *Nature (London)* **268,** 654 (1977).
133. A. Derouiche and T. Rauen, *J. Neurosci. Res.* **42,** 131 (1995).
134. S. Kentroti and A. Vernadakis, *GLIA* **18,** 79 (1996).
135. D. L. Broussard, P. G. Bannerman, C. M. Tang, M. Hardy, and D. Pleasure, *J. Neurosci. Res.* **34,** 24 (1993).
136. H. Kato, T. Yamamoto, H. Yamamoto, R. Chi, N. So, and Y. Iwasaki, *J. Pediatr. Surg.* **25,** 514 (1990).
137. K. R. Jessen and R. Mirsky, *J. Neurosci.* **3,** 2206 (1983).
138. S. Berl and D. D. Clarke, in "Glutamine Metabolism in Mammalian Tissues" (D. Häussinger and H. Sies, eds.), p. 223. Springer-Verlag, Berlin and New York, 1984.
139. A. H. Lockwood, J. M. McDonald, R. E. Reiman, A. S. Gelbard, J. S. Laughlin, T. E. Duffy, and F. Plum, *J. Clin. Invest.* **63,** 449 (1979).
140. A. J. Patel, A. Hunt, R. D. Gordon, and R. Balázs, *Dev. Brain Res.* **4,** 3 (1982).
141. Y. Aizenman and J. deVellis, *Brain Res.* **414,** 301 (1987).
142. D. Chatterjee and P. K. Sarkar, *Int. J. Dev. Neurosci.* **2,** 55 (1984).
143. K. Hallermayer, C. Harmening, and B. Hamprecht, *J. Neurochem.* **37,** 43 (1981).
144. M. J. Jackson, H. R. Zielke, and S. R. Max, *Neurochem. Res.* **20,** 201 (1995).
145. C. Arcuri, M. Tardy, B. Rolland, R. Armellini, A. R. Menghini, and V. Bocchini, *Neurochem. Res.* **20,** 1133 (1995).
146. B. H. J. Juurlink, A. Schousboe, O. S. Jorgensen, and L. Hertz, *J. Neurochem.* **36,** 136 (1981).
147. M. Khelil, B. Rolland, C. Fages, and M. Tardy, *GLIA* **3,** 75 (1990).
148. M. K. O'Banion, D. A. Young, and M. C. Bohn, *Mol. Brain Res.* **22,** 57 (1994).
149. H. H. Samuels, D. Klein, F. Stanley, and J. Casanova, *J. Biol. Chem.* **253,** 5895 (1978).
150. J. Ruel and J. H. Dussault, *Dev. Brain Res.* **21,** 83 (1985).
151. P. J. Andres-Barquin, C. Fages, G. Le Prince, B. Rolland, and M. Tardy, *Neurochem. Res.* **19,** 65 (1994).
152. A. J. Patel, M. Hayashi, and A. Hunt, *J. Neurochem.* **50,** 803 (1988).
153. A. J. Patel, A. Hunt, and S. M. Tahourdin, *Dev. Brain Res.* **10,** 83 (1983).
154. C. Loret, T. Janet, G. Labourdette, H. Schneid, and M. Binoux, *GLIA* **4,** 378 (1991).
155. M. Pomerance, J. M. Gavaret, M. Breton, and M. Pierre, *J. Neurosci. Res.* **40,** 737 (1995).
156. F. Perraud, F. Besnard, B. Pettman, M. Sensenbrenner, and G. Labourdette, *GLIA* **1,** 124 (1988).
157. P. Honegger and M. Tenot-Sparti, *J. Neuroimmunol.* **40,** 295 (1992).
158. B. Guentert-Lauber and P. Honegger, *Dev. Neurosci.* **7,** 286 (1985).
159. P. Honegger and B. Guentert-Lauber, *Dev. Brain Res.* **11,** 245 (1983).
160. S. R. Max, M. E. Landry, and H. R. Zielke, *Neurochem. Res.* **15,** 583 (1990).
161. H. R. Zielke, J. T. Tildon, M. E. Landry, and S. R. Max, *Neurochem. Res.* **15,** 1115 (1990).
162. H. Hama, T. Sakurai, Y. Kasuya, M. Fujiki, T. Masaki, and K. Goto, *Biochem. Biophys. Res. Commun.* **186,** 355 (1992).
163. I. Kronfeld, A. Zsukerman, G. Kazimirsky, and C. Brodie, *J. Neurochem.* **65,** 1505 (1995).
164. P. Honegger, *J. Neurochem.* **46,** 1561 (1986).
165. S. Kentroti, R. Baker, K. Lee, C. Bruce, and A. Vernadakis, *J. Neurosci. Res.* **28,** 497 (1991).

166. F. Perraud, F. Besnard, M. Sensenbrenner, and G. Labourdette, *Int. J. Dev. Neurosci.* **5,** 181 (1987).
167. R. B. Nelson and R. Siman, *Brain Res.* **54,** 93 (1990).
168. C. Brodie and N. Goldreich, *J. Neuroimmunol.* **55,** 91 (1994).
169. C. C. Chao, S. Hu, M. Tsang, J. Weatherbee, T. W. Molitor, W. R. Anderson and P. K. Peterson, *J. Clin. Invest.* **90,** 1786 (1992).
170. K. G. Low, R. G. Allen, and M. H. Melner, *Endocrinology* **131,** 1908 (1992).
171. G. Labourdette, T. Janet, P. Laeng, F. Perraud, D. Lawrence, and B. Pettman, *J. Cell. Physiol.* **144,** 473 (1990).
172. B. H. J. Juurlink and L. Hertz, *Dev. Neurosci.* **7,** 263 (1985).
173. S. Berl and D. D. Clarke, in "Amino Acids as Chemical Transmitters" (F. Fonnum, ed.), p. 691 Plenum, New York, 1978.
174. C. J. van den Berg, D. F. Matheson, and W. C. Nijenmanting, in "Amino Acids as Chemical Transmitters" (F. Fonnum, ed.), p. 709. New York, 1978.
175. L. Peng, L. Hertz, R. Huang, U. Sonnewald, S. B. Petersen, N. Westergaard, O. Larsson, and A. Schousboe, *Dev. Neurosci.* **15,** 367 (1993).
176. E. A. Newsholme and M. Board, *Adv. Enzyme Regul.* **31,** 225 (1991).
177. J. H. Laake, T. A. Slyngstad, F. M. S. Haug, and O. P. Ottersen, *J. Neurochem.* **65,** 871 (1995).
178. D. V. Pow and D. K. Crook, *Neuroscience* **70,** 295 (1995).
179. I. Torgner and E. Kvamme, *Mol. Chem. Neuropathol.* **12,** 11 (1990).
180. A. Schousboe, N. Westergaard, U. Sonnewald, S. B. Petersen, A. C. H. Yu, and L. Hertz, *Prog. Brain Res.* **94,** 199 (1992).
181. K. Kanamori and B. D. Ross, *Biochem. J.* **305,** 329 (1995).
182. K. Kanamori, and B. D. Ross, *J. Biol. Chem.* **270,** 24805 (1995).
183. K. Kanamori, B. D. Ross, J. C. Chung, and E. L. Kuo, *J. Neurochem.* **67,** 1584 (1996).
184. K. Kanamori and B. D. Ross, *Biochem. J.* **293,** 461 (1993).
185. K. Kanamori, B. D. Ross, and E. L. Kuo, *Biochem. J.* **311,** 681 (1995).
186. R. P. Shank, G. C. Leo, and H. R. Zielke, *J. Neurochem.* **61,** 315 (1993).
187. N. Westergaard, J. Drejer, A. Schousboe, and U. Sonnewald, *GLIA* **17,** 160 (1996).
188. N. R. Sibson, A. Dhankhar, G. F. Mason, K. L. Behar, D. L. Rothman, and R. G. Shulman, *Proc. Natl. Acad. Sci. U.S.A.* **94,** 2699 (1997).
189. A. J. Cooper and F. Plum, *Physiol. Rev.* **67,** 440 (1987).
190. N. Zhang, J. H. Laake, E. A. Nagelhus, J. Storm-Mathisen, and O. P. Ottersen, *Anat. Embryol.* **184,** 213 (1991).
191. J. Storm-Mathisen, N. C. Danbolt, F. Rothe, R. Torp, N. Zhang, J. E. Aas, B. I. Kanner, I. Langmoen, and O. P. Ottersen, *Prog. Brain Res.* **94,** 225 (1992).
192. O. P. Ottersen, N. Zhang, and F. Walberg, *Neuroscience* **46,** 519 (1992).
193. R. L. Veech, R. L. Harris, D. Veloso, and E. H. Veech, *J. Neurochem.* **46,** 519 (1992).
194. J. C. Szerb and R. F. Butterworth, *Prog. Neurobiol.* **39,** 135 (1992).
195. H. Laursen, *Acta Neurol. Scand.* **65,** 381 (1982).
196. A. T. Blei, S. Olafsson, G. Therrien, and R. F. Butterworth, *Hepatology* **19,** 1437 (1994).
197. A. M. Mans, M. R. DeJoseph, and R. A. Hawkins, *J. Neurochem.* **63,** 1829 (1994).
198. R. A. Hawkins, J. Jessy, A. M. Mans, and M. R. De Joseph, *J. Neurochem.* **60,** 1000 (1993).
199. H. Takahashi, R. C. Koehler, S. W. Brusilow, and R. J. Traystman, *Am. J. Physiol.* **261,** H825 (1991).
200. C. L. Willard-Mack, R. C. Koehler, T. Hirata, L. C. Cork, H. Takahashi, R. J. Traystman, and S. W. Brusilow, *Neuroscience* **71,** 589 (1996).
201. E. E. Mena and C. W. Cotman, *Exp. Neurol.* **89,** 259 (1985).
202. J. D. Rothstein, L. Martin, A. I. Levey, M. Dykes-Hoberg, L. Jin, D. Wu, N. Nash, and R. W. Kuncl, *Neuron* **13,** 713 (1994).

203. C. Hermenegildo, G. Marcaida, G. Montoliu, S. Grisolía, M. D. Minana, and V. Felipo, *Neurochem. Res.* **21,** 1237 (1996).
204. D. D. Mousseau and R. F. Butterworth, *Proc. Soc. Exp. Biol. Med.* **206,** 329 (1994).
205. G. Lemieux, G. Baverel, P. Vinay, and P. Wadoux, *Am. J. Physiol.* **231,** 1068 (1976).
206. A. G. Craan, G. Lemieux, P. Vinay, and A. Gougoux, *Kidney Int.* **22,** 103 (1982).
207. H. B. Burch, S. Choi, W. Z. McCarthy, P. Y. Wong, and O. H. Lowry, *Biochem. Biophys. Res. Commun.* **82,** 498 (1978).
208. S. R. DiGiovanni, K. M. Madsen, A. D. Luther, and M. A. Knepper, *Am. J. Physiol.* **267,** F407 (1994).
209. A. C. Schoolwerth, P. A. J. de Boer, A. F. M. Moorman, and W. H. Lamers, *Am. J. Physiol.* **267,** F400 (1994).
210. H. Lie-Venema, P. A. J. de Boer, A. F. M. Moorman, and W. H. Lamers, *Biochem. J.* **323,** 611 (1997).
211. G. Lemieux, M. R. Aranda, P. Fournel, and C. Lemieux, *Can. J. Physiol. Pharmacol.* **62,** 70 (1984).
212. M. Domenech, F. J. Lopez-Soriano, M. Marzabal, and J. M. Argiles, *Cell. Mol. Biol.* **39,** 405 (1993).
213. M. Watford, *Contrib. Nephrol.* **92,** 211 (1991).
214. T. R. Austgen, M. K. Chen, W. Moore, and W. W. Souba, *Arch. Surg.* **126,** 23 (1991).
215. S. Dugelay and G. Baverel, *Biochim. Biophys. Acta* **1075,** 191 (1991).
216. D. D. Van Slyke, R. A. Phillips, P. B. Hamilton, R. M. Archibald, P. H. Futcher, and A. Miller, *J. Biol. Chem.* **150,** 481 (1943).
217. T. C. Welbourne and V. Phromphetcharat, *in* "Glutamine Metabolism in Mammalian Tissues" (D. Häussinger and H. Sies, eds.), p. 161. Springer-Verlag, Berlin and New York, 1984.
218. T. C. Welbourne, *Sem. Nephrol.* **10,** 339 (1990).
219. B. K. Tamarappoo, S. Joshi, and T. C. Welbourne, *Mineral Electrolyte Metab.* **16,** 322 (1990).
220. T. C. Welbourne and M. J. Cronin, *Am. J. Physiol.* **260,** R1036 (1991).
221. N. P. Curthoys and O. H. Lowry, *J. Biol. Chem.* **248,** 162 (1973).
222. D. W. Good and M. A. Knepper, *Am. J. Physiol.* **248,** F459 (1985).
223. R. L. Tannen and A. Sahai, *Mineral Electrolyte Metab.* **16,** 249 (1990).
224. D. M. Parry and J. T. Brosnan, *Biochem. J.* **174,** 387 (1978).
225. R. P. Hughey, B. B. Rankin, and N. P. Curthoys, *Am. J. Physiol.* **238,** F199 (1980).
226. M. Lowry and B. D. Ross, *Biochem. J.* **190,** 771 (1980).
227. C. Michoudet, M. F. Chauvin, and G. Baverel, *Biochem. J.* **297,** 69 (1994).
228. M. F. Chauvin, F. Megnin-Chanet, G. Martin, J. M. Lhoste, and G. Baverel, *J. Biol. Chem.* **269,** 26025 (1994).
229. M. J. Rennie, P. A. MacLennan, H. S. Hundal, B. Weryk, K. Smith, P. M. Taylor, C. Egan, and P. W. Watt, *Metabolism* **38,** 47 (1989).
230. E. A. Newsholme and M. Parry-Billings, *J. Parent. Enteral Nutr.* **14,** 63S (1990).
231. X. Remesar, L. Arola, A. Palou, and M. Alemany, *Horm. Metab. Res.* **14,** 419 (1982).
232. M. T. Falduto, A. P. Young, and R. C. Hickson, *Am. J. Physiol.* **262,** 214 (1992).
233. R. C. Hickson, L. E. Wegrzyn, D. F. Osborne, and I. E. Karl, *Am. J. Physiol.* **270,** E912 (1996).
234. M. S. M. Ardawi and M. F. Majzoub, *Metabolism* **40,** 155 (1991).
235. M. S. M. Ardawi, *Clin. Sci.* **74,** 165 (1988).
236. M. S. M. Ardawi and Y. S. Jamal, *Clin. Sci.* **79,** 139 (1990).
237. K. Iqbal and J. H. Ottaway, *Biochem. J.* **119,** 145 (1970).
238. W. B. Rowe, *Methods Enzymol.* **113,** 199 (1985).

239. S. F. Abcouwer, B. P. Bode, and W. W. Souba, *J. Surg. Res.* **59,** 59 (1995).
240. M. T. Falduto, A. P. Young, and R. C. Hickson, *Am. J. Physiol.* **263,** 1157 (1992).
241. M. T. Falduto, A. P. Young, G. Smyrniotis, and R. C. Hickson, *Am. J. Physiol.* **262,** 1131 (1992).
242. P. A. King, L. Goldstein, and E. A. Newsholme, *Biochem. J.* **216,** 523 (1983).
243. R. C. Hickson, L. E. Wegrzyn, D. F. Osborne, and I. E. Karl, *Am. J. Physiol.* **271,** R1165 (1996).
244. M. S. M. Ardawi, *Biochem. J.* **270,** 829 (1990).
245. B. Feng, D. C. Hilt, and S. R. Max, *J. Biol. Chem.* **265,** 18702 (1990).
246. S. R. Max, J. Mill, K. M. Mearow, Y. Konagaya, J. W. Thomas, C. Banner, and L. Vitkovic, *Am. J. Physiol.* **255,** E397 (1988).
247. S. R. Max, *Med. Sci. Sports Exerc.* **22,** 325 (1990).
248. M. E. Tischler, E. J. Henriksen, and P. H. Cook, *Muscle Nerve* **11,** 752 (1988).
249. T. R. Austgen, R. Chakrabarti, M. K. Chen, and W. W. Souba, *J. Trauma* **32,** 600 (1992).
250. S. F. Abcouwer, J. Norman, G. Fink, G. Carter, R. J. Lustig, and W. W. Souba, *Surgery* **120,** 255 (1996).
251. T. B. Kelso, C. R. Shear, and S. R. Max *Am. J. Physiol.* **257,** E885 (1989).
252. L. Arola, A. Palou, X. Remesar, and M. Alemany, *Horm. Metab. Res.* **13,** 189 (1981).
253. B. Feng, M. Konagaya, Y. Konagaya, J. W. Thomas, C. Banner, J. Mill, and S. R. Max, *Am. J. Physiol.* **258,** E757 (1990).
254. E. B. Marliss, T. T. Aoki, T. Pozefsky, A. S. Most, and G. F. Cahill, Jr., *J. Clin. Invest.* **50,** 814 (1971).
255. E. A. Newsholme, P. Newsholme, R. Curi, E. Challoner, and M. S. M. Ardawi, *Nutrition* **4,** 261 (1988).
256. M. C. Herrero, N. Angles, X. Remesar, L. Arola, and C. Blade, *J. Obs. Rel. Metab. Dis.* **18,** 255 (1994).
257. J. Salvadó and L. Arola, *Comp. Biochem. Physiol.* **103A,** 817 (1997).
258. G. Hod, M. Chaouat, Y. Haskel, O. Z. Lernau, S. Nissan, and M. Mayer, *Eur. J. Clin. Invest.* **12,** 445 (1982).
259. O. P. Ganda and N. B. Ruderman, *Metabolism* **25,** 427 (1976).
260. A. H. Lockwood, J. M. McDonald, R. E. Reiman, A. S. Gelbard, J. S. Laughlin, T. E. Duffy, and F. Plum, *J. Clin. Invest.* **63,** 449 (1979).
261. C. H. Dejong, N. E. P. Deutz, and P. B. Soeters, *J. Hepatol.* **21,** 299 (1994).
262. J. E. Fischer and P. O. Hasselgren, *Am. J. Surg.* **161,** 266 (1991).
263. G. M. Vaughan, R. A. Becker, J. P. Allen, C. W. Goodwin, B. A. Pruitt, and A. D. Mason, *J. Trauma* **22,** 263 (1982).
264. M. F. Dallman, A. M. Strack, S. F. Skana, M. J. Bradbury, E. S. Hanson, K. A. Scribner, and M. Smith, *Front. Neuroendocrinol.* **14,** 303 (1993).
265. E. Roth, J. Funovics, F. Mühlbacher, M. Schemper, W. Mauritz, P. Sporn, and A. Fritsch, *Clin. Nutr.* **1,** 25 (1982).
266. H. S. Hundal, M. J. Rennie, and P. W. Watt, *J. Physiol.* **393,** 283 (1987).
267. H. S. Hundal, P. Babij, P. W. Watt, M. R. Ward, and M. J. Rennie, *Am. J. Physiol.* **259,** E148 (1990).
268. H. S. Hundal, P. Babij, P. M. Taylor, P. W. Watt, and M. J. Rennie, *Biochim. Biophys. Acta* **1092,** 376 (1991).
269. M. J. Rennie, L. Tadros, S. Khogali, A. Ahmed, and P. M. Taylor, *J. Nutr.* **124** (Suppl. 8), 1503S (1994).
270. M. J. Rennie, A. Ahmed, S. E. Khogali, S. Y. Low, H. S. Hundal, and P. M. Taylor, *J. Nutr.* **126** (Suppl. 4), 1142S (1996).

271. R. R. Roy, P. F. Gardiner, D. R. Simpson, and V. R. Edgerton, *Arch. Oral Biol.* **28,** 639 (1983).
272. T. Seene and A. Viru, *J. Steroid Biochem.* **16,** 349 (1982).
273. R. R. Almon and D. C. Dubois, *Med. Sci. Sports Exerc.* **22,** 304 (1990).
274. P. A. MacLennan, R. A. Brown, and M. J. Rennie, *FEBS Lett.* **215,** 187 (1987).
275. P. A. MacLennan, K. Smith, B. Weryk, P. W. Watt, and M. J. Rennie, *FEBS Lett.* **237,** 133 (1988).
276. V. Hack, O. Stutz, R. Kinscherf, M. Schykowski, M. Kellerer, E. Holm, and W. Droge, *J. Mol. Med.* **74,** 337 (1996).
277. A. D. Fox, S. A. Kripke, J. M. Berman, R. M. McGintey, R. G. Settle, and J. L. Rombeau, *J. Surg. Res.* **44,** 391 (1988).
278. X. Remesar, L. Arola, A. Palou, and M. Alemany, *Reprod. Nutr. Dev.* **25,** 861 (1985).
279. K. Holovska, V. Lenartova, and I. Havassy, *Physiol. Bohemoslovaca* **28,** 145 (1978).
280. H. Lie-Venema, P. A. J. de Boer, A. F. M. Moorman, and W. H. Lamers, *Eur. J. Biochem.* **248,** 644 (1997).
281. C. J. F. van Noorden, I. M. C. Vogels, G. Fronik, and R. D. Bahattacharya, *Exp. Cell Res.* **155,** 381 (1984).
282. J. C. Roig, V. Shenoy, R. Chakrabarti, J. Y. N. Lau, and J. J. Neu, *Parent. Enteral Nutr.* **19,** 179 (1995).
283. P. Sarantos, A. Abouhamze, R. Chakrabarti, and W. W. Souba, *Arch. Surg.* **129,** 59 (1994).
284. P. Sarantos, R. Chakrabarti, E. M. Copeland, and W. W. Souba, *Am. J. Surg.* **167,** 8 (1994).
285. M. S. M. Ardawa and E. A. Newsholme, *Biochem. J.* **231,** 713 (1985).
286. S. F. Abcouwer, G. C. Lukascewicz, U. S. Ryan, and W. W. Souba, *Surgery* **118,** 325 (1995).
287. J. D. Crapo, M. Peters-Golden, and J. Marsh-Sahlin, *Lab. Invest.* **39,** 640 (1987).
288. M. S. M. Ardawi, *Clin. Sci.* **81,** 603 (1991).
289. T. R. Austgen, M. K. Chen, R. M. Salloum, and W. W. Souba, *J. Trauma* **31,** 1068 (1991).
290. P. Sarantos, D. Howard, and W. W. Souba, *Metabolism* **42,** 795 (1993).
291. S. F. Abcouwer, G. C. Lukaszewicz, and W. W. Souba, *Am. J. Physiol.* **14,** L 141 (1996).
292. D. A. Plumley, T. R. Austgen, R. M. Salloum, and W. W. Souba, *J. Parent. Enteral Nutr.* **14,** 569 (1990).
293. W. W. Souba, D. A. Plumley, R. M. Salloum, and E. M. Copeland, *Surgery* **108,** 213 (1990).
294. D. A. Plumley, W. W. Souba, R. D. Hautamaki, T. D. Martin, T. C. Flynn, W. R. Rout, and E. M. Copeland, *Arch. Surg.* **125,** 57 (1990).
295. R. J. Smith, *J. Parent. Enteral Nutr.* **14,** 40S (1990).
296. A. Fine, *Kidney Int.* **21,** 439 (1982).
297. F. C. Fowler, R. K. Banks, and M. E. Mailliard, *Hepatology* **16,** 1187 (1992).
298. F. J. Lopez-Soriano and M. Alemany, *Biochim. Biophys. Acta* **925,** 265 (1987).
299. R. E. Miller and D. A. Carrino, *J. Biol. Chem.* **255,** 5490 (1980).
300. R. E. Miller, S. R. Pope, J. E. DeWille, and D. M. Burns, *J. Biol. Chem.* **258,** 5405 (1983).
301. R. E. Miller and D. M. Burns, *Curr. Top. Cell. Regul.* **26,** 65 (1985).
302. R. L. Stone and D. A. Bernlohr, *Differentiation* **45,** 119 (1990).
303. V. R. Martins and N. M. Brentani, *J. Steroid Biochem. Molec. Biol.* **37,** 183 (1990).
304. B. Bhandari and R. E. Miller, *Mol. Cell Endocrinol.* **51,** 7 (1987).
305. R. Brandes, R. Arad, N. Benvenisty, S. Weil, and J. Bar-Tana, *Biochim. Biophys. Acta* **1054,** 219 (1990).
306. N. P. Curthoys and M. Watford, *Annu. Rev. Nutr.* **15,** 133 (1995).
307. M. D. Kvidera and G. B. Carey, *Proc. Soc. Exp. Biol. Med.* **206,** 360 (1994).
308. B. T. Hinton and M. A. Palladino, *Microsc. Res. Technol.* **30,** 67 (1995).
309. M. W. Fornes and J. C. De Rosas, *Anat. Rec.* **231,** 193 (1991).

310. J. A. Holash, S. I. Harik, G. Perry, and P. A. Stewart, *Proc. Natl. Acad. Sci. U.S.A.* **90,** 11069 (1993).
311. B. Bhandari, R. H. Wilson, and R. E. Miller, *Mol. Endocrinol.* **1,** 403 (1987).
312. J. M. Harmon and E. B. Thompson, *J. Cell. Physiol.* **110,** 155 (1982).
313. P. Marchetti, F. O. Ranelletti, V. Natoli, G. Sica, G. De Rossi, and S. Iacobelli, *J. Steroid Biochem.* **19,** 1665 (1983).
314. T. J. Schmidt and E. B. Thompson, *Cancer Res.* **39,** 376 (1979).
315. M. S. M. Ardawi and E. A. Newsholme, *Biochem. J.* **212,** 835 (1983).
316. E. A. Newsholme and A. L. Carrié, *Gut* **35** (Suppl. 1), S13 (1994).
317. W. L. McKeehan, *Cell Biol. Int. Rep.* **6,** 635 (1982).
318. W. W. Souba, *Ann. Surg.* **218,** 715 (1993).
319. C. Wu, *Arch. Biochem. Biophys.* **106,** 394 (1964).
320. W. E. Knox, H. Z. Kupchik, and L. P. Liu, *Enzyme* **12,** 88 (1971).
321. L. Arola, A. Palou, X. Remesar, and M. Alemany, *Arch. Int. Physiol. Biochim.* **90,** 163 (1982).
322. L. Arola, A. Palou, X. Remesar, and M. Alemany, *Arch. Int. Physiol. Biochim.* **89,** 189 (1981).
323. H. Lie-Venema, W. T. Labruyère, M. A. van Roon, P. A. J. de Boer, A. F. M. Moorman, A. J. M. Berns, and W. H. Lamers, *J. Biol. Chem.* **270,** 28251 (1995).
324. V. Shenoy, J. C. Roig, R. Chakrabarti, P. Kubilis, and J. Neu, *Pediatr. Res.* **39,** 643 (1996).
325. A. Palou, X. Remesar, L. Arola, and M. Alemany, *Arch. Int. Physiol. Biochim.* **91,** 43 (1983).
326. C. E. Martin, M. H. Cake, P. E. Hartman, and I. F. Cook, *Acta Endocrinol.* **84,** 167 (1977).
327. S. J. Henning, *Am. J. Physiol.* **235,** E451 (1978).
328. S. R. Magnuson and A. P. Young, *Dev. Biol.* **130,** 536 (1988).
329. K. M. Mearow, J. F. Mill, and L. Vitkovic, *Mol. Brain Res.* **6,** 223 (1989).
330. B. Bhandari, W. J. Roesler, K. D. De Lisio, D. J. Klemm, N. S. Ross, and R. E. Miller, *J. Biol. Chem.* **266,** 7784 (1991).
331. R. G. E. Notenboom, A. F. M. Moorman, and W. H. Lamers, *Microsc. Res. Technol.* **39,** 413 (1997).
332. M. Caldani, B. Rolland, C. Fages, and M. Tardy, *Experientia* **38,** 1199 (1982).
333. A. J. Patel, A. Hunt, and S. M. Tahourdin, *Dev. Brain Res.* **8,** 31 (1983).
334. K. M. Mearow, J. F. Mill, and E. Freese, *GLIA* **3,** 385 (1990).
335. J. Akimoto, H. Itoh, T. Miwa, and K. Ikeda, *Dev. Brain Res.* **72,** 9 (1993).
336. F. Rosier, D. Lambert, and J. Mertens-Strijthagen, *Biochem. J.* **315,** 607 (1996).
337. G. Arad and R. G. Kulka, *Biochim. Biophys. Acta* **544,** 153 (1978).
338. G. Arad, A. Freikopf, and R. G. Kulka, *Cell* **8,** 95 (1976).
339. R. G. Kulka and H. Cohen, *J. Biol. Chem.* **248,** 6738 (1973).
340. A. Freikopf-Cassel and R. G. Kulka, *FEBS Lett.* **124,** 27 (1981).
341. R. B. Crook, M. Louie, T. F. Deuel, and G. M. Tomkins, *J. Biol. Chem.* **253,** 6125 (1978).
342. B. Feng, S. K. Shiber, and S. R. Max, *J. Cell. Physiol.* **145,** 376 (1990).
343. R. J. Smith, S. Larson, S. E. Stred, and R. A. Durschlag, *J. Cell. Physiol.* **120,** 197 (1984).
344. L. B. Tadros, N. M. Willhoft, P. M. Taylor, and M. J. Rennie, *Am. J. Physiol.* **265,** E935 (1993).
345. T. Kitoh, M. Kubota, M. Takimoto, H. Hashimoto, T. Shimizu, H. Sano, Y. Akiyama, and H. Mikawa, *J. Cell. Physiol.* **143,** 150 (1990).
346. A. Colquhoun and E. A. Newsholme, *Biochem. Mol. Biol. Int.* **41,** 583 (1997).
347. D. C. Tiemeyer and G. Milman, *J. Biol. Chem.* **247,** 5722 (1972).
348. L. Lacoste, K. D. Chaudhary, and J. Lapointe, *J. Neurochem.* **39,** 78 (1982).
349. A. Michalak and R. F. Butterworth, *Hepatology* **25,** 631 (1997).

350. G. Milman, L. S. Portnoff, and D. C. Tiemeier, *J. Biol. Chem.* **250,** 1393 (1975).
351. R. L. Levine, *J. Biol. Chem.* **258,** 11828 (1983).
352. N. F. Schor, *Brain Res.* **456,** 17 (1988).
353. C. N. Oliver, P. E. Starke-Reed, E. R. Stadtman, G. J. Liu, J. M. Carney, and R. A. Floyd, *Proc. Natl. Acad. Sci. U.S.A.* **87,** 5144 (1990).
354. J. M. Farber and R. L. Levine, *J. Biol. Chem.* **261,** 4574 (1986).
355. E. R. Stadtman, P. E. Starke-Reed, C. N. Oliver, J. M. Carney, and R. A. Floyd, *in* "Free Radicals and Aging" (I. Emerit and B. Chance, eds.), p. 64. Birkhauser Verlag, Basel, Switzerland, 1992.
356. E. Pinteaux, J. C. Copin, M. Ledig, and G. Tholey, *Dev. Neurosci.* **18,** 397 (1996).
357. J. A. Sakahian, L. I. Szweda, B. Friguet, K. Kitani, and R. L. Levine, *Arch. Biochem. Biophys.* **318,** 411 (1995).
358. K. Hensley, J. M. Carney, M. P. Mattson, M. Aksenova, M. Harris, and R. A. Floyd, *Proc. Natl. Acad. Sci. U.S.A.* **91,** 3270 (1994).
359. G. Le Prince, P. Delaere, C. Fages, L. Lefrançois, M. Touret, M. Salanon, and M. Tardy, *Neurochem. Res.* **20,** 859 (1995).
360. C. E. Finch and D. M. Cohen, *Exp. Neurol.* **143,** 82 (1997).
361. M. E. Gibbs, B. S. Odowd, L. Hertz, S. R. Robinson, G. L. Sedman, and K. T. Ng, *Cogn. Brain Res.* **4,** 57 (1996).
362. C. J. Pike, N. Ramezan-Arab, S. Miller, and C. W. Cotman, *Exp. Neurol.* **139,** 167 (1996).
363. A. P. Young and G. M. Ringold, *J. Biol. Chem.* **258,** 11260 (1983).
364. S. J. Bulera, R. B. Birge, S. D. Cohen, and E. A. Khairallah, *Toxicol. Appl. Pharmacol.* **134,** 313 (1995).
365. C. J. de Groot, D. Zonneveld, R. T. M. de Laaf, M. A. Dingemanse, P. G. Mooren, A. F. M. Moorman, W. H. Lamers, and R. Charles, *Biochim. Biophys. Acta* **866,** 61 (1986).
366. A. T. Das, J. Salvadó, L. Boon, G. Biharie, A. F. M. Moorman, and W. H. Lamers, *Eur. J. Biochem.* **235,** 677 (1996).
367. A. S. Spirin, J. B. W. Hershey, M. B. Mathews, and N. Sonenberg, (eds.), *In* "Translational Control," Vol. 1, p. 319. Cold Spring Harbor Laboratory, Cold Spring Harbor, New York, 1996.
368. L. P. W. G. van de Zande, W. T. Labruyère, M. M. Smaling, A. F. M. Moorman, R. H. Wilson, R. Charles, and W. H. Lamers, *Nucleic Acids Res.* **16,** 7726 (1988).
369. J. W. Hershey, *Annu. Rev. Biochem.* **60,** 717 (1991).
370. C. Santos and J. P. Ballesta, *J. Biol. Chem.* **269,** 15689 (1994).
371. K. Saini, P. Thomas, and B. Bhandari, *Biochem. J.* **267,** 241 (1990).
372. C. M. Berlin and R. T. Schimke, *Mol. Pharmacol.* **1,** 149 (1965).
373. I. J. McEwan, A. P. H. Wright, and J. A. Gustafsson, *BioEssays* **19,** 153 (1997).
374. A. P. Wolffe, *Nature (London)* **387,** 16 (1997).
375. M. A. Cahill, W. H. Ernst, and A. Nordheim, *FEBS Lett.* **344,** 105 (1994).
376. M. F. White, *Phil. Trans. R. Soc. Lond., Ser. B.* **351,** 181 (1996).
377. L. S. Argetsinger and C. Carter-Su, *Physiol. Rev.* **76,** 1089 (1996).
378. J. C. Chow, P. R. Ling, Z. Qu, L. Laviola, A. Ciccarone, B. R. Bristian, and R. J. Smith, *Endocrinology* **137,** 2880 (1996).
379. Y. Kumada, D. R. Benson, D. Hillemann, T. J. Hosted, D. A. Rochefort, C. J. Thompson, W. Wohlleben, and Y. Tateno, *Proc. Natl. Acad. Sci. U.S.A.* **90,** 3009 (1993).
380. G. Pesole, C. Gissi, C. Lanave, and C. Saccone, *Mol. Biol. Evol.* **12,** 189 (1995).
381. J. R. Brown, Y. Masuchi, F. T. Robb, and W. F. Doolittle, *J. Mol. Evol.* **38,** 566 (1994).
382. C. Saccone, C. Gissi, C. Lanave, and G. Pesole, *J. Mol. Evol.* **40,** 273 (1995).
383. P. R. Laud and J. W. Campbell, *J. Mol. Evol.* **39,** 93 (1994).
384. J. W. Campbell and D. D. Smith, *Mol. Biol. Evol.* **9,** 787 (1992).

385. J. W. Campbell, J. E. Vorhaben, and D. D. Smith, *J. Exp. Zoo.* **243,** 349 (1987).
386. G. Pesole, M. P. Bozzetti, C. Lanave, G. Preparata, and C. Saccone, *Proc. Natl. Acad. Sci. U.S.A.* **88,** 522 (1991).
387. Y. Tateno, *Jpn. J. Genet.* **69,** 489 (1994).
388. C. Caggese, P. Barsanti, L. Viggiano, M. P. Bozzetti, and R. Caizzi, *Genetica* **94,** 275 (1994).
389. R. Caizzi, M. P. Bozzetti, C. Caggese, and F. Ritossa, *J. Mol. Biol.* **212,** 17 (1990).
390. H. G. Trapido-Rosenthal, P. J. Linser, R. M. Greenberg, R. A. Gleeson, and W. E. S. Carr, *Gene* **129,** 275 (1993).
391. L. Fucci, A. Piscopo, F. Aniello, M. Branno, A. Di Gregorio, R. Calogero, and G. Geraci, *Gene* **152,** 205 (1995).
392. D. D. Smith, N. M. Ritter, and J. W. Campbell, *J. Biol. Chem.* **262,** 198 (1987).
393. S. Hatada, M. Kinoshita, M. Noda, and M. Asashima, *FEBS Lett.* **371,** 287 (1995).
394. D. D. Smith and J. W. Campbell, *Comp. Biochem. Physiol. B Mol. Biol.* **86B,** 755 (1987).
395. H. Pu and A. P. Young, *Gene* **81,** 169 (1989).
396. S. S. Tate, F. Y. Leu, and A. Meister, *J. Biol. Chem.* **247,** 5312 (1972).
397. V. Pamiljans, P. R. Krishnaswamy, G. Dumville, and A. Meister, *Biochemistry* **1,** 153 (1962).
398. S. Seyama, Y. Kuroda, and N. Katunuma, *J. Biochem.* **72,** 1017 (1972).
399. T. F. Deuel, M. Louie, and A. Lerner, *J. Biol. Chem.* **253,** 6111 (1978).
400. K. Iqbal and C. Wu, *Enzyme* **12,** 553 (1971).
401. D. C. Tiemeyer and G. Milman, *J. Biol. Chem.* **247,** 2272 (1972).
402. C. Wu, *Can. J. Biochem.* **55,** 332 (1977).
403. H. Yamamoto, H. Konno, T. Yamamoto, K. Ito, M. Mizugaki, and Y. Iwasaki, *J. Neurochem.* **49,** 603 (1987).
404. H. Tumani, S. Q. Shen, and J. B. Peter, *J. Immunol. Meth.* **188,** 155 (1995).
405. L. Jaenicke and W. Berson, *Hoppe-Seyler's Z. Physiol. Chem.* **358,** 883 (1977).
406. R. A. Ronzio, W. B. Rowe, S. Wilk, and A. Meister, *Biochemistry* **8,** 2670 (1969).
407. R. B. Denman and F. Wedler, *Arch. Biochem. Biophys.* **232,** 427 (1984).
408. P. K. Sarkar, D. A. Fischman, E. Goldwasser, and A. A. Moscona, *J. Biol. Chem.* **247,** 7743 (1972).
409. J. E. Vorhaben, S. S. Smith, and J. W. Campbell, *Int. J. Biochem.* **14,** 747 (1982).
410. P. Lund, *Biochem. J.* **118,** 35 (1970).
410a. L. Levintow, *J. Natl. Cancer Inst.* **15,** 347 (1954).
411. S. Goenner, C. Cosson, A. Boutron, A. Legrand, N. Moatti, and A. Lemonnier, *Med. Sci. Res* **23,** 483 (1995).
412. H. S. Hundal, P. M. Taylor, N. M. Willhoft, B. Mackenzie, S. Y. Low, M. R. Ward, and M. J. Rennie, *Biochim. Biophys. Acta* **1180,** 137 (1992).
413. R. E. Miller, *Diabetes* **24,** 416 (1975).
414. B. E. Hayward, A. Hussain, R. H. Wilson, A. Lyons, V. Woodcock, B. McIntosh, and T. J. B. Harris, *Nucleic Acids Res.* **14,** 999 (1986).
415. D. M. Burns, B. Bhandari, J. M. Short, P. G. Sanders, R. H. Wilson, and R. E. Miller, *Biochem. Biophys. Res. Commun.* **134,** 146 (1986).
416. C. S. Gibbs, K. E. Campbell, and R. H. Wilson, *Nucleic Acids Res.* **15,** 6293 (1987).
417. M. J. B. van den Hoff, W. J. C. Geerts, A. T. Das, A. F. M. Moorman, and W. H. Lamers, *Biochim. Biophys. Acta* **1090,** 249 (1991).
418. G. Patejunas and A. P. Young, *Mol. Cell. Biol.* **7,** 1070 (1987).
419. B. Bhandari, K. D. Beckwith, and R. E. Miller, *Proc. Natl. Acad. Sci. U.S.A.* **85,** 5789 (1988).
420. R. Chakrabarti, J. B. McCracken, D. Chakrabarti, and W. W. Souba, *Gene* **153,** 163 (1995).
421. Y. Wang, J. Kudoh, J. Kubota, S. Asakawa, S. Minoshima, and N. Shimizu, *Genomics* **37,** 195 (1996).
422. K. P. Clancy, R. Berger, M. Cox, J. Bleskan, K. A. Walton, I. Hart, and D. Patterson, *Genomics* **38,** 418 (1996).

423. K. Helou, A. T. Das, W. H. Lamers, J. M. N. Hoovers, C. Szpirer, J. Szpirer, K. Klinga-Levan, and G. Levan, *Mammal. Genome* **8**, 362 (1997).
424. R. H. Wilson and R. E. Kellems, (eds.), *in* "Gene Amplification in Mammalian Cells: A Comprehensive Guide," p. 301. Dekker, Inc., New York, 1993.
425. N. J. Proudfoot and G. G. Brownlee, *Nature (London)* **263**, 211 (1976).
426. J. Fahrner, W. T. Labruyère, C. Gaunitz, A. F. M. Moorman, R. Gebhardt, and W. H. Lamers, *Eur. J. Biochem.* **213**, 1067 (1993).
427. F. Gaunitz, C. Gaunitz, M. Papke, and R. Gebhardt, *Biol. Chem.* **378**, 11 (1997).
428. F. Gaunitz, M. Papke, L. Scheja, G. Beckers, and R. Gebhardt, *Z. Gastroenterol.* **33**, 54 (1995).
429. H. Akaoka, H. Hardin-Pouzet, A. Bernard, B. Verrier, M. F. Belin, and P. Giraudon, *J. Virol.* **70**, 8727 (1996).
430. D. Ghosh, *Nucleic Acids Res.* **20** (Suppl.), 2091 (1992).
431. D. Ghosh, *Nucleic Acids Res.* **21**, 3117 (1993).
432. D. Jin and K. Jeang, *Nucleic Acids Res.* **25**, 379 (1997).
433. S. Wagner and M. R. Green, *Science* **262**, 395 (1997).
434. A. M. Baranger, C. R. Palmer, M. K. Hamm, H. A. Glebler, A. Brauweller, J. K. Nyborg, and A. Schepart, *Nature (London)* **376**, 606 (1997).
435. G. Perini, S. Wagner, and M. R. Green, *Nature (London)* **376**, 602 (1997).
436. V. M. Christoffels, M. J. B. van den Hoff, M. C. Lamers, M. A. van Roon, P. A. J. de Boer, A. F. M. Moorman, and W. H. Lamers, *J. Biol. Chem.* **271**, 31243 (1996).
437. M. M. McGrane, J. de Vente, J. Yun, J. Bloom, E. Park, A. Wynshaw-Boris, T. Wagner, F. M. Rottman, and R. W. Hanson, *J. Biol. Chem.* **263**, 11443 (1988).
438. V. M. Christoffels, M. J. B. van den Hoff, A. F. M. Moorman, and W. H. Lamers, *J. Biol. Chem.* **270**, 24932 (1995).
439. C. J. Decker and P. Parker, *Curr. Opin. Cell Biol.* **7**, 386 (1997).
440. A. Jacobson and S. W. Peltz, *Annu. Rev. Biochem.* **65**, 693 (1996).
441. J. Ross, *Trends Genet.* **12**, 171 (1996).
442. C. Y. Chen and A. B. Shyu, *Trends Biochem. Sci.* **20**, 465 (1995).
443. T. Kastelic, J. Schnyder, A. Lentwiler, R. Traber, B. Streit, H. Niggli, A. MacKenzie, and D. Cheneval, *Cytokine* **8**, 751 (1996).
444. P. V. Riccelli, J. Hilario, F. J. Gallo, A. P. Young, and A. S. Benight, *Biochemistry* **35**, 15364 (1996).
445. W. Su, S. Jackson, R. Tjian, and H. Echols, *Gene Dev.* **5**, 820 (1991).
446. P. J. Linser and A. A. Moscona, *Proc. Natl. Acad. Sci. U.S.A.* **76**, 6476 (1979).
447. Y.-C. Li, D. Beard, S. Hayes, and A. P. Young, *J. Mol. Neurosci.* **6**, 169 (1995).
448. R. Grossman, L. E. Fox, L. Gorovits, I. Ben-Dror, S. Reisfeld, and L. Vardimon, *Mol. Brain Res.* **21**, 312 (1994).
449. A. A. Moscona and P. J. Linser, *Curr. Top. Dev. Biol.* **18**, 155 (1983).
450. A. Herzfeld and S. M. Raper, *Biol. Neonate* **163** (1975).
451. M. Moscona and A. A. Moscona, *Differentiation* **13**, 165 (1979).
452. L. Vardimon, L. E. Fox, and A. A. Moscona, *Proc. Natl. Acad. Sci. U.S.A.* **83**, 9060 (1986).
453. A. A. Moscona, N. Osborne, and G. Chader, (eds.), *in* "Progress in Retina Research," p. 111. Pergamon, Oxford, 1983.
454. G. Patejunas and A. P. Young, *J. Biol. Chem.* **265**, 15280 (1990).
455. A. A. Moscona, *FEBS Lett.* **24**, 1 (1972).
456. C. Marie, *J. Endocrinol.* **90**, 193 (1981).
457. S. Ben-Or and S. Okret, *Mol. Cell. Biol.* **13**, 331 (1993).
458. D. E. Koehler and A. A. Moscona, *Arch. Biochem. Biophys.* **170**, 102 (1975).
459. A. D. Saad and A. A. Moscona, *Cell Differ.* 16, 241 (1985).

460. M. E. Lippman, B. O. Wiggert, G. J. Chader, and E. B. Thompson, *J. Biol. Chem.* **249,** 5916 (1974).
461. H. Zhang and A. P. Young, *J. Biol. Chem.* **268,** 2850 (1993).
462. L. Vardimon, I. Ben-Dror, N. Havazelet, and L. E. Fox, *Dev. Dyn.* **196,** 276 (1993).
463. D. Dutting, A. Gierer, and G. Hansmann, *Brain Res.* **312,** 21 (1983).
464. S. G. Spence and J. A. Robson, *Neurosci.* **32,** 801 (1989).
465. J. B. Piperberg and L. Reif-Lehrer, *Cell Biophys.* **6,** 131 (1984).
466. R. Gorovits, A. Yakir, L. E. Fox, and L. Vardimon, *Mol. Brain Res.* **43,** 321 (1996).
467. H. Zhang and A. P. Young, *J. Biol. Chem.* **266,** 24332 (1991).
468. H. Pu and Young, A. P. *Gene* **89,** 259 (1990).
469. I. Ben-Dror, N. Havazelet, and L. Vardimon, *Proc. Natl. Acad. Sci. U.S.A.* **90,** 1117 (1993).
470. M. Beato, *Cell* **56,** 335 (1989).
471. R. Gorovits, I. Ben-Dror, L. E. Fox, H. M. Westphal, and L. Vardimon, *Proc. Natl. Acad. Sci. U.S.A.* **91,** 4786 (1994).
472. G. G. Kuiper and A. O. Brinkman, *Mol. Cell. Endocrinol.* **100,** 103 (1994).
473. M. L. Moyer, K. C. Borror, B. J. Bona, D. B. DeFranco, and S. K. Nordeen, *J. Biol. Chem.* **268,** 22933 (1993).
474. Y. Berko-Flint, G. Levkowitz, and L. Vardimon, *EMBO J.* **13,** 646 (1994).
475. P. J. Linser and M. Perkins, *Dev. Brain Res.* **31,** 277 (1986).
476. M. F. Notter, M. del Cerro, and P. C. Balduzzi, *J. Neurosci. Res.* **29,** 326 (1990).
477. L. Vardimon, L. L. Fox, L. Degenstein, and A. A. Moscona, *Proc. Natl. Acad. Sci. U.S.A.* **85,** 5981 (1988).
478. L. Vardimon, L. E. Fox, R. Cohen-Kupiec, L. Degenstein, and A. A. Moscona, *Mol. Cell. Biol.* **11,** 5275 (1991).
479. P. J. Linser and A. A. Moscona, *Dev. Biol.* **96,** 529 (1983).
480. S. Reisfeld and L. Vardimon, *Mol. Endocrinol.* **8,** 1224 (1994).
481. C. M. Yoshihara, J. D. Lee, and J. B. Dodgson, *Nucleic Acids Res.* **15,** 753 (1987).
482. C. I. Cheeseman, *Prog. Biophys. Mol. Biol.* **55,** 71 (1991).
483. M. S. Kilberg, B. R. Stevens, and D. A. Novak, *Annu. Rev. Nutr.* **13,** 137 (1993).
484. J. D. McGivan and M. Pastor-Anglada, *Biochem. J.* **299,** 321 (1994).
485. M. E. Mailliard, B. R. Stevens, and G. E. Mann, *Gastroenterology* **108,** 888 (1995).
486. M. S. Malandro and M. S. Kilberg, *Annu. Rev. Biochem.* **65,** 305 (1996).
487. L. Kaczmarek, M. Kossut, and J. Skangiel-Kramska, *Physiol. Rev.* **77,** 217 (1997).
488. A. L. Buchman, *J. Am. Coll. Nutr.* **15,** 199 (1996).
489. J. Neu, V. Shenoy, and R. Chakrabarti, *FASEB J.* **10,** 829 (1996).
490. J. C. Hall, K. Heel, and R. McCauley, *Br. J. Surg.* **83,** 305 (1995).
491. J. E. Blume and D. J. Shapiro, *Nucleic Acids Res.* **17,** 9003 (1989).
492. B. Van der Zeijst and H. J. P. Bloemers, (G. D. Fasman, ed.), *in* "Handbook of Biochemistry and Molecular Biology. Physical and Chemical Data" Vol. 3, p. 426. CRC Press, Cleveland, Ohio, 1976.
493. Z. Krawczyk and C. Wu, *Anal. Biochem.* **165,** 20 (1987).
494. J. Felsenstein, *Cladistics* **5,** 164 (1986).
495. W. E. Knox, (ed.), *in* "Enzyme Patterns in Fetal, Adult and Neoplastic Rat Tissues," p. 254. S. Karger, New York, 1972.
496. C. Wu, *Comp. Biochem. Physiol. A Physiol.* **8,** 335 (1963).
497. F. J. López-Soriano and M. Alemany, *Biochem. Int.* **12,** 471 (1986).
498. A. Herzfeld and S. M. Raper, *Enzyme* **21,** 471 (1976).
499. M. K. Chen, N. J. Espat, K. I. Bland, E. M. Copeland, and W. W. Souba, *Ann. Surg.* **217,** 655 (1993).

Structural Organization and Transcription Regulation of Nuclear Genes Encoding the Mammalian Cytochrome c Oxidase Complex

NIBEDITA LENKA,
C. VIJAYASARATHY,
JAYATI MULLICK, AND
NARAYAN G. AVADHANI[1]

Laboratories of Biochemistry
Department of Animal Biology
School of Veterinary Medicine
University of Pennsylvania
Philadelphia, Pennsylvania 19104

I. Introduction	311
A. Subunit Composition and General Function	311
B. Tissue-Specific and Species-Specific Isologs	311
II. Sequence Properties of Nuclear-Encoded Subunits	313
A. COX IV	313
B. COX Va	314
C. COX Vb	315
D. COX VIa	315
E. COX VIb	316
F. COX VIc	318
G. COX VIIa	318
H. COX VIIb	319
I. COX VIIc	319
J. COX VIII	320
III. Structure of Nuclear-Encoded Genes	320
A. Multiple Transcription Start Sites and Their Sequence Properties	321
B. Regulation of Ubiquitous and Muscle-Specific Genes by Upstream Transcription Activators and Negative Enhancers	329
IV. Mechanisms of Coordinate Regulation	332
A. Intergenic Regulation of Gene Expression	333
B. Coordinate Regulation of Nuclear Genes	335
V. Summary and Future Direction	337
References	338

[1] To whom correspondence may be addressed (Fax: 215-898-9923; e-mail: narayan@vet.upenn.edu).

Cytochrome c Oxidase (COX) is the terminal component of the bacterial as well as the mitochondrial respiratory chain complex that catalyzes the conversion of redox energy to ATP. In eukaryotes, the oligomeric enzyme is bound to mitochondrial innermembrane with subunits ranging from 7 to 13. Thus, its biosynthesis involves a coordinate interplay between nuclear and mitochondrial genomes. The largest subunits, I, II, and III, which represent the catalytic core of the enzyme, are encoded by the mitochondrial DNA and are synthesized within the mitochondria. The rest of the smaller subunits implicated in the regulatory function are encoded on the nuclear DNA and imported into mitochondria following their synthesis in the cytosol. Some of the nuclear coded subunits are expressed in tissue and developmental specific isologs. The ubiquitous subunits IV, Va, Vb, VIb, VIc, VIIb, VIIc, and VIII (L) are detected in all the tissues, although the mRNA levels for the individual subunits vary in different tissues. The tissue specific isologs VIa (H), VIIa (H), and VIII (H) are exclusive to heart and skeletal muscle. cDNA sequence analysis of nuclear coded subunits reveals 60 to 90% conservation among species both at the amino acid and nucleotide level, with the exception of subunit VIII, which exhibits 40 to 80% interspecies homology. Functional genes for COX subunits IV, Vb, VIa 'L' & 'H', VIIa 'L' & 'H', VIIc and VIII (H) from different mammalian species and their 5' flanking putative promoter regions have been sequenced and extensively characterized. The size of the genes range from 2 to 10 kb in length. Although the number of introns and exons are identical between different species for a given gene, the size varies across the species. A majority of COX genes investigated, with the exception of muscle-specific COXVIII(H) gene, lack the canonical 'TATAA' sequence and contain GC-rich sequences at the immediate upstream region of transcription start site(s). In this respect, the promoter structure of COX genes resemble those of many housekeeping genes. The ubiquitous COX genes show extensive 5' heterogeneity with multiple transcription initiation sites that bind to both general as well as specialized transcription factors such as YY1 and GABP (NRF2/ets). The transcription activity of the promoter in most of the ubiquitous genes is regulated by factors binding to the 5' upstream Sp1, NRF1, GABP (NRF2), and YY1 sites. Additionally, the murine COXVb promoter contains a negative regulatory region that encompasses the binding motifs with partial or full consensus to YY1, GTG, CArG, and ets. Interestingly, the muscle-specific COX genes contain a number of striated muscle-specific regulatory motifs such as E box, CArG, and MEF2 at the proximal promoter regions. While the regulation of COXVIa (H) gene involves factors binding to both MEF2 and E box in a skeletal muscle-specific fashion, the COXVIII (H) gene is regulated by factors binding to two tandomly duplicated E boxes in both skeletal and cardiac myocytes. The cardiac-specific factor has been suggested to be a novel bHLH protein. Mammalian COX genes provide a valuable system to study mechanisms of coordinated regulation of nuclear and mitochondrial genes. The presence of conserved sequence motifs common to several of the nuclear genes, which encode mitochondrial proteins, suggest a possible regulatory function by common physiological factors like heme/O2/carbon source. Thus, a well-orchestrated regulatory control and cross talks between the nuclear and mitochondrial genomes in response to changes in the mitochondrial metabolic conditions are key factors in the overall regulation of mitochondrial biogenesis.
© 1998 Academic Press

I. Introduction

A. Subunit Composition and General Function

Cytochrome c Oxidase (COX) is the terminal oxidase of the bacterial and mitochondrial respiratory chain complexes. In eukaryotes the oligomeric enzyme represents a mitochondrial inner membrane-bound structure with subunits ranging in number from 7 in slime molds, ~12 in yeast, and up to 13 in mammalian cells (1–4), suggesting a possible correlation between the number of subunits with the evolutionary stage of the organism. The known catalytic functions of the enzyme include (1) the transfer of electrons from reduced cytochrome c to O_2 through the bimetalic CuA–heme a–heme a3–CuB centers to form H_2O, and (2) proton pumping from the matrix side across the mitochondrial inner membrane to set up the electrochemical potential, which is used to drive the synthesis of ATP by the mitochondrial ATP synthase complex. The three largest mitochondrially encoded subunits (I–III) of the complex form the catalytic centers of the enzyme, because they carry the heme and Cu^{2+} redox centers. Subunit III has also been suggested to have a role in coupling electron transfer to oxidative phosphorylation by participating in proton translocation (1, 5–11). Though the precise functional and structural roles of the nuclear-encoded subunits are still unclear, it is generally believed that they may modulate the overall activity of the complex by way of binding to nucleotides, hormones, second messengers, free fatty acids, Zn^{2+}, etc. (12–15). An interesting feature of the mammalian enzyme, distinguishing it from the bacterial or the lower eukaryotic forms, is that some of the nuclear gene-coded subunits are expressed as tissue- and developmental-specific isologs (16). The structure, function, and activities of the COX complexes from different sources comprise an intensely pursued area of research, and some of these aspects have been reviewed extensively (1, 17, 18). The main objective of this review is to discuss more recent information on the structure and activity of the mammalian enzyme, with special reference to transcription regulation of different ubiquitous and muscle-specific COX genes.

B. Tissue-Specific and Species-Specific Isologs

Of the 10 nuclear-encoded subunits in mammals, subunits VIa, VIIa, and VIII are expressed as heart-type (H) and liver-type (L) isoforms (1, 19–21). The H isologs are predominantly expressed in the heart and skeletal muscle tissues whereas the L isologs are ubiquitous, in that they are also expressed in the heart and skeletal muscle, albeit at low levels. Exceptions to this general rule have been noted in some species as follows: rodent muscle and heart tissues appear to lack detectable VIIa (H) mRNA, but rat heart and skeletal muscle express both L and H forms of VIa mRNAs. In addition, the human system has been shown to express only the subunit VIII (L) form. Similarly, in the bovine smooth muscle the H form of subunit VII and L form of subunits VIa and VIII are expressed (22). These studies clearly suggest a marked species-specific variation in the regulation of cytochrome oxidase in different tissues.

Studies with the human COX also show that the steady-state levels of mRNAs for different subunits vary with the developmental stage, ranging from fetus to the adult (23). The COX mRNA levels in relation to the cytoplasmic 28S rRNA in the liver show minimal changes from the fetal to adult stage, whereas the heart and skeletal muscle show a 2- to 20-fold increase from the fetal to adult stage. Skeletal muscle and the heart, from both embryonic and postnatal stages, contain mRNAs for the L and H forms of COX VIa and VIIa subunits. The levels of both VIa(H) and VIIa(H)mRNA in adult skeletal muscle are increased by 5- to 10-fold over the fetal level; the L form of VIa and VIIa transcripts declines after birth (23). Similarly in adult heart, the H form transcripts (both VIa and VIIa) show 7-fold increase over the fetal level and the L form remains either at the same level (VIa) or increases by 3-fold (VIIa). Developmental switching in the expression of VIa isoforms has also been reported by Ewart et al. (24). During bovine skeletal muscle development, switching occurs from a mixture of VIa (H) + VIa (L) mRNAs in the fetus to exclusively VIa (H) isoform in the adult. Studies on the stage-dependent disappearance of liver-specific VIa and appearance of muscle-specific VIII (H) in the developing heart suggest that thyroid hormone may directly or indirectly influence the developmental stage-related switching of tissue-specific isologs (25).

The heart and liver have different energy requirements and COX complexes from these tissues are believed to exhibit different kinetic properties and response to adenine nucleotides (26–29). For these reasons the observed developmental shift in the expression of tissue-specific isologs is often interpreted as indirect evidence for their role in the regulation of holoenzyme activity. This possibility gains support from elegant studies using yeast (30–32), demonstrating a regulated expression of subunit 5b and 5a isologs (homologs of mammalian subunit IV) under aerobic and anaerobic conditions, respectively. Both genetic and biochemical analyses showed that these two isologs modulate the catalytic activity of the complex (17, 33–36). Subsequent studies using unfractionated mitochondrial membranes or submitochondrial particles from rat tissues showed that COX complexes from tissues with different oxidative capacities show variations in subunits IV and Vb stoichiometries and kinetic parameters (29). Additionally, COX from different compartments of the heart, which have different work loads and O_2 tensions, exhibits differences in subunit content and enzyme kinetics (29). The possibility of nuclear-encoded subunits regulating enzyme activity is supported by experimental analyses of the COX complex from human patients with mitochondrial myopathies. To date a number of diseases with varied clinical symptoms have been associated with COX deficiency. These include fatal infantile and benign myopathy as well as cardiomyopathy (37–41). Immunohistochemical analysis showed a specific deficiency of VIIa and VIIb in both groups of patients (42) and also a deficiency of mtDNA-encoded subunit II. Leigh's syndrome, a severe encephalopathy that is associated with a decrease in COX activity (43, 44), appears to have an incompletely assembled enzyme complex (45). Other COX deficiency diseases include encephalomyopathy (46), MELAS (mitochondrial encephalomyopathy, lactic acidosis, and stroke-like episodes), MERRF (myoclonus epilepsy and ragged red fibers), PEO (progressive external ophthalmoplegia), and Kearns–Sayre syndrome (40, 47–51). A thorough understanding of the structure and regulation of genes for various COX subunits would be of great benefit in the molecular genetic analysis of these fatal diseases.

II. Sequence Properties of Nuclear-Encoded Subunits

Based on their tissue distribution in different vertebrates, subunits IV, Va, Vb, VIb, VIc, VIIb, and VIIc and the liver-specific isoforms VIa, VIIa, and VIII of the mammalian COX complex are classified as ubiquitous subunits and are detected in all tissues, though the relative mRNA levels for individual subunits vary in different tissues. The tissue-specific subunits of the mammalian COX enzyme include the H isologs VIa, VIIa, and VIII, which are exclusive to the heart and skeletal muscle with no apparent detection in any other tissues.

A. COX IV

The bovine subunit IV was the first nuclear-encoded subunit of the mammalian complex to be characterized by cDNA cloning and sequencing (52). Since then, sequence properties of subunit IV from primates, humans (53), rats (54), and mice (55) have been reported. COX IV mRNA is a ubiquitously expressed transcript of about 700–750 nucleotides, which is translated into a 17-kDa [169 amino acid (aa) residues] polypeptide. Based on DNA sequence analysis, the subunit from four different mammalian species (mouse, rat, bovine and human) contains a 22-aa-long N-terminal mitochondrial targeting sequence, which is cleaved off during importation. The mature polypeptide is 147 amino acids long and begins with an N-terminal Ala residue. The length of the cDNA varies with the variable 5' and 3' untranslated regions in different species. The hydropathy profile of the mature protein shows a single membrane-spanning amphipathic alpha-helical structure, and the possible transmembrane organization of the subunit suggests its role in membrane anchoring and/or assembly of the complex, and some studies postulate a role in transmembrane proton translocation. The yeast homolog of this subunit (COX 5) is expressed as 5a and 5b isologs that play important roles in modulating the enzyme activity under aerobic and anaerobic conditions, respectively (17, 30–32, 35, 36, 56). Despite efforts, however, the identification of such tissue-specific isologs of COX IV in the mammalian systems has been unsuccessful.

The protein-coding regions of the human, bovine, rat, and mouse subunits show 77–95% conservation at the amino acid level. A higher level of homology (20 out of 220 is observed between the human and bovine presequence as compared to that of the mature protein (120 out of 147). On the other hand, comparing these two species, there is a uniform 85–87% homology at the nucleotide sequence level in the mature protein as well as the N-terminal presequence coding regions of the cDNAs. On the whole, there are 29 positional nucleotide replacements in the entire coding region, out of which 20 (69%) are single nucleotide substitutions, whereas 9 (31%) show double nucleotide substitutions.

B. COX Va

Based on the predicted amino acid sequence from the cDNA (57–59), the bovine and the human Va subunits appear to be the most conserved polypeptides among all the identified COX subunits of bovine–human pairs, with 95% nucleotide sequence identity (58). The mRNA for this ubiquitously expressed subunit is 750 nucleotides long and encodes a 12.4-kDa protein. The deduced human cDNA contains a 19-nucleotide-long 5′ untranslated region (UTR) with a highly G + C-rich (84%) repetitive trinucleotide motif (GCC) and contains a 150-aa-long reading frame, including a 41-aa-long N-terminal presequence. The presequence is rich in both basic and hydroxylated amino acids (6 out of 41 each) and contains an Asp residue at position -24 with respect to the N-terminal Ser $(+1)$ of the mature polypeptide. Thus, the presequence of subunit Va differs from those of other COX subunits in having a negatively charged residue. Additionally, the subunit Va presequence lacks the typical amphipathic arrangement of the basic residues on one side of the helix that is the characteristic of N-terminal signal sequences of a large number of mitochondrial imported proteins (60, 61). Some studies suggest that the negatively charged residue in the presequence may help in intramitochondrial polypeptide sorting (62, 63), whereas other studies (64) suggest no role for the acidic residues in transportation or intramitochondrial sorting. Thus, the significance of this single Asp residue in the COX Va presequence is not clear. The mature peptide in both the bovine and human systems is 109 amino acids long. In the bovine species, the protein contains 48.6% hydrophobic, 17.4% neutral, and 34% charged (18.4% negatively charged and 15.6% positively charged) residues. The 3' UTR of the human mRNA contains an A + T-rich ATTTA as well as a polyadenylation signal AATAAA. The former sequence motif is thought to have a role in the mRNA stability (58, 65).

C. COX Vb

The COX Vb mRNA species, 500–600 nucleotides long, have been detected in various tissues with varying abundance, and encode a protein with an apparent molecular mass of 10.6 kDa (66, 67). The cDNA consists of a 129-aa-long open reading frame that includes a 31-residue-long presequence and a 98-aa-long mature protein both in the rat (68) and in humans (69). Out of 31 amino acid residues of the human presequence, 17 are hydrophobic, 9 are uncharged polar, and the remaining 5 are basic residues. The first amino acid of the mature peptide is Ala in all four species studied (mouse, rat, bovine, and human), similar to that found in the COX IV subunit. The mouse subunit shares 81% positional identity with the human subunit in the entire reading frame, whereas at the nucleotide level it is 78%. Similarly, a 91% amino acid conservation between the rat and mouse is observed. The bovine cDNA sequence shows 85% identity with the human sequence at the nucleotide level; at the 3' UTR only 77% identity is detected (69). In both mouse (66) and rat (68), the 3' UTR contains a consensus polyadenylation signal, AATAAA. The human and bovine cDNA sequences show a less common putative polyadenylation signal, AGTAAA.

Northern blot hybridization of RNA from different tissues (70) shows that the heart and kidney contain nearly 10-fold higher levels of COX Vb mRNA as compared to the liver. Immunoblot analysis of mitochondrial proteins from different tissues shows widely varying levels of subunit Vb (29). Based on equal amounts of protein and heme a/a3 contents, the liver mitochondrial COX complex shows 8-fold lower subunit Vb content as compared to the heart and kidney mitochondrial complexes. These results suggest a variable COX Vb stoichiometry in different tissues. Furthermore, cytochrome oxidase from liver also shows a higher enzyme activity (turnover number of <750/sec); in contrast, the kidney and heart enzymes, with lower turnover numbers (340 to 490/sec), contain higher molar contents of subunit Vb. In support of the reported differences in catalytic activities of the enzyme from different tissues, and under different physiological conditions (20), these results suggest a role for the murine COX Vb in modulating enzyme activity.

D. COX VIa

In the bovine, rat, and human systems, mRNAs for the L type of subunit are 580–620 nucleotides long, whereas mRNAs for the H subunit form are significantly smaller and range from 370 to 520 nucleotides (24, 71–74). The species-specific and isolog-specific differences in the sizes of mRNAs appear largely due to differences in the 3' UTR sequence. Results of Northern blot hybridization with RNA from bovine tissues showed that the VIa (H + L) mRNAs occur at >2-fold higher levels in adult tissues relative to the fetal stage, suggesting a developmental variation in their pattern of expression (24). Additionally, the ratios of each of these forms vary with the developmental stage of the organism (75). Ewart et al. (24) reported that mRNA for the L-specific isolog in 100- to 120-day-old fetal bovine heart and skeletal muscle represented nearly 20–25% of the total L + H pool, which is reduced to 5 and 1% in adult heart and skeletal muscle tissues, respectively. The occurrence of mRNA species for both of the isologs during different stages of evolution has been ascribed to a possible gene duplication event resulting in the presence of two distinct forms of the same gene during the mammalian evolution (76). Additionally, a developmental pattern of gene switching suggests a possible regulatory role for the tissue-predominant isologs of this subunit.

Based on the cDNA sequence analysis, the VIa (H) subunit is predicted to contain a 12-aa-long leader sequence and a 255-aa-long mature protein. The leader peptide shows 67–83% sequence identity between different species, but the mature protein shows 80% sequence conservation. As shown for subunits IV and Vb, Ala is the first amino acid of the mature VIa (H) protein in all the three mammalian species studied. In the case of the VIa (L) subunit, no leader peptide has been detected in the bovine and human species (71), but the rat subunit (73) has been reported to contain a 26-aa-long cleavable presequence. The mature H subunit contains Pro at the N terminus and the L subunit contains a Gln residue. The amino acid sequence of the L subunit from different mammalian species shows 90% conservation as compared to 80% intraspecies identity in the H form.

Most of the sequence similarity between the two isoforms from different species is apparent in the C-terminal half of the protein, mostly beyond the Glu residue at position 45, prompting suggestions on the existence of tissue-specific functional domains in the N-terminal region of the VIa (H) and (L) isoforms (24) (see sequence alignment in Fig. 1). The H and L isologs within a given species show a low level (60%) of sequence homology. However, the individual isologs show high interspecies homology (80–90%), suggesting different tissue-specific roles for the two isologs of this subunit.

E. COX VIb

The mRNA encoding this subunit, 500–550 nucleotides long, is ubiquitously expressed, though it is detected at much higher abundance in the heart and skeletal muscle than in the liver. The mature protein deduced from the bovine (77) and human (78, 79) cDNA sequences is 86 amino acids long with no cleavable presequence. The sequence homology shared between the

```
BM: malplkslsrgl ASAAKGDHGGTGARTWRFLTFGLALPSVALCTLNSWLHS--GHRERPAFIPYHHLRIRTKPFSWGDGNHTFFHNPRVNPLPTGYEKP
RM: [---plrvlsrxm]ASASKGDHGGAGANTWRLLTFVLALPSVALCSLNCWMHA--GHHERPEFIPYHHLRIRTKPFSWGDGNHTLFHNPHVNPLPTGYEQP
HM: malplrplsrgl ASAAKGGHGGAGARTWRLLTFVLALPSVALCTFNSYLHS--GHRPRPEFRPYQHLRIRTKPYPWGDGNHTLFHNSHVNPLPTGYEHP
HL:             SSGAHGEEG--SARMWKTLTFFVALPGVAVSMLDVYLKSHHGEHERPEFIAYPHLRIRTKPFPWGDGNHTLFHNPHVNPLPTGYEDE
BL:             SSGAHGEEG--SARMWKALTLFVALPGVGVSMLNVFMKSHHGEEERPEVAYPHLRIRSKPFPWGDGNHTLFHNPHVNPLPTGYEDE
RL:             SSGAHGEEG--SARIWKALTYFVALPGVGVSMLNVFLKSRHEEHERPEFVAYPHLRIRTKPFPWGDGNHTLFHIPHMNPLPTGYEDE
```

FIG. 1. Amino acid alignment of the COX VIa subunit (muscle and liver isoforms) from different species. Here M and L denote the muscle and liver specific isoforms from the bovine (B), human (H), and rat (R) species). Presequences are in lowercase and mature peptides are in uppercase letters. Asterisks denote the sequence conservation among species and isoforms. Reprinted from *Gene* **119**, G. M. Fabrizi, J. Sadlock, M. Hirano, S. Mita, Y. Koga, R. Rizzuto, M. Zeviani, and E. A. Schon; Differential expression of genes specifying two isoforms of subunit VIa of human cytochrome c oxidase; 307–312, with permission from Elsevier Science. (From Ref. 72.)

two species is 85% in terms of amino acid sequence, whereas at the nucleotide level, the homology is 89%. The nucleotide sequences of the human heart and skeletal muscle cDNAs show absolute homology in the entire coding and noncoding regions, except within the 5' end of the poly(A) tract. In the skeletal muscle, an AAA triplet at the beginning of the poly(A) tract is replaced by a GTG sequence, resulting in the cleavage and polyadenylation of the transcript six bases further downstream from that of the heart mRNA (79). The mature subunit from the bovine heart contains an acetylated form of Ala residue, suggesting that the N-terminal Met might have been removed posttranslationally (80). A similar removal of the terminal Met residue has been commonly observed for polypeptides imported to mitochondria that do not contain cleavable presequences (19, 81–83).

F. COX VIc

This is another ubiquitous COX subunit without a cleavable presequence. In the human, the N terminus of the mature protein contains basic amino acid residues and exhibits the properties of an uncleaved signal sequence (84). In the rat, there is a TAG stop codon 18 nucleotides upstream of, and inframe with, the initiation codon ATG, suggesting the absence of a processed presequence (83). The 8.5-kDa polypeptide is encoded by an mRNA 450 to 510 nucleotides long. The mature subunit from the rat is 75 amino acids long, as compared with 76 and 77 residues for the human and bovine subunits, respectively. Both the rat and human subunits contain Met as the first residue, compared to the bovine subunit, which contains a Ser residue (85). The rat COX VIc protein shows 80% homology with the bovine subunit and 69% homology with the human protein. Similarly, between the bovine and human, a 73% positional conservation is observed.

G. COX VIIa

There are two different isologs (H and L) for this subunit. The mRNA for the H form (420–460 nucleotides long) is detected exclusively in the heart and skeletal muscle tissues, whereas the L form (530 nucleotides long) is ubiquitously expressed in all tissues. As shown for the COXVIa isologs, mRNAs for both the H and L forms of COX VIIa are expressed in heart and skeletal muscle. The human uterus expresses a novel form of COX VIIa mRNA, about 470 nucleotides long, which is distinct from either the H or the L form found in other tissues. Unlike the H and L isologs, the uterine form cross-hybridizes with both of the H- and L-specific probes, suggesting that it is a novel hybrid form (86). No further characteristics of this form have been reported. A notable difference, unique to the rat, is that no VIIa (H) mRNA has been detected in any of the tissues investigated; irrespective of the tissue, the rat appears to contain only the VIIa (L) form.

The deduced amino acid sequence of the COX VIIa (H) subunit in the human (86) and bovine systems (87), based on cDNA sequences, reveals the presence of 58- and 59-aa-long mature proteins, respectively, in addition to a 21-aa leader peptide. The mature VIIa (L) protein in all three mammalian species, rat (88), human (86, 89), and bovine (90), on the other hand, is 60 amino acids long. The COX VIIa (H) mRNA contains a smaller 3' UTR as compared to mRNA for the L form, accounting for the larger transcript size in the latter case. The N-terminal presequences of both the VIIa (L) and (H) forms show a high degree of homology and show the characteristics of a two-step processing signal (91). Thus, the transport/maturation of this isoform may involve a two-step cleavage of the precursor by distinct matrix proteases (87, 91). The interspecies conservation with the presequence is generally higher than that detected in the mature protein for a given isoform. On the contrary, the homology within the presequence region is drastically reduced between the two isoforms either in a single species or among species, compared to that in the mature peptide. In the bovine species, the conservation in the N-terminal regions of the tissue-specific isologs is higher than in the C-terminal half. Interestingly, both the H and the L isologs from different species contain a 9-aa-long conserved sequence stretch (EKQKLFQED) at an identical N-terminal position, from position 7 to position 15 of the mature protein, implying that its has an important functional role (87).

H. COX VIIb

This is a ubiquitously expressed subunit and currently sequence information is available only for the bovine (92) and human subunit VIIb (93) species, where it is expressed as an 83- to 88-aa-long preprotein with a 27- or 32-aa-long cleavable presequence. Another noteworthy feature of the VIIb subunit compared to other subunits is that mRNA contains an unusual 79-nucleotide-long inverted repeat. The 5' end repeat sequence is located in the unusually long 5' UTR (156 nucleotides long), where the 3' end repeat spans the 23 nucleotides from the C-terminal end of the reading frame and the 3' UTR. The human polypeptide is 80 amino acids long, including a stretch of 24 amino acids (leader sequence) plus the 56 amino acids of the mature peptide. In the mature protein, there is 82% sequence homology between the two species.

I. COX VIIc

This is another ubiquitous subunit, characterized from the mouse (94), bovine (95), and human systems (96). The cDNA sequence shows the presence of an open reading frame that includes a presequence (16 amino acids) and a 47-aa-long mature protein. The presequence from different species shows 100% amino acid identity. However, comparing the mature peptide from mouse–bovine systems and mouse–human systems, there is 94 and 81% sequence identity, respectively. In the bovine and human systems, 84% amino acid sequence conservation is noted. The mature protein starts with a His residue in all three species studied.

J. COX VIII

This is the smallest of all the nuclear-encoded subunits and it is expressed in a tissue-specific manner, in different species, as H and L isoforms. In bovine, pig, mouse, and rat tissues, both isoforms are prevalent, depending on the tissue types. In the human and primate (monkey) species, however, only the COX VIII (L) form is detectable in all tissues. This differential expression of the COX VIII subunit in different mammals suggests that the putative gene duplication event leading to the formation of the two isoform-specific genes might have occurred prior to the divergence of primates from other mammals. It is likely that the heart form might have been silenced in the human during the evolutionary process (*97*). mRNA for the VIII (L) isoform is about 600 nucleotides long (*21, 83, 97–99*) and the size of the VIII (H) mRNA in different species is about 400 to 500 nucleotides long (*21, 98, 100*). Some of the size differences between the two isoform-specific mRNAs can be attributed to the longer 3′ UTR of the L form. In bovines (*21*), the L form mRNA is more abundant in the kidney, brain, and liver as opposed to low abundance in skeletal muscle and the heart, which express high levels of the H form. In contrast, in humans (*97*), the L form is detected in high abundance in the heart and skeletal muscle as well, where no H form is detected.

The presequences of both the L and H isoforms are about 25–26 amino acid long and are rich in basic amino acids. The mature proteins of the H and L forms are also similar in size and range from 44 to 46 amino acids long. The 3′ UTR of the VIII (L) mRNA sequences from different species is equally conserved as that of the protein-coding region, suggesting an important role in both transcriptional and translational control of gene expression (*21*). In the rat and human species, a high nucleotide conservation (81%) is observed at the 3′ UTR. In the coding region the homology ranges from 40 to 80% among isoforms and species. Compared to other COX subunits, subunit VIII shows the least conservation among species, indicating a rapid evolution of this particular subunit of the multisubunit complex.

III. Structure of Nuclear-Encoded Genes

Characterization of the functional genes for some of the COX subunits in different mammalian organisms is made complicated by the presence of multiple pseudogenes and retroinserts. The genes for COX IV, Vb, VIa (L and H), VIIa (L and H), VIIc, and VIII (H) subunits from different mammalian sources, including the human, mouse, rat, and bovine species, have been characterized. The 5' flanking putative promoter regions have been sequenced. The genome size for different subunits ranges from 2 kb for the rat (101) and bovine (102) COX VIII (H), gene, with only two exons, to about 10 kb, with six exons for the rat (103), mouse (55), and bovine (104) COX IV genes. Although the numbers of introns and exons are identical among different species for a given gene, the sizes of introns and exons vary among the species (see Table I). In most cases, the exon sequence encoding the presequence of the protein is separated by an intron from that encoding the N terminus of the mature protein. A detailed structural analysis of genes for different COX subunits is presented in Table I and Figs. 2 and 3. The 5' upstream regions of many COX genes, including the COX IV, Vb, VIa, (L and H), VIIa (L and H), VIIC, and VIII (H) genes, contain highly G + C-rich sequences. Most of them are either placed in CpG islands (COX VIIa) or contain multiple GC boxes that bind Sp1 factor (Table I). The CpG islands usually remain undermethylated or nonmethylated. Thus genes containing these islands in their 5' proximal region or in the introns are, in general, transcriptionally more active (105). This feature is also common to most of the housekeeping genes. Among the ubiquitous genes, the COX IV (106–109), Vb (110), and VIIc (111) genes from different sources have been extensively investigated. Under the tissue-specific category, the information on transcription regulation has been reported for the H form of both VIa (112) and VIII (101) as well as the L form of VIa (73) and VIIa (113).

A. Multiple Transcription Start Sites and Their Sequence Properties

Based on the sequence properties, transcription initiation sites of diverse tissue-specific and ubiquitous COX genes fall into two distinct classes. A large number of specialized genes contain a canonical TATAA or a modification thereof at about 29 nucleotides upstream of transcription initiation sites. A number of TATAA-containing genes also contain a pyrimidine-rich CAPyPyPy initiator (Inr) motif that marks the transcriptional start points (114–118). Another class of genes, mostly belonging to the housekeeping family, lack the canonical TATAA sequence, and instead contain G + C-rich sequences in the immediate upstream regions of transcription start sites. In most cases investigated, the mRNAs encoded by the TATAA containing genes do not show 5' heterogeneity because transcription is initiated at single or restricted promoter sites. Genes belonging to the second category lack the canonical TATAA sequence and transcription initiation on these genes is driven by the Inr motifs. A variety of sequences that show lose consensus to

TABLE I
STRUCTURAL FEATURES OF NUCLEAR-ENCODED COX GENES AND CONSERVED
PROTEIN-BINDING MOTIFS AT THE 5′ PROMOTER REGIONS

Subunits and species	Transcription initiation site(s)	Exon number/ size (bp)	Intron number/ size (bp)	Promoter elements
IV				
Mouse	Multiple	I/92 II/74 III/168 IV/132 V/134	I/950 II/3000 III/126 IV/288	Sp1, GAPB, Hap2/Hap3
Rat	Multiple	I/44 II/74 III/168 IV/132 V/235	I/1000 II/3200 III/400 IV/900	Sp1, AP4, Hap2/Hap3, GABP
Bovine	Two	I/112 II/74 III/168 IV/132 V/232	I/1171 II/2500 III/400 IV/600	Sp1, GABP, ATF/CREB
Vb				
Mouse	Multiple	I/149 II/73 III/99 IV/189	I/520 II/165 III/648	Sp1, GABP, YY-1, AP1, AP4, GTG, CArG, UAS2, Hap2/Hap3, NRF1
Rat	Multiple	I/125 II/74 III/100 IV/193	I/523 II/162 III/656	Sp1, AP1, AP4, NRF1, Hap2/Hap3, GABP, Mt1, Mt2, Mt3
Human	—	II/103 III/74 IV/100 V/191	II/876 III/200 IV/552	Sp1, AP1, AP2, NRF1, GABP, Mt1, Mt3, Mt4
VIa (H)				
Mouse	Multiple	I/103 II/137 III/186	I/186 II/77	E box, MEF2, GATA I
Rat	Multiple	I/276 II/137 III/195	I/188 II/84	Sp1, AP4, AP1/GCN4, Oxbox, Mt1, Mt2, Mt3, Mt4, UAS2, CArG, E box
Bovine	Multiple	I/253 II/137 III/116	I/146 II/96	TATA, CCAAT, E box, NRF1, AP4, Mt1, Mt2, Mt3, Mt4

(continues)

TABLE I (Continued)

Subunits and species	Transcription initiation site(s)	Exon number/ size (bp)	Intron number/ size (bp)	Promoter elements
VIa (L)				
Rat	Multiple	I/167 II/143 III/276	I/115 II/2359	AP1/GCN4, Sp1, Oxbox, UAS2, Mt1, Mt2, Mt3, Mt4, CArG, E box
VIc				
Rat	Multiple	I/96 II/130 III/129 IV/97	I/843 II/3200 III/1000	TATA, CCAAT, AP1
VIIa (H)				
Bovine	Two differ by 1 nucleotide	I/61 II/87 III/85 IV/138	I/696 II/134 III/380	Sp1, CArG, E box, Oxbox, Hap2/Hap3, CpG islands
VIIa (L)				
Bovine	Two major	I/97 II/90 III/85 IV/203	I/1180 II/2150 III/1580	Sp1, NRF1, NRF2, (GABP), CpG islands
VIIc				
Bovine	Multiple	I/185 II/129 III/218	I/838 II/915	YY-1, NRF2 (GABP)
VIII (H)				
Rat	Single	I/178 II/151	I/1200	Sp1, TATA, E box, AP1, AP2, Ψ CArG, PEA3, GRE, GCN4, NF-κB
Bovine	Single	I/73 II/65	I/1227	TATA, E box, Ψ CArG, NRF1

the classical Inr, initially reported for the mouse TdT promoter, have been shown to function as "initiators" under both *in vivo* and *in vitro* conditions (*114, 117, 118*). The evidence suggests that a putative Inr binding factor helps in the recruitment of TBP (TFIIB) and/or other components of the basal transcription complex by an unknown mechanism, and positions the initiation complex at the correct start site (*117, 119–121*). The mRNAs encoded by these TATAA-lacking genes often show 5′ heterogeneity, although little is known about mechanisms involved in multiple-site initiations. A majority of the COX genes investigated, with the exception of the muscle-specific COX VIII (H) gene, lack the canonical TATAA sequence, and therefore resemble the housekeeping genes. Additionally, mRNAs for some of COX genes, particu-

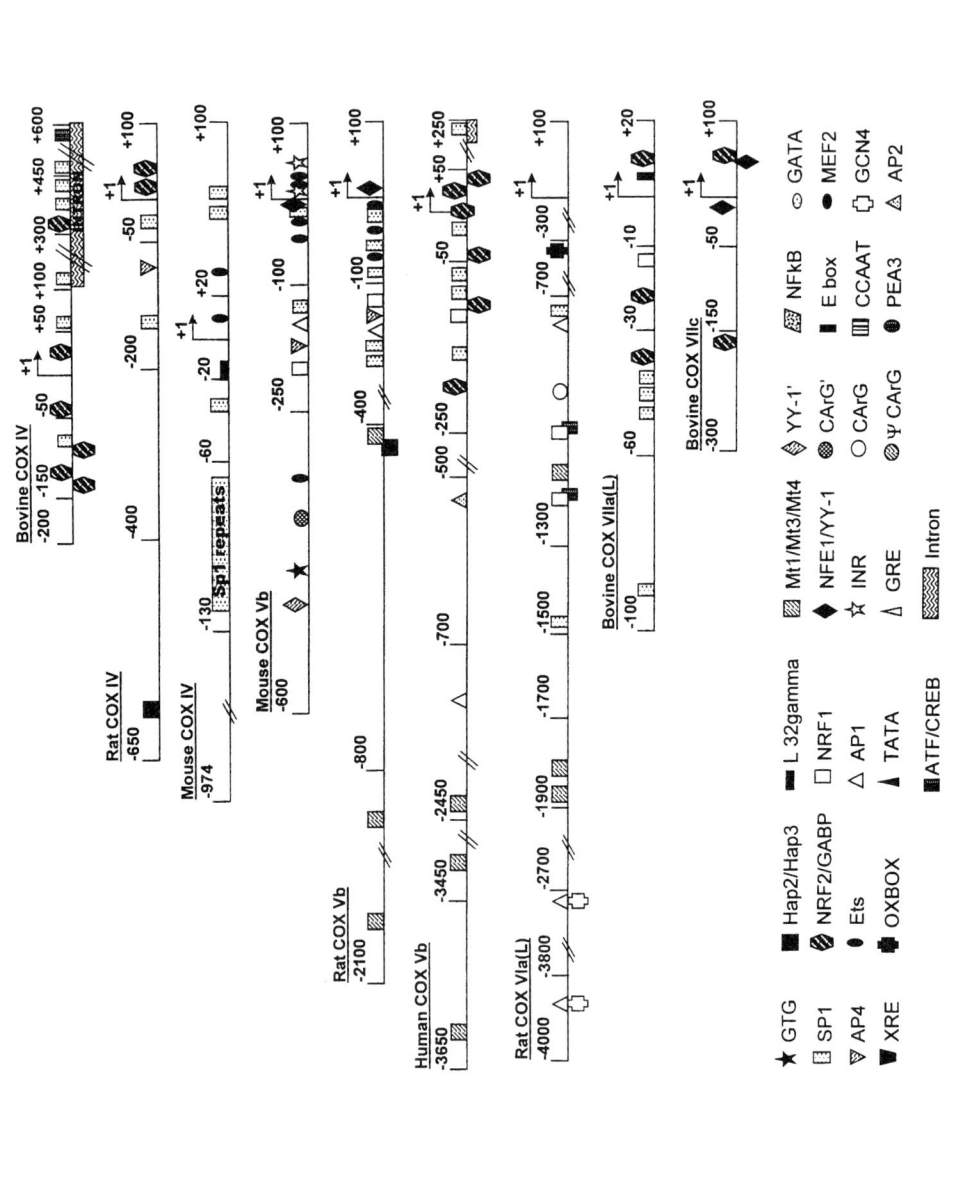

larly those of the ubiquitous types, show extensive 5' heterogeneity as described below.

The transcription initiation sites of the mouse COX IV and Vb promoters have been extensively studied by mapping the *in vivo* and *in vitro* transcripts and by mutational analyses (67, 108–110, 122). In the case of both mouse and rat COX IV promoters, two major clusters of mRNA 5' ends map to 21-base-long tandem duplicated sequences (103, 106, 108, 123). Each of these repeats contains a CGGAAG core sequence that binds to an ets family transcription factor GABP (124, 125), alternatively referred to as β, E4TF1, and NRF2 (106, 126, 127). Studies with the mouse COX IV promoter showed that the 33-bp initiation region sequence lacking the TATAA binding or any consensus Inr sequence was sufficient to direct site-specific transcription initiation under the *in vitro* and *in vivo* conditions. GABP-α and -β heteromeric factor binding to the 33-bp tandemly duplicated ets site was essential for transcription initiation using this minimal promoter. RNA mapping studies showed that the RNA start sites with the minimal promoter under the *in vivo* and *in vitro* conditions resemble the steady-state COX IV mRNA 5' ends. Mutations targeted to the upstream or downstream ets sites abolished transcripts mapping to these respective sites, while mutations targeted to both ets sites abolished transcripts mapping to the +1 (upstream ets) and +22 (downstream ets) sites of the mouse COX IV promoter. Interestingly, the COX IV mRNA in the rat also shows similar 5' end heterogeneity (106) and maps to two tandemly duplicated ets sites. In the bovine system, however, the major cluster of 5' ends map to CTCGAT, a sequence that resembles the YY-1 binding site. Currently the nature of 5' heterogeneity and the protein-binding properties of the initiation site sequence motif of bovine COX IV remain to be elucidated.

Of the nuclear gene-encoded COX mRNAs so far investigated, the mouse Vb mRNA probably shows the most extensive 5' heterogeneity, with six different clusters of mRNA termini mapping to +1, +6, +12, +17–22, +24–29, and +32–36 sequences of the gene (67). Interestingly, all of these six clusters map to sequence motifs that bind to known (GABP) or yet undefined protein-binding sites. By *in vivo* transcription in COS and 3T3 cells and *in vitro* transcription in factor-supplemented and -depleted extracts, and also by mutational analyses, Sucharov *et al.* (67) showed that transcription initiation at the +1 position of the mouse COX Vb gene is highly dependent on the presence of factor YY-1, alternatively referred to as NF-E1, gamma fac-

---

FIG. 2. Promoter structure and putative cis DNA elements of the ubiquitous COX genes from different mammalian species. Representative symbols of various protein-binding motifs are given at the bottom of the figure.

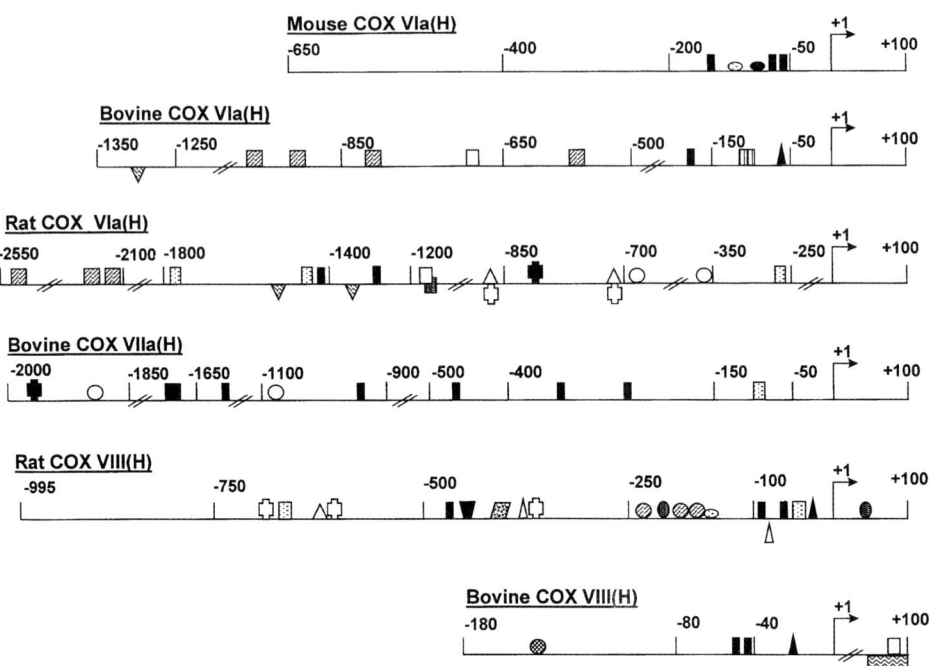

FIG. 3. Promoter structure and putative cis DNA elements of the muscle-specific COX genes from different mammalian species. Representative symbols of various protein-binding motifs are as shown in Fig. 2.

tor, and delta/UCRBP (110, 128–132). In another study, it was also shown that GABP factor binding to the inverted repeat ets motif at the +16 to +26 sequence of the mouse COX Vb promoter was essential for the initiation of two clusters of transcripts mapping to the +17–22 and +24–29 positions under both in vivo and in vitro conditions (67). In support of these observations, a study using promoter constructs containing upstream Sp1 factor-binding sites showed that two tandemly spaced GABP-binding ets sites facing the same side of the helix can indeed function as Inr under in vitro conditions (133). In the case of the mouse COX Vb gene, transcripts originating at the +6, +12, and +32–36 positions of the promoter map to an unknown factor-binding motif, and two initiator-like sequences, respectively. As shown in Fig. 4, promoter site selection mapped by S1 nuclease analysis of RNA from COS cells transfected with various promoter constructs and also in vitro transcribed RNA closely resemble the map positions of endogenous mRNA. It should be noted that depletion of YY-1 factor exclusively affected transcripts originating at +1 and depletion of GABP affected only transcripts

A: COX Vb ENDOGENOUS mRNA 5' ENDS

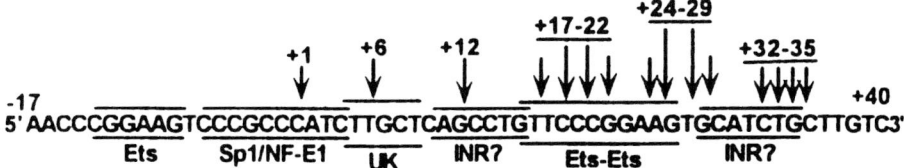

B: START SITE SELECTION IN VIVO

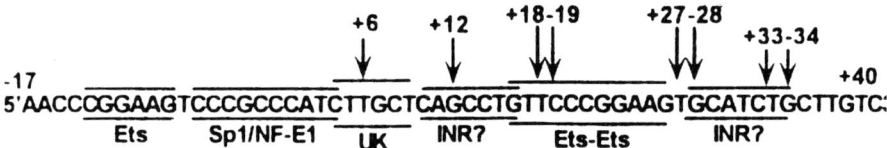

C: START SITE SELECTION IN VITRO

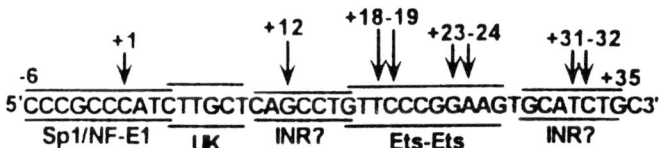

UK=Unknown

FIG. 4. Transcription start sites of the murine COX Vb gene mapped by *in vivo* and *in vitro* transcription analyses. (A) The endogenous COX Vb mRNA 5' map positions are based on S1 nuclease protection assay. (B) The 5' terminal positions of RNA from COS cells transfected with −17CAT, −6CAT, and ets–ets constructs as determined by S1 nuclease protection assay. (C) Transcription start site selection under *in vitro* transcription conditions using −6CAT (−13 to +35) and ets–ets (+12 to +31) templates. The nucleotide sequences of −17CAT DNA, −6CAT DNA, and various protein-binding motifs are presented. (From Ref. 67.)

mapping to the +17–22 and +24–29 positions, further suggesting that the 5' heterogeneity of the COX Vb mRNA is largely due to independent transcription initiation at multiple initiator motifs that bind to some of the general transcription factors, such as YY-1 and GABP. Interestingly, the major mRNA 5' end in the rat system also maps to a putative YY-1-binding site, CATCTG (68). Similarly, the rat basal promoter region contains an inverted ets repeat, identical to that found in the mouse gene, excepting that it occurs at the immediate upstream position of the putative YY-1 site. In the human COX Vb promoter, the major transcript maps to a single GABP-binding ets site fused to an immediate upstream Sp1 site (*134*). Similar to that observed in the mouse COX Vb promoter, the human promoter contains an inverted ets repeat motif at about 22 nucleotides downstream of the major transcription start site at +1. Additionally, the human promoter also contains a putative YY-1-binding site at about the +15 position (*134*). Thus, although the details of the 5' heterogeneity and factor binding properties of different initiation site sequence motifs remain unclear, there is a high degree of interspecies similarity with respect to initiation site sequences and conserved sequence motifs (see Fig. 2).

In the case of the recently reported VIIc gene, the basal promoter contains a YY-1 site flanking the major transcription start site at +1 and also a YY-1 site fused to a GABP site at about the +15 position of the promoter (*111*). Interestingly, as shown for the mouse COX Vb gene, mutations targeted to the upstream YY-1 site reduced the level of transcripts mapping to the +1 position, suggesting its role in transcription initiation at the +1 site. The ubiquitous mRNA for COX VIIa (L) exhibits multiple 5' ends that map to a duplicated GABP-binding ets motif in a manner analogous to the mouse and rat COX IV genes.

Both the muscle-specific rat COX VIa (H) and bovine VIIa (H) genes are TATA-less genes, whereas both the rat and bovine COX VIII (H) genes contain TATAA sequences at about 28 nucleotides upstream of major transcription start sites. In contrast to the ubiquitous COX genes, these muscle-specific genes lack any of the general transcription factor binding motifs, such as GABP or YY-1, as part of their basal promoters. Thus, a unifying theme in all of the ubiquitous COX genes is that the major transcription start sites are positioned to sites that bind to general transcription factors such as GABP-binding ets sites and/or the YY-1-binding sites. The initiation regions of these genes, therefore, resemble the ribosomal protein genes and other housekeeping genes (*115, 126, 135, 136*).

The initiation region sequence motifs GABP and YY-1 from the ubiquitous COX genes are similar to the sequence elements classified as initiator motifs by the following criteria: (1) they yield a basal level of promoter activity in the absence of other promoter elements, (2) they specify the position of transcriptional initiation *in vivo* and *in vitro*, (3) they function synergistically with the upstream activator motifs, such as Sp1 element, to produce strong promoter activity, and (4) in one of these cases, i.e., YY-1, the factor has been shown to interact with TBP and other components of TFIID (*120, 121*), suggesting that it is able to recruit the basal transcription machinery needed for the initiator function. However, currently it is not clear if the GABP-β subunit has any role in recruiting TBP and other TAFs to the basal transcription complex. Results emerging from these studies, however, suggest that a number of general transcription factor binding motifs, including YY-1 and GABP, are mutifunctional in that they can function both as transcription activators and initiators, depending on their location within the promoter.

B. Regulation of Ubiquitous and Muscle-Specific Genes by Upstream Transcription Activators and Negative Enhancers

As stated before, all but one of the nuclear-encoded COX genes lack putative TATA and CAAT motifs and most of the ubiquitously expressed genes contain multiple GC boxes and sequence motifs for binding to various general as well as specialized transcription factors.

In the ubiquitous COX genes, in addition to the transcription initiation region YY-1 and GABP factor-binding sites, 5' upstream NRF1, GABP (NRF2), and Sp1 sites appear to play a dominant role in the transcription activity of the promoter. An extensive 5' deletion analysis of the mouse COX IV gene shows that the -5 to $+36$ sequence of the basal promoter that contains two tandem GABP sites, placed 15 nucleotides apart, shows only about 6–7% $in\ vivo$ transcription activity of the more intact -142CAT promoter. These results show that multiple Sp1 sites within the -5 to -142 region of the mouse COX IV promoter play a major role in transcription activation of the promoter. A similar effect of the upstream Sp1 sites (at sequence -28 to -52) on the transcriptional activity of the rat COX IV promoter has been reported. The rat gene contains additional Sp1 sites at further upstream regions, which might also contribute to the transcription activation of the promoter. Thus a 15- to 20-fold stimulation of the basal transcription activity by sequences from the distal regions of the COX IV and Vb genes in both the mouse and the rat depends on Sp1 factor-binding sites as well as sequence motifs targeted by a relative of the ets oncogene family heteromeric transcription factor, GABP (67, 106, 108, 110). Additionally, site-specific mutational analyses showed that the transcriptional activities of the mouse COX Vb promoter and rat COX VIIa (L) promoters are positively modulated by upstream NRF1 sites. In the murine COX Vb promoter the functional NRF1 site is located at about the -175 region (137) and in the bovine COX VIIa (L) gene it is located immediately upstream (the -100 region) of the transcription initiation site (113, 138). The activity of the bovine COX VIIc promoter is also positively modulated by an immediate upstream YY-1 (18, 111). Thus, the overall promoter architecture and transcription factor binding sites of the ubiquitously expressed mammalian COX genes resemble those of the housekeeping genes.

The deletion analysis of the murine COXVb promoter also showed the presence of a negative regulatory region mapping to the −481 to −320 sequence of the promoter (70). This region encompasses binding sites exhibiting partial or full consensus to YY-1 (YY-1' site), GTG, CArG (CArG' site), and ets. Studies on positional alterations and site-specific mutations suggest that interaction between the protein factors binding to the YY-1', GTG, and CArG' elements is essential for the function of the negative enhancer. Individually the three sequence motifs act as transcription activators when placed upstream of the COX Vb basal promoter or the heterologous TK CAT promoters. However, when placed in a tandem array in an order as they exist in the COX Vb promoter the sequence motif down-regulates the activity of the homologous and heterologous (mouse COX IV promoter and TK CAT) promoters by five- to sevenfold of the control promoter activities. The 90-bp DNA containing the three motifs forms a number of protein complexes and shows tissue-specific differences in terms of number and migration patterns of complexes. Gel-shift analysis and DNase I footprinting studies show that the CArG' motif binds to factor YY-1, and the YY-1' and GTG motifs bind to unidentified protein components that show tissue-specific differences. Interestingly, the activity of the negative enhancer is down-modulated during the induced differentiation of C2C12 myocytes, resulting in a five- to sevenfold higher transcription rates in myotubes as compared to undifferentiated myoblasts. The results of this study suggest that the activity of the negative enhancer plays an important role in the tissue-specific variations in the level of COX Vb gene expression. Although the comparable 5' deletion data for the rat and human COX Vb promoters are not available, a comparison of DNA sequences suggests the occurrence of YY-1'-, GTG-, and CArG'-like motifs in these genes. Additionally, 5' deletion results show the presence of a negative enhancer activity at sequences −974 to −142 of the mouse COX IV promoter (108), although the details with respect to sequence and protein-binding properties remain unknown. Thus the overall rates of mouse COX Vb expression appear to be a result of a coordinated interplay between the positive- and negative-acting enhancer elements.

A distinct feature of the promoters of heart/muscle-specific COX genes is that they contain a number of striated muscle-specific regulatory motifs, e.g., E box, CArG, MEF2, Oxbox, and GATA (Table I). Interestingly, both COX VIa (H) and VII (H) promoters in bovines show the presence of putative NRF1-binding sites, although they have not yet been functionally characterized (102, 139). The rat COX VIa (H) and COX VIII (H) promoters that have been characterized for basal- and muscle-specific transcription activity do not contain canonical NRF1- or GABP-binding sites (101, 112). The functional domain of the murine COX VIa (H) gene has been localized to a 300-bp 5' flanking minimal promoter region (112). The sequence motifs required for the skeletal muscle-specific transcription activity include three potential

E boxes (CANNTG) and a single MEF2 site located with the 300-bp region. Mutations targeted to the MEF2-binding site abolished inducible transcription in differentiated skeletal C2C12 myocytes. Mutations targeted to the E box motifs, however, showed a general decline in transcription activity in both differentiated and undifferentiated myocytes, as well as in 3T3 fibroblast cells. In view of reports showing increased levels of MEF2 in induced C2C12 myotubes (*140*), it was suggested that MEF2 binding is important in transcriptional regulation of the COX VIa (H) promoter in differentiated skeletal muscle cells (*112*). However, a possible interaction between the MEF2-binding elements and the bHLH factors binding to the distal E box motif in regulated muscle-specific COXVIa (H) gene expression cannot be ruled out, because a significant decrease (65%) in the promoter activity is observed with the mutation at the distal E box region.

In the rat COX VIII (H) gene, the basal promoter activity maps to a 44-bp sequence upstream of the transcription initiation site at +1, which contains a TATA element at position −29/−22 and an upstream binding site for Sp1 factor (−42/−33) (*101*). In both skeletal C2C12 myocytes and cardiac-specific H9C2 myocytes, sequences up to position 158 of the promoter are required for muscle-specific transcription activation. This region, between −44 and −158, includes two tandemly duplicated E boxes and a single GRE consensus motif. The E boxes, with the core sequence of CANNTG, are the potential binding sites for the bHLH superfamily of proteins that include the skeletal muscle-specific myogenic factors such as MyoD, MRF4, Myf5, and myogenin (*141–147*). The transcription activity of the rat (COX VIII (H) promoter construct (−158 to +50) is induced 12- to 15-fold during induced differentiation of skeletal and cardiac muscle cell lines. Mutations, targeted to a single E box, reduced the transcription activity to that of the basal promoter and abolished differentiation-specific inducibility in the cardiac and skeletal myocytes, suggesting functional cooperativity between factors binding to the two E boxes. These mutational studies suggest the involvement of a bHLH protein factor in the cardiac muscle-specific activation of the COX VIII (H) promoter. The possible involvement of a bHLH factor in the cardiac muscle-specific activation of the COX VIII (H) promoter was supported by two different lines of experiments. First, coexpression with Id suppressed the transcription activity in differentiated H9C2 cardiomyocytes in a dose-dependent fashion. Id, a bHLH factor, is known to down-regulate the activity of myogenic bHLH factors such as MyoD and myogenin by heterodimerization (*148, 149*). Second, a polyclonal antibody to human E2A caused a supershift of E-box-specific complexes formed with nuclear extract from differentiated H9C2 cardiomyocytes. Interestingly, extract from undifferentiated myocytes yielded differently migrating complexes and characteristic supershifting with the antibody, which was different from the pattern observed with the differentiated nuclear extract (*101*). This suggests that the induced putative bHLH factor (s) might play a regulatory role during the induced differentiation of cardiomyocytes. Additionally, gel mobility-shift analysis using nuclear extract from E2A-overexpressing cells and purified bacterially expressed E47 protein showed that the antibody-supershifting complex is unlikely to be an E2A homodimer. Results of cDNA expression also suggest that the putative bHLH factors induced during H9C2 cardiomyogenesis may be different from the e-HAND and d-HAND bHLH proteins that are transiently expressed during early stages of cardiogenesis, but not detected in the adult heart (*150, 151*). Thus results of COX VIII (H) promoter analysis suggest the involvement of the same promoter/enhancer elements in the regulation of gene expression in two different striated muscle types. This is probably the first demonstration of cardiac muscle-specific expression of a gene without the involvement of previously characterized cardiac factors, such as the GATA4, CArG (SRE), MCAT, or MEF2.

Thus, the results indicate no commonality in terms of promoter structure or regulatory elements between the ubiquitous and tissue-specific COX genes (see Figs. 2 and 3). Specifically, as shown in Fig. 3, the muscle-specific promoters do not contain the initiation region YY-1 or GABP (NRF2) sites and show some species-specific variations with respect to the presence of Sp1 and NRF1 sites. Nevertheless, the muscle-specific promoters contain enhancer elements, such as E box, MEF2, and CArG, although their functional significance in some cases remains to be elucidated.

IV. Mechanisms of Coordinate Regulation

Mitochondrial membrane biogenesis involves a well-orchestrated interplay between protein subunits encoded by the nuclear and mitochondrial genes. Nuclear genes contribute to the majority of the mitochondrial enzymes, structural proteins, and proteins involved in mitochondrial transcription, translation, and DNA replication. In vertebrates, the mitochondrial genomes encode only 13 proteins that are associated with four different oligomeric complexes involved in mitochondrial electron transport-coupled oxidative phosphorylation. Although unicellular eukaryotes and higher plants contain significantly larger organelle genomes, the number of protein subunits contributed by these mitochondrial genomes may not far exceed those encoded by mammalian DNA. Though smaller in number, mitochondrial-encoded proteins play an important role in the functional maturation of these cytoplasmic organelles; they are essential either in catalytic activity, as in the COX complex (*1*) or in assembly, as in the case of F0–F1 ATPase (*152*). Thus, two major types of regulatory controls can be envisioned in the overall regulation of mitochondrial biogenesis: (1) A mechanism to coordinate the level of expression of well over 100 genes encoding ubiquitous and tissue-specific proteins that have mitochondrial destination, and (2) intergenic coordination between the spatially separated nuclear and mitochondrial genes in response to changes in the mitochondrial metabolic and physiological conditions. This could be accomplished either by using the same or related transcription factor(s) regulating both nuclear and mitochondrial gene expression, or by genetic or metabolic signals originating in the mitochondria that can modulate nuclear gene expression. Although these two levels of regulation may be closely interconnected, for the sake of presentation these two aspects will be covered separately as follows.

A. Intergenic Regulation of Gene Expression

Based on mutational studies in yeast and *Neurospora*, it was hypothesized that nuclear and mitochondrial genes coding for the various mitochondrial membrane proteins may be regulated in a highly coordinated fashion (*153*). However, the evidence for coordinate expression of nuclear and mitochondrial genes in the vertebrate systems is mixed. Some studies (*154*) show that induced mitochondrial biogenesis in skeletal muscle by electrical stimuli or other treatments results in a concomitant increase in the transcription of both mitochondrial and nuclear genes. A number of additional studies, by contrast, show that mRNAs encoding various mitochondrial membrane/matrix proteins, including different subunits of the same complex, are differently regulated by various physiological factors, such as thyroid hormone, growth factors, and retinoic acid (*155–159*). In many of these cases, increased nuclear transcription is not always accompanied by a parallel increase in mitochondrial transcription (*158*).

It is quite likely that nuclear genes may control the level of mitochondrial gene expression because they contribute proteins and RNA (only in some cases) necessary for mitochondrial DNA replication, transcription, mRNA processing, mRNA stability, translation, etc. These genes could regulate mitochondrial genetic and biosynthetic processes as part of the nucleus-to-mitochondrion control pathway. There is also evidence suggesting that regulation in the other direction, namely, from mitochondrion to nucleus, occurs, though the precise nature of this communication, particularly in the metazoan systems, remains unknown. The possibility of a mitochondrial factor regulating the expression of some nuclear genes was initially suggested based on observation in *Neurospora* that cells treated with chloramphenicol, an inhibitor of mitochondrial translation, contained elevated levels of nuclear-encoded RNA polymerase and tRNA synthetase within the organelle compartment (*160*). Similarly, somatic cell hybridization and genetic analysis studies (*161–164*) showed that one of the histocompatibility complex class I genes in the mouse is under the regulation of a maternally inherited, putative mitochondrial gene-encoded factor. It was also observed in *Neurospora* that when mitochondrial function is disrupted due to cytoplasmic "poky" mutations or when cells are grown in the presence of chloramphenicol, the mRNA level of a nuclear-encoded mitochondrial ribosomal protein was substantially elevated (*165*). These studies lead to the hypothesis that a mitochondrial-encoded factor, translocated to the nuclear compartment, might function as a transcription repressor (*160*). Despite intensive efforts, however, translocation of the putative mitochondrial-encoded protein or RNA factor outside the mitochondrial membrane matrix compartment has not been shown in any cell system.

In an important series of experiments, the Butow lab (*166*) showed that yeast respiratory mutants lacking all (ρ^0) or specific regions (Mit$^-$ and rho$^-$) of the mitochondrial genome have increased levels of nuclear-specified RNAs mapping to different regions of the genome. These investigators described two classes of RNAs that were elevated 10- to 20-fold in various mutants. Class I transcripts that include mRNA encoded by the CIT2 gene respond to mitochondrial respiratory deficiency either due to mitochondrial DNA mutations or treatment with respiratory poisons such as antimycin (*160, 167, 168*). The second class of transcripts (class II) responds to mitochondrial genotype or specific regions of the mitochondrial genome, because they are overexpressed only in certain deletion mutants and are not affected by respiratory inhibitors (*17*). These transcripts include mRNAs encoded by the spacer region of the nuclear ribosomal gene repeat, PUT1, MRP13, and ORF-D (*166, 169–171*). Farrell *et al.* (*172*) described a third class of transcripts that were down-regulated in respiratory-deficient yeast cells that lack mitochondrial DNA or contain large deletions. These transcripts respond to only certain mitochondrial genotypes and are unaffected by respiratory inhibitors. Class III transcripts include nuclear COX 4, COX 5a, COX 6, COX 8, and COX 9 genes (*17*). Regulation of the CIT2 gene, which codes for a member of class I transcripts in response to respiratory deficiency, has been extensively studied. The promoter region responsible for up-regulation in respiratory-deficient cells has been mapped to a 76-bp region designated as UASr, the activity of which appears to depend on the protein products of three genes, RTG1, RTG2, and RTG3 (*168, 173–175*). This mode of up-regulation, termed "retrograde regulation," appears to be controlled by a signaling mechanism set up in response to mitochondrial metabolic and respiratory activities. Interestingly, Rtg1p and Rtg3p are bHLH leucine zipper proteins that bind to specific sequence motifs on nuclear genes that respond to mitochondrial retrograde signaling (*174–176*). The down-regulation of genes for class III transcripts, on the other hand, is mediated by O_2 tension, heme level, and carbon source. The best studied members of this group are

COX 6 and COX 5a (*17*). Currently, there is limited information on the vertebrate nuclear genes that are up- or down-regulated in response to mitochondrial DNA damage or oxidative stress. Thus, the extent of retrograde regulation and the nature of the signal(s) involved in the mitochondrion-to-nuclear regulation in the vertebrate systems remain unclear.

B. Coordinate Regulation of Nuclear Genes

A possible global regulation of nuclear genes encoding various mitochondrial proteins was indicated initially in genetic analysis studies in yeast (*35, 56, 177–179*). A more detailed analysis indicates that the nuclear COX genes in yeast are regulated independently through different pathways or involving specific transcription factors (*177, 180–182*), although many of them respond to heme/O_2 tension. Various cytochrome-encoding genes in yeast contain common cis-acting elements, initially suggesting a possible means of coordinated expression through a common or a closely interrelated mechanism. There is a general consensus that heme/O_2 may be a common physiological factor involved in regulation. Heme is required by transcription factor HAP1 for binding and activation of cis DNA element UAS1 found in upstream promoter regions of cytochrome c_1, iso-1, iso-2, and a number of COX genes (*180–182*). Similarly, another cis DNA element, UAS2, found in the yeast cytochrome c_1, iso-1, cytochrome c, and cytochrome oxidase 4 and 5a genes, causes activation through binding to a heme-dependent HAP2/HAP3/HAP4 complex (*183–187*). Biosynthesis of a limiting factor, HAP4, is induced when yeast cells are shifted from glucose to lactate, resulting in subsequent activation of UAS2-bearing genes (*188, 189*). The genes coding for COX 5a and 5b isotypes are particularly sensitive to O_2 tension. Under aerobic conditions the 5a gene is up-regulated through heme, REO1, and HAP1. Under anaerobic conditions the 5b gene is activated and the gene for 5a is down-regulated by a heme/O_2-dependent mechanism (*30–32, 35, 36, 56, 190*).

Analyses of a wide spectrum of vertebrate genes for the mitochondrial respiratory proteins have resulted in the discovery of DNA elements common to several of the nuclear genes encoding the mitochondrial proteins. This has led to a popular hypothesis that developmental and physiological stimuli modulating cellular responses to respiratory demand exert their effects through regulatory proteins targeted to various cis-acting DNA elements common to the nuclear genes. Scarpulla and co-workers described two distinct elements, designated NRF1 and NRF2 (*107, 137*). Since its initial discovery as an activator of the cytochrome c gene, NRF1 consensus motif, a palindromic sequence of T/CGCGCAT/CGCGCA/G, has been found to occur in a large number of vertebrate genes coding for various mitochondrial membrane matrix proteins, including that coding for the mitochondrial transcription factor, MtTF1 (*137, 191*). Additional studies on cDNA cloning by Scarpulla's group have demonstrated that NRF1 activity resides in a single 68-kDa protein with a novel DNA-binding domain. Interestingly, NRF1 is structurally related to the developmentally regulated factors such as the sea urchin P3A2 and *Drosophila* erect wing (EWG) proteins (*137*). NRF2 is an ets-like element initially found in the rat COX IV promoter (*106*). Working with mouse COX IV and Vb promoters (*108, 110*), we independently found that the NRF2 motif binds an ets family heterodimeric factor, GABP, originally discovered by McKnight's group (*192–194*) as an activator of HSV immediate-early genes. Studies by various groups have shown that the motif CGGAAG(A) is a high-affinity site and that A(T)GGAAT(C) is a low-affinity site for binding to GABP (NRF2) (*111, 194–196*). Extensive studies by Scarpulla's group and McKnight's group have also led to the discovery of multiple isoforms of DNA-binding subunit α and the non-DNA-binding β subunits that are generated by a combination of differential splicing and expression of multiple genes (*107, 124, 194, 197–199*). Furthermore, the β subunit, which binds to the complex by protein–protein interactions, contains the transcription activation domain. An important difference between the isoforms of β subunit appears to reside within the N-terminal ankyrin repeat, which is believed to be involved in protein dimerization and may influence the ability of the heteromeric factor binding to the target motifs as a dimer or a tetramer. The precise biological roles of these individual isoforms remain to be elucidated.

It is becoming increasingly clear that functionally important NRF1 and GABP (NRF2) sites are also detected in the promoter regions of many nonmitochondrial housekeeping genes (*196, 200, 201*), indicating that the function of these factors is not limited to mitochondrial genes. Additionally, available evidence suggests that NRF1 and GABP may not play any significant role in the regulation of tissue-specific genes, such as the muscle-specific COX genes. These factors also seem to play minimal roles in the regulation of a spectrum of hormone-regulated genes that code for mitochondrial cytochrome P450 heme proteins and associated electron transfer proteins (*202–205*) (see Fig. 5). It is therefore unlikely that NRF1 and GABP (NRF2) have a global role in the regulation of mitochondrial-destined genes. Nevertheless, these transcription factors seem to play an important role in the regulation of a large majority of the genes encoding mitochondrial proteins associated with the OXPHOS function. Of particular interest is the observation that NRF1 and GABP (NRF2) are involved in the regulation of the gene encoding the RNA component of MtTF1 (*191*), suggesting that they may represent an important connecting link for coordinating the nuclear and mitochondrial transcription levels. This mode of coordinate regulation is also supported by findings that in yeast, factors such as the carbon source, which

CYTOCHROME c OXIDASE COMPLEX

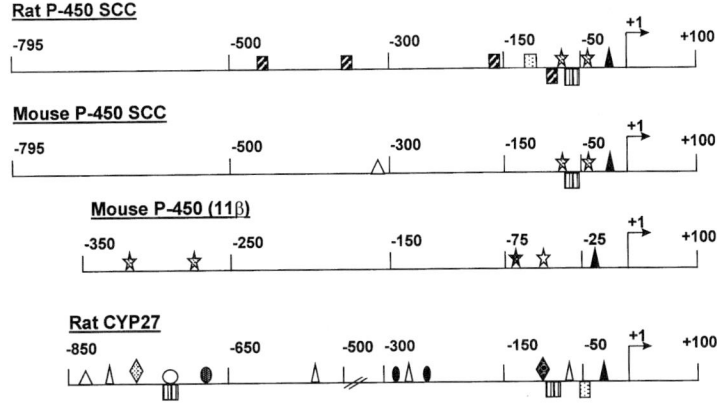

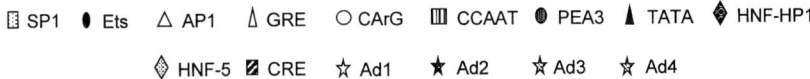

FIG. 5. Promoter structure and the cis DNA elements involved in regulation of mitochondrial-targeted P450 genes.

will affect the expression of a number of nuclear-encoded OXPHOS genes, also affect the expression of the catalytic subunits of mitochondrial RNA polymerase and MtTF1 (17, 206).

Additional conserved sequence motifs that might be involved in the co-ordinated regulation include Oxbox and Mt3/Mt4 consensus sequences. Oxbox was reported to act as an enhancer of two muscle-specific genes involved in the ADP/ATP exchange (207–209). Mt3/Mt4 motifs were initially found in the upstream regions of cytochrome c, ATPase-β, and ubiquinol-binding protein, and also in the D-loop regions of the mouse, rat, human, and bovine mitochondrial genomes. These motifs have also been reported to occur in the 5′ upstream regions of many mitochondrial-destined genes. Although these motifs were shown to bind a protein from rat liver nuclear extract (210), their precise roles in the transcription regulation of nuclear genes, and/or coordination, remain unclear.

V. Summary and Future Direction

It is becoming increasingly apparent that nuclear-encoded subunits may have an important regulatory role in modulating the catalytic activity of mammalian COX. Although there is no dramatic isolog switching as shown for the yeast enzyme, the mechanism appears to involve subtle variation in the relative levels of ubiquitous and tissue-specific isologs (29). A case in point is the observation that thyroid hormone alters the catalytic activity of the enzyme by directly binding to the complex (15).

A comparison of 5' distal and proximal promoters of ubiquitous and tissue-specific COX genes, and genes encoding the mitochondrial cytochrome P450 proteins, reveals fundamental differences in both the overall architecture and transcription activation mechanisms. The ubiquitous genes resemble the housekeeping genes in terms of cis DNA elements and trans-acting factors involved in the activation, whereas genes for the tissue-specific isologs seem to closely resemble the muscle-specific genes and use some of the myogenic factors. The three well-characterized P450 genes that respond to glucocorticoid hormone and cyclic AMP are regulated through a distinct set of DNA elements and factors, including GR and CRE. The emerging picture of the mode of regulation of mitochondrial-destined genes appears to be that of a microcosm of cellular genes, with different functions and cell specificities. Isolation and characterization of additional COX and other OXPHOS genes will help in completing this emerging picture.

Finally, extensive analysis of yeast COX has provided interesting insights into the mechanisms of hypoxia-mediated changes in enzyme activity and transcription regulation of individual genes. However, currently very little is known about the effects of hypoxia and other physiological factors on the mammalian COX genes. It is well known that mitochondria from different tissues with different oxidative demands and work loads exhibit different structure and membrane physiology, therefore it would be interesting to see if the COX and other OXPHOS genes are regulated differently, and more important, if they are subject to retrograde regulation by O_2 tension and other physiological factors.

Acknowledgments

We are thankful to Aruna Basu, Robert Carter, and Carmen Sucharov for their help and suggestions. The work from our laboratory is supported by in part by NIH Grant GM-49683.

References

1. R. A. Capaldi, *Annu. Rev. Biochem.* **59**, 569 (1990).
2. S. I. Chan and P. M. Li, Biochemistry **29**, 1 (1990).
3. J. W. Taanman and R. A. Capaldi, *J. Biol. Chem.* **267**, 22481 (1992).

4. A. E. LaMarche, M. I. Abate, S. H. Chan, and B. L. Trumpower, *J. Biol. Chem.* **267,** 22473 (1992).
5. R. P. Casey, M. Thelen, and A. Azzi, *Biochem. Biophys. Res. Comm.* **87,** 1044 (1979).
6. R. P. Casey, M. Thelen, and A. Azzi, *J. Biol. Chem.* **255,** 3994 (1980).
7. L. J. Prochaska, R. Bisson, R. A. Capaldi, G. C. Steffens, and G. Buse, *Biochim. Biophys. Acta* **637,** 360 (1981).
8. L. J. Prochaska and K. A. Reynolds, *Biochemistry* **25,** 781 (1986).
9. F. Malatesta, G. Georgevich, and R. A. Capaldi, in "Structure and Function of Membrane Proteins" (E. Quagliariello and F. Palmieri, eds.), p. 223. Elsevier, Amsterdam, 1983.
10. B. C. Hill and N. C. Robinson, *J. Biol. Chem.* **261,** 15356 (1986).
11. S. Wu, R. Moreno-Sanchez, and H. Rottenberg, *Biochemistry* **34,** 16298 (1995).
12. W. Dowhan, C. R. Bibus, and G. Schatz, *EMBO J.* **4,** 179 (1985).
13. F. Malatesta, G. Antonini, P. Sarti, and M. Brunori, *Biochem. J.* **248,** 161 (1987).
14. F. J. Huther and B. Kadenbach, *Biochem. Biophys. Res. Comm.* **153,** 525 (1988).
15. F. Goglia, A. Lanni, J. Barth, and B. Kadenbach, *FEBS Lett.* **346,** 295 (1994).
16. B. Kadenbach, L. Kuhn-Nentwig, and U. Buge, *Curr. Top. Bioenerg.* **15,** 113 (1987).
17. R. O. Poyton and J. E. McEwen, *Annu. Rev. Biochem.* **65,** 563 (1996).
18. L. I. Grossman and M. I. Lomax, *Biochim. Biophys. Acta* **1352,** 174 (1997).
19. A. Schlerf, M. Droste, M. Winter, and B. Kadenbach, *EMBO J.* **7,** 2387 (1988).
20. B. Kadenbach, A. Stroh, A. Becker, C. Eckerskorn, and F. Lottspeich, *Biochim. Biophys. Acta* **1015,** 368 (1990).
21. R. Lightowlers, G. Ewart, R. Aggeler, Y. Z. Zhang, L. Calavetta, and R. A. Capaldi, *J. Biol. Chem.* **265,** 2677 (1990).
22. G. Anthony, A. Stroh, F. Lottspeich, and B. Kadenbach, *FEBS Lett.* **277,** 97 (1990).
23. G. Bonne, P. Seibel, S. Possekel, C. Marsac, and B. Kadenbach, *Eur. J. Biochem.* **217,** 1099 (1993).
24. G. D. Ewart, Y. Z. Zhang, and R. A. Capaldi, *FEBS Lett.* **292,** 79 (1991).
25. J. Meehan and J. F. Kennedy, *Biochem. J.* **327,** 155 (1997).
26. B. Kadenbach, A. Stroh, F. J. Huther, A. Reimann, and D. Steverding, *J. Bioenerg. Biomembr.* **23,** 321 (1991).
27. G. Anthony, A. Reimann, and B. Kadenbach, *Proc. Natl. Acad. Sci. U.S.A.* **90,** 1652 (1993).
28. F. Rohdich and B. Kadenbach, *Biochemistry* **32,** 8499 (1993).
29. C. Vijayasarathy, I. Biunno, N. Lenka, Y. Ming, A. Basu, I. P. Hall, and N. G. Avadhani, *Biochim. Biophys. Acta* **1371,** 71 (1998).
30. C. E. Trueblood and R. O. Poyton, *Genetics* **120,** 671 (1988).
31. C. E. Trueblood, R. M. Wright, and R. O. Poyton, *Mol. Cell. Biol.* **8,** 4537 (1988).
32. M. R. Hodge, G. Kim, K. Singh, and M. G. Cumsky, *Mol. Cell. Biol.* **9,** 1958 (1989).
33. R. A. Waterland, A. Basu, B. Chance, and R. O. Poyton, *J. Biol. Chem.* **266,** 4180 (1991).
34. L. A. Allen, X. J. Zhao, W. Caughey, and R. O. Poyton, *J. Biol. Chem.* **270,** 110 (1995).
35. H. F. Bunn and R. O. Poyton, *Physiol. Rev.* **76,** 839 (1996).
36. P. V. Burke, D. C. Raitt, L. A. Allen, E. A. Kellogg, and R. O. Poyton, *J. Biol. Chem.* **272,** 14705 (1997).
37. M. A. Johnson, B. Kadenbach, M. Droste, S. L. Old, and D. M. Turnbull, *J. Neurol. Sci.* **87,** 75 (1988).
38. S. DiMauro, A. Lombes, H. Nakase, S. Mita, G. M. Fabrizi, H. J. Tritschler, E. Bonilla, A. F. Miranda, D. C. DeVivo, and E. A. Schon, *Pediatr. Res.* **28,** 536 (1990).
39. D. C. Wallace, *Annu. Rev. Biochem.* **61,** 1175 (1992).
40. J. Muller-Hocker, G. Hubner, K. Bise, C. Forster, S. Hauck, I. Paetzke, D. Pongratz, and B. Kadenbach, *Arch. Pathol. Lab. Med.* **117,** 202 (1993).

41. S. Possekel, A. Lombes, H. Ogier de Baulny, M. A. Cheval, M. Fardeau, B. Kadenbach, and N. B. Romero, *Histochem. Cell Biol.* **103,** 59 (1995).
42. H. J. Tritschler, E. Bonilla, A. Lombes, F. Andreetta, S. Servidei, B. Schneyder, A. F. Miranda, E. A. Schon, B. Kadenbach, and S. DiMauro, *Neurology* **41,** 300 (1991).
43. P. Zimmermann and B. Kadenbach, *Biochim. Biophys. Acta* **1180,** 99 (1992).
44. V. Tiranti, M. Munaro, D. Sandona, E. Lamantea, M. Rimoldi, S. Di Donato, R. Bisson, and M. Zeviani, *Human Mol. Genet.* **4,** 2017 (1995).
45. B. Kadenbach, J. Barth, R. Akgun, R. Freund, D. Linder, and S. Possekel, *Biochim. Biophys. Acta* **1271,** 103 (1995).
46. C. Mariotti, G. Uziel, F. Carrara, M. Mora, A. Prelle, V. Tiranti, S. Di Donato, and M. Zeviani, *J. Neurol.* **242,** 547 (1995).
47. C. T. Moraes, S. DiMauro, M. Zeviani, A. Lombes, S. Shanske, and A. F. Miranda, *New Engl. J. Med.* **320,** 1293 (1989).
48. Y. Goto, I. Nonaka, and S. Horai, *Nature (London)* **348,** 651 (1990).
49. J. M. Shoffner, M. T. Lott, A. M. Lezza, P. Seibel, S. W. Ballinger, and D. C. Wallace, *Cell* **61,** 931 (1990).
50. E. A. Schon, M. Hirano, and S. DiMauro, *J. Bioenerg. Biomembr.* **26,** 291 (1994).
51. L. I. Grossman, and E. A. Shoubridge, *BioEssays* **18,** 983 (1996).
52. M. I. Lomax, N. J. Bachman, M. S. Nasoff, M. H. Caruthers, and L. I. Grossman, *Proc. Natl. Acad. Sci. U.S.A.* **81,** 6295 (1984).
53. M. Zeviani, M. Nakagawa, J. Herbert, M. I. Lomax, L. I. Grossman, A. A. Sherbany, A. F. Miranda, S. DiMauro, and E. A. Schon, *Gene* **55,** 205 (1987).
54. Y. Goto, N. Amuro, and T. Okazaki, *Nucleic Acids Res.* **17,** 2851 (1989).
55. R. S. Carter and N. G. Avadhani, *Arch. Biochem. Biophys.* **288,** 97 (1991).
56. R. O. Poyton and P. V. Burke, *Biochim. Biophys. Acta* **1101,** 252 (1992).
57. M. Tanaka, H. Mitsuru, K. T. Yasunobu, C. A. Yu, Y. H. Wei, and T. E. King, *J. Biol. Chem.,* **254,** 3879 (1979).
58. R. Rizzuto, H. Nakase, M. Zeviani, S. DiMauro, and E. A. Schon, *Gene* **69,** 245 (1988).
59. M. Droste, E. Schon, and B. Kadenbach, *Nucleic Acids Res.* **17,** 4375 (1989).
60. D. Roise, S. J. Horvath, J. M. Tomich, J. H. Richards, and G. Schatz, *EMBO J.* **5,** 1327 (1986).
61. G. Von Heijne, *EMBO J.* **5,** 1335 (1986).
62. A. P. Van Loon, A. W. Brandli, B. Pesold-Hurt, D. Blank, and G. Schatz, *EMBO J.* **6,** 2433 (1987).
63. A. P. Van Loon and G. Schatz, *EMBO J.* **6,** 2441 (1987).
64. F. U. Hartl, J. Ostermann, B. Guiard, and W. Neupert, *Cell* **51,** 1027 (1987).
65. G. Shaw and R. Kamen, *Cell* **46,** 659 (1986).
66. A. Basu, and N. G. Avadhani, *Biochim. Biophys. Acta* **1087,** 98 (1990).
67. C. Sucharov, A. Basu, R. S. Carter, and N. G. Avadhani, *Gene Exp.* **5,** 93 (1995).
68. H. Hoshinaga, N. Amuro, Y. Goto, and T. Okazaki, *J. Biochem.* **15,** 194 (1994).
69. M. Zeviani, S. Sakoda, A. A. Sherbany, H. Nakase, R. Rizzuto, C. E. Samitt, S. DiMauro, and E. A. Schon, *Gene* **65,** 1 (1988).
70. A. Basu, N. Lenka, J. Mullick, and N. G. Avadhani, *J. Biol. Chem.* **272,** 5899 (1997).
71. G. M. Fabrizi, R. Rizzuto, H. Nakase, S. Mita, B. Kadenbach, and E. A. Schon, *Nucleic Acids Res.* **17,** 6409 (1989).
72. G. M. Fabrizi, J. Sadlock, M. Hirano, S. Mita, Y. Koga, R. Rizzuto, M. Zeviani, and E. A. Schon, *Gene* **119,** 307 (1992).
73. O. C. Mell, P. Seibel, and B. Kadenbach, *Gene* **140,** 179 (1994).
74. E. O. Smith, D. M. BeMent, L. I. Grossman, and M. I. Lomax, *Biochim. Biophys. Acta* **1089,** 266 (1991).

75. R. Schillace, T. Preiss, R. N. Lightowlers, and R. A. Capaldi, *Biochim. Biophys. Acta* **1188**, 391 (1994).
76. C. Saccone, G. Pesole, and B. Kadenbach, B. *Eur. J. Biochem.* **195**, 151 (1991).
77. R. N. Lightowlers and R. A. Capaldi, *Nucleic Acids Res.* **17**, 5845 (1989).
78. J. W. Taanman, C. Schrage, N. J. Ponne, A. T. Das, P. A. Bolhuis, H. de Vries, and E. Agsteribbe, *Gene* **93**, 285 (1990).
79. R. D. Carrero-Valenzuela, F. Quan, R. Lightowlers, N. G. Kennaway, M. Litt, and M. Forte, *Gene* **102**, 229 (1991).
80. G. C. Steffens, G. J. Steffens, and G. Buse, *Hoppe-Seyler Z. Physiol. Chem.* **360**, 1641 (1979).
81. M. Smith, D. W. Leung, S. Gillam, C. R. Astell, D. L. Montgomery and B. D. Hall, *Cell* **16**, 753 (1979).
82. R. M. Wright, L. K. Dirks, and R. O. Poyton, *J. Biol. Chem.* **261**, 17183 (1986).
83. G. Suske, T. Mengel, M. Cordingley, and B. Kadenbach, *Eur. J. Biochem.* **168**, 233 (1987).
84. M. Otsuka, Y. Mizuno, M. Yoshida, Y. Kagawa, and S. Ohta, *Nucleic Acids Res.* **16**, 10916 (1988).
85. M. Erdweg, and G. Buse, *Biol. Chem. Hoppe-Seyler* **366**, 257 (1985).
86. E. Arnaudo, M. Hirano, R. S. Seelan, A. Milatovich, C. L. Hsieh, G. M. Fabrizi, L. I. Grossman, U. Francke, and E. A. Schon, *Gene* **119**, 299 (1992).
87. R. S. Seelan and L. I. Grossman, *J. Biol. Chem.*, **266**, 19752 (1991).
88. C. Enders, A. Schlerf, O. Mell, L. I. Grossman, and B. Kadenbach, *Nucleic Acids Res.* **18**, 7143 (1990).
89. G. M. Fabrizi, R. Rizzuto, H. Nakase, S. Mita, M. I. Lomax, L. I. Grossman, and E. A. Schon, *Nucleic Acids Res.*, **17**, 7107 (1989).
90. R. S. Seelan, D. Scheuner, M. I. Lomax, and L. I. Grossman, *Nucleic Acids Res.*, **17**, 6410 (1989).
91. J. P. Hendrick, P. E. Hodges, and L. E. Rosenberg, *Proc. Natl. Acad. Sci. U.S.A.* **86**, 4056 (1989).
92. R. Lightowlers, S. Takamiya, R. Wessling, M. Lindorfer, and R. A. Capaldi, *J. Biol. Chem.* **264**, 16858 (1989).
93. J. E. Sadlock, R. N. Lightowlers, R. A. Capaldi, and E. A. Schon, *Biochim. Biophys. Acta* **1172**, 223 (1993).
94. M. Akamatsu and L. I. Grossman, *Nucleic Acids Res.* **18**, 3645 (1990).
95. M. S. Aqua, N. J. Bachman, M. I. Lomax, and L. I. Grossman, *Gene* **104**, 211 (1991).
96. Y. Koga, G. M. Fabrizi, S. Mita, E. Arnaudo, M. I. Lomax, M. S. Aqua, L. I. Grossman, and E. A. Schon, *Nucleic Acids Res.* **18**, 684 (1990).
97. R. Rizzuto, H. Nakase, B. Darras, U. Francke, G. M. Fabrizi, T. Mengel, F. Walsh, B. Kadenbach, S. DiMauro, and E. A. Schon, *J. Biol. Chem.*, **264**, 10595 (1989).
98. K. Scheja and B. Kadenbach, *Biochim. Biophys. Acta* **1132**, 91 (1992).
99. G. J. Makris and M. I. Lomax, *Biochim. Biophys. Acta* **1308**, 197 (1996).
100. A. D. Hegeman, J. S. Brown, and M. I. Lomax, *Biochim. Biophys. Acta* **1261**, 311 (1995).
101. N. Lenka, A. Basu, J. Mullick, and N. G. Avadhani, *J. Biol. Chem.* **271**, 30281 (1996).
102. M. I. Lomax, P. K. Riggs, and J. E. Womack, *Mammal. Genome* **6**, 118 (1995).
103. M. Yamada, N. Amuro, Y. Goto, and T. Okazaki, *J. Biol. Chem.*, **265**, 7687 (1990).
104. N. J. Bachman, *Gene* **162**, 313 (1995).
105. A. P. Bird, *Nature (London)* **321**, 209 (1986).
106. J. V. Virbasius and R. C. Scarpulla, *Mol. Cell. Biol.* **11**, 5631 (1991).
107. J. V. Virbasius, C. A. Virbasius, and R. C. Scarpulla, *Genes Dev.* **7**, 380 (1993).
108. R. S. Carter, N. K. Bhat, A. Basu, and N. G. Avadhani, *J. Biol. Chem.*, **267**, 23418 (1992).
109. R. S. Carter and N. G. Avadhani, *J. Biol. Chem.* **269**, 4381 (1994).

110. A. Basu, K. Park, M. L. Atchison, R. S. Carter, and N. G. Avadhani, *J. Biol. Chem.* **268,** 4188 (1993).
111. R. S. Seelan and L. I. Grossman, *J. Biol. Chem.* **272,** 10175 (1997).
112. B. Wan, and R. W. Moreadith, *J. Biol. Chem.* **270,** 26433 (1995).
113. R. S. Seelan, L. Gopalakrishnan, R. C. Scarpulla, and L. I. Grossman, *J. Biol. Chem.* **271,** 2112 (1996).
114. S. T. Smale and D. Baltimore, 1989, *Cell* **57,** 103 (1989).
115. A. O'Shea-Greenfield and S. T. Smale, *J. Biol. Chem.* **267,** 1391 (1992).
116. L. Weis and D. Reinberg, *FASEB J.* **6,** 3300 (1992).
117. L. Zawel and D. Reinberg, *Annu. Rev. Biochem.* **64,** 533 (1995).
118. S. K. Burley and R. G. Roeder, *Annu. Rev. Biochem.* **65,** 769 (1996).
119. E. Martinez, C. M. Chiang, H. Ge, and R. G. Roeder, *EMBO J.* **13,** 3115 (1994).
120. A. Usheva and T. Shenk, *Cell* **76,** 1115 (1994).
121. A. Usheva and T. Shenk, *Proc. Natl. Acad. Sci. U.S.A.* **93,** 13571 (1996).
122. A. Basu and N. G. Avadhani, *J. Biol. Chem.* **266,** 15450 (1991).
123. J. V. Virbasius and R. C. Scarpulla, *Nucleic Acids Res.* **18,** 6581 (1990).
124. C. C. Thompson, T. A. Brown, and S. L. McKnight, *Science* **253,** 762 (1991).
125. R. Janknecht and A. Nordheim, *Biochim. Biophys. Acta* **1155,** 346 (1993).
126. N. Hariharan, D. E. Kelley, and R. P. Perry, *Genes Dev.* **3,** 1789 (1989).
127. H. Watanabe, T. Imai, P. A. Sharp, and H. Handa, *Mol. Cell. Biol.* **8,** 1290 (1988).
128. N. Hariharan, D. E. Kelley, and R. P. Perry, *Proc. Natl. Acad. Sci. U.S.A.* **88,** 9799 (1991).
129. K. Park, and M. L. Atchison, *Proc. Natl. Acad. Sci. U.S.A.* **88,** 9804 (1991).
130. Y. Shi, E. Seto, L. S. Chang, and T. Shenk, *Cell* **67,** 377 (1991).
131. J. R. Flanagan, K. G. Becker, D. L. Ennist, S. L. Gleason, P. H. Driggers, B. Z. Levi, E. Appella, and K. Ozato, *Mol. Cell. Biol.*, **12,** 38 (1992).
132. S. Hahn, *Curr. Biol.* **2,** 152 (1992).
133. M. Yu, X.-Y. Yang, T. Schmidt, Y. Chinenov, R. Wang, and M. Martin, *J. Biol. Chem.* **272,** 29060 (1997).
134. N. J. Bachman, T. L. Yang, J. S. Dasen, R. E. Ernst, and M. I. Lomax, *Arch. Biochem. Biophys.* **333,** 152 (1996).
135. M. L. Atchison, O. Meyuhas, and R. P. Perry, *Mol. Cell. Biol.* **9,** 2067 (1989).
136. N. Hariharan and R. P. Perry, *Nucleic Acids Res.* **17,** 5323 (1989).
137. C. A. Virbasius, J. V. Virbasius, and R. C. Scarpulla, *Genes Dev.* **7,** 2431 (1993).
138. R. S. Seelan and L. I. Grossman, *Genomics* **18,** 527 (1993).
139. E. O. Smith and M. I. Lomax, *Biochim. Biophys. Acta* **1174,** 63 (1993).
140. M. S. Parmacek, H. S. Ip, F. Jung, T. Shen, J. F. Martin, A. J. Vora, E. N. Olson, and J. M. Leiden, *Mol. Cell. Biol.* **14,** 1870 (1994).
141. C. P. Emerson, *Curr. Opin. Cell Biol.* **2,** 1065 (1990).
142. E. N. Olson, *Genes Dev.* **4,** 1454 (1990).
143. A. B. Lassar, R. L. Davis, W. E. Wright, T. Kadesch, C. Murre, A. Voronova, D. Baltimore, and H. Weintraub, *Cell* **66,** 305 (1991).
144. H. Weintraub, R. Davis, S. Tapscott, M. Thayer, M. Krause, R. Benezra, T. K. Blackwell, D. Turner, R. Rupp, and S. Hollenberg, *Science* **251,** 761 (1991).
145. H. Weintraub, V. J. Dwarki, I. Verma, R. Davis, S. Hollenberg, L. Snider, A. Lassar, and S. J. Tapscott, *Genes Dev.* **5,** 1377 (1991).
146. T. Kadesch, *Cell Growth Differ.* **4,** 49 (1993).
147. C. Murre, G. Bain, M. A. van Dijk, I. Engel, B. A. Furnari, M. E. Massari, J. R. Matthews, M. W. Quong, R. R. Rivera, and M. H. Stuiver, *Biochim. Biophys. Acta* **1218,** 129 (1994).
148. R. Benezra, R. L. Davis, A. Lassar, S. Tapscott, M. Thayer, D. Lockshon, and H. Weintraub, *Ann. N.Y. Acad. Sci.* **599,** 1 (1990).

149. R. Benezra, R. L. Davis, D. Lockshon, D. L. Turner, and H. Weintraub, *Cell* **61,** 49 (1990).
150. S. M. Hollenberg, R. Sternglanz, P. F. Cheng, and H. Weintraub, *Mol. Cell. Biol.* **15,** 3813 (1995).
151. D. Srivastava, P. Cserjesi, and E. N. Olson, *Science* **270,** 1995 (1995).
152. L. G. Nijtmans, P. Klement, J. Houstek, and C. van den Bogert, *Biochim. Biophys. Acta* **1272,** 190 (1995).
153. P. Nagley, *Trends Genet.* **7,** 1 (1991).
154. R. S. Williams, M. Garcia-Moll, J. Mellor, S. Salmons, and W. Harlan, *J. Biol. Chem.* **262,** 2764 (1987).
155. K. Luciakova, R. Li, and B. D. Nelson, *Eur. J. Biochem.* **207,** 253 (1992).
156. K. Luciakova and B. D. Nelson, *Eur. J. Biochem.* **207,** 247 (1992).
157. R. J. Wiesner, *Trends Genet.* **8,** 264 (1992).
158. R. J. Wiesner, T. T. Kurowski, and R. Zak, *Mol. Endocrinol.* **6,** 1458 (1992).
159. R. Sewards, B. Wiseman, and H. T. Jacobs, *Mol. Gen. Genet.* **245,** 760 (1994).
160. Z. Barath and H. Kuntzel, *Proc. Natl. Acad. Sci. U.S.A.* **69,** 1371 (1972).
161. R. Smith, M. M. Huston, R. N. Jenkins, D. P. Huston, and R. R. Rich, *Nature (London)* **306,** 599 (1983).
162. K. F. Lindahl and B. Hausmann, *Genetics* **103,** 483 (1983).
163. K. F. Lindahl, B. Hausmann, and V. M. Chapman, *Nature (London)* **306,** 383 (1983).
164. K. F. Lindahl, B. Hausmann, P. J. Robinson, J. L. Guenet, D. C. Wharton, and H. Winking, *J. Exp. Med.* **163,** 334 (1986).
165. M. T. Kuiper, R. A. Akins, M. Holtrop, H. de Vries, and A. M. Lambowitz, *J. Biol. Chem.* **263,** 2840 (1988).
166. V. S. Parikh, M. M. Morgan, R. Scott, L. S. Clements, and R. A. Butow, *Science* **235,** 576 (1987).
167. X. S. Liao, W. C. Small, P. A. Srere, and R. A. Butow, *Mol. Cell. Biol.* **11,** 38 (1991).
168. X. Liao and R. A. Butow, *Cell* **72,** 61 (1993).
169. V. S. Parikh, H. Conrad-Webb, R. Docherty, and R. A. Butow, *Mol. Cell. Biol.* **9,** 1897 (1989).
170. S. S. Wang, and M. C. Brandriss, *Mol. Cell. Biol.* **7,** 4431 (1987).
171. J. A. Partaledis and T. L. Mason, *Mol. Cell. Biol.* **8,** 3647 (1988).
172. L. E. Farrell, J. D. Trawick, and R. O. Poyton, "Structure, Function, and Biogenesis of Energy Transfer Systems" (E. Quagliariollo, S. Papa, F. Palmiery, and C. Saccone, eds.), p. 131. Elsevier, Amsterdam, 1990.
173. A. Chelstowska and R. A. Butow, *J. Biol. Chem.* **270,** 18141 (1995).
174. B. A. Rothermel, A. W. Shyjan, J. L. Etheredge, and R. A. Butow, *J. Biol. Chem.* **270,** 29476 (1995).
175. Y. Jia, B. Rothermel, J. Thornton, and R. A. Butow, *Mol. Cell. Biol.* **17,** 1110 (1997).
176. B. Rothermel, J. Thornton, and R. A. Butow, *J. Biol. Chem.* **272,** 19801 (1997).
177. R. O. Poyton, C. E. Trueblood, R. M. Wright, and L. E. Farrell, *Ann. N.Y. Acad. Sci.* **550,** 289 (1988).
178. J. H. de Winde and L. A. Grivell, *Prog. Nucleic Acid Res. Mol. Biol.* **46,** 51 (1993).
179. C. L. Dieckmann and R. R. Staples, *Int. Rev. Cytol.* **152,** 145 (1994).
180. L. Guarente, *Annu. Rev. Genet.* **21,** 425 (1987).
181. J. Olesen, S. Hahn, and L. Guarente, *Cell* **51,** 953 (1987).
182. K. Pfeifer, T. Prezant, and L. Guarente, *Cell* **49,** 19 (1987).
183. S. Hahn and L. Guarente, *Science* **240,** 317 (1988).
184. J. D. Trawick, C. Rogness, and R. O. Poyton, *Mol. Cell. Biol.* **9,** 5350 (1989).
185. J. D. Trawick, R. M. Wright, and R. O. Poyton, *J. Biol. Chem.* **264,** 7005 (1989).
186. J. D. Trawick, N. Kraut, F. R. Simon, and R. O. Poyton, *Mol. Cell. Biol.* **12,** 2302 (1992).

187. J. T. Olesen and L. Guarente, *Genes Dev.* **4,** 1714 (1990).
188. S. L. Forsburg and L. Guarente, *Genes. Dev.* **3,** 1166 (1989).
189. F. H. MacIver, I. W. Dawes, and C. M. Grant, *Curr. Genet.* **31,** 119 (1997).
190. M. R. Hodge, K. Singh, and M. G. Cumsky, *Mol. Cell. Biol.* **10,** 5510 (1990).
191. J. V. Virbasius and R. C. Scarpulla, *Proc. Natl. Acad. Sci. U.S.A.* **91,** 1309 (1994).
192. K. L. LaMarco and S. L. McKnight, *Genes Dev.* **3,** 1372 (1989).
193. K. LaMarco, C. C. Thompson, B. P. Byers, E. M. Walton, and S. L. McKnight, *Science* **253,** 789 (1991).
194. F. C. de la Brousse, E. H. Birkenmeier, D. S. King, L. B. Rowe, and S. L. McKnight, *Genes Dev.* **8,** 1853 (1994).
195. A. Seth, R. Ascione, R. J. Fisher, G. J. Mavrothalassitis, N. K. Bhat, and T. S. Papas, *Cell Growth Differ.* **3,** 327 (1992).
196. R. R. Genuario, D. E. Kelley, and R. P. Perry, *Gene Expr.* **3,** 279 (1993).
197. T. A. Brown and S. L. McKnight, *Gene Dev.* **6,** 2502 (1992).
198. S. Gugneja, J. V. Virbasius, and R. C. Scarpulla, *Mol. Cell. Biol.,* **15,** 102 (1995).
199. S. Gugneja, C. M. Virbasius, and R. C. Scarpulla, *Mol. Cell. Biol.* **16,** 5708 (1996).
200. T. Yoganathan, N. K. Bhat, and B. H. Sells, *Biochem. J.* **287,** 349 (1992).
201. E. Sadasivan, M. M. Cedeno, and S. P. Rothenberg, *J. Biol. Chem.* **269,** 4725 (1994).
202. A. R. Mouw, D. A. Rice, J. C. Meade, S. C. Chua, P. C. White, B. P. Schimmer, and K. L. Parker, *J. Biol. Chem.* **264,** 1305 (1989).
203. R. B. Oonk, K. L. Parker, J. L. Gibson, and J. S. Richards, *J. Biol. Chem.,* **265,** 22392 (1990).
204. D. A. Rice, M. S. Kirkman, L. D. Aitken, A. R. Mouw, B. P. Schimmer, and K. L. Parker, *J. Biol. Chem.* **265,** 11713 (1990).
205. J. Mullick, S. Addya, C. Sucharov, and N. G. Avadhani, *Biochemistry* **34,** 13729 (1995).
206. T. L. Ulery, S. H. Jang, and J. A. Jaehning, *Mol. Cell. Biol.* **14,** 1160 (1994).
207. K. Li, C. K. Warner, J. A. Hodge, S. Minoshima, J. Kudoh, R. Fukuyama, M. Maekawa, Y. Shimizu, N. Shimizu, and D. C. Wallace, *J. Biol. Chem.* **264,** 13998 (1989).
208. K. Li, J. A. Hodge, and D. C. Wallace, *J. Biol. Chem.* **265,** 20585 (1990).
209. N. Neckelmann, C. K. Warner, A. Chung, J. Kudoh, S. Minoshima, R. Fukuyama, M. Maekawa, Y. Shimizu, N. Shimizu, and J. D. Liu, *Genomics,* **5,** 829 (1989).
210. H. Suzuki, Y. Hosokawa, M. Nishikimi, and T. Ozawa, *J. Biol. Chem.* **266,** 2333 (1991).

Control of Meiotic Recombination in *Schizosaccharomyces pombe*

MARY E. FOX AND
GERALD R. SMITH

*Division of Basic Sciences
Fred Hutchinson Cancer Research Center
Seattle, Washington 98109*

I. Control of Entry into Meiosis	348
A. Physiological Control	348
B. Genetic Control	348
C. Induction of Meiosis in *pat1–114* Mutant Cells	351
D. Timing of Commitment to Meiosis and Recombination	351
II. Nuclear Cytology during Meiosis	352
A. Nuclear Reorganization and Movement	352
B. Linear Elements Instead of Synaptonemal Complex	353
III. Meiotic Recombination Hotspots	355
A. Genetic Properties of *M26*	355
B. The Question of Meiotic DSBs in *Schizosaccharomyces pombe*	357
C. The *M26* Heptamer Sequence and Associated Binding Proteins	357
D. *M26* and Transcription	359
E. Chromosomal Context Requirements	359
F. *M26* and Chromatin Structure	360
G. The *ura4–aim* Hotspot	361
IV. Recombination-Deficient Mutants and Genes	361
A. Isolation of Recombination-Deficient (Rec$^-$) Mutants	361
B. Cloning of *rec* Genes	363
C. General Features of *Schizosaccharomyces pombe* Rec$^-$ Mutants and the Corresponding Genes	364
D. Features of Specific *Schizosaccharomyces pombe* Rec$^-$ Mutants	365
V. Region-Specific Control of Meiotic Recombination	369
A. The Rec8, Rec10, and Rec11 Activators	369
B. The Rik1, Swi6, and Clr Repressors	372
VI. Conclusion and Perspective	373
Note Added in Proof	375
References	375

Homologous recombination occurs at high frequency during meiosis and is essential for the proper segregation of chromosomes and the generation of genetic diversity. Meiotic recombination is controlled in numerous ways. In the fission yeast *Schizosaccharomyces pombe* nutritional starvation induces meiosis and high-level expression of many genes, including numerous recombination (*rec*) genes, whose products are required for recombination. Accompanying the two meiotic divisions are profound changes in nuclear and chromosomal structure and movement, which may play an important role in meiotic recombination. Although recombination occurs throughout the genome, it occurs at high frequency in some intervals (hotspots) and at low frequency in others (coldspots). The well-characterized hotspot *M26* is activated by the Mts1/Mts2 protein; this site and its binding proteins interact with the local chromosomal structure to enhance recombination. A coldspot between the silent mating-type loci is repressed by identified proteins, which may also alter local chromatin. We discuss in detail the *rec* genes and the possible functions of their products, some but not all of which share homology with other identified proteins. Although some of the *rec* gene products are required for recombination throughout the genome, others demonstrate regional specificity and are required in certain genomic regions but not in others. Throughout the review contrasts are made with meiotic recombination in the more thoroughly studied budding yeast *Saccharomyces cerevisiae*. © 1998 Academic Press

Meiosis is an essential part of the life cycle of sexually reproducing eukaryotes. During meiosis a diploid progenitor cell forms four haploid gametes; fusion of compatible gametes regenerates a diploid cell of the next generation. Faithful reproduction requires that the haploid gametes have exactly one copy of each chromosome. This is ensured by two special features of meiosis. A single DNA replication is followed by two meiotic cell divisions; this feature halves the number of chromosomes per cell. Prior to the first division the replicated homologous chromosomes pair and segregate to opposite sides of the cell; this feature ensures that each cell receives a representative of each chromosome. During the second division the duplicate copies of each homolog segregate to form the haploid gametes.

The pairing and segregation of homologs at the first division are unique to meiosis and are intimately associated with genetic recombination between the homologs. Homologous recombination occurs about 10^2 to 10^3-fold more frequently during meiosis than during mitosis and is essential for ensuring correct chromosome segregation in meiosis (*1*). In numerous organisms meiotic recombination-deficient (Rec⁻) mutants display faulty homolog segregation, and consequently such mutants frequently produce aneuploid gametes and inviable progeny. In addition, recombination can generate new allelic combinations in the gametes and thereby increase genetic diversity and the potential for natural selection. Recombination is therefore crucial for life and evolution.

The fission yeast *Schizosaccharomyces pombe* and other fungi are suitable for the study of meiotic recombination because of the ease of culturing large numbers of cells, extensive and facile genetics, and the ability to grow both haploids and diploids (for reviews see Ref. *2*). The inclusion of all four meiotic products (spores) in a sac (ascus) allows, through microscopic dissection, the determination of the fate of the four chromatids in an individual meiosis. Recombination events can therefore be classified as reciprocal or nonreciprocal. The descent of all commonly used S. *pombe* strains from a single isolate (3) facilitates the construction of isogenic strains. Temperature-sensitive mutations in the S. *pombe pat1* (also called *ran1*) gene, which encodes a repressor of meiosis, allow the nearly synchronous induction of meiosis in large cultures ($>10^{10}$ cells) (Section I,C) (4, 5).

Several features of meiosis and recombination distinguish S. *pombe* from the distantly related budding yeast *Saccharomyces cerevisiae*, in which meiotic recombination has been extensively studied (6). *Schizosaccharomyces pombe* has three chromosomes (3), whereas S. *cerevisiae* has 16 (7). In Rec$^-$ mutants random segregation of homologs at the first meiotic division would be expected to produce 2^{-3} (= 12.5%) as many true haploids as in wild-type S. *pombe*, but only 2^{-16} (= 0.0015%) in S. *cerevisiae*. The small number of S. *pombe* chromosomes thus favors recovery of Rec$^-$ mutants and further study of them (Section IV).

Unlike most eukaryotes, including S. *cerevisiae*, S. *pombe* does not form a complete synaptonemal complex, a large proteinaceous structure containing aligned homologs that becomes visible by electron microscopy during the first meiotic division (Section II,B) (8). *Schizosaccharomyces pombe* also lacks genetic interference, the lower than random frequency of two reciprocal recombinational exchanges (crossovers) occurring in nearby intervals along homologs (9). Both S. *pombe* and S. *cerevisiae* have ~50 crossovers per genome in each meiosis. In S. *cerevisiae* these crossovers are distributed among the 16 chromosomes. Without crossover interference helping to distribute these crossovers, the shortest chromosomes might frequently receive no crossovers and hence fail to segregate properly; the synaptonemal complex may be required for interference (Section II). In S. *pombe* the crossovers are randomly distributed among only three chromosomes; even the shortest, chromosome III, receives ~10 crossovers and the probability of zero crossovers and hence faulty segregation is extremely low. Its relative simplicity makes S. *pombe* meiotic recombination a good model for studying recombination without the complexities of the synaptonemal complex and interference.

The intensity of recombination (frequency of exchange per kilobase of DNA) is not constant; rather, there are hotspots and coldspots in both prokaryotes and eukaryotes (10). In S. *pombe* the *M26* hotspot (Section III) and the *mat2–mat3* coldspot (Section V,B) are well characterized. The *M26* hotspot is the only eukaryotic hotspot with a unique nucleotide sequence identified to date; this feature permits a detailed genetic and physical analysis of this hotspot.

Many Rec⁻ mutants of S. *pombe* have been isolated (*11, 12*), and the identification of the corresponding genes and their products will help elucidate the mechanism of meiotic recombination (Section IV). The analysis of these Rec⁻ mutants identified genes whose products are required for recombination in some genomic regions but not others (Section V,A). These novel region-specific activators of meiotic recombination may give new insights into the control of recombination.

This study reviews the genetic and biochemical control of S. *pombe* meiotic recombination. To put this work into context, we present in the next two sections an overview of the physiological and genetic control of the entry into meiosis and the cytological changes of nuclei and chromosomes during meiosis.

I. Control of Entry into Meiosis

We summarize here the genetic and physiological factors influencing entry of cells into meiosis, with emphasis on those factors used to study the function of meiotic recombination genes (Section IV). More comprehensive reviews of the control of meiosis in S. *pombe* are provided by Nielsen (*13*) and Yamamoto (*14*).

A. Physiological Control

Entry of wild-type cells into meiosis requires both diploidy (heterozygosity at the mating-type locus) and nutritional starvation, especially of nitrogen. Under conditions of nitrogen starvation haploid cells of opposite mating types (h^+ and h^-) arrest in the G_1 phase of the cell cycle and can mate to give a diploid zygote. Under continued starvation conjugation is immediately followed by nuclear fusion (karyogamy), premeiotic DNA synthesis, and two divisions to yield four haploid spores (Fig. 1). This represents a zygotic meiosis and differs from that in S. *cerevisiae*, in which nitrogen starvation is required for meiosis but not for mating (*15*). If nutrients are restored to S. *pombe* following mating, cells reenter vegetative growth as stable diploids. Starvation of such diploids induces entry into meiosis; this is termed azygotic meiosis (Fig. 1). The life cycle of S. *pombe* is described further by Gutz *et al.* (*15*).

B. Genetic Control

Mating and meiosis involve a cascade of protein interactions regulated by a complex network of control mechanisms in which the Pat1 (Ran1) kinase plays a central role (Fig. 2). The Pat1 kinase is a critical negative regulator of meiosis; inhibition of Pat1 triggers the switch from vegetative growth into

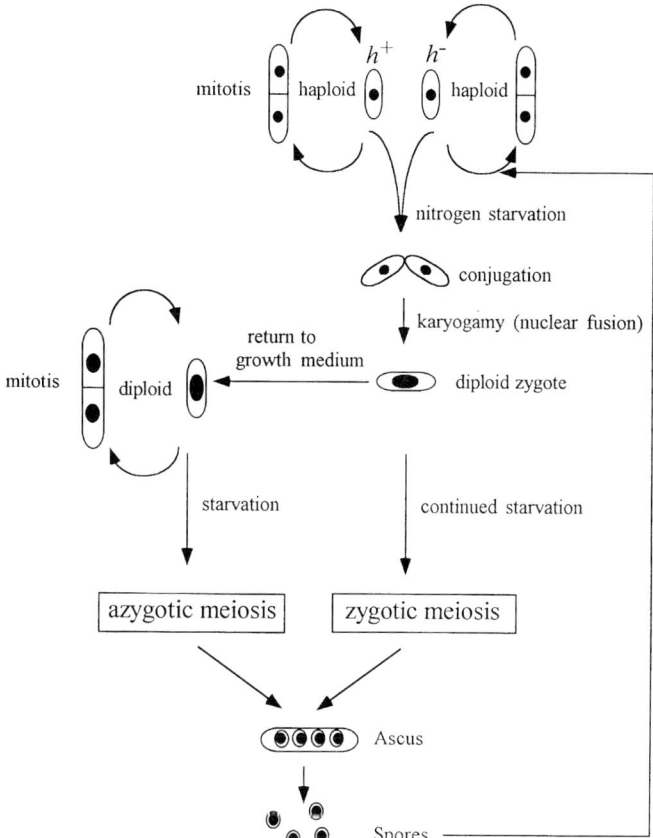

FIG. 1. The life cycle of *Schizosaccharomyces pombe*. Vegetative haploid cells enter zygotic meiosis following starvation-induced conjugation of cells of opposite mating type. Stable diploids can be obtained if zygotes are returned to growth medium following nuclear fusion. Nitrogen starvation of diploids induces azygotic meiosis. (Adapted from Ref. 15.)

meiosis (*4, 5, 16*). Active Pat1 kinase suppresses meiosis directly through phosphorylation of the Mei2 protein (*17*) and indirectly through phosphorylation of Ste11, the transcription factor required for expression of *mei2* (*18*). Mei2 is an RNA-binding protein that activates premeiotic DNA synthesis and meiosis I through interaction with different RNA species, including meiRNA (*19*).

Initiation of meiosis requires the interaction of two pathways: the first results in expression of *mei2*, and the second inactivates Pat1. Both pathways are controlled by Ste11 (Aff1), a transcription factor belonging to the high-

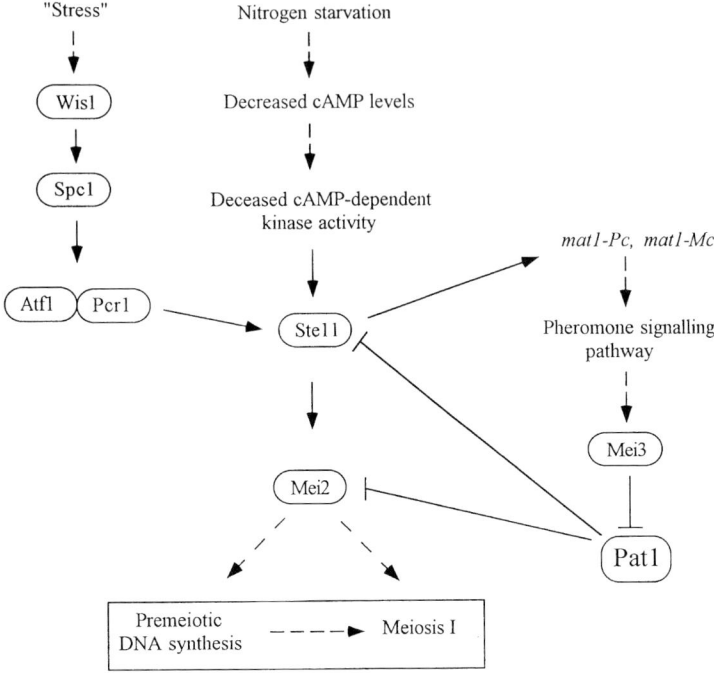

Fig. 2. Genetic control of meiosis in *Schizosaccharomyces pombe*. Major genes and pathways involved are summarized. Solid arrows indicate direct interaction between the proteins indicated in boxes; an arrowhead indicates activation, and a straight line indicates inhibition. Dashed arrows indicate that one process or component leads to another. "Stress" indicates that the MAP kinase cascade can be activated by a variety of agents, including osmotic stress, heat shock, oxidative stress, and DNA damage.

mobility-group (HMG) family (20). Ste11 activates expression of several genes required for mating and meiosis, including *mei2*, through binding to a specific motif in the 5′ region of target genes called the T-rich (TR) box (20). Ste11 can be activated by two different pathways. The first involves decreased cAMP-dependent kinase activity resulting from decreased intracellular levels of cAMP (20–22). The second involves activation by the transcription factor Atf1 (Gad7), which functions as a heterodimer with another transcription factor Pcr1 (23–25). Atf1 is activated in response to a variety of stresses through a mitogen-activated protein (MAP) kinase cascade involving the Wis1 and Spc1 (Sty1) kinases (26–31).

Pat1 kinase is inactivated by the Mei3 protein, which is produced in response to expression of the mating-type genes in an h^+/h^- diploid. In S. *pombe* mating type is determined by expression of plus (P) or minus (M) information at the *mat1* locus (Section V,B). The *mat1-P* and *mat1-M* genes can be differentially transcribed to give four products, *mat1-Pc* and *mat1-Pi*, *mat1-Mc* and *mat1-Mi* (constitutive; inducible) (32). Following nitrogen starvation of haploid cells Ste11 further increases expression of *mat1-Pc* and *mat1-Mc*, which in turn induce expression of the genes for mating-type pheromones and their receptors. This activates the pheromone signaling pathway through a MAP kinase cascade that leads to mating and expression of *mat1-Pi* and *mat1-Mi*. Coexpression of these genes in an h^+/h^- diploid results in expression of the *mei3* gene (33, 34). Mei3 binds to Pat1 kinase and inactivates its kinase activity (35), thus triggering entry of cells into meiosis through activation of Mei2.

C. Induction of Meiosis in *pat1–114* Mutant Cells

Cells carrying the temperature-sensitive *pat1–114* allele stop vegetative growth and initiate meiosis at the restrictive temperature, irrespective of the availability of a nitrogen source or ploidy of the cell (4, 5, 16). Thus, synchronous meiosis can be rapidly induced in diploid or haploid cells simply by shifting *pat1* mutant cells to the restrictive temperature. Meiotic events induced in both diploid and haploid *pat1–114* cells generally mimic those of meiosis induced by starvation of *pat1*$^+$ diploids. Commitment to meiosis, meiotic recombination, and chromosome segregation (in diploids) occur normally in *pat1–114* cells at the restrictive temperature, resulting in abundant formation of viable spores (36, 37). Recombinant frequencies have been reported to be slightly reduced compared with *pat1*$^+$ cells in some studies (37), but comparable to *pat1*$^+$ in others (38). The *pat1–114* mutation has been vital in the identification of several genes involved in meiosis; for example, *mei2* and *ste11* were identified as suppressors of the growth defect of a temperature-sensitive *pat1* mutant (19, 20). The ability to induce synchronous meioses through inactivation of *pat1* has allowed study of the temporal regulation of many meiotic events, for example the expression of *rec* genes required for meiotic recombination (Section IV).

D. Timing of Commitment to Meiosis and Recombination

With the exception of the timing of commitment to meiosis, the timing of events following meiotic induction is similar in *S. pombe* and *S. cerevisiae*. In appropriate strains of both yeasts premeiotic DNA synthesis begins ~2 hr after induction, prophase I begins at 3–4 hr, the first and second meiotic divisions at 5 and 6 hr, respectively, and spores are observed by 7 hr (37). Commitment to recombination occurs coincident with, or just after, initiation of premeiotic DNA synthesis in both *S. pombe* and *S. cerevisiae*. In *S. pombe* cells become committed to meiosis at about this time, but in *S. cerevisiae* commitment to meiosis occurs later (after the spindle pole bodies have separated just prior to the first meiotic division). Thus, S. *cerevisiae* cells, but not S. *pombe* cells, can be returned to mitotic growth after commitment to high levels of recombination (37, 39).

II. Nuclear Cytology during Meiosis

A. Nuclear Reorganization and Movement

The switch from mitotic growth to mating and the entry of cells into meiosis involve major reorganization of nuclear structure and of the spatial arrangement of chromosomes within the nucleus. During mitotic interphase the three chromosomes of S. *pombe* have a polarized arrangement within the nucleus typical of that in other eukaryotes: the centromeres are clustered at a single site near the spindle pole body (SPB), and the telomeres are located at the nuclear periphery (40). During nuclear division the centromeres interact with the mitotic spindle to lead chromosome movement to opposite poles.

In S. *pombe* major nuclear reorganization occurs following conjugation of haploid cells and karyogamy. Chikashige *et al.* (41) used fluorescence microscopy to study chromosome movement in living S. *pombe* cells during mating and meiosis. Following conjugation the two nuclei fuse and assume an elongated morphology, called a "horse tail" nucleus due to its shape (42). Following karyogamy horse tail nuclei migrate repeatedly back and forth from one end of the cell to the other. This movement continues throughout premeiotic DNA synthesis and prophase until just prior to the first meiotic division, when movement stops with the nucleus in the center of the cell (41). Within the horse tail nuclei the telomeres are clustered at the SPB, an arrangement unusual in eukaryotes, and lead movement of the nucleus, while the centromeres are separated from the SPB. This is in contrast to mitotic prophase in which the centromeres remain clustered at the SPB and lead chromosome movement. Thus, the telomeres and centromeres switch positions on conjugation of haploid cells. This switching occurs in two stages. First, in response to mating pheromone, the telomeres become associated with the SPB and the centromeres, and, second, following conjugation, the centromeres detach from the SPB (43). Once the first meiotic division is underway, the centromeres resume the leading position, moving the chromosomes toward the poles.

During the first meiotic prophase homologous chromosomes pair and become aligned. Extensive recombination occurs between paired homologs prior to chromosome separation and the first meiotic division. The mechanism by which homologs recognize each other and pair is not known (*44*). Telomeres may play a role in directing the alignment of homologous chromosomes in meiotic prophase. During prophase chromosomes are arranged in a "bouquet" configuration in which the telomeres are clustered at the nuclear surface and the chromosomes protrude as loops into the nuclear lumen (*41*). This bouquet arrangement of chromosomes is seen transiently during meiotic prophase in most eukaryotes (*45*). Fluorescence *in situ* hybridization (FISH) analysis of diploid cells undergoing azgyotic meiosis indicates that homologous chromosomes are close to each other during mitotic growth, entry into meiosis, and meiotic prophase (*46*). This close proximity presumably aids pairing of homologs, and the clustering of telomeres may bring homologs closer together and allow recognition of homologous DNA sequences. Pairing may thus initiate at the telomeres and subsequently spread along the chromosome (*41*). Movement of the nucleus may further aid chromosome pairing by constantly moving the chromosomes relative to each other and through stretching out of the chromosomes due to tension applied to their ends.

A role for telomeres in chromosome pairing is further supported by studies with an artificial minichromosome III, which contains 390 kb (about 10%) of chromosome III spanning the centromere and flanked by telomeric sequences (*47*). This minichromosome recombines rarely with full-length chromosome III despite the extensive homology (*47, 48*). FISH analysis revealed that the telomeres of the minichromosome remain associated with the telomeres of the endogenous chromosome, while the centromeres separate, resulting in spatial separation of the homologous DNA and presumably preventing recombination between them (*13*). Thus, alignment of homologous chromosomes may be initiated at the telomeres.

B. Linear Elements Instead of Synaptonemal Complex

In most eukaryotes the aligned chromosomes are held in a proteinaceous structure called the synaptonemal complex (SC) (reviewed in Ref. 49). The SC is a tripartite structure consisting of two lateral elements to which the chromatin is attached and which are closely synapsed ~100 nm apart, and a central core that extends between the lateral elements. This structure is highly conserved among most eukaryotes. During prophase I axial elements form along each set of sister chromatids. As the chromosomes pair, these elements synapse and become the lateral elements of the SC.

Schizosaccharomyces pombe lacks this classical SC but does possess long filaments called linear elements (*8, 50, 51*) (Fig. 3). Bähler *et al.* (*8*) and Scherthan *et al.* (*46*) examined nuclei during meiosis by silver staining and electron microscopy. Short linear elements are first observed following DNA replication, at about the time of chromosomal pairing. During prophase these

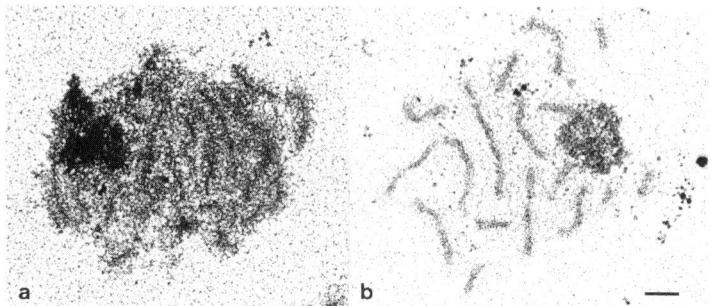

FIG. 3. Cytology of spread nuclei from *Schizosaccharomyces pombe* and *Saccharomyces cerevisiae*. Spread and silver-stained meiotic nuclei are shown for S. *pombe* (a) and S. *cerevisiae* (b). In S. *pombe* many linear elements with associated chromatin loops are visible. For S. *cerevisiae* the classical tripartite synaptonemal complex is observed associated with paired homologs. The bar represents 1 micron. (Reproduced from Ref. 8: *The Journal of Cell Biology*, 1993, **121**, 241; by copyright permission of the Rockefeller University Press.)

elements aggregate into networks and bundles, then into long filaments that are degraded shortly before the first meiotic division. In haploids, diploids, and tetraploids the number of linear elements is proportional to, but always greater than, the number of chromosomes, suggesting that linear elements are not formed continuously along the length of the chromosome. Linear elements appear to connect the two sister chromatids and may correspond to the axial elements seen in the classical SC of other eukaryotes.

Although it was originally proposed that the SC may play a role in pairing and recombination (*49*), it is obviously not essential because S. *pombe* can perform both functions in the absence of a true SC. This has also been shown for S. *cerevisiae* mutants that lack complete SCs (*52–55*).

It is, however, likely that the SC plays a role in crossover interference, the mechanism by which crossovers are distributed evenly throughout the genome. Crossover interference is abolished in S. *cerevisiae* mutants that lack a functional SC (*56*). Both S. *pombe* and the filamentous fungus *Aspergillus nidulans* lack a classical SC and do not show crossover interference (*9, 57, 58*). This is consistent with the discontinuous nature of linear elements observed in S. *pombe*, because crossover interference would presumably require communication along the length of the chromosome.

Studies on the recombination-deficient *rec8* mutant (Section IV) indicate that the *rec8* gene product may be required for linear element formation (*59*). A *rec8* mutant forms aberrant linear elements (*59*) and shows a region-specific reduction in meiotic recombination (Section V,A). The enhanced level of precocious sister chromatid separation seen in *rec8* mutants is consistent with the observed location of linear elements between sister chromatids (8). Mutants of S. *cerevisiae* that lack a full SC but still have pairing of homologs and/or high levels of recombination form fragmented axial elements similar to the linear elements of S. *pombe* (55). Together, these observations support the proposal that linear elements represent the minimal structures needed to perform pairing, recombination, and chromosome segregation (46, 57).

III. Meiotic Recombination Hotspots

The frequency of meiotic recombination does not occur evenly throughout the genome but is greater than average in some regions (hotspots) and lower than average in other regions (coldspots). In S. *pombe* a specific mutation in the *ade6* gene creates a meiotic recombination hotspot called *M26*. We discuss in Sections III,A–III,F the genetic characteristics of this hotspot and the mechanism of *M26* hotspot activity. We discuss in Section III,G a hotspot created by transplacement of the *ura4* gene and in Section V,B a coldspot between *mat2* and *mat3*.

A. Genetic Properties of M26

The *ade6–M26* mutation is one of 394 mutations in the *ade6* gene isolated and characterized by Gutz (60). Among these mutations *M26* is unique: when crossed with other mutations within *ade6* it increases the frequency of intragenic recombination 10- to 15-fold compared with the nearby mutation *ade6–M375*. *M26* is a C → T transversion that creates a nonsense codon near the 5' end of the coding region, whereas *M375* is a G → T nonsense mutation in the preceding codon and thus serves as an excellent control for *M26* (61, 62). With one exception noted in Section III,B, *M26* is active only in meiosis (62, 63) and stimulates recombination when homozygous or heterozygous, indicating that hotspot activity is not due to mismatch correction at the *M26* site (62, 64).

The *M26* mutation also undergoes gene conversion (nonreciprocal exchange) approximately 10 times more frequently than does *M375*. In such cases *M26* demonstrates an unusual disparity of conversion, with *M26* being preferentially converted to wild type (60). *M26* thus acts preferentially as a recipient of genetic information. In addition to conversion events, *M26* also simulates crossover events between duplicated *ade6* genes (63) and between two markers flanking *ade6* (65). The conversion of *M26* is accompanied by conversion of other *ade6* mutations located on either side or both sides of *M26*. The disparity of coconversion occurs in the same direction as that of

M26 and occurs frequently for mutations close to *M26*, but decreases in frequency for more distant markers. These observations led Gutz (60) to propose that recombination is initiated in the vicinity of *M26* and that conversion subsequently extends in both directions.

This proposal is supported by more detailed analysis of "coconversion" tracts in the presence and absence of *M26* in strains carrying silent restriction site polymorphisms in the region of *ade6* (65). Among random spores from appropriate crosses Ade$^+$ recombinants were selected and scored for flanking markers. Those retaining the flanking markers of the *M26* (or *ade6–706* control) parent were deemed to be convertants for *M26* or *706*. However, in the absence of tetrad analysis one cannot be certain that the events were nonreciprocal. Analysis by Southern blot hybridization of the DNA from these selected recombinants revealed the extent of coconversion of the restriction site polymorphisms. Nearly all of the conversion tracts were continuous, and their lengths were nearly the same for *M26* and *706:* the frequency of conversion of a restriction site decreased by a factor of about 4 for each kilobase from *M26* or *706*. Grimm *et al.* (65) concluded that *M26* enhances the frequency of recombination rather than stimulating novel events. It should be noted, however, that only events having an exchange between the selected *ade6* markers were analyzed. The authors propose a role for *M26* in the initiation of recombination, although these data do not rule out *M26* acting at other stages of recombination, such as processing of recombination intermediates or termination.

Genetic evidence indicates that hybrid DNA is frequently formed near the *M26* site. When crossed with wild type, a G → C transversion mutant can give C:C mismatches, which are repaired at reduced frequency compared with other mismatches and thus show higher postmeiotic segregation (PMS) than other mutations (66). Schär and Kohli (67) studied the segregation patterns of such markers located 2 bp to the left and 6 bp to the right of *M26* and found significant PMS, at a level equivalent to that of G → C transversions at other loci, indicating that hybrid DNA is frequently formed on either side of *M26*. Furthermore, strand transfer is highly biased, with preferential transfer of the transcribed strand (67). The *M26* mutation also demonstrates PMS when crossed with wild type in a *pms1* mutant background deficient for mismatch repair (68). These data provide evidence for the presence of heteroduplex DNA at *M26* as a recombination intermediate and argue against the presence of double-strand gaps at *M26*. The data are consistent with the presence near *M26* of a single-strand break, or a double-strand break (DSB) that is subsequently processed to give single-stranded ends, leading to invasion of the homolog by the transcribed strand and heteroduplex formation.

B. The Question of Meiotic DSBs in Schizosaccharomyces pombe

In *S. cerevisiae* there is strong evidence that recombination at hotspots is initiated through the formation of DSBs that are processed to give single-stranded 3' ends that invade the homologous duplex (reviewed in Ref. 69). There is no direct evidence that DSBs initiate recombination in *S. pombe*; searches for DSBs in the vicinity of *M26* have failed to detect any breakage (A. Ponticelli, J. Virgin, M. Cervantes, and J. Kohli, personal communication). These findings may reflect limitations in the techniques used to detect breaks, which may be transient or induced at low levels. There is, however, indirect evidence for the presence of DSBs in *S. pombe* meiosis. The introduction of a DSB 0.4 or 4.5 kb from *M26* using the HO endonuclease of *S. cerevisiae* results in mitotic activity of *M26* in recombination events between duplicated *ade6* genes (70). This result suggests that DSBs can initiate recombination in *S. pombe*. The presence of heteroduplex DNA at *M26* and the PMS of the *M26* mutation observed in a *pms1* mutant, noted above, suggest that such breaks, if they occur, are not at *M26* but are some distance from *M26*. The normal lack of *M26* activity in mitosis may be due to a lack of DSBs.

A special DSB can promote meiotic recombination in *S. pombe*. A DSB located in the *mat1* locus is essential for the directed gene conversion that switches the cell's mating-type during mitotic growth (see Section V,B for further description). In diploid cells deleted for the silent *mat2* and *mat3* loci, conversion occurs by interhomolog interaction between *mat1-P* and *mat1-M* in 20% of the meioses (71). Many of these convertants have an associated reciprocal exchange. A small deletion, *smt-s*, near *mat1* or a mutation in the unlinked *swi3* gene reduces both the level of the DSB and the frequency of meiotic conversion by at least a factor of 5. The high level of conversion is seen for *mat1* (near the DSB) but not for the more remote genes *leu1* or *his2* or the unlinked *ade6* gene. The DSB at *mat1* thus leads to meiotic gene conversion specifically near the DSB. Whether meiosis-specific DSBs lead to recombination elsewhere in the genome remains to be determined.

C. The M26 Heptamer Sequence and Associated Binding Proteins

A distinctive feature of the *M26* hotspot of *S. pombe*, compared with other eukaryotic hotspots, is that hotspot activity is known to depend on a unique DNA sequence, the heptamer 5'-A$\underline{T}$GACGT-3' (the underlined T is the site of the G → T transversion in *ade6–M26*) (72). Extensive site-directed mutagenesis demonstrated the requirement for each of these seven bases for hotspot activity, whereas mutations of bases outside of the heptamer have little or no effect on activity. The heteromeric protein Mts1/Mts2 binds specifically to the *M26* heptamer sequence (73). Binding of the protein to mutated DNA substrates correlates well with hotspot activity of these mutations (Fig. 4) (73), and both proteins are required for hotspot activity, but not for basal recombination involving *M375* (73a). However, *M26* normally functions as a hotspot only in meiosis (62, 63), whereas Mts1/Mts2 binding activity is present in extracts from both meiotic and mitotic cells (73). Other factor(s), as yet unidentified, must therefore confer meiotic specificity, possibly through the induction of DSBs. The genes encoding Mts1 and Mts2 have recently been cloned and sequenced. Surprisingly, Mts1 is identical to the transcription factor Atf1, and Mts2 is identical to its binding partner Pcr1 (73a). As described in Section I,B, Atf1 activates expression of several genes, including

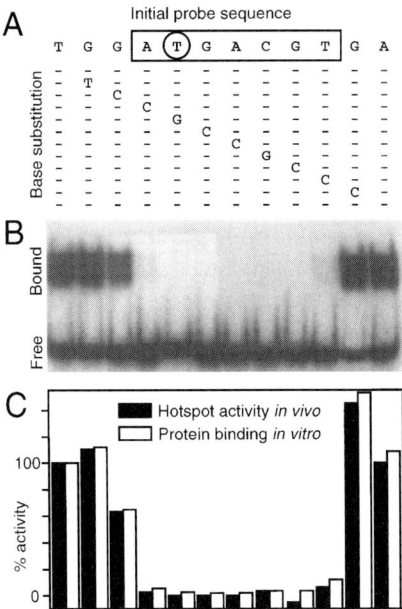

FIG. 4. Hotspot activity and Mts1/Mts2 binding to the *M26* site and to mutant derivatives. (A) The 7-bp nucleotide sequence (5'-ATGACGT-3') required for hotspot activity in meiotic crosses (72) is boxed; the site of the G → T *ade6–M26* mutation is circled (61). Single base pair substitutions analyzed below are indicated. (B) Binding of purified Mts1/Mts2 protein to DNA fragments bearing the sequences indicated in panel A was determined by a gel retardation assay (73). (C) Hotspot activity was measured by the frequency of intragenic *ade*$^+$ recombinants compared with that in a cross with the *ade6–M375* standard (72). Binding of Mts1/Mts2 was measured by quantitation of the gel in panel B. The hotspot and binding activities of *M26* were set at 100%, and others are expressed relative to *M26* (after subtracting the *M375 ade*$^+$ recombinant frequency). (Reproduced, with permission, from Ref. 73.)

ste11, whose product is required for entry into meiosis (Fig. 2). Mts1 (Atf1) thus plays a role in both meiotic recombination hotspot activity and transcription; the relationship between these two activities remains to be determined.

D. M26 and Transcription

It has been suggested that hotspot activity of *M26* requires transcription of the *ade6* gene. Deletion of ~500 bp containing the putative promoter region of *ade6* abolishes hotspot activity when in cis position to *M26* (74), whereas an identical deletion in *trans* position has no effect (74, 75). Replacement of the weak *ade6* promoter with the strong *adh1* promoter increases both meiotic and mitotic recombination at *ade6* (76). The presence of *M26* increases meiotic recombinant frequencies further, although the additional enhancement due to *M26* is small (about twofold) (76). The deleterious effect of the promoter deletion on *M26* function indicates that the heptamer sequence alone is not sufficient to confer hotspot activity. The abolition of *M26* hotspot activity could be due to decreased transcriptional activity or to other factors, for example, disruption of gross chromosomal structure or deletion of a second DNA element required for hotspot function (74) (Section III,E). *M26* does not significantly change *ade6* transcript levels (76; N. Kon and W. Wahls, personal communication); this result argues against a simple connection between transcription and *M26*-activated recombination.

E. Chromosomal Context Requirements

The *M26* heptamer sequence requires a particular chromosomal context for activity. In plasmid-by-chromosome crosses *M26* functions when on the chromosome but not when on the plasmid (77, 78). *M26* is also inactive in certain transplacements of *ade6–M26* to novel chromosomal locations but retains full activity in others (77, 78). These results indicate that the heptamer can function as a hotspot in certain other genomic locations but may require a specific chromatin structure or a second particular DNA sequence for hotspot activity. Because large pieces of DNA (usually >3 kb) were transplaced, with >1 kb on either side of *M26*, such chromosomal context effects must act over a distance of at least 1 kb.

Interpretation of the above data is complicated by the fact that large DNA fragments were transplaced, which may disrupt gross chromosomal structure. When the heptamer sequence was made at novel locations in the *ade6* and *ura4* genes by site-directed mutagenesis of only 1–4 bp, the heptamer was active at all locations tested, independent of its orientation (79). These results indicate that, provided gross chromosomal structure is not changed, the heptamer functions as a hotspot in a position- and orientation-independent manner and does not require a fixed relation to any other DNA element.

We infer that the heptamer functions as a hotspot in several chromosomal contexts in the presence of a wild-type chromatin structure and is inactivated only when certain large transplacements are made.

Although *M26* is the result of a mutation, the heptamer sequence occurs in the wild-type S. *pombe* genome. In the S. *pombe* sequences deposited to date (GenBank Release 100) the heptamer is underrepresented by approximately sixfold compared with scrambled heptamers of identical nucleotide composition. The results described above predict that naturally occurring heptamers are recombination hotspots distributed throughout the S. *pombe* genome; however, this prediction has not yet been confirmed.

F. *M26* and Chromatin Structure

The mechanism by which *M26* acts as a recombination hotspot presumably involves enhancement of a rate-limiting step of recombination. In S. *cerevisiae* meiotic recombination hotspots colocalize with sites of DNA double-strand breaks and sites of hypersensitivity of chromatin to nucleases (69). It is proposed that opening of the chromatin at these regions allows access to the DNA of proteins required for recombination.

Meiosis-specific micrococcal nuclease-hypersensitive sites are present at the site of *M26* and in the promoter region of *ade6* (80). In wild-type and *M375* chromatin, discrete, regularly spaced hypersensitive sites in mitotic and meiotic chromatin indicate that nucleosomes are phased throughout the *ade6* promoter and coding region; the wild-type site corresponding to the *M26* mutation is covered by a phased nucleosome. In the *M26* chromatin nucleosome phasing is disrupted over the 5′ end of the gene, and the *M26* site is hypersensitive to nuclease digestion. Nuclease sensitivity at *M26* and in the promoter region increases as cells enter meiosis. These hypersensitive sites are abolished in mutants lacking Mts1 (K. Ohta and W. Wahls, personal communication). This raises the possibility of a "two-site" mechanism in which the *M26* heptamer works in conjunction with a second site, possibly the recombination initiation site, in the promoter region (80). Although this proposal may seem to contradict the hotspot activity of the heptamer when created in several locations in *ade6* or *ura4* (79), there may be flexibility between *M26* and the second site analogous to that between enhancers of transcription and promoters.

The nuclease hypersensitivity at *M26* appears to be determined by the heptamer sequence. When the heptamer sequence is moved 747 bp downstream of the *M26* site, the nuclease-sensitive site is also shifted from the *M26* site to the new location of the heptamer; the nuclease-hypersensitive site in the promoter region remains (K. Ohta, M. Fox, J. Virgin, and G. Smith, unpublished observations). It will be interesting to determine whether the site of exchange is at the site of the heptamer or at a distinct fixed site, for example, in the promoter.

Although the M26 heptamer and associated binding proteins, including Mts1/Mts2, interact with chromatin structure and enhance recombination, a mechanistic connection between chromatin structural changes and enhanced recombination has not been made. The binding of Mts1/Mts2 may induce an appropriate chromatin structure, or binding may follow induction of a favorable chromatin structure at the heptamer sequence. Either way, recombination must involve additional proteins that confer meiotic specificity; these may include the products of the rec genes described in Section IV.

G. The ura4–aim Hotspot

The mutation ura4–aim (artificially introduced marker) is a 1.8-kb insertion of the ura4 gene into a site ~15 kb upstream of ade6 (65). The inserted ura4 gene demonstrates hotspot activity: ura4–aim converts ~7 times more frequently than does ade6–M375. This hotspot activity results from the transplacement of ura4 to a novel chromosomal location, because the endogenous ura4 locus does not demonstrate hotspot activity (74). The ura4–aim hotspot is influenced in a complex manner by M26 and by a ~500-bp deletion of the ade6 promoter region (74). Additional factors influencing this hotspot, such as particular DNA sequences and gene products, have not been identified.

IV. Recombination-Deficient Mutants and Genes

A. Isolation of Recombination-Deficient (Rec⁻) Mutants

To identify genes, and hence components, required for meiotic recombination, Ponticelli and Smith (11) developed a screen for meiotic Rec⁻ mutants. This screen was designed to yield both dominant and recessive mutations as well as mutations inactivating the M26 hotspot. Cells with the genotype h^{90} ade6–M26 ura4–294 (pade6–469) are mutagenized and plated to give isolated colonies on a minimal medium that selects for maintenance of the plasmid. During growth of the colonies the cells switch their mating type, owing to the h^{90} homothallism allele. When nutrients are depleted, the cells mate to form diploids homozygous for any induced mutation. During the ensuing meiosis recombination between the chromosomal ade6–M26 allele and the plasmid-borne ade6–469 allele can generate ade^+ recombinants. These occur at high frequency ($\sim 2 \times 10^{-2}$ per viable spore) in rec^+ cells but at low or undetectable frequency in Rec⁻ mutants. Spore suspensions free of vegetative cells are spotted onto solid media to estimate the titers of ade^+ recombinants and of total viable spores. Candidates with low ade^+ recombinant frequency are analyzed further. Subsequent crosses with appropriate heterothallic (h^+ and h^-) derivatives allow the determination of *ade6* interchromosomal recombination in standard heterothallic matings. The *rec* mutants described below are recombination deficient by this measure.

This screen identified one dominant and 39 recessive *rec* mutations that were studied further (*11, 12*). Complementation analyses placed the recessive mutations into 16 groups, designated *rec6–rec21*. [The previously described *rec1, rec2, rec3*, and *rec5* mutations affect mitotic or ectopic meiotic recombination, but not homologous meiotic recombination (*81, 82*).] The dominant *rec* mutation, *rec-101*, reduces *ade6, his7*, and *ura4* intragenic recombination by factors of 3–10, and *mat1–his5* and *lys7–leu2* intergenic recombination by a factor of ~2; similar reductions are observed whether *rec-101* is homozygous or heterozygous (Y. F. Li, personal communication). This mutation has not been studied further. The recessive *rec* mutations are discussed further below.

Because the screen described above demands that viable spore yields be ≳ 1% of that of rec^+ cells, certain classes of *rec* mutants would not be expected. These include mutants defective in mitotic growth, mating-type switching, mating, meiosis, sporulation, or germination. Although Rec⁻ mutants yielding 10–15% as many viable spores as wild type (as expected from random chromosome segregation at the first meiotic division; see Section I) are readily recovered, those with much lower viable spore yields would not be. In S. *cerevisiae* some Rec⁻ mutations have been classified as "early" or "late," depending on whether they produce high or low viable spore yields (in a *spo13* background, which results in one meiotic division and two diploid spores) (*6*). Early mutations are epistatic to late mutations. A plausible interpretation is that in the early mutants recombination never begins and the chromosomes remain intact but separate and therefore free to segregate, whereas in the late mutants recombination is initiated but not completed, and the chromosomes are damaged or intertwined and therefore unable to segregate. If a similar situation exists in S. *pombe*, the mutants isolated in the screen described above may be early mutants.

Support for the notion of early and late Rec⁻ S. *pombe* mutants comes from studies of the *rad32* mutant. This mutant was isolated on the basis of its hypersensitivity to gamma rays during mitotic growth (*83*). Further tests showed that only 0.5% of the spores from a *rad32* mutant meiosis are viable, compared to essentially 100% from a rad^+ meiosis. Among the few viable spores produced the *ade6* intragenic recombinant frequency is reduced by a factor of about 10. These properties suggest that *rad32* is a late meiotic Rec⁻ mutant. The putative early *rec6* mutation increases the viable spore yield of the *rec6 rad32* double mutant to that of the *rec6* single mutant, about 20% of the *rec⁺* level (J. Bedoyan, personal communication). A new set of S. *pombe* Rec⁻ mutants might be isolated as those that yield few viable spores but are "rescued" by a *rec6* (or other early) mutation. This type of screen identified S. *cerevisiae* mutants that are rescued by a *spo11* mutation (84–86). The converse of this screen, selection for mutations that rescue the late *rad52* mutation in S. *cerevisiae*, identified several new early *rec* genes in that organism (87).

Additional meiotic Rec⁻ mutants were sought by testing S. *pombe* mutants isolated on other bases, but these tests have revealed only two additional mutants, one of which is *rad32*, described above. The *swi5* mutation reduces mitotic mating-type switching and meiotic *ade6* intragenic recombination by a factor of about 5–10 (12, 88); mutations in seven other *swi* genes are not significantly meiotic Rec⁻. Similarly, mitotic radiation-sensitive mutants (*rad*), other than *rad32*, and mitotic hyper- or hyporecombination mutants (*rec50–rec60*) are not significantly meiotic Rec⁻ (89).

These and other observations discussed below indicate that S. *pombe* possesses many genes acting separately in meiotic recombination, mitotic recombination, mating-type switching, and recovery from DNA damage. Additional genes acting in two or more of these events may not have been identified because of limitations in the screens used. It may, therefore, be too early to tell whether the recombination and repair pathways of S. *pombe* have less overlap, as suggested so far, than do those of S. *cerevisiae*, which overlap extensively (6).

B. Cloning of *rec* Genes

To date nine *rec* genes have been cloned: *6, 7, 8, 10, 11, 12, 14, 15,* and *16* (Table I). Of these all but *rec10* and *rec16* have introns. For *rec6, rec7,* and *rec8* the complete intron–exon structure has not been determined. Approximately half of the S. *pombe* genes analyzed to date have introns, generally of ≤100 bp (90). Thus, in this regard the *rec* genes do not appear extraordinary although *rec11* has an unusually high number of introns (eight) (91). Comparisons with previously reported genes, including the entire S. *cerevisiae* genome, reveal limited homology between only four of the predicted polypeptides (Rec7, Rec12, Rec14 and Rad32) and those of other genes (see Table I and Section IV,D). This outcome suggests differences between the mechanism or control (or both) of meiotic recombination in S. *pombe* and in S. *cerevisiae*. However, it should be noted that the different methods for isolating Rec⁻ mutants in the two organisms may have yielded largely nonoverlapping sets of mutants.

TABLE I
Schizosaccharomyces pombe Meiotic Recombination-Deficient Mutants

Gene	Map location[a]	Size of predicted polypeptide[b]	Possible homolog in S. cerevisiae	Reduction of ade6 recombination[c]	Mitotic phenotype[d]	Ref.
rec6	II (tps13)	—	—	1000	—	11, 96
rec7	III (ade6)	~340	Rec114	1000	—	11, 95
rec8	II (mat1)	—	—	1000	—	11, 59, 92, 95
rec9	I	—	—	3	UV^s, MMS^s	11
rec10	I (his1)	791	—	100	—	11, 92, 97
rec11	III (ade6)	923	—	300	—	11, 91, 92
rec12	I	345	Spo11	1000	—	12, 96
rec13	—	—	—	8	—	12
rec14	II	302	Rec103 (Ski8)	1000	slow growth	12, 94
rec15	II (mat1)	180	—	1000	—	12, 98
rec16	II	472	—	100	—	12, 38, 108, 110a
rec17	—	—	—	5	MMS^s	12
rec18	I	—	—	5	—	12
rec19	I	—	—	5	MMS^s	12
rec20	I	—	—	5	—	12
rec21	II	—	—	5	—	12
swi5	II (ade7)	—	—	10	UV^s, IR^s	12, 88, 93, 114, 115
rad32	I (cdc25)	648	Mre11	10	UV^s, IR^s, MMS^s	83

[a]Chromosome and, if known, locus linked to the gene in meiotic crosses.
[b]Number of amino acids. Sizes for Rec 6, 7, and 8 are uncertain, as their complete intron–exon structure has not been determined.
[c]Factor by which ade^+ recombinant frequency, in crosses of ade6–M26 × ade6–52, is reduced relative to rec^+ (swi^+ or rad^+) crosses.
[d]UV^s, MMS^s, IR^s: hypersensitivity to ultraviolet light, methyl methanesulfonate, and ionizing radiation, respectively.

C. General Features of *Schizosaccharomyces pombe* Rec⁻ Mutants and the Corresponding Genes

The *rec* mutants were isolated on the basis of their reduced *ade6* meiotic recombinant frequency (Section IV,A). The factor of reduction varies from ~3 to >1000 (Table I), indicating that some *rec* gene products play a much greater role in meiotic recombination than do others. Previously, the mutants were placed into three classes based on the factor of reduction: class I (*rec6*,

7, 8, 12, 14, and 15), ~1000; class II (rec10, 11, and 16), ~100; and class III (rec9, 13, 17, 18, 19, 20, and 21, swi5, and rad32), ~3–10 (12, 83). Subsequent analyses of the mutants have shown phenotypic heterogeneity in these classes, described below, and question the utility of this classification.

By testing meiotic recombination in intervals other than ade6 a regional specificity of rec8, rec10, and rec11 was discovered (92). This specificity is discussed in Section V. Mutations in the other rec genes tested (rec6, 7, 12, 14, 15, and 16) reduce recombination strongly in all the intervals tested and presumably throughout the genome.

Only some of the rec mutations appear to have a mitotic phenotype. The rec9, rec17, and rec19 (and rad32) mutants are sensitive to the DNA-damaging agent methyl methanesulfonate (MMS) during mitotic growth, and rec9 (as well as swi5 and rad32) mutants are UV sensitive (11, 12, 83, 93). The rec14 mutants have slightly slower mitotic growth than do rec^+ cells (94). To the extent tested, none of the rec mutants have altered frequencies of mitotic ade6 plasmid-by-chromosome recombination (94; K. L. Larson, N. Hollingsworth, and J. Virgin, personal communication). These observations suggest a limited overlap in S. pombe functions for meiotic recombination and mitotic repair, as noted earlier.

The lack of mitotic phenotypes for many of the rec mutants is consistent with the corresponding rec gene transcript abundance. Transcripts of rec6, 7, 8, 10, 11, 12, 15, and 16 are barely, if at all, detectable during mitotic growth; transcripts of these genes accumulate to high levels at 2 and 3 hr after induction of meiosis, and then disappear (38, 91, 95–98; R. Ding, D. Evans, Y. F. Li, Y. Lin, and G. R. Smith, unpublished observations). The rec9, rec13, and rec17 genes, which have mitotic phenotypes, have not been cloned, and their transcript patterns are unknown. However, rec14 is an unusual case: although rec14 mutants grow slowly, they have wild-type levels of mitotic plasmid-by-chromosome recombination and resistance to DNA-damaging agents (94). The rec14 transcripts are readily detectable in mitotic cells and are induced about three- to fivefold during meiosis (94,110a). Rec14 thus appears to play a role in mitotic growth other than in recombination or repair, in addition to its strong role in meiotic recombination.

D. Features of Specific *Schizosaccharomyces pombe* Rec⁻ Mutants

1. rec7

The predicted Rec7 polypeptide has limited similarity to the predicted S. cerevisiae Rec114 polypeptide (99). An alignment with FASTA (Version 2) (100) reveals 20% amino acid identity. The rec114 mutants were isolated as suppressors of the low-viable-spore-yield phenotype of a rad52 mutant (87);

it is therefore considered an early Rec⁻ mutant, as is *rec7* (Section IV,A). Both *rec7* and *rec114* mutations strongly reduce meiotic recombination but have no known mitotic phenotype and the corresponding transcripts are strongly induced during meiosis (95, 99). Rec7 and Rec114 may therefore be functional homologs.

2. *rec8*, *rec10*, AND *rec11*

Mutations in these genes reduce recombination more strongly in intervals on chromosome III than in the intervals tested on the other two chromosomes. This regional (or perhaps chromosomal) specificity is discussed in Section V,A. The *rec8* mutants have aberrant linear elements and frequent aberrant chromosome segregation during meiosis (Section II,B).

3. *rec12*

Rec12 appears to be a functional homolog of the S. *cerevisiae* Spo11 protein. The *rec12* gene has four introns (B. Baum, personal communication; *100a*), and the predicted polypeptide encoded by the spliced *rec12* transcript contains 345 amino acids, 89 (26%) of which can be aligned with identical amino acids in the 398 amino acids of Spo11 (with the inclusion of 13 gaps). The *spo11* and *rec12* mutants have similar phenotypes: they are strongly deficient in meiotic recombination but have no detectable mitotic phenotypes (*12, 101*). Their transcripts are barely detectable during mitotic growth, are strongly induced early in meiosis, and are degraded late in meiosis (*96, 102*).

Spo11 appears to form double-strand breaks that initiate S. *cerevisiae* meiotic recombination. Spo11 becomes covalently linked to DNA at the site of a DSB, probably through a tyrosine–DNA phosphodiester bond (*103*). Spo11 shares amino acid identities not just with Rec12 but also with a subunit of a type II DNA topoisomerase of the archaeon *Sulfolobus shibatae* (*104*). Substitution of one of the conserved tyrosines in Spo11 with phenylalanine inactivates Spo11, presumably by abolishing DSB formation (*104*). Substitution of the corresponding tyrosine in Rec12 with phenylalanine inactivates Rec12 as measured by *ade6* intragenic recombination (J. Farah, personal communication).

These observations suggest that Rec12 and Spo11 play similar roles in meiotic recombination in S. *pombe* and S. *cerevisiae*, respectively. If so, DSBs may initiate meiotic recombination in S. *pombe* as in S. *cerevisiae*. However, arguments based on homology, as between Rec12 and Spo11, may be misleading: for example, the S. *pombe* Rad32 protein shares extensive identity with the S. *cerevisiae* Mre11 protein (*83*), yet the two appear to function differently (Section IV,D,6).

4. rec14

As noted in Section IV,C, *rec14* is an unusual *rec* mutation: it is the only one with a strong meiotic recombination deficiency and an observable mitotic phenotype (slow growth). Unlike those of other *rec* genes tested, *rec14* transcripts are readily detectable during mitotic growth. The predicted Rec14 polypeptide contains six copies of the loosely defined WD (or β-transducin) repeat, which is found in many proteins having a wide variety of activities and is thought to play a role in protein–protein interactions (*94*). Because of these repeats, Rec14 shares homologies with many proteins, including the *S. cerevisiae* Ski8 protein. *SKI8* is identical to *REC103*, mutations which were identified as early Rec⁻ mutations by the criteria discussed in Section IV,A (*87, 105*). The phenotypes of *rec14* and *rec103* mutants are similar. The *ski8* mutations were initially identified as suppressors of *mak* mutations that reduce the maintenance of the double-stranded RNA virus-like killer particle present in some *S. cerevisiae* strains (*106*). The Ski8 protein appears to inhibit the translation of nonpolyadenylylated RNAs, such as those of the killer particle (*107*). The role of such a protein in meiotic recombination is unclear. Gardiner *et al.* (*105*) hypothesize an inhibitor of meiotic recombination encoded by a nonpolyadenylylated RNA; this inhibitor would accumulate in *ski8* mutants, making them Rec⁻. Further analysis of *rec14* and *SKI8* (*REC103*), and of suppressors of *rec14* and *ski8* mutations, may test this hypothesis and shed light on how these proteins promote meiotic recombination.

5. rec16

In a physiological analysis of meiotic events, such as DNA replication, transcription, and splicing, the *rec16* mutant was unique among the *rec* mutants tested: this mutant is deficient not just in meiotic recombination, but also in meiotic replication and transcription (*38*). In a *rec16* mutant premeiotic DNA synthesis is delayed and, in some cases, reduced in amount. Accumulation of the transcripts of certain other *rec* genes during meiosis is also reduced. In contrast, meiotic replication and transcription appear unaltered in the 10 other *rec* mutants examined. These results implicated Rec16 as a major regulator of multiple meiotic events.

Cloning and sequencing of *rec16* revealed it to be identical to *rep1*, which may be a transcriptional activator (*110a*). The *rep1*⁺ gene was isolated as a high-level suppressor of the *cdc10–129* temperature-sensitive mutation (*108*). Cdc10 is a transcriptional activator of many genes required for DNA metabolism in both mitosis and meiosis (*5, 109*). Several high-level suppressors of *cdc10–129* have been isolated; some of the corresponding gene products form complexes with Cdc10 (*110*), although no Rep1–Cdc10 complex has been reported. The *rec16* (*rep1*) gene is induced early in meiosis (*108, 110a*), and its product is required for the meiotic induction of several *rec* genes and *cdc22* (*38, 108*), which is required for DNA synthesis (*109*). Rec16 (Rep1) may, with Cdc10, form a meiosis-specific transcriptional activator. The Rec$^-$ phenotype of *rec16* (*rep1*) mutants can thus be accounted for by the failure to induce some of the *rec* genes. Additional roles are not excluded, however: delay (*38*) or abolition (*108*) of meiotic replication in *rec16* (*rep1*) mutants may block recombination because the two processes are coupled (e.g., by break-copy recombination; *111*).

6. *rad32*

The *rad32* mutant was isolated as a gamma-ray-sensitive mutant; subsequent analysis showed it to be deficient in the mitotic repair of gamma-ray-induced DSBs (*83*). Rad32 is also crucial in meiosis: in a *rad32* mutant meiosis only 0.5% of the spores are viable, and among these the *ade6* intragenic recombinant frequency is reduced by a factor of about 10 (*83*). The predicted Rad32 polypeptide shares extensive homology with the S. cerevisiae Mre11 protein: in the amino-terminal three-quarters of the proteins, 50% of the amino acids are identical (with the inclusion of nine gaps for alignment). Mre11 is an early Rec$^-$ mutant of S. cerevisiae and, like a *rad32* mutant, is sensitive to MMS during mitotic growth, suggesting that both proteins are required for repair of mitotic DSBs (*107*).

In spite of the extensive homology and their common requirement in meiotic recombination, the two proteins appear to function differently in meiosis. By inference from epistasis with *rec6*, *rad32* appears to be a late Rec$^-$ mutant (see Section IV,A), whereas *mre11* appears to be an early Rec$^-$ mutant and is required for the formation of meiotic DSBs (*112, 113*). The two proteins may have different activities, or they may be components of protein complexes that act in both the formation and the repair of DSBs.

7. *swi5*

The *swi* mutants were isolated based on their deficiency in mating-type switching (*114*). Subsequent analysis showed *swi5* mutants to be UV sensitive and gamma ray sensitive and to be reduced by a factor of 5–10 in intragenic and intergenic meiotic recombination (*12, 88, 93, 115*). The *swi5* mutation does not reduce spore viability and is useful for demonstrating meiotic linkage of markers unlinked in wild-type meiosis (*115*). DSBs at the mating-type locus are present at the same level in *swi*$^+$ and *swi5* mutants (*116*), but these breaks are evidently inefficiently utilized for mating-type switching in *swi5* mutants. The role of Swi5 in meiotic recombination is unclear. Within the context of the arguments in Section IV,A the high spore viability of *swi5* mutants suggests that Swi5 acts early. The inefficient utilization of DSBs in mating-type switching, however, suggests that Swi5 acts late. Further epistasis analyses may help clarify the role of Swi5 in meiotic recombination.

V. Region-Specific Control of Meiotic Recombination

In Section III we discussed a special site, *M26*, that enhances meiotic recombination within ~1 kb of itself. Here, we describe genes whose products control recombination, either positively or negatively, over much larger intervals (up to ~3 Mb).

A. The Rec8, Rec10, and Rec11 Activators

The *rec* mutants described in Section IV were isolated on the basis of their reduced frequency of *ade6* intragenic recombination. Further analyses showed that the *rec8*, *rec10*, and *rec11* mutations reduce meiotic recombination to a greater extent in some intervals, including *ade6*, than in others (92). The most strongly affected intervals are on chromosome *III* (Fig. 5). Because the *rec8*, *rec10*, and *rec11* mutations are recessive (11) and, for *rec8* and *rec10*, map on chromosomes *II* and *I*, respectively (92), these genes encode diffusible activators of recombination specific for the affected region, which may be the entire chromosome *III*. There may be activators of recombination specific to other regions, e.g., chromosomes *I* and *II*. Such activators would not have been revealed by the screen for mutants deficient in *ade6* recombination if they are highly region specific like Rec10.

Rec10 is the clearest example of a region-specific activator. The *rec10* mutations reduce recombination in the eight intervals tested on chromosome *III* by factors of 3 to 100, but less than a factor of 2 in the eight intervals tested on chromosomes *I* and *II* (92; G. R. Smith, unpublished data). Rec10-dependent intervals span nearly the entirety of chromosome *III*. Thus, Rec10 appears to be required for high levels of meiotic recombination on *III* but dispensable for recombination on *I* and *II*. There may, however, be untested regions of *I* and *II* that do require Rec10.

Rec8 and Rec11 also activate recombination more strongly in some genetic intervals than in others. For example, *rec8* mutations reduce intragenic recombination by factors ranging from 2 (at *lys7* on chromosome *I*) to 1000 (at *ade6* on *III*) and intergenic recombination by factors ranging from 2 (*ura2-leu2* on *I*) to 300 (*ade7-his3* on *II* and *ade6-arg1* on *III*) (92; J. Kohli, personal communication). The intervals activated by Rec8 and Rec11 for recombination appear to be region specific, rather than chromosome specific as inferred for Rec10. This may reflect a difference between the function of Rec8 and Rec11 and that of Rec10.

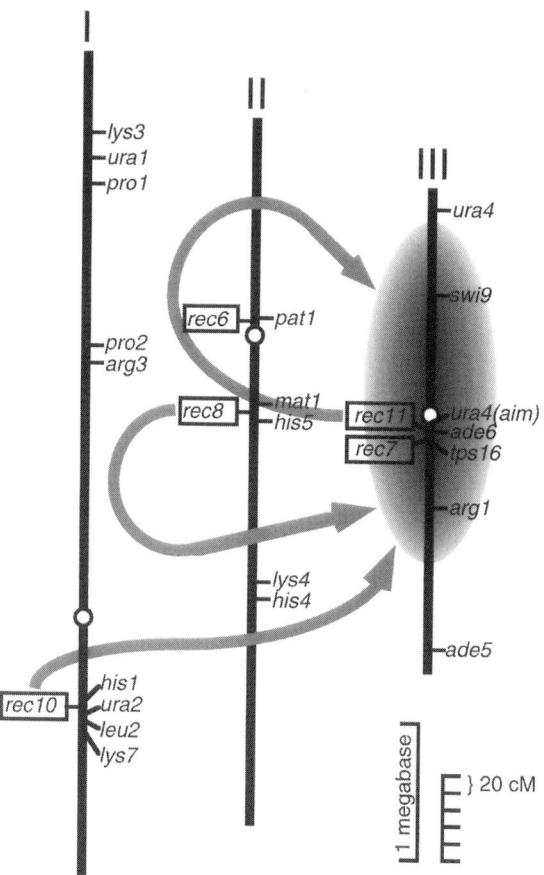

FIG. 5. Regional specificity of Rec8, Rec10, and Rec11. Shown are the three S. pombe chromosomes and markers used to measure recombination. Open circles indicate centromeres. Scales indicate physical distances (in base pairs) and meiotic genetic distances (in centimorgans). The arrows from *rec8*, *rec10*, and *rec11* indicate the promotion of recombination by their gene products preferentially in the region indicated on chromosome III. The precise extent of this region has not been determined. (Reproduced, with permission, from Ref. 92.)

The specificities described above appear to reflect the region of the genome tested, not the type of recombination or the particular markers used. Intragenic recombination (principally nonreciprocal exchange or gene conversion) and intergenic recombination (principally reciprocal exchange or crossing-over) are affected approximately equally. Multiple *ade6* alleles behave similarly, arguing against differential mismatch correction being responsible for the effects. Furthermore, studies with transplacements of 3-kb

DNA fragments bearing *ade6* alleles into *ura4* showed that the degree of reduction of recombination by *rec10* and *rec11* mutations corresponds to the location of *ade6*: recombination of the transplaced *ade6* alleles within *ura4* is reduced by factors similar to that of *ura4* alleles and less than that of endogenous *ade6* alleles.

The mechanism by which Rec8, Rec10, and Rec11 act is unclear. The phenotypes of the mutations imply that there are one or more special chromosomal site(s) that allow Rec8, Rec10, and Rec11, directly or indirectly, to distinguish the affected region (perhaps chromosome *III*) from the unaffected regions. Such sites have not been identified. The nucleotide sequences of the genes indicate that they encode novel proteins and give no clue to their function (*91, 95, 97*). One possibility is that they promote pairing of chromosome *III*, which may be required for its recombination. If pairing depends on sites present on each homolog, however, this hypothesis does not readily account for the dependence of plasmid-by-chromosome recombination on Rec8, Rec10, and Rec11 (*11*), unless the 3-kb *ade6* DNA fragment in the plasmid used for the screen to isolate the *rec* mutants (see Section IV,A) contains such a site. Further studies of the proteins and searches for the sites at which they act may help elucidate their mechanism.

Chromosome-specific activation of meiotic recombination occurs also in the worm *Caenorhabditis elegans* (*117, 118*; P. Meneely, personal communication). The *him-1, him-5,* and *him-8* mutations reduce recombination on the X chromosome by factors as great as 50 but have much less or no significant effect on autosomal recombination. The *him-1, him-5,* and *him-8* mutations are recessive and autosomal; the genes thus encode activators of recombination. Reduced X chromosome recombination results in a high frequency of XX nondisjunction at the first meiotic division and a <u>h</u>igh <u>i</u>ncidence of <u>m</u>ales (XO animals) from XX hermaphrodites.

Pairing of X homologs may initiate at one end of the X chromosome, and the *him-1, him-5,* and *him-8* gene products may be required for the extension of pairing along the homologs. In the *him-1, him-5,* and *him-8* mutants recombination, relative to that in *him*$^+$ animals, is elevated up to twofold near one end of the X chromosome and is reduced progressively farther from this end (*118*). A large deletion mutation at this end also reduces X, but not autosomal, recombination by a factor as great as 50 (*119*). Curiously, this deletion has little effect when hemizygous; in other words the putative site deleted in the mutant can promote recombination when the site is present on only one homolog. One possibility is that the *him-1, him-5,* and *him-8* gene products load onto the X chromosome at this site, move along the chromosome, and promote pairing of subsequently encountered regions of homology. These observations encourage the search for sites at which the *S. pombe* Rec8, Rec10, and Rec11 proteins act.

B. The Rik1, Swi6, and Clr Repressors

There is a coldspot of meiotic recombination located between the *mat2* and *mat3* loci; reduction of recombination in this interval, designated K, requires the *rik1*, *swi6*, and *clr* gene products (and perhaps others) as described below. These gene products appear to be repressors of recombination specific for the K region.

The gene cluster *mat1–L–mat2–K–mat3* on chromosome *II* determines the mating type of S. *pombe* cells (reviewed in Refs. *120* and *121*). The *mat1* locus can contain either P or M information, and *mat1* expression results in cells of either plus or minus mating type, respectively. The *mat2* and *mat3* loci contain P and M information, respectively, but are not expressed. In wild-type (h^{90}) cells repeated transfer of information from *mat2* or *mat3* to *mat1* by unidirectional gene conversion results in switching of mating type and the ability of cells in a culture derived from a single cell to mate (homothallism). Complex duplications and deletions in this region produce cells with stable plus (h^+) or minus (h^-) mating type (heterothallism).

Meiotic recombination in the *mat2–mat3* interval (K) rarely, if ever, occurs. The physical size of the K interval, 15 kb (*122*), would predict a genetic distance of about 2 cM, based on the average frequency of meiotic recombination in the S. *pombe* genome. The measured genetic distance, however, is <0.002 cM (*123*), and the K region is therefore designated a meiotic coldspot. Homologous substitution of exogenous linear DNA during transformation of mitotic cells also appears to occur at >50-fold lower frequency in the K–*mat3* region than at a control locus (*ura4*) (*124*). The *rik1*, *swi6*, and *clr* mutations, which allow recombination in the *mat2–K–mat3* region, also allow expression of the *mat2* and *mat3* genes (or of genes such as *ura4* inserted into this region) (*125–127*). This region may have a special chromatin structure, perhaps designed to allow the directed gene conversion of mating-type switching and to prevent expression of *mat2* and *mat3* and consequent sterility (*128*).

Genes whose products repress *mat2–mat3* recombination were found in various ways. The *rik1* (recombination in K) mutation was serendipitously found in a mutagenized strain containing the *end1* mutation (*129*); *rik1* reduces the abundance of spores and their viability in an h^{90} strain. The *swi6* (switching) mutation was identified in a set of mutants deficient in mating-type switching (*114*). The *clr1, 2, 3*, and *4* (cryptic loci regulator) mutants were selected for their ability to express *mat2* and *mat3* or the *ura4*$^+$ gene after it was inserted near *mat3*, where its transcription is repressed in *clr*$^+$ cells (*126, 127, 130*).

The pleiotropic effect of these mutations on mating-type switching, gene expression, and meiotic recombination suggest that they control the chromatin structure of the mating-type locus. The *swi6* gene encodes a 328-amino acid polypeptide containing the "chromodomain" motif found in proteins associated with chromatin (*125*). The mechanism by which chromatin structure controls recombination and gene expression in the *mat2–mat3* region remains to be elucidated.

Mutations in the *rik1*, *swi6*, *clr1*, *clr2*, *clr3*, and *clr4* genes increase meiotic recombination between *mat2* and *mat3* from <0.002 cM to 2–8 cM (*125–129*). This level is near that for genes elsewhere in the genome separated by a similar physical distance. The mutations thus appear to remove completely the repression of meiotic recombination in the coldspot. Effects on recombination elsewhere in the genome have not been extensively reported, but *swi6* does not affect *ade6* intragenic recombination (cited in Ref. *125*; G. R. Smith, unpublished data) or *leu1–mat1* intergenic recombination (*128*). The repressors may thus be specific for the *mat2–mat3* region.

Region-specific repressors of meiotic recombination are present in certain strains of the distantly related fungus *Neurospora crassa* (see Refs. *131* and *132* for reviews). Dominant alleles of *rec* loci reduce meiotic recombination by a factor of ~10 at specific target intervals, but not at others. For example, *rec1* represses *his-1* intergenic recombination by a factor of ~15 but has no significant effect on recombination in several other intervals. Three *rec* genes, each acting at two or three target intervals, have been identified by comparisons of recombination in different strains of *N. crassa* and their hybrids. Typically, the *rec* genes are distant from their targets, or even on separate chromosomes. This result implies that the *rec* genes encode diffusible products that, because of the dominance of the low-frequency alleles, repress recombination in specific regions. Although sites near the target intervals have been inferred, neither they nor the *rec* genes have been further characterized.

The identification of region-specific meiotic recombination activators in S. *pombe* and C. *elegans*, and repressors in S. *pombe* and *N. crassa*, suggests that the control of meiotic recombination by such products may be widespread in eukaryotes.

VI. Conclusion and Perspective

Homologous recombination in S. *pombe* is controlled in numerous ways. Recombination occurs at high levels between paired chromosomes during meiosis. In S. *pombe* meiosis is induced by nutritional starvation, which activates a complex network of proteins controlling and coordinating the activity of many genes and their products. Among these are the *rec* genes, whose products are required for recombination and most of which are specifically induced in meiosis. Accompanying the induction of these genes are profound changes in nuclear morphology and movement, which may aid chromosome pairing, and in chromatin structure, which may control access to the DNA of recombination-promoting enzymes. These enzymes interact, directly or indirectly, with chromosomal sites that control the frequency of recombination. The recombination hotspot *M26*, a 7-bp sequence, together with its binding protein Mts1/Mts2, acts over relatively short distances (a few kilobases) to increase recombination. Region-specific *rec* genes act over relatively long distances (a few megabases) and may promote recombination on specific chromosomes. Although some aspects of the control of meiotic recombination are known, they have not yet been put into a unified picture.

The mechanism of meiotic recombination in S. *pombe* is still largely unknown. It is currently unclear how similar it is to that in S. *cerevisiae*, in which the mechanism is more thoroughly understood. *Schizosaccharomyces pombe* Rec12 is similar to S. *cerevisiae* Spo11, which appears to be a central component of a complex that makes DSBs that initiate meiotic recombination. This similarity suggests that DSBs may initiate S. *pombe* meiotic recombination, but direct evidence for such DSBs in S. *pombe* is lacking. Some other S. *pombe* and S. *cerevisiae* proteins required for meiotic recombination share homology: Rec7 ≈ Rec114, Rec 14 ≈ Rec103 (Ski8), and Rad32 ≈ Mre11. However, many S. *pombe* Rec proteins do not appear to have homologs in S. *cerevisiae*. It is not clear whether the mechanism of meiotic recombination is different in the two organisms, or whether the mechanism is similar but involves a largely different set of proteins.

Key advances in elucidating the mechanism in S. *pombe* would include the identification of enzymatic activities required for meiotic recombination and the identification of DNA intermediates in the process. The former might stem from further study of the *rec* genes and their products, and the latter from physical analysis of DNA isolated during meiosis of *rec* mutants blocked at intermediate stages of recombination. New screens for late *rec* mutants may be essential for this approach.

Understanding the coordinated control of meiotic recombination remains an additional challenge. This is likely to require knowledge of gene expression, enzyme activity, protein complex formation, changes in chromatin structure, interaction of proteins with chromosomal sites, and movement of chromosomes within nuclei. This knowledge will broaden our understanding of recombination and its vital role in promoting proper chromosome segregation and generating genetic diversity during meiosis.

Acknowledgments

We are grateful to Sue Amundsen, Jirair Bedoyan, Amar Klar, Jürg Kohli, Ramsay McFarlane, Phil Meneely, Henning Schmidt, Andrew Taylor, Jeff Virgin, and Wayne Wahls for helpful comments on the manuscript. We thank Bobby Baum, Jirair Bedoyan, Marcella Cervantes, Joe Farah, Nancy Hollingsworth, Jürg Kohli, Karen Larson, Feng Li, Yukang Lin, Kunihiro Ohta, Fred Ponticelli, Jeff Virgin, and Wayne Wahls for allowing us to quote unpublished information. We thank Karen Brighton for skillfully preparing the manuscript. Research in our laboratory is supported by NIH Grants GM31693 and GM32194.

NOTE ADDED IN PROOF: Rec8 shares amino acid sequence homology with S. cerevisiae Scc1, a protein required for mitotic sister chromatid cohesion (C. Michaelis, R. Ciosk, and K. Nasmyth, Cell **91,** 35–45, 1997), and with S. pombe Rad21 (J. Kohli, personal communication). Rec8,. and perhaps Rec11, may be involved in pairing (and hence recombination) in specific regions of the genome.

References

1. B. S. Baker, A. T. C. Carpenter, M. S. Esposito, R. E. Esposito, and L. Sandler, *Annu. Rev. Genet.* **10,** 53 (1976).
2. A. Nasim, P. Young, and B. F. Johnson, "Molecular Biology of the Fission Yeast." Academic Press, San Diego, 1989.
3. P. Munz, K. Wolf, J. Kohli, and U. Leupold, *in* "Molecular Biology of the Fission Yeast" (A. Nasim, P. Young, and B. F. Johnson, eds.), p. 1. Academic Press, San Diego, 1989.
4. Y. Iino and M. Yamamoto, *Mol. Gen. Genet.* **198,** 416 (1985).
5. D. Beach, L. Rodgers, and J. Gould, *Curr. Genet.* **10,** 297 (1985).
6. T. D. Petes, R. E. Malone, and L. S. Symington, *in* "The Molecular and Cellular Biology of the Yeast *Saccharomyces*" (J. Broach, E. Jones, and J. Pringle, eds.), p. 407. Cold Spring Harbor Laboratory, Cold Spring Harbor, New York, 1991.
7. M. V. Olson, *in* "The Molecular and Cellular Biology of the Yeast *Saccharomyces*" (J. R. Broach, J. R. Pringle, and E. W. Jones, eds.), Vol. 1, p. 1. Cold Spring Harbor Laboratory, Cold Spring Harbor, New York, 1991.
8. J. Bähler, T. Wyler, J. Loidl, and J. Kohli, *J. Cell Biol.* **121,** 241 (1993).
9. P. Munz, *Genetics* **137,** 701 (1994).
10. G. R. Smith, *Experientia* **50,** 234 (1994).
11. A. S. Ponticelli and G. R. Smith, *Genetics* **123,** 45 (1989).
12. L. C. DeVeaux, N. A. Hoagland, and G. R. Smith, *Genetics* **130,** 251 (1992).
13. O. Nielsen, *Trends Cell Biol.* **3,** 60 (1993).
14. M. Yamamoto, *Trends Biochem.* **21,** 18 (1996).
15. H. Gutz, H. Heslot, U. Leupold, and N. Loprieno, *in* "Handbook of Genetics" (R. C. King, ed.), vol. 1, p. 395. Plenum, New York, 1974.
16. P. Nurse, *Mol. Gen. Genet.* **198,** 497 (1985).
17. Y. Watanabe, S. Shinozaki-Yabana, Y. Chikashige, Y. Hiraoka, and M. Yamamoto, *Nature (London)* **386,** 187 (1997).
18. P. Li and M. McLeod, *Cell* **87,** 869 (1996).
19. Y. Watanabe and M. Yamamoto, *Cell* **78,** 487 (1994).
20. A. Sugimoto, Y. Iino, T. Maeda, Y. Watanabe, and M. Yamamoto, *Genes Dev.* **5,** 1990 (1990).

21. Y. Watanabe, Y. Iino, K. Furuhata, C. Shimoda, and M. Yamamoto, *EMBO J.* **7**, 761 (1988).
22. J. DeVoti, G. Seydoux, D. Beach, and M. McLeod, *EMBO J.* **10**, 3759 (1991).
23. T. Takeda, T. Toda, K. Kominami, A. Kohnosu, M. Yanagida, and N. Jones, *EMBO J.* **14**, 6193 (1995).
24. J. Kanoh, Y. Watanabe, M. Ohsugi, Y. Iino, and M. Yamamoto, *Genes Cells* **1**, 391 (1996).
25. Y. Watanabe and M. Yamamoto, *Mol. Cell. Biol.* **16**, 704 (1996).
26. E. Warbrick and P. A. Fantes, *EMBO J.* **10**, 4291 (1991).
27. K. Shiozaki and P. Russell, *Nature (London)* **378**, 739 (1995).
28. K. Shiozaki and P. Russell, *Genes Dev.* **10**, 2276 (1996).
29. M. G. Wilkinson, M. Samuels, T. Takeda, W. M. Toone, J.-C. Shieh, T. Toda, J. B. A. Millar, and N. Jones, *Genes Dev.* **10**, 2289 (1996).
30. G. Degols, K. Shiozaki, and P. Russell, *Mol. Cell. Biol.* **16**, 2870 (1996).
31. J.-C. Shieh, M. G. Wilkinson, V. Buck, B. A. Morgan, K. Makino, and J. B. A. Millar, *Genes Dev.* **11**, 1008 (1997).
32. R. Egel, O. Nielsen, and D. Weilguny, *Trends Genet.* **6**, 369 (1990).
33. M. McLeod, M. Stein, and D. Beach, *EMBO J.* **6**, 729 (1987).
34. M. Willer, L. Hoffmann, U. Styrkarsdottir, R. Egel, J. Davey, and O. Nielsen, *Mol. Cell. Biol.* **15**, 4964 (1995).
35. M. McLeod and D. Beach, *Nature (London)* **332**, 509 (1988).
36. Y. Iino and M. Yamamoto, *Proc. Natl. Acad. Sci. U.S.A.* **82**, 2447 (1985).
37. J. Bähler, P. Schuchert, C. Grimm, and J. Kohli, *Curr. Genet.* **19**, 445 (1991).
38. Y. F. Li and G. R. Smith, *Genetics* **146**, 57 (1997).
39. R. E. Esposito and S. Klapholtz, in "The Molecular Biology of the Yeast *Saccharomyces*: Life Cycle and Inheritance" (J. N. Strathern, E. W. Jones, and J. R. Broach, eds.), p. 211. Cold Spring Harbor Laboratory, Cold Spring Harbor, New York, 1981.
40. H. Funabiki, I. Hagan, S. Uzawa, and M. Yanagida, *J. Cell Biol.* **121**, 961 (1993).
41. Y. Chikashige, D.-Q. Ding, H. Funabiki, T. Haraguchi, S. Mashiko, M. Yanagida, and Y. Hiraoka, *Science* **264**, 270 (1994).
42. C. F. Robinow, *Genetics* **87**, 491 (1977).
43. Y. Chikashige, D.-Q. Ding, Y. Imai, M. Yamamoto, T. Haraguchi, and Y. Hiraoka, *EMBO J.* **16**, 193 (1997).
44. P. B. Moens, *BioEssays* **16**, 101 (1994).
45. C. P. Fussell, in "Meiosis" (P. B. Moens, ed.), p. 275. Academic Press, San Diego, 1987.
46. H. Scherthan, J. Bähler, and J. Kohli, *J. Cell Biol.* **127**, 273 (1994).
47. O. Niwa, T. Matsumoto, and M. Yanagida, *Mol. Gen. Genet.* **203**, 397 (1986).
48. O. Niwa, T. Matsumoto, Y. Chikashige, and M. Yanagida, *EMBO J.* **8**, 3045 (1989).
49. D. von Wettstein, S. W. Rasmussen, and P. B. Holm, *Annu. Rev. Genet.* **18**, 331 (1984).
50. C. F. Robinow and J. S. Hyams, in "Molecular Biology of the Fission Yeast" (A. Nasim, P. Young, and B. F. Johnson, eds.), p. 273. Academic Press, San Diego, 1989.
51. L. W. Olson, U. Eden, M. Egel-Mitani, and R. Egel, *Hereditas* **89**, 189 (1978).
52. G. S. Roeder, *Trends Genet.* **6**, 385 (1990).
53. R. Padmore, L. Cao, and N. Kleckner, *Cell* **66**, 1239 (1991).
54. B. M. Weiner and N. Kleckner, *Cell* **77**, 977 (1994).
55. J. Loidl, F. Klein, and H. Scherthan, *J. Cell Biol.* **125**, 1191 (1994).
56. M. Sym and G. S. Roeder, *Cell* **79**, 283 (1994).
57. J. Kohli and J. Bähler, *Experientia* **50**, 295 (1994).
58. M. Egel-Mitani, L. W. Olson, and R. Egel, *Hereditas* **97**, 179 (1982).
59. M. Molnar, J. Bähler, M. Sipiczki, and J. Kohli, *Genetics* **141**, 61 (1995).
60. H. Gutz, *Genetics* **69**, 317 (1971).
61. P. Szankasi, W. D. Heyer, P. Schuchert, and J. Kohli, *J. Mol. Biol.* **204**, 917 (1988).

62. A. S. Ponticelli, E. P. Sena, and G. R. Smith, *Genetics* **119,** 491 (1988).
63. P. Schuchert and J. Kohli, *Genetics* **119,** 507 (1988).
64. C. Grimm, P. Munz, and J. Kohli, *Curr. Genet.* **18,** 193 (1990).
65. C. Grimm, J. Bähler, and J. Kohli, *Genetics* **136,** 41 (1994).
66. P. Schär and J. Kohli, *Genetics* **133,** 825 (1993).
67. P. Schär and J. Kohli, *EMBO J.* **13,** 5212 (1994).
68. P. Schär, M. Baur, C. Schneider, and J. Kohli, *Genetics* **146,** 1275 (1997).
69. M. Lichten and A. S. H. Goldman, *Annu. Rev. Genet.* **29,** 423 (1995).
70. F. Osman, E. Fortunato, and S. Subramani, *Genetics* **142,** 341 (1996).
71. A. J. S. Klar and L. M. Miglio, *Cell* **46,** 725 (1986).
72. P. Schuchert, M. Langsford, E. Käslin, and J. Kohli, *EMBO J.* **10,** 2157 (1991).
73. W. P. Wahls and G. R. Smith, *Genes Dev.* **8,** 1693 (1994).
73a. N. Kon, M. D. Krawchuk, B. G. Warren, G. R. Smith, and W. P. Wahls, *Proc. Natl. Acad. Sci. U.S.A.* **94,** 13765 (1997).
74. M. Zahn-Zabal, E. Lehmann, and J. Kohli, *Genetics* **140,** 469 (1995).
75. A. S. Ponticelli, Ph.D. Thesis. University of Washington, Seattle, Washington, 1988.
76. C. Grimm, P. Schaer, P. Munz, and J. Kohli, *Mol. Cell. Biol.* **11,** 289 (1991).
77. A. S. Ponticelli and G. R. Smith, *Proc. Natl. Acad. Sci. U.S.A.* **89,** 227 (1992).
78. J. B. Virgin, J. Metzger, and G. R. Smith, *Genetics* **141,** 33 (1995).
79. M. E. Fox, J. B. Virgin, J. Metzger, and G. R. Smith, *Proc. Natl. Acad. Sci. U.S.A.* **94,** 7446 (1997).
80. K. Muzuno, Y. Emura, M. Baur, J. Kohli, K. Ohta, and T. Shibata, *Genes Dev.* **11,** 876 (1997).
81. S. L. Goldman and H. Gutz, in "Mechanisms in Recombination" (R. F. Grell, ed.), p. 317. Plenum, New York, 1974.
82. P. Thuriaux, *Mol. Gen. Genet.* **199,** 365 (1985).
83. M. Tavassoli, M. Shayeghi, A. Nasim, and F. Z. Watts, *Nucleic Acids Res.* **23,** 383 (1995).
84. S. Prinz, A. Amon, and F. Klein, *Genetics* **146,** 781 (1997).
85. A. H. Z. McKee and N. Kleckner, *Genetics* **146,** 797 (1997).
86. A. H. Z. McKee and N. Kleckner, *Genetics* **146,** 817 (1997).
87. R. E. Malone, S. Bullard, M. Hermiston, R. Rieger, M. Cool, and A. Galbraith, *Genetics* **128,** 79 (1991).
88. H. Schmidt, P. Kapitza, and H. Gutz, *Curr. Genet.* **11,** 303 (1987).
89. A. Gysler-Junker, Z. Bodi, and J. Kohli, *Genetics* **128,** 495 (1991).
90. M. Q. Zhang and T. G. Marr, *Nucleic Acids Res.* **22,** 1750 (1994).
91. Y. F. Li, M. Numata, W. P. Wahls, and G. R. Smith, *Mol. Microbiol.* **5,** 869 (1997).
92. L. C. DeVeaux and G. R. Smith, *Genes Dev.* **8,** 203 (1994).
93. H. Schmidt, P. Kapitza-Fecke, E. R. Stephen, and H. Gutz, *Curr. Genet.* **16,** 89 (1989).
94. D. H. Evans, Y. F. Li, M. E. Fox, and G. R. Smith, *Genetics* **146,** 1253 (1997).
95. Y. Lin, K. L. Larson, R. Dorer, and G. R. Smith, *Genetics* **132,** 75 (1992).
96. Y. Lin and G. R. Smith, *Genetics* **136,** 769 (1994).
97. Y. Lin and G. R. Smith, *Curr. Genet.* **27,** 440 (1995).
98. Y. Lin and G. R. Smith, *Mol. Microbiol.* **17,** 439 (1995).
99. D. Pittman, W. Lu, and R. E. Malone, *Curr. Genet.* **23,** 295 (1993).
100. R. F. Smith, B. A. Wiese, M. K. Wojzynski, D. B. Davison, and K. C. Worley, *Genome Res.* **6,** 454 (1996).
100a. Y. Lin and G. R. Smith, unpublished data.
101. S. Klapholz, C. S. Waddell, and R. E. Esposito, *Genetics* **110,** 187 (1985).
102. C. L. Atcheson, B. DiDomenico, S. Frackman, R. E. Esposito, and R. T. Elder, *Proc. Natl. Acad. Sci. U.S.A.* **84,** 8035 (1987).
103. S. Keeney, C. N. Giroux, and N. Kleckner, *Cell* **88,** 375 (1997).

104. A. Bergerat, B. de Massy, D. Gadelle, P.-C. Varoutas, A. Nicolas, and P. Forterre, *Nature (London)* **386,** 414 (1997).
105. J. M. Gardiner, S. A. Bullard, C. Chrome, and R. E. Malone, *Genetics* **146,** 1265 (1997).
106. R. B. Wickner, *Microbiol. Rev.* **60,** 250 (1996).
107. D. C. Masison, A. Blanc, J. C. Ribas, K. Carroll, N. Sonenberg, and R. B. Wickner, *Mol. Cell. Biol.* **15,** 2763 (1995).
108. A. Sugiyama, K. Tanaka, K. Okazaki, H. Nojima, and H. Okayama, *EMBO J.* **13,** 1881 (1994).
109. N. F. Lowndes, C. J. McInerny, A. L. Johnson, P. A. Fantes, and L. H. Johnston, *Nature (London)* **355,** 449 (1992).
110. N. Nakashima, K. Tanaka, S. Strum, and H. Okayama, *EMBO J.* **14,** 4794 (1995).
110a. R. Ding and G. R. Smith, *Mol. Gen. Genet.*, in press.
111. T. Kogoma, *Microbiol. Mol. Biol. Rev.* **61,** 212 (1997).
112. K. Johzuka and H. Ogawa, *Genetics* **139,** 1521 (1995).
113. M. Ajimura, S.-H. Leem, and H. Ogawa, *Genetics* **133,** 51 (1993).
114. H. Gutz and H. Schmidt, *Curr. Genet.* **9,** 325 (1985).
115. H. Schmidt, *Curr. Genet.* **24,** 271 (1993).
116. R. Egel, D. H. Beach, and A. J. S. Klar, *Proc. Natl. Acad. Sci. U.S.A.* **81,** 3481 (1984).
117. J. Hodgkin, H. R. Horvitz, and S. Brenner, *Genetics* **91,** 67 (1979).
118. S. A. Broverman and P. M. Meneely, *Genetics* **136,** 119 (1994).
119. A. M. Villeneuve, *Genetics* **136,** 887 (1994).
120. A. J. S. Klar, *in* "The Molecular Biology of the Yeast *Saccharomyces:* Gene Expression" (E. W. Jones, J. R. Pringle, and J. R. Broach, eds.), p. 745. Cold Spring Harbor Laboratory, Cold Spring Harbor, New York, 1992.
121. H. Schmidt and H. Gutz, *in* "The Mycotia I: Growth Differentiation and Sexuality" (Wessels and Meinhardt, eds.), p. 283. Springer-Verlag, Berlin and New York, 1994.
122. D. H. Beach and A. J. S. Klar, *EMBO J.* **3,** 603 (1984).
123. R. Egel, *Curr. Genet.* **8,** 199 (1984).
124. G. Thon and A. J. S. Klar, *Genetics* **134,** 1045 (1993).
125. A. Lorentz, L. Heim, and H. Schmidt, *Mol. Gen. Genet.* **233,** 436 (1992).
126. G. Thon and A. J. S. Klar, *Genetics* **131,** 287 (1992).
127. G. Thon, A. Cohen, and A. J. Klar, *Genetics* **138,** 29 (1994).
128. A. J. S. Klar and M. J. Bonaduce, *Genetics* **129,** 1033 (1991).
129. R. Egel, M. Willer, and O. Nielsen, *Curr. Genet.* **15,** 407 (1989).
130. K. Ekwall and T. Ruusala, *Genetics* **136,** 53 (1994).
131. D. G. Catcheside, "The Genetics of Recombination." University Park Press, Baltimore, Maryland, 1977.
132. G. R. Smith, *in* "The Recombination of Genetic Material" (B. Low, ed.), p. 115. Academic Press, New York, 1988.

The Nucleosome: A Powerful Regulator of Transcription

ALAN P. WOLFFE AND
HITOSHI KURUMIZAKA

Laboratory of Molecular Embryology
National Institute of Child Health and
  Human Development
National Institutes of Health
Bethesda, Maryland 20892

I. The Nucleosome Core 381
II. The Nucleosome 390
III. Transcription Factor Access to Nucleosomal DNA 400
   A. TATA-Binding Proteins Do Not Bind to DNA in the Nucleosome Core 401
   B. TFIIIA Access to Nucleosomal DNA 402
   C. Functional Studies of Dinucleosomal Templates—Nucleosome Mobility and Linker Histones 406
   D. Functional Studies of Nucleosomes Using Nuclear and Steroid Hormone Receptors 409
   E. Nucleosomal Arrays: Higher Order Structure and Transcription Repression 417
IV. Concluding Remarks 418
   References 418

Nucleosomes provide the architectural framework for transcription. Histones, DNA elements, and transcription factors are organized into precise regulatory complexes. Positioned nucleosomes can facilitate or impede the transcription process. These structures are dynamic, reflecting the capacity of chromatin to adopt different functional states. Histones are mobile with respect to DNA sequence. Individual histone domains are targeted for posttranslational modifications. Histone acetylation promotes transcription factor access to nucleosomal DNA and relieves inhibitory effects on transcriptional initiation and elongation. The nucleosomal infrastructure emerges as powerful contributor to the regulation of gene activity.

An understanding of the structural and functional properties of the nucleosome is essential for any scientist investigating the biology of the nucleus. The molecular machines controlling transcription, replication, recombination, and repair all utilize DNA that is packaged into nucleosomes.

All of these events occur efficiently within a restrictive environment in which DNA is constrained on the surface of the histones and then progressively wrapped into ever more complex structures. The fact that chromatin is not an impediment to the regulatory machinery reflects the coevolution of DNA with both the architectural proteins that package the double helix and those that utilize it as a substrate. We now know that the histones, nucleosomes, and nucleosomal arrays function as integral components of many regulatory processes. It is probable that the constraint of DNA within chromatin is essential for the control of DNA function during transcription, replication, and repair.

In this review we discuss an approach to nucleosome structure and function that relies on the assembly of specific nucleoprotein structures containing the histones. This line of experimentation builds directly on seminal work from the laboratory of Robert T. Simpson reviewed in this series several years ago (*1*). The aim is to define as accurately as possible important features of these regulatory nucleoprotein structures in terms of the path of DNA, the site and extent of histone–DNA interactions, and their stability and dynamics. It is then possible to alter the components present and to define the structural and functional consequences. Such alterations include removal of histones and the introduction both of posttranslational modification and of mutant histones into nucleosomes. All of these manipulations lead to defined structural transitions and influence the access of nucleosomal templates to transcription factors and the transcriptional machinery.

With experience, it has proved possible to extend the analysis of chromatin structures from mononucleosomal particles to dinucleosomes and nucleosomal arrays. Among other discoveries, this approach has led to the development of a new asymmetric model of the nucleosome (*2*), to the recognition that histones can stimulate transcription (*3*), and that core histone acetylation can facilitate transcription factor access to nucleosomal DNA (*4*). Importantly, histone acetylation relieves inhibitory effects on transcription initiation and elongation (*5*). Other findings include the spontaneous mobility of histone octamers under physiological conditions (*6*). This mobility is restricted by linker histones or HMG1 leading to transcriptional repression (*7, 8*). Internucleosomal interactions also influence transcription with chromatin compaction exerting a strong repressive influence on RNA polymerase initiation and elongation (*9, 10*). These observations provide a structural and functional foundation for understanding the significance of genetic experiments delineating histone and other structural chromosomal proteins as important for gene activation and repression.

I. The Nucleosome Core

The available structures for the core histones (*11–13*), linker histones (*14, 15*) and for the core histones complexed with DNA (*16, 17*) provide an important framework for considering nucleosome structure in solution. The analysis of chromatin structure with nucleases provided the first footprints of nucleoprotein structures. Of particular importance was the definition of a kinetic intermediate in the digestion of chromatin with micrococcal nuclease, known as the nucleosome core, which contains 146 bp of DNA wrapped around the histone octamer. Information about this structure has been obtained largely from crystallographic analysis (*11–13, 16, 17*) and from protein–DNA cross-linking together with chemical and enzymatic cleavage of DNA.

The four core histones, H2A, H2B, H3, and H4, have very selective interactions with each other. Each core histone has a flexible N-terminal tail domain that reaches outside the two superhelical turns of DNA within the nucleosome (see later) and a C-terminal domain that is involved in histone–histone interactions inside nucleosomal DNA. Histone H2A forms a heterodimer with H2B, and H3 forms a heterodimer with H4. The interface between the histones in each heterodimer is very similar and is described as a 'handshake' motif (Fig. 1A). The C-terminal domains of each core histone are predominantly α-helical, with a long central helix bordered on each side by a loop segment and a shorter helix. This overall structure has been termed the 'histone fold' (*11–13*). The helices participate in extensive intermolecular contacts and each of the loops forms a segment of intermolecular β-structure (Fig. 1A) (*12*).

Previous work had found that two molecules of each of the four core histones were present within each nucleosome and that this histone octamer had a tripartite organization, in which a central histone tetramer (H3, H4)$_2$ interacts with two histone dimers (*18*). The tetramer (H3, H4)$_2$ and dimer (H2A, H2B) are stable at physiological ionic strength, whereas the octamer structure is stable only at high ionic strength (~2 M NaCl) or in the presence of polyanions such as DNA. Although the area of the interface between the two (H3, H4) heterodimers is less extensive than that between the (H3, H4) and H2A, H2B) heterodimers (*11*), the latter contacts break apart first in the absence of counterions. The interface between the dimer and the tetramer is more accessible to solvent and is consequently less stable. Disruption of these contacts within the nucleosome has important consequences for transcription through a nucleosome and for the access of trans-acting factors to DNA in the nucleosome (*10, 19*).

The shape of the octamer is that of a wedge, more narrow at the end

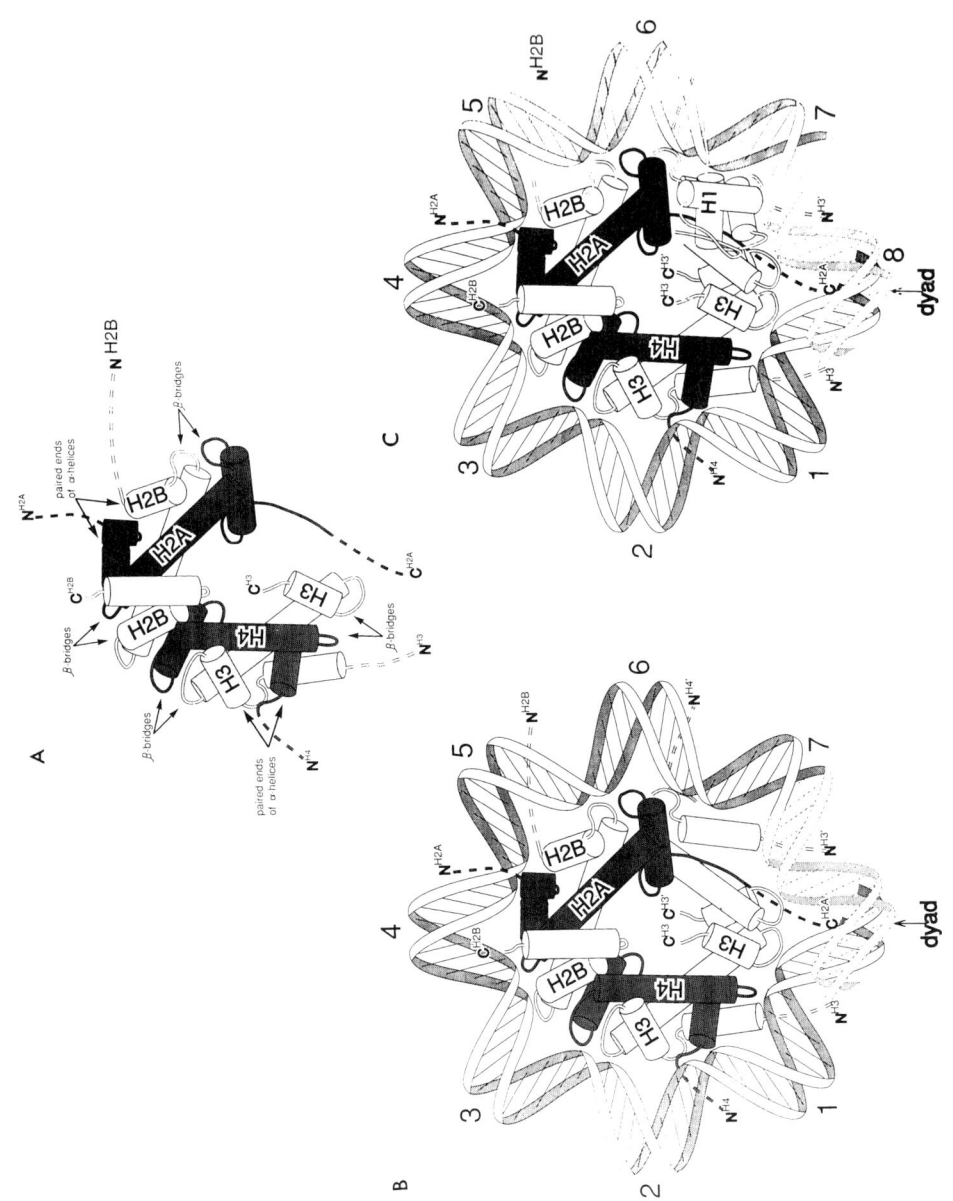

formed by a V-shaped tetramer $(H3, H4)_2$ and more wide where two flattened spheres of (H2A, H2B) dimers are attached. The resolved portion of the octamer structure has several grooves and ridges on its surface. These make a left-handed helical ramp onto which DNA maybe wrapped (12). Within this ramp are eight histone fold motifs containing 16 loop segments (Fig. 1B). Following dimerization of the histones, loop segments are paired to form eight parallel β-bridge segments, two of which are found within each of the histone dimers (H3, H4) and (H2A, H2B). Each β-bridge segment is associated with at least two positively charged amino acids, which are available to make contact with DNA on the surface of the histone octamer (12). The second repeating motif within the nucleosome is assembled from the pairing of the N-terminal ends of the first helical domain of each of the histones in the heterodimers (Fig. 1A). These four "paired-ends-of-helices" motifs also appear to contact DNA (12). Thus each of the four heterodimers within the core can make at least three pseudosymmetrical contiguous contacts with three consecutive inward-facing minor grooves of DNA. The eight parallel β-bridges and the four paired-ends-of-helices motifs provide potential contacts sites for 12 out of 14 helical turns of the nucleosomal DNA, regularly arranged along the ramp on which the DNA is wound; the two remaining DNA turns are bound by additional helix-loop segments of histone H3 on the flanks of this superhelical ramp (Fig. 1B).

Although the crystallization of DNA on the surface of the histone octamer shows exactly where the double helix will make contact with the C-terminal domains of the core histones, it does not provide exact information about the position of the N-terminal tail domains (17). Mirzabekov and colleagues have pioneered protein–DNA cross-linking techniques that both confirm the predictions of C-terminal domain contacts with DNA made from the crystal structure and provide additional information about the N-terminal tails (20–22). Important features of nucleosome structure determined from those studies include the presence of histone H4 cross-linking to DNA over 30 bp to either side of the nucleosomal dyad axis. The N-terminal tail of histone H4 can be cross-linked to DNA at ±1.5 turns from the dyad axis, coincident with the sites of strong DNA deformation (23). A second feature of histone–DNA interaction around the dyad axis of the nucleosome core is strong histone H3

← FIG. 1. Nucleosomal architecture. (A) The histone fold and DNA-binding motifs. The relative juxtaposition of the two histone heterodimers as viewed from the top (i.e., along the superhelical axis of the DNA) is shown. The appropriate positions of the flexible histone tails are shown by broken lines. Note the six regularly spaced domains (double arrows) predicted to be involved in DNA binding. (B) A view down the superhelical axis of the nucleosome core. The helical turns of DNA are numbered relative to the dyad axis. (C) One potential position for the linker histone H1 globular domain within the nucleosome.

cross-linking (*24*). Toward the periphery of the nucleosome, histone H4 contacts DNA over 60 bp from the dyad (*20*), this demonstrates that the (H3/H4)$_2$ tetramer contacts DNA as it exits and enters the nucleosome core.

The (H2A, H2B) dimer interacts with DNA both around the dyad axis and at the periphery of the nucleosome. Histone H2A is unique among the core histones in having both an N- and C-terminal basic tail. The C-terminal tail binds to DNA around the dyad axis (*25, 26*). The N-terminal tails of histone H2B and the tail of histone H2A contact DNA at the periphery of the nucleosome, as far as 80 bp from the dyad axis (*20*).

It is important to recognize that the contacts made by the histone tail domains as determined using a mononucleosome core may change following inclusion of additional proteins such as linker histones and HMGs, and in the presence of adjacent nucleosomes within an array (see later). The organization of DNA in the nucleosome core has been the subject of much speculation (*27*). Fortunately the development of the hydroxyl radical as a DNA cleavage reagent by Tullius and colleagues allowed a direct approach to resolving points of controversy (*28*). Hydroxyl radical cleavage of nucleosome cores reveals that the structure of DNA is different when it is wrapped around the histone octamer compared to when it is free in solution (*29–31*). Importantly, on a longer DNA fragment associated with a histone octamer, histone–DNA contacts extend over at least 160 bp, suggesting that two full turns of DNA may be wrapped around the core histones. Within the nucleosome core, three turns of DNA at the center of nucleosomal DNA (dyad axis) have a helical period of 10.7 bp/turn, whereas outside this region the helical period is 10.0 bp/turn. This DNA structure is ideally suited to match the regularly spaced sites of potential contact on the histone octamer surface, including the three turns of 10.7 bp/turn at the dyad axis (*12*). Thus DNA in the nucleosome core has a tripartite organization comparable to that of the histone octamer. A similar conclusion has been made with the use of UV-induced pyrimidine dimerization as a probe (*32*). These differences in helical periodicity of DNA in the nucleosome may help contribute to determining exactly where along a DNA molecule the histone octamer will prefer to bind (but see Ref. 33). Removal of the N-terminal tail domains of the core histones has no influence on the extent or organization of DNA as detected by hydroxyl radical cleavage (*31*). The (H3, H4)$_2$ tetramer in isolation has the capacity organize the central 120 bp of DNA bound by the histone octamer in an identical way (Fig. 2) (*31*).

Hydroxyl radical cleavage of nucleosome cores containing specific DNA sequences provides information on two concepts that had been proposed from constraints imposed on DNA by the histone core. First, DNA curvature was suggested to be a major determinant of nucleosome positioning (*34*). This is presumed to be a consequence of the fact that DNA is wrapped

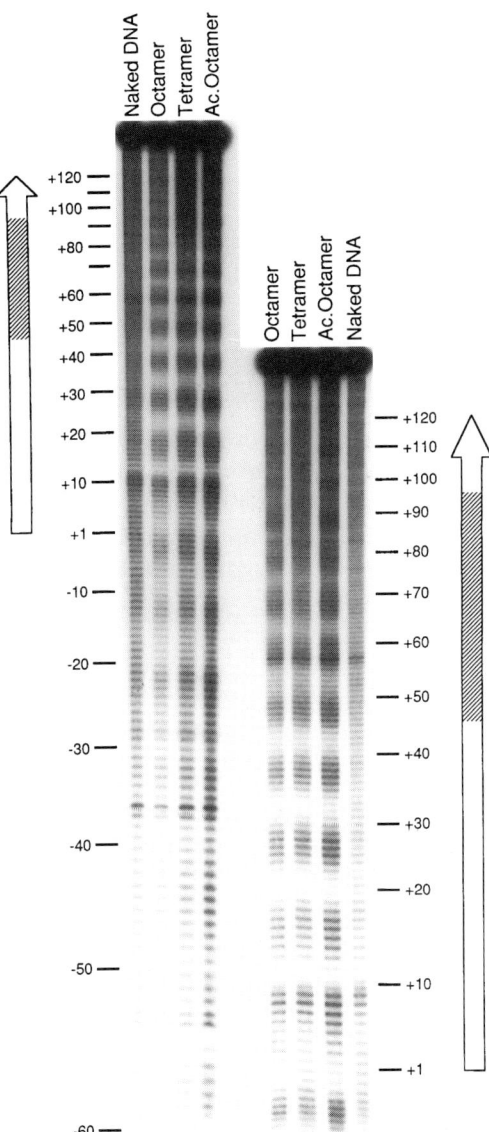

FIG. 2. Hydroxyl radical footprinting of a single histone octamer, a tetramer, and a hyperacetylated octamer from HeLa cells reconstituted by salt dialysis with the *X. borealis* 5S DNA nucleosome positioning sequence. The arrow indicates the position of the 5S RNA gene, and the hatched box the site of TFIIIA binding; the numbers indicate the distance from the start site of transcription +1. An autoradiogram is shown of cleaved DNA fragments as indicated after resolution on a 6% polyacrylamide gel containing 7 M urea in 90 mM Tris borate, EDTA (pH 8.3) buffer. (See Ref. *112* for details.)

around the histone core in a shallow helical path with one complete turn comprising only 80 bp. Second, the helical periodicity of nucleosome-wrapped DNA is different from that in free DNA. Such a difference in helical period might resolve the so-called linking-number paradox (27). This apparent discrepancy, that 1.75 turns of DNA around the nucleosome lead to only one DNA supercoil, had been explained by a change in helical periodicity from 10.5 bp/turn in solution to 10.0 bp/turn over the 146 bp of DNA believed to be associated with the histone core (27). Both of these concepts require alterations to the structure of DNA in a nucleosome compared to its state free in solution.

DNA curvature is clearly seen to have a role in directing the positioning of the histone octamer with respect to specific sequences found within both the *Xenopus borealis* 5S rRNA gene (Fig. 3) and the *Xenopus laevis* vitellogenin B1 promoter (Fig. 4). The hydroxyl radical detects variations in minor groove width (35). Such variations are apparent in the hydroxyl radical cleavage pattern of naked DNA containing either the 5S RNA gene from *X. borealis* or the vitellogenin B1 gene from *X. laevis*. Quantitative analysis indicates a periodic reduction in minor groove width every 10–11 bp (Fig. 3;

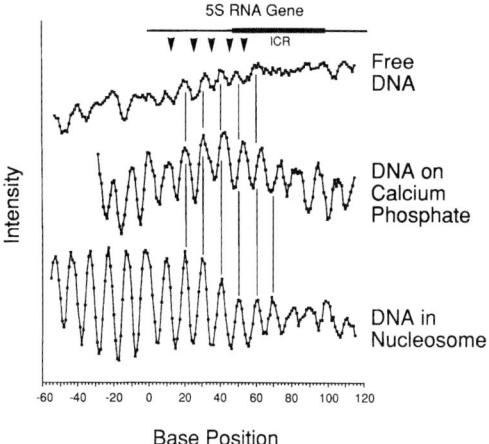

FIG. 3. Structure of DNA in a nucleosome, on a crystal surface (calcium phosphate), and free in solution as determined by hydroxyl radical cleavage. Densitometry scans are shown for 5S DNA cleaved with hydroxyl radical under the conditions indicated. Vertical lines are to indicate common structural features over 5S DNA (the 5S RNA gene is shown schematically at the top of the figure). Arrowheads indicate regions of narrow minor groove width (detected in free DNA) that have a periodicity of 10–11 bp, indicative of intrinsic curvature. This structural feature is maintained and exaggerated in nucleosomal DNA and when DNA is bound to calcium phosphate.

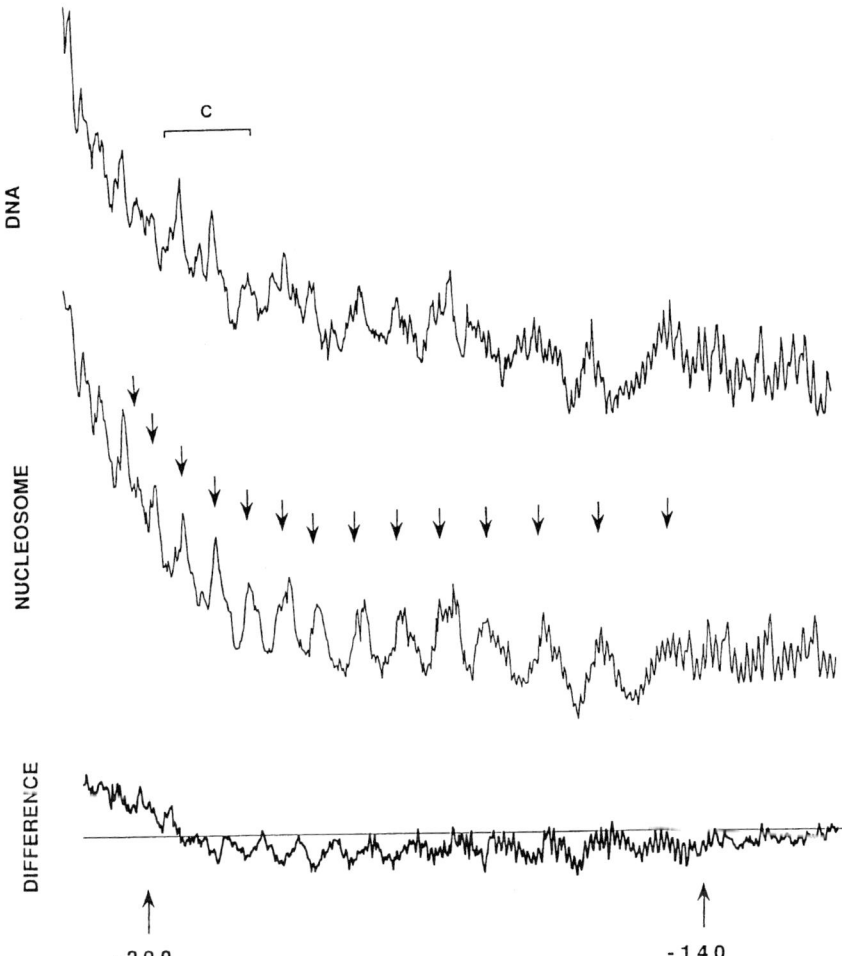

FIG. 4. Densitometric analysis of the hydroxyl radical analysis. Cleavage pattern for the lower strand is shown. The first two traces represent the cleavage pattern for free and reconstituted DNA (0.8, octamer), respectively. The lower trace represents the difference between the two previous scans. The small arrows mark the position of the maximal hydroxyl radical cleavage of the nucleosomal DNA. The borders of the nucleosome are indicated by the arrows on the lower scan. The bracket on the DNA scan (labeled C) indicates the highly conserved sequence between the four *Xenopus* and the chicken vitellogenin II genes.

free DNA trace) and indicates that these DNA sequences bend toward the narrowed grooves. More importantly, the features present in naked DNA are retained and exaggerated in the hydroxyl radical cleavage pattern of the nucleosome (Figs. 3 and 4, compare nucleosome to DNA data sets). Thus, positions of reduced minor groove width in the naked DNA, located where sequence-directed positioning elements are expected, are also places where the minor groove is narrowed when the DNA is bent around the histone core. Comparison of sequences found at these positions reveals few similarities, except for some A + T-rich character. Our results imply that these different sequences can adopt similar secondary structural conformations. The hydroxyl radical provides a simple means to detect structural elements in free DNA that are likely to be functional in the positioning of a nucleosome (2, 30). Experimental tests demonstrate the capacity of regions of DNA curvature to move the translational position of the histone octamer with direct consequences for the access of DNA-binding proteins to nucleosomal templates (36).

The linking-number paradox refers to the fact that DNA is wrapped in approximately two superhelical turns around the core histones in the nucleosome core, although measurements of topological change would be more consistent with a single superhelical turn (27, 37). With the detailed information hydroxyl radical cleavage provides on the helical periodicity of DNA in the nucleosome and in solution, the structural changes that occur in DNA on nucleosome formation can be evaluated. Using mathematical formalism, the linking-number paradox represents the discrepancy in the change in linking number (ΔLk) observed on nucleosome formation (≈ 1) and that expected from knowledge of the structure of the nucleosome (≈ 2). The equation most often used to understand this process is

$$\Delta Lk = \Delta Tw + \Delta Wr, \qquad (1)$$

where ΔTw is the change in twist and ΔWr is the change in writhe experienced by the DNA (38). The experimental determination of hydroxyl radical cleavage rate gives information on the helical repeat in the local frame of reference [i.e., with respect to the surface of the nucleosome core (38)]. The helical repeat of DNA in the local frame is related to the quantity Φ, called the winding number, defined as the number of times that a vector associated with the DNA backbone rotates past the surface that is the reference frame (38). The helical repeat h, defined as the number of base pairs per unit winding number, or $h = N/\Phi$, is the quantity experimentally measured. Here, Φ is not the same as the quantity Tw in Eq. (1). They are simply related, though, by Eq. (2) (39):

$$Tw = STw + \Phi, \qquad (2)$$

where STw, the surface twist, is calculated from the geometry of the system. The corresponding equation for relating the changes in these quantities that occur on nucleosome formation from DNA in solution is

$$\Delta Tw = \Delta STw + \Delta \Phi. \tag{3}$$

ΔSTw, the change in surface twist on forming the nucleosome, is calculated to be -0.19 (38).

The average helical repeat of DNA bound in a nucleosome is 10.18 ± 0.05 bp/turn, a value that agrees very well with the average helical periodicity of random sequence nucleosomes inferred by sequence analysis (40). Substituting this value into Eq. (3) along with the value for ΔSTw for the nucleosome and the value we determined for the helical repeat of DNA bound to a precipitate of calcium phosphate, and considering that there are 146 bp of DNA in the nucleosome core, we obtain

$$\Delta Tw = \Delta STw + \Delta(N/h) = 0.19 + (146/10.18 - 146/10.49) = 0.23. \tag{4}$$

The change in writhe calculated for DNA assembling onto the nucleosome is -1.65 (38), so, using Eq. (1), the change in linking number for formation of a nucleosome is 1.42 ± 0.13 (assuming an error of ± 0.1 in the calculated change in writhe).

Although this calculation was originally performed for DNA in a nucleosome containing the *X. borealis* somatic 5S rRNA gene (30), subsequent experiments have established that the histone octamer exerts a dominant constraint on the structure of DNA in the nucleosome core (31, 41). Thus the calculated change of linking number of 1.42 still represents a major discrepancy from the experimentally measured value of 1.01 found in nucleosomal array (37).

The observed alteration in the helical periodicity of DNA on nucleosome formation does not, by itself, completely explain the linking-number paradox as currently formulated (38). This is entirely due to the difference in the helical periodicity of nucleosomal DNA of 10.0 bp/helical turn assumed in theoretical work (38) and the actual helical periodicity of 10.18 bp/turn measured directly. One reason for this discrepancy may be that in nucleosomal arrays (37) sources other than nucleosomal writhe and surface twist exist. Moreover, it has been suggested that the nucleosome might not be best represented as a simple cylinder and that alternative paths of DNA may contribute to the experimentally observed change in linker number (42). The calculation above considers only the length of DNA found in the nucleosome core particle as isolated by micrococcal nuclease digestion of chromatin. More than 146 bp of DNA may be in contact with the histone core in nucleosome arrays. Inspection of the hydroxyl radical footprints (Figs. 2 and 3) reveals a modulation of cleavage extending more than 80 bp from the center of dyad symmetry of the nucleosome. These extended interactions are consistent with histone–DNA cross-linking (20) and controlled micrococcal nuclease digestion of chromatin (24), but of course are not resolved in crystallographic studies of nucleosome cores containing only 146 bp of DNA (*17*).

The path of DNA as it wraps around the core histones is not uniform. The crystal structure of the nucleosome core (*16, 17*) reveals that the 146 bp of DNA within it are bent in 1.75 superhelical turns around the histones. At the dyad axis, the minor groove of DNA is orientated toward solution. DNA is more severely distorted about ±1.5 and ±3.5 turns from the dyad axis (Fig. 5). These sites are preferentially reactive when probed by enzymes sensitive to DNA distortion, such as the integrase protein of human immunodeficiency virus (Figs. 5 and 6) (*43*). The sites ± 1.5 turns from the dyad axis are also sensitive to chemical cleavage by reagents that recognize unstacked base pairs, such as singlet oxygen (*44*). These distortions delimit the different regions of distinct helical structure within the nucleosome core. It might be anticipated that the necessity to distort DNA at these sites might also contribute to the positioning of histones with respect to DNA sequence.

Finally, it is clear that we are still far from a full appreciation of the determinants of core histone interaction with DNA. The triplet repeat sequences $(CTG)_n$ and $(C\text{-methyl}CG)_n$ are associated with the human diseases myotonic dystrophy and Fragile X mental retardation, respectively (*45, 46*). Short repeats of these triplets containing 6–10 reiterations generate the strongest known interactions between core histones and DNA in the nucleosome core (*47, 48*). Why this should occur is not yet known. However, two current hypotheses are that (1) there are sequence-specific interactions between the core histones and these triplet repeats, and (2) that these particular triplets can best accommodate the writhe imposed by the histones on DNA at the dyad axis. Further experiments will explore these possibilities.

II. The Nucleosome

Micrococcal nuclease cleavage sites are initially spaced at an average distance of 180–190 bp within most chromatin isolated from somatic cells; this distance is described as the "nucleosomal repeat." Characterization of the proteins associated with DNA in discrete histone–DNA complexes during digestion of chromatin with micrococcal nuclease revealed that a fifth histone, known as the linker histone (e.g., H1, H5, H1°), was lost when DNA was progressively reduced in length from that of the nucleosomal repeat to the 146 bp within the nucleosome core (*49*). Detailed analysis revealed that linker histone inclusion leads to the more stable protection of an additional 20 bp of linker DNA immediately contiguous to the 146 bp in the nucleosome core (Fig. 7), and because only a single molecule of linker histone existed in the nucleosome, the particle could not be genuinely symmetric (*50*). Neverthe-

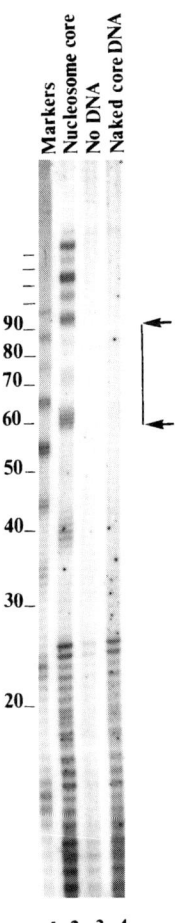

FIG. 5. HIV integrase directs integration to sites of severe DNA distortion within the nucleosome core. Integration products made with or without a heterologous target (nucleosome cores or naked core DNA) were resolved in a sequencing gel. Lane 1 shows markers of DNA fragments end-labeled with [γ^{32}P]ATP and T4 polynucleotide kinase following DNase I digestion of chicken erythrocyte chromatin. Lane 2 shows integration into nucleosome cores; lane 3, self-integration products in the absence of target DNA; and lane 4, integration into naked DNA extracted from nucleosome cores. The numbers represent actual distance between the integration site and the DNA 3' terminus. The arrows indicate the sites of preferred integration 1.5 turns to either side of the dyad axis of the nucleosome core at which integration is favored; the line between them indicates the 32 bp of DNA that includes the dyad axis at which integration occurs less frequently.

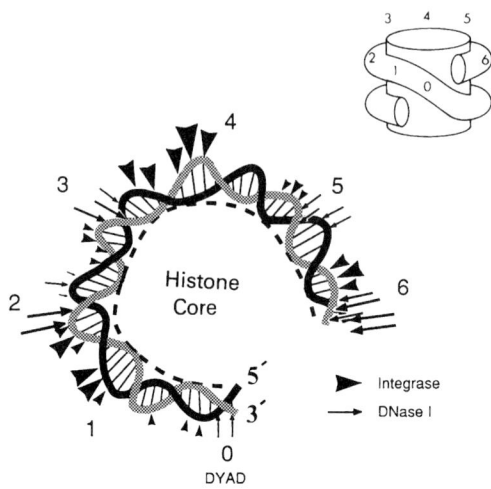

Fig. 6. HIV integrase directs integration to sites of severe DNA distortion within the nucleosome core. Sites of integration (arrowheads) are shown for a single superhelical turn of DNA within the nucleosome. Arrows indicate sites of DNase I cleavage. The inset shows the positions of the numbers on the surface of the nucleosome.

less, it was suggested that this additional nuclease protection occurred symmetrically, with 10 bp of linker being protected to either side of the nucleosome core and with the single molecule of linker histone lying quasi-symmetrically across the dyad axis (51–53). The exact role of the linker histone in generating the 166 bp of protection from micrococcal nuclease is questionable because (1) the core histones alone can generate patterns of protection extending to 166 bp (24, 54) and (2) other proteins structurally unrelated to linker histones, such as HMG1, can generate patterns of protection from micrococcal nuclease digestion identical to those of histone H1 (8, 55). It is therefore quite possible that allosteric changes in core histone interactions with DNA induced by linker histones or other proteins that bind to nucleosomes might lead to protection from micrococcal nuclease cleavage.

Linker histones have three domains: N- and C-terminal tails flank a central globular domain. The globular domain alone is sufficient to direct the protection of linker DNA (44). Therefore a model for linker histone interaction was proposed that would require the globular domain to make three contacts with DNA, one with each DNA strand entering and exiting the nucleosome and one with the minor groove of the DNA, which is orientated directly away from the nucleosomal surface at the dyad axis (52). Consistent with this hypothesis, DNase I footprinting of dinucleosomes containing linker histones revealed protection from cleavage at the nucleosomal dyad (53). It was not,

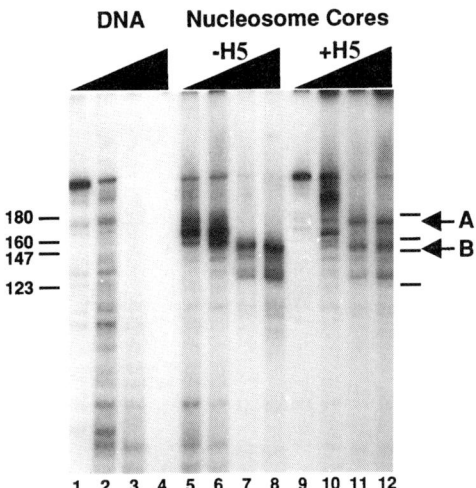

FIG. 7. Association of linker histone with reconstituted 5S nucleosome cores protects an extra 20 bp of linker DNA. Naked 5S DNA (0.5 μg, lanes 1–4) or reconstituted 5S nucleosome cores (0.125 μg of DNA, lanes 5–12) in the absence (lanes 5–8) or presence (lanes 9–12) of 40 ng of histone H5 were digested with 0.075, 0.15, 0.3, and 0.6 unit of micrococcal nuclease (5 min, 22°C) as indicated by the filled triangles above the lanes. Products of digestion were labeled with [γ-^{32}P]ATP and analyzed by native gel electrophoresis. Horizontal bars indicate size markers, and arrows A and B indicate the 168- and 147-bp products of digestion, respectively.

however, clear whether this protection comes from the linker histone molecule or from the second nucleosomal unit of the dinucleosome.

The structure of the globular domain of linker histone H5 (GH5) shows remarkable similarity to that of the DNA–binding domain of transcription factor HNF-3 (56, 57). HNF-3 binds across the major groove of DNA as a monomer bending DNA toward itself. Both HNF-3 and GH5 consist of a bundle of three helices attached to a three-stranded antiparallel β-sheet (Fig. 8). Moreover, the key DNA-binding amino acid residues of HNF-3 are either conserved between HNF-3 and H5, or changed to positively charged residues in the latter, suggesting that the globular domain of H5 will contact nucleosomal DNA in the same way. This is consistent with the known contact of His-25 of histone H5 with DNA in the nucleosome (58) and with the protease sensitivity of a conserved aromatic residue on the opposite side of the linker histone globular domain in the nucleosome (59). Because the globular domain of H1/H5 is expected to interact with at least one other DNA molecule (52), it was proposed that it should have an auxiliary DNA-binding site at a distance from α-helix 3 (which corresponds to the principal DNA-binding motif of HNF-3) (60). Such an auxiliary site could include a loop join-

FIG. 8. Putative DNA-binding element of the globular domain of histone H5. A ribbon scheme of the protein structure and a conserved aromatic residue anchoring the β-structure are shown. The three-dimensional structure matches the structure of DNA-binding protein HNF-3 extremely well. Elements shared by both proteins are implicated in HNF-3-DNA binding. α-Helix 3 is the principal DNA-binding element of HNF-3; throughout the entire H1/H5 family this helix invariably harbors three positively charged residues (Lys-69, Arg-73, and Arg-74 in H5), shown here as dark stars. Residues shown by white stars are as follows: (C) Ser-92 corresponds to phosphate-binding Ser-191 of HNF-3 and is conserved throughout most of the linker histone family, although in a few cases it is substituted to arginine; (B) Arg-47 (in H5s and H1°s) is substituted to leucine in the other members of the family and corresponds to the DNA backbone-binding Leu-142 residue of HNF-3; (D) Arg-94 (lysine in some other linker histones) corresponds to the DNA phosphate-binding Trp-193 residue of HNF-3; (G) a triangle marks residue His-25 (in the immediate vicinity of Tyr-28, which is cross-linked to DNA.

ing the α-helices 1 and 2 (*61*). The strongest support for the idea of multiple DNA-binding sites comes from the observation that the linker histones (either intact or globular domains alone) bind to *naked* DNA molecules cooperatively by constraining two double helices next to each other and stacking linker histones between them like the sleepers between railway tracks (*62*). It is, however, possible to have selective association of linker histones with nucleosomal DNA in the absence of this second putative DNA-binding site (*63*). Thus, it remains unclear whether it is a single linker histone molecule or a dimer that ties the two DNA molecules together. Indeed, in crystals GH5 dimerizes and express its putative DNA-binding α-helical domains toward the outside. The spacing between these domains is compatible with that between the two DNA molecules in a binary GH5–DNA complex (*14, 62*). The possibility of further interactions of linker histones with linker histone molecules within adjacent nucleosomes, or with the DNA in adjacent nucleosomes, may contribute to the stabilization of the higher order chromatin structures assembled by nucleosomal arrays.

The new information on the structure of the globular domain of a linker histone imposes several constraints on existing models of the chromatosome. None of the new structural data (*14, 56*) suggest a strong interaction of the globular domain with the minor groove of DNA, such as must be envisaged if the linker histone were bound directly over the dyad. One could hypothesize that the globular domain interacts with both DNA linkers at a distance from the central DNA superhelical turn and protects two 10-bp segments of DNA in a quasisymmetric fashion. However, this would place the globular domain too far away from the core histone octamer to be compatible with neutron scattering (*64*) and core histone–linker histone cross-linking data (*65*).

An alternative hypothesis is that the single molecule of the linker histone interacts asymmetrically with DNA in the nucleosome. New evidence supporting this model comes from studies of nucleosomes incorporating specific DNA sequences (*2, 7, 8, 55, 66–70*). Linker histones bind preferentially to a *X. borealis* somatic 5S rRNA gene associated with an octamer of core histones rather than to naked DNA (Fig. 9) (*53*). This preferential binding requires free linker DNA and results in the protection of an additional 20 bp of the linker DNA from micrococcal nuclease digestion. Importantly, this additional linker DNA is asymmetrically distributed to either side of the nucleosome core. There are 15 bp protected to one side of the 5S nucleosome by H5, whereas on the opposite side of the nucleosome only 5 bp are protected. This evidence alone could be accounted for by changes in core-histone DNA interactions or simply by sequence selectivity in DNA cleavage by micrococcal nuclease. Moreover, formally, asymmetric protection of linker DNA could still occur through interaction of the histone H5 over the nucleosomal dyad. However, incorporation of linker histones cause no additional protection of DNA in the 5S nucleosome from cleavage by hydroxyl radical or DNase I cleavage (*66*), i.e., outward-facing minor grooves of the 5S nucleosome (including the one on the dyad axis) remain exposed. The asymmetric protection of linker DNA and lack of protection at the dyad are consistent with the asymmetric linker histone molecule interacting predominantly with one end of nucleosomal 5S DNA close to the surface of the histone octamer. More recent observations have shown that the linker histone requires the (H2A, H2B) dimer to be present before stable association with nucleosomal DNA can occur (*67*). This is consistent with the existence of close linker histone contacts with histone H2A (*65*), and indicates a lack of strong independent interactions of the linker histone with the DNA–(H3, H4)$_2$ tetramer complex at the dyad axis.

The globular domain of histone H5 alone has also been shown to confer

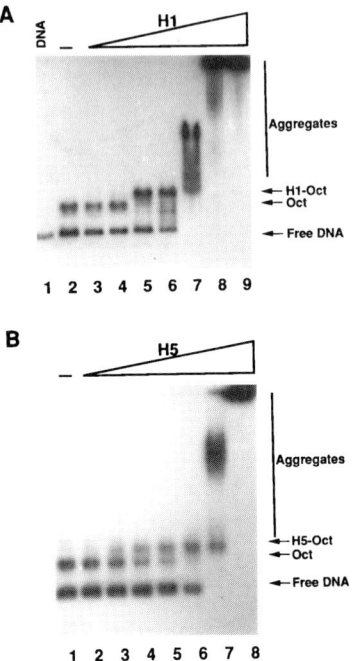

FIG. 9. Preferential binding of linker histones to reconstituted 5S nucleosome cores, rather than to naked DNA. (A) Binding of calf thymus histone H1 to 5S nucleosome cores. Reconstituted nucleosome cores were mixed with various amounts of H1 and analyzed by nucleoprotein gel electrophoresis. Each lane contained 50 ng of DNA (~0.3 pmol), half of which was in the core particle. Lane 1, Naked HpaII–DdeI 5S DNA fragment only; lanes 2–9, 50 ng (DNA content) of reconstituted nucleosome cores (~0.3 pmol) in linker histone binding buffer with 0, 0.1, 0.2, 0.4, 0.8, 1.5, 3.0, and 5.0 pmol of H1, respectively. Oct, octamer. (B) Binding of chicken erythrocyte H5 to reconstituted 5S nucleosome cores. Lanes 1–8, 50 ng (DNA content) of reconstituted nucleosome cores (~0.3 pmol) were mixed with 0, 0.05, 0.1, 0.25, 0.5, 1.0, 2.0, and 5.0 pmol of H5 before resolution in a nondenaturing gel.

asymmetric protection of linker DNA in the 5S nucleosome (67). Moreover, histone–DNA cross-linking by two independent methodologies, including the use of photocross-linking reagents introduced at specific sites within the nucleosome (Fig. 10), reveals a single major contact around 65 bp from the nucleosomal dyad to be made by the globular domain with 5S DNA (Fig. 11). Modeling indicates that the major groove of this portion of nucleosomal DNA is largely devoid of spatial constraints imposed by the core histones, hence the globular domain of the linker histone could adopt a variety of positions between 64 and 70 bp from dyad axis (2, 67). Our favored model fits the globular domain close to the dyad axis (but from the inner side of the DNA

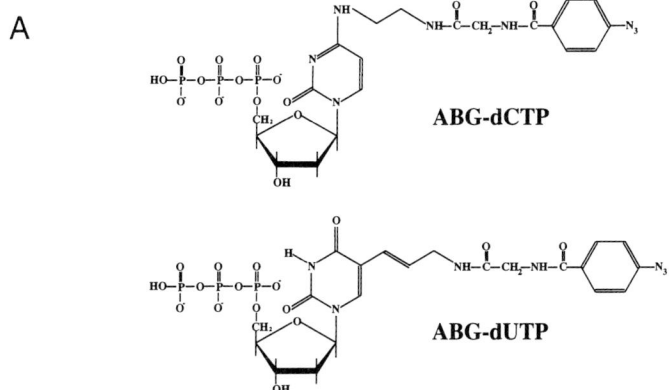

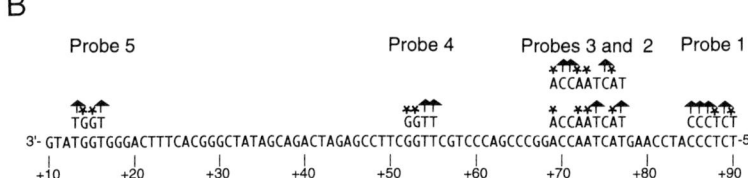

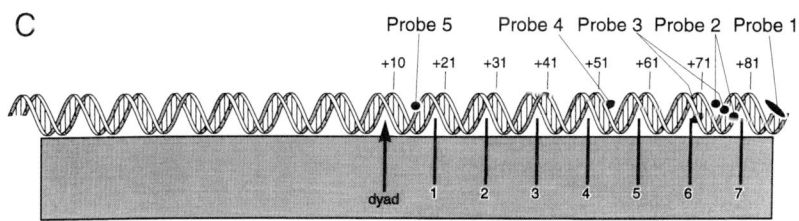

FIG. 10. Design of the photoactive probes. (A) Photoactive DNA precursors. (B) Positions of the photoactive bases (arrows) and ^{32}P-labeled nucleotides (asterisks) within the bottom strand of the 5S RNA gene. The photoreactive group is attached to the bottom of the major groove by a 1.3-nm linker. (C) Approximate positions of the photoactive azido moieties with respect to the surface of the histone octamer (shaded) within nucleosomes.

superhelix). The "hole" in the octamer filled by the globular domain is framed by the H2A and H3 polypeptide chains (Fig. 1C). An alternative location would have the globular domain right at the top edge of the nucleosome. Any of these positions is consistent with the protein–protein interactions of linker histones with H2A (65), the lack of protection by linker histones across the dyad axis of the 5S nucleosome (66), and the known dimensions and the

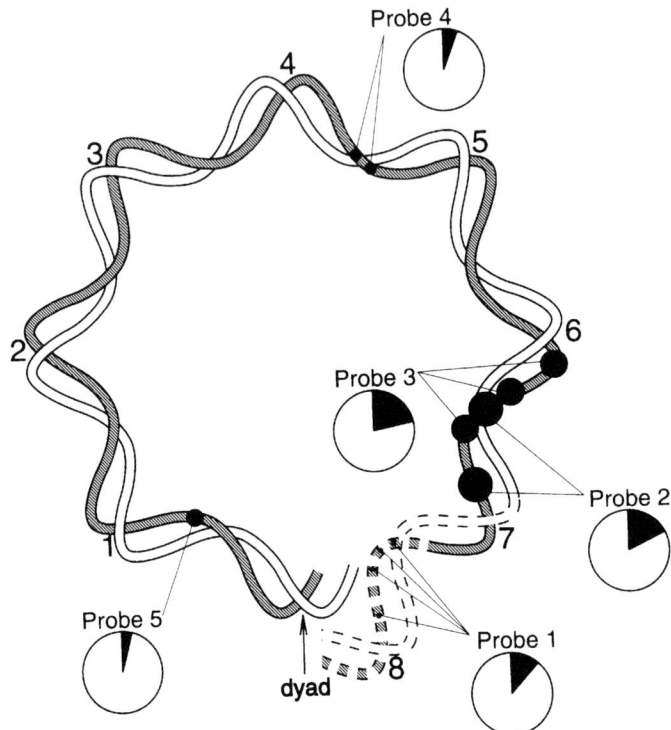

FIG. 11. Relative efficiency of photocross-linking. Showing positions of photoactive nucleotides with the 5S nucleosome (●). The sizes of the circles correspond to the average GH5 cross-linking yield per photoactive base of the probe. The pie diagrams show the ratio of GH5 (black) and total core histone (white) cross-linking.

putative HNF-3-like mode of interaction of the globular domain of histone H5 with DNA (56). An inward location seems to be compatible with the extensive protection on the linker DNA from micrococcal nuclease in the mononucleosome (67), the extended wrapping of linker DNA released by DNase I and hydroxyl radical cleavage of dinucleosome templates containing linker histones (Fig. 12) (7), with the reduction of the Stokes radius of the nucleosome on binding of the globular domain (71), and with the capacity of HNF-3 to bend DNA toward the protein (56). Although most of the published evidence at this time utilizes mono- or dinucleosomal templates containing the X. borealis somatic 5S rRNA gene, other nucleosomes reconstituted on specific DNA sequences show an asymmetric association of linker histones (70). Nevertheless, it remains possible that linker histone association

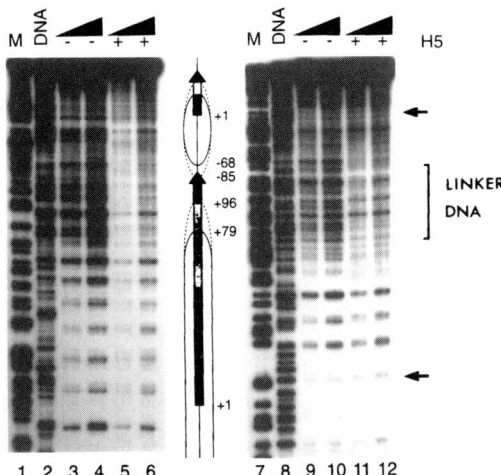

FIG. 12. DNase I footprinting of dinucleosomes containing histone H5. A bound or unbound H5 dinucleosome was prepared and digested with DNase I. Individual complexes were isolated by nucleoprotein agarose (0.7%) gel electrophoresis. DNA from these complexes was isolated and analyzed by denaturing polyacrylamide (6%) gel electrophoresis. Lanes 1–6, The 5' end-radiolabeled noncoding strand of the 5S RNA gene is used as a template. Lane 1 shows the G-specific cleavage reactions used as markers. Digestion of naked DNA (lane 2), of dinucleosoms (lanes 3 and 4), and of dinucleosomes containing H5 (lanes 5 and 6) is shown, as indicated at the top. Filled triangles indicate increasing DNase I digestion. Small arrows indicate the position of the axis of dyad symmetry of nucleosome. The large vertical arrows show the location and orientation of the 5S RNA gene. Gray boxes show internal control region. Solid and dotted ovals indicate the predominant regions contacted by the nucleosome cores and chromatosome, respectively. The position of linker DNA is indicated. Lanes 7–12 are the same as in 1–6, except the coding strand is labeled.

relative to the nucleosome core may well change as long nucleosomal arrays are assembled and compacted into higher order structures (24).

An explanation for the protection from nuclease cleavage by linker histones within mixed-sequence dinucleosomes (53) comes from experiments examining the compaction of linker DNA within dinucleosomes (72, 73). These results indicate that dinucleosomes containing linker histones can fold together more readily than linker histone-deficient particles. Thus the adjacent nucleosome could occlude access to the dyad axis by DNase I in the presence of histone H1. The effective compaction of adjacent dinucleosomes (72, 73), the extended interaction of 160 bp of DNA with the histone octamer (20, 30), and the asymmetric placement of the linker histone (2, 67), all favor the linker DNA being supercoiled between adjacent nucleosomes (74). This linker DNA organization would support a solenoid model for chromatin higher order folding, in which the core DNA together with the linker DNA form a continuous superhelix (24). Oligonucleosomes with proteolytically removed tails do not fold efficiently into compacted chromatin structures (75), implying that the core histone tail domains have a role in the formation of higher order chromatin structures. More specifically, variants of histone H2B exist in which the N terminus of H2B is considerably extended in length and interacts stably with linker DNA over 80 bp from the nucleosomal dyad (76), likewise, a variant of histone H2A with an extended C-terminal tail interacts with linker DNA (77). The interaction of histone H2B with linker DNA is of particular interest because spatial modeling shows that this requires the linker DNA to be folded or supercoiled between adjacent nucleosomes. Thus a role for the basic histone tails in shielding some of the negative charge of the linker DNA and folding of linker DNA in the chromatin fiber seems probable (78).

The preferential association of the linker histone to one end of the nucleosome generates an asymmetrical particle that may impart a directionality to the folding of the chromatin fiber. This might be propagated by a polar, head-to-tail arrangement of linker histone molecules along the nucleosomal array or chromatin fiber (79). However, the proximity of the linker histone to the surface of the histone octamer within the chromatin fiber (80) makes it unlikely that this type of arrangement could be propagated through direct interactions between the globular domains. The asymmetric positioning of the linker histone is also likely to favor interaction of the extended basic C-terminal tail of the linker histone with the linker DNA. The charge neutralization of the phosphodiester backbone of linker DNA by the basic C-terminal tail of the linker histone has a major role in chromatin compaction (78). Moreover, the removal of the linker histone away from the dyad axis of the nucleosome would eliminate the need for the DNA at the entry and exit of the nucleosome to come into close proximity. This also eliminates a central requirement of certain models of the chromatin fiber that have histone H1 uniquely localized in the center of fiber (81). There is still much remaining to be accomplished in defining the structure of nucleosomal arrays. It is an important task for the future.

III. Transcription Factor Access to Nucleosomal DNA

The assembly of nucleosomes potentially impedes transcription factor access to regulatory DNA. Attempts to examine this issue *in vitro* are complicated by the methodologies used to reconstitute nucleosomes and the typical use of radiolabeled substrates at high dilution (82). A large excess of DNA-binding proteins might also be expected to displace histones from DNA

(83, 84). In the examples discussed here the histones remain associated with DNA in the presence of the transcription factors. In certain instances the presence of the histones actually potentiates transcription (3).

A. TATA-Binding Proteins Do Not Bind to DNA in the Nucleosome Core

The analysis of TATA-binding protein (TBP) association with nucleosomal DNA is of special interest due to the general requirement for this protein for transcription (85, 86). *In vivo*, the access of the basal transcriptional machinery to DNA is impeded by nucleosomes. Most of these studies have made effective use of the yeast, *Saccharomyces cerevisiae*. Histone depletion leads to the derepression of several genes, including the *PHO5* gene (87). However the exact structural basis for transcriptional inhibition has not been determined. In a systematic analysis Patterton and Simpson (88) found that movement of the TATA box into a linker DNA region between two nucleosomes would not derepress transcription of the STE6 gene. In this case the nucleosome spacing was 160 bp, therefore the linker length was short (15 bp). However, other proteins such as micrococcal nuclease or Dam methyltransferase find access to DNA between nucleosomes (89). Patterton and Simpson speculated that the known distortion of DNA in the TBP–DNA cocrystal structure (90, 91) might not be accommodated by the internucleosomal linker and particularly by the flanking nucleosomes. Consistent with these *in vivo* observations, nucleosome assembly represses the transcription process *in vitro* (92–94). However, the prior association of TBP, either alone or together with the basal transcriptional machinery, with a promoter is reported to prevent nucleosome-mediated transcriptional repression under these *in vitro* conditions (94–96). Importantly, the subsequent addition of TBP and the basal transcriptional machinery to a template that has been assembled into nucleosomes will not lead to the recovery of transcriptional competence, suggesting that TBP (as TFIID) cannot bind to a TATA box sequence on the surface of a nucleosome (94).

Nucleosome positioning sequences have been used to rotate the adenovirus major late promoter TATA box relative to the surface of the histone octamer at a single translational position approximately at the nucleosomal dyad (97). In general, association with the core histones under these conditions prevents the association of TBP with the TATA box. In fact, TBP in the presence of TFIIA has been reported to bind to only one rotational position of the TATA box dependent on the presence of a human chromatin remodeling complex SWI/SNF (under conditions such that a 100-fold molar excess of SWI/SNF relative to nucleosomes was present) and ATP (97). We examined TBP access in the presence of TFIIA to a TATA box at defined positions within the nucleosome, making use of nucleosome positioning sequences contained within the X. borealis 5S ribosomal RNA gene (30), to determine the rotational and translational positioning of the TATA box. We made use of both intact core histones and those from which the N-terminal tails have been removed in order to mimic the structural consequences of hyperacetylation of the amino-terminal tails (98, 99). Functionally, either histone hyperacetylation or proteolytic removal of the N-terminal tails can facilitate transcription factor binding to nucleosomal DNA (see later). TBP/TFIIA can bind only to the TATA box within the linker DNA at the edge of the nucleosome following removal of the N-terminal tails or depletion of histones H2A and H2B. This result is entirely consistent with that of Patterton and Simpson (88), and emphasizes the important role of the core histone tails in transcriptional regulation and in controlling the properties of linker DNA. All other sites internal to the nucleosome core prevented TBP binding. More recently we have examined the influence of mutagenesis of histone H3 at sites known to suppress the requirement for the SWI/SNF complex in facilitating transcriptional activation *in vivo* (100, 101). Surprisingly although these H3 mutations destabilize histone DNA interactions within the nucleosome, they do not lead to enhanced binding of TBP/TFIIA to the TATA box within the nucleosome (102). We conclude that the distortion of DNA inherent to TBP binding (90, 91) is incompatible with continued stable association of that DNA on the surface of the core histone octamer. This result is consistent with *in vivo* data (88) yet is in contrast to earlier conclusions from *in vitro* experiments (97). We do not yet understand the basis for the discrepancy.

B. TFIIIA Access to Nucleosomal DNA

Among the best studied examples of trans-acting factors known to interact with chromatin is the association of transcription factor TFIIIA with a 5S ribosomal RNA gene assembled into a nucleosome or subnucleosomal particle. Substantial information exists concerning the essential sequences recognized by the zinc-finger protein TFIIIA within the 5S RNA gene (103). The crystallization of a zinc finger protein complexed with DNA (104) and missing nucleoside experiments (105) have led to these significant insights. These latter results demonstrate that two 10-bp segments within 5S DNA, from +51 to +61 and from +81 to +91 (relative to the start of transcription at +1), are continuously contacted by TFIIIA on both strands of the double helix. However, between these elements, TFIIIA makes contact with only one side of the DNA helix (positions +74 and +75 on the coding strand and +69 on the noncoding strand). These results support models in which TFIIIA is envisioned to wrap around the DNA helix, following the major groove at either end of the complex and then lying on one face of the DNA helix between these contacts (106). TFIIIA has nine zinc fingers, of which three fingers are believed to contact DNA over the 10-bp regions at either end of the complex and three fingers are believed to act as a linker associated with one side of the DNA helix (Fig. 13A). Such an extensive interaction with DNA would appear incompatible with the simultaneous wrapping of this DNA around the histone core.

A large number of functional studies (107–110) demonstrated that prior association of the 5S RNA gene with a histone octamer prevented transcription. These functional studies are in agreement with the observation that the prior assembly of a *X. laevis* 5S RNA gene with a complete octamer of core histones prevented the subsequent binding of TFIIIA (111).

Hydroxyl radical footprinting of the *X. borealis* 5S RNA gene associated with a histone octamer reveals that key contacts that should be made by TFI-

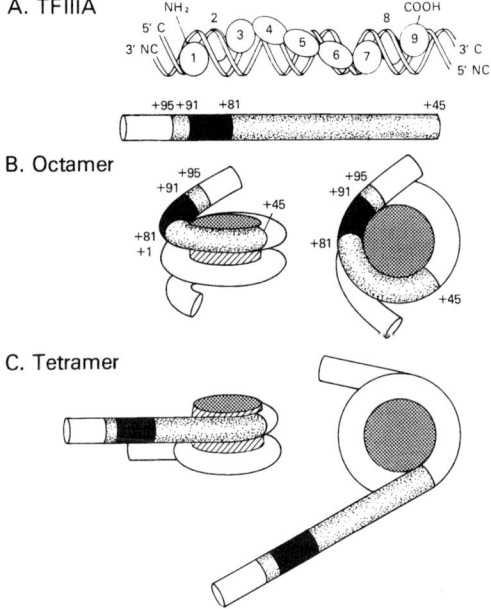

FIG. 13. HISTONE AND TFIIIA BINDING TO A *Xenopus borealis* somatic 5S RNA gene. (A) A representation of TFIIIA bound to a 5S RNA gene is shown. TFIIIA is believed to consist of nine domains, or zinc fingers (see text for details). The binding site for TFIIIA is also represented as a cylinder, and the speckled region is protected from DNase I cleavage. Missing contact analysis reveals the region from +81 to +91 to be an essential contact for TFIIIA (solid cylinder). (B) When associated with a complete octamer of core histones [(H2A/H2B/H3/H4)$_2$] the key contacts from +81 to +91 are in contact with the histone core, and TFIIIA cannot bind to the gene. (C) When associated with a tetramer of histones [(H3/H4)$_2$] the key contacts from +81 to +91 are accessible to TFIIIA, which forms a triple complex with the gene.

IIA in these nucleosomes are bound by histone (Fig. 13B). TFIIIA would not bind to the 5S RNA gene under these conditions. In contrast, the key contacts between +81 and +91 are exposed when the *X. borealis* gene is associated with a histone tetramer (Fig. 13C). TFIIIA interacts quite readily with the 5S RNA gene with equivalent affinity whether or not it is associated with a histone tetramer (Fig. 14) (*19*). Movement of the tetramer–5S DNA contacts to include those at the 3' end of the 5S RNA gene prevents TFIIIA from forming a tertiary complex (*4*). Thus, a histone tetramer is also capable of preventing transcription factor access to DNA.

Acetylated histone octamers assembled on both the *X. borealis* and *X. laevis* 5S RNA gene have contacts with the entire TFIIIA binding site (*112*). The extent of interaction is identical to that of unmodified histone octamers, yet TFIIIA still binds to the genes associated with acetylated histones (Fig. 14). Therefore, the nature of histone–DNA contacts with the acetylated octamer, rather than the length of DNA helix associated with histones, must change. DNase I footprinting shows that TFIIIA binding to the complex of the acetylated octamer with 5S DNA occurs at the 3' end of the binding site between the DNase I hypersensitive sites at +62/+63 and at +95 (Fig. 15). These encompass the key recognition elements for stable protein binding to naked DNA. This is true for the complexes formed by TFIIIA with the acetylated histone octamers and for octamers from which the histone tails have been removed from nucleosomes associated with the *X. borealis* 5S RNA gene (Fig. 16). The footprint of the acetylated octamer outside of the TFIIIA binding site does not change as a consequence of TFIIIA binding. This result, together with the clear differences in mobility on nondenaturing gels (Fig. 14), strongly suggests that no histone displacement from the complex occurs as a

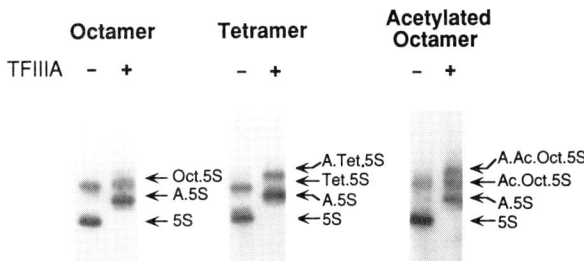

FIG. 14. TFIIIA binds to the *Xenopus borealis* 5S RNA gene associated with an acetylated octamer or an unmodified tetramer, but not to the gene associated with an unmodified octamer. An autoradiograph of mobility-shifted complexes is shown. The presence (+) or absence (−) of TFIIIA (10-fold molar excess) is indicated. All lanes are from the same gel. Histones were prepared from HeLa cells. The octamer and acetylated octamer were reconstituted onto 5S DNA by exchange from long chromatin; the tetramer was assembled onto 5S DNA through salt-urea dialysis.

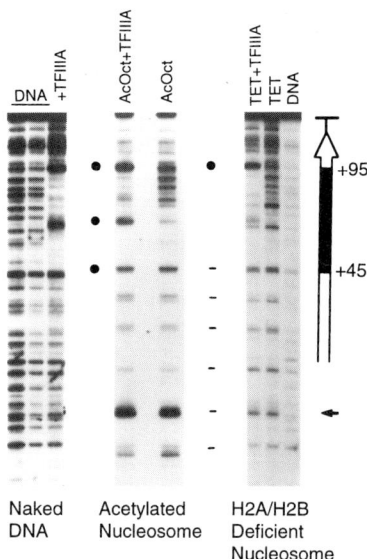

FIG. 15. Histone acetylation of N-terminal tails or H2A/H2B deficiency allows the coexistence of TFIIIA and the acetylated octamer on the 5S RNA gene. Lanes 1, 2 and 8 show naked DNA cleaved with DNase I. Lane 3 shows cleavage of DNA in the presence of TFIIIA; lane 4, in the presence of TFIIIA and an octamer of acetylated histones (AcOct); lane 5, in the presence of an octamer of acetylated histones; lane 6, in the presence of a tetramer [TET; $(H3/H4)_2$] and TFIIIA; and lane 7, DNase I cleavage in the presence of a tetramer alone.

result of TFIIIA binding. Potentially the other components of the transcription complex that bind subsequent to TFIIIA might displace the histones.

The biological relevance of these observations follows from the hypothesis (4) that the molecular machines that remodel chromatin function either by displacing histones H2A/H2B from the nucleosome (113) or through acetyltransferase activities that acetylate the histones (114, 115). Thus deficiency of histones H2A/H2B or acetylation of the core histones might provide a general mechanism of facilitating trans-acting factor access to DNA. A proven event that disrupts both chromatin structure and transcription complexes is transit of the DNA replication fork (116, 117). Each cell division event will lead to competition between chromatin structural proteins and transcription factors for binding to DNA and a new opportunity for reestablishing or altering the state of gene expression. *In vivo* and *in vitro*, chromatin assembly on replicating DNA has been found to occur in stages, with a tetramer of histones H3 and acetylated H4 first rapidly associating with the newly replicated DNA (108, 118). Next, histones H2A and H2B are assembled into the nascent chromatin to form ordered arrays of nucleosomes. Thus

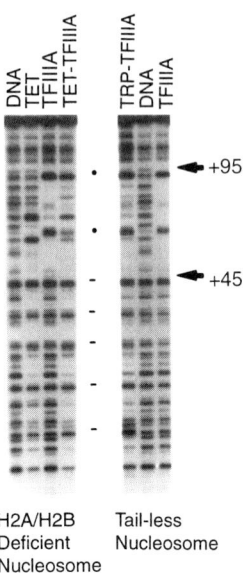

FIG. 16. Removal of the amino-terminal tails or depletion of histones H2A/H2B facilitates TFIIIA binding to nucleosomal DNA. Lanes 1 and 6 show DNase I cleavage of naked DNA. Lanes 3 and 7 show DNase I cleavage of a TFIIIA–DNA complex, lane 2 shows a tetramer [TET; (H3/H4)$_2$] and DNA complex. Lane 4 shows a complex of TFIIIA, the tetramer, and DNA (TET–TFIIIA) and lane 5 shows a complex of TFIIIA, the trypsinized octamer lacking the N-terminal tails, and TFIIIA (TRP–TFIIIA).

the staging of chromatin assembly, coupled with the positioning of nucleosomes on the 5S RNA gene promoter will present a window of opportunity for TFIIIA to bind to the gene, and thus the potential will exist for the eventual assembly of a functional transcription complex (119).

C. Functional Studies of Dinucleosomal Templates—Nucleosome Mobility and Linker Histones

A useful approach to interrelate chromatin structure with transcription has been to make use of short DNA fragments that are long enough to be competent for transcription, but short enough to allow aspects of their nucleoprotein organization to be determined (120–122). A synthetic dinucleosome containing two 5S rRNA genes was used to demonstrate that physiologically spaced histone octamers would partially repress transcription, but only by ~70% relative to naked DNA (Fig. 17) (7). The reason for this relative accessibility of this chromatin template to the transcriptional machinery appears to be the mobility of the histone octamers with respect to DNA sequence (6).

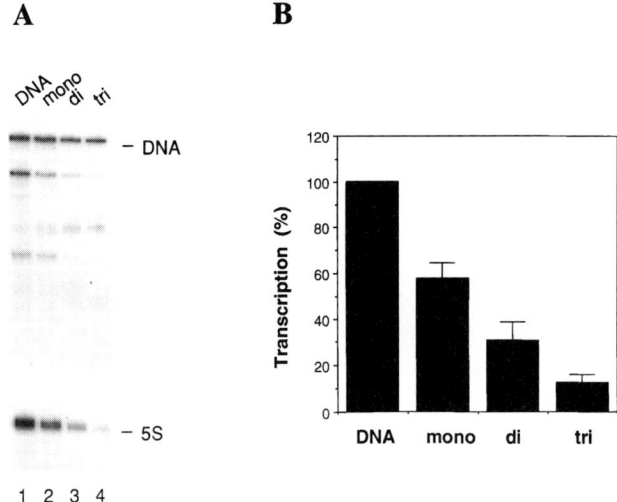

FIG. 17. Repression of 5S RNA gene transcription depends on the numbers of histone octamers reconstituted on the short, linear, 414-bp X5S 197–2 DNA template. (A) The radiolabeled DNA template was reconstituted with histone octamers; and mono-, di-, and trinucleosome products were separated by sucrose gradient centrifugation. These complexes were then used as templates for transcription in an extract from *Xenopus* oocyte nuclei. Lane 1, naked DNA; lane 2, mononucleosome; lane 3, dinucleosome; lane 4, trinucleosome template. Transcripts were analyzed by electrophoresis in a 6% denaturing polyacrylamide gel. (B) The level of 5S RNA transcription in panel A was quantitated by PhosphorImager and the data are shown as a bar graph using radiolabeled DNA as an internal control. The transcriptional activity of the Naked DNA template is taken as 100%.

Surprisingly, unlike the mononucleosomes we had previously studied (66), the histone octamers in the dinucleosome are mobile and show multiple translational positions. Because TFIIIA will not bind to an unmodified histone octamer assembled on a shorter (211 bp) DNA fragment, we interpret octamer mobility on the longer DNA fragment (424 bp) as an essential requirement for TFIIIA binding and the subsequent assembly of a functional transcription complex. We have found the difference between the immobile nucleosome previously studied and the mobile dinucleosome to be a consequence of DNA length. There is evidence that the presence of two adjacent histone octamers will influence their translational positioning (123). The major consequence of the mobile histone octamers is that the mapping of the boundaries of octamer–DNA association reveal that the binding site for TFIIIA will be accessible in a substantial fraction of the reconstitutes (>50%). Thus, a transcriptionally competent template can be assembled in the presence of a physiological density of histone octamers. The consequences of linker histone association on transcription could now be directly assessed.

Stable incorporation of linker histones into dinucleosomal templates leads to the disappearance of nucleosome mobility. Linker histones direct the nucleosomes to adopt a single predominant position that brings all of the TFIIIA binding site (+45 to +95) into contact with the histones (Fig. 18). A dinucleosomal template with the nucleosomes immobile and positioned over the TFIIIA binding site is transcriptionally silent. Therefore nucleosome mobility is an important characteristic of transcriptionally competent templates. Fixation of core histone–DNA contacts through incorporation of linker histones can determine transcriptional repression at a dinucleosomal level. The addition of one molecule of histone H1 or H5 per histone octamer directs the establishment of a completely repressive chromatin structure. The restriction of nucleosomal mobility by linker histones therefore provides a potential ex-

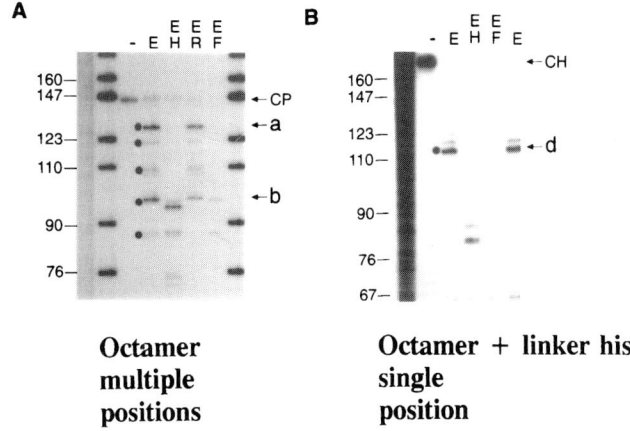

FIG. 18. Micrococcal nuclease mapping of core and chromatosome positions on reconstituted dinucleosome complexes. DNA from the nucleosome core particle and the chromatosome was recovered from an acrylamide gel and digested with restriction enzymes to determine the positions of the boundaries of histone–DNA complexes. (A) Core particle positions. Predominant products of EcoRV (E) digestion of core particle DNA were a 129- and a 100-bp fragment, labeled a and b, respectively (lane 4). However, a ladder of bands is visible below the 129-bp band (a) separated by 10–11 bp (dots). These represent the multiple positions occupied by the histone octamer on the 5' 5S RNA gene repeat. Band lengths were determined using MspI-digested pBR322 size markers (lanes 2 and 8) and DNA fragments from a hydroxyl radical cleavage reaction (lane 1) which provide a more accurate indication of fragment sizes with the intermediate size markers. (B) Chromatosome positions. EcoRV digestion of chromatosome DNA produced a 117- and a 63-bp fragment, labeled d (marked with a dot). Lane 1, DNA fragments from a hydroxyl radical cleavage reaction.

planation for the transcriptional competence of templates assembled only with histone octamers and for the repressive influence of histone H1 on transcription (124–128).

In vivo histone H1 has a causal role in repressing *Xenopus* 5S rRNA genes during development (129). The sequence selective interactions of the globular domain of histone H1 revealed in the positioning of histone–DNA contacts on 5S DNA may have an important role in determining which particular genes are controlled by linker histones in vivo (130). Becker and colleagues (131) have established nucleosome mobility as an important aspect of transcriptional competence within chromatin templates assembled in *Drosophila* embryonic extracts. In their studies, ATP hydrolysis is necessary to generate mobile nucleosomes and addition of histone H1 is without effect. Differences in our results might be related to chromatin composition: embryonic chromatin might contain unusual histone variants and be enriched in nonhistone HMG proteins that differ from the somatic core histones and linker histones used in this study (132–134).

Core histone acetylation influences transcription of the dinucleosomal template, substantially relieving transcriptional repression in the absence of linker histones (Fig. 19) (5). This increase in transcriptional efficiency occurs without any increase in nucleosome mobility, thus the effects of acetylation on trans-acting factor access to nucleosomal DNA (4) most probably rely on events involving a single nucleosome and not changes in interactions between nucleosomes. Histone H1 binds efficiently to nucleosomes reconstituted with either acetylated or control core histones and represses transcription (Fig. 20) (135). Incorporation of histone H1 restricts the translational mobility of both acetylated and control nucleosomes, without influencing the rotational positioning of DNA on the surface of the histones. Restriction of the mobility of histone octamers by histone H1 can again account for the observed repression of transcription (7).

The repression of transcription mediated by histone H1 on acetylated templates implies that removal of H1 will still be a necessary component of any transcriptional activation process in which targeted core histone acetylation is involved. How this might be accomplished is yet to be determined, however targeted linker histone phosphorylation is an attractive possibility.

D. Functional Studies of Nucleosomes Using Nuclear and Steroid Hormone Receptors

The vitellogenin genes of *Xenopus* and the chicken provide an attractive model system for investigating the molecular mechanisms of inducible gene expression by the hormone estrogen (136). In vitro and in vivo experiments have suggested roles for several trans-acting factors, including the estrogen

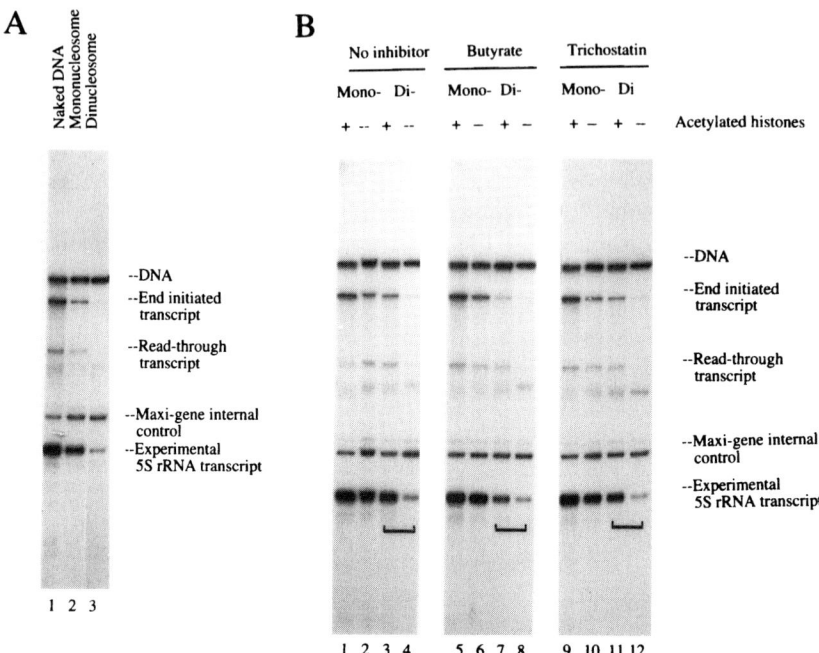

FIG. 19. The influence of histone acetylation on transcription of dinucleosome templates reconstituted with histone octamers. (A) Repression of transcription depends on the number of control histone octamers with minimal histone acetylation reconstituted on the X5S 197-2 DNA template. The radiolabeled DNA template was reconstituted with histone octamers, and mono- and dinucleosome products were separated by sucrose gradient centrifugation. These complexes were then used as templates for transcription together with a naked plasmid template encoding a maxi-5S RNA gene as an internal control in an extract of *Xenopus* oocyte nuclei. Lane 1, Naked DNA; lane 2, mononucleosome; lane 3, dinucleosome. Transcripts were analyzed by electrophoresis in a 6% denaturing polyacrylamidel gel. The positions of the 5S rRNA (Experimental 5S rRNA transcript), the maxi-gene 5S transcript (Maxi-gene internal control), the transcripts initiated at the first 5S rRNA gene that read through the first termination signal (Read-through transcript), and the transcripts initiated at either end of the DNA fragment that read through the length of the DNA sequence (End-initiated transcript) are indicated. (B) Influence of histone acetylation on transcription of the X5S 197-2 DNA template reconstituted with a single (mono-) or two (di-) histone octamers. Radiolabeled DNA was reconstituted with octamers containing acetylated (+) or control histones with minimal acetylation (−) together with a naked plasmid template encoding a maxi-5S RNA gene as an internal control, in an extract of *Xenopus* oocyte nuclei. These transcription reactions were carried out with no additions (lanes 1–4), or with addition of 10 mM sodium butyrate (lanes 5–8) or 10 μM Trichostatin A (lanes 9–12). Transcripts are indicated as in panel A. Transcripts derived from the dinucleosomal templates in the presence (+) or absence (-) of histone acetylation are indicated by brackets.

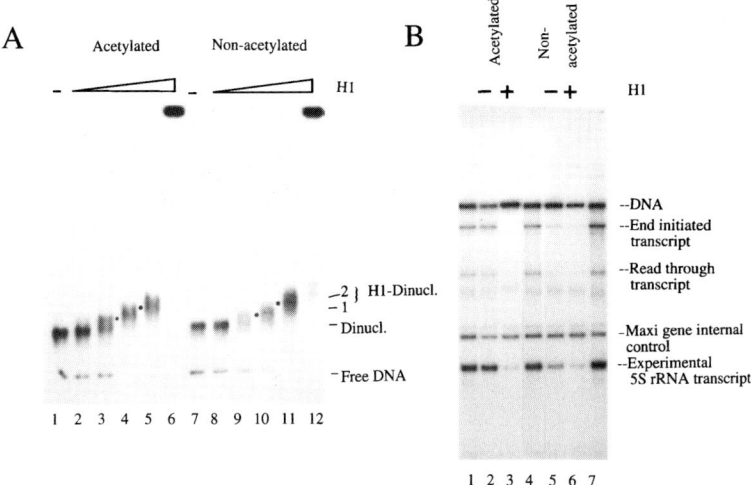

FIG. 20. Binding of histone H1 to reconstituted acetylated and control dinucleosomes represses transcription. (A) Binding of H1 to reconstituted dinucleosomes. Reconstituted dinucleosomes containing acetylated (lanes 1–6) or control core histones (lanes 7–12) were mixed with free DNA before various amounts of histone H1 were added and analyzed by nucleoprotein agarose (0.7%) gel electrophoresis. Lanes 1–6, 100 ng (DNA content) of reconstituted acetylated dinucleosomes (= 0.6 pmol) was mixed with 0, 0.3, 0.6, 1.2, 2.4, and 4.8 pmol of H1, respectively. Lanes 7–12 are as for lanes 1–6, except that control core histones with minimal acetylation were used. Dinucleosomes containing one (1) and (2) molecules of histone H1 per dinucleosome are indicated. Note that nucleosome mobility interferes with the resolution of the nucleoprotein complexes. (B) Binding of histone H1 represses transcription. The X5S 197-2 DNA template reconstituted with two histone octamers containing acetylation histones (lanes 2 and 3) or control histones with minimal acetylation (lanes 5 and 6) were separated on sucrose gradients. To 100 ng of reconstituted dinucleosome (0.6 pmol) was added 1.2 pmol of histone H1 (lanes 3 and 6) or H1 was not added (lanes 2 and 5) before 10 ng of the template was transcribed in the oocyte nuclear extract. As a control, 10 ng of naked X5S 197-2 DNA fragment was transcribed (lanes 1, 4, and 7). To all transcription reactions a naked plasmid template encoding a maxi-5S rRNA gene as an internal control was added.

receptor (137), a nuclear factor 1-like activity (138), a leucine zipper protein VBP (139), a USF-like activity (140), a liver-specific repressor (141), and the basal transcriptional machinery (142) in transcriptional control of the vitellogenin gene promoter. Alterations in chromatin structure over the promoter elements of the vitellogenin genes following the estrogen-dependent induction of transcription have also been described (143–145). Surprisingly, *in vitro* experiments have suggested that the concomitant assembly of chromatin during transcription complex formation potentiates transcription from the *Xenopus* vitellogenin B1 gene promoter (146).

We found that prior nucleosome assembly on the vitellogenin B1 promoter led to transcriptional potentiation in an estrogen-responsive transcription extract (Fig. 21) (3). Nuclease digestion and hydroxyl radical cleavage indicate that strong, DNA sequence-directed positioning of a nucleosome occurs between −300 and −140 relative to the start site of transcription (Fig. 4). Deletion of this DNA sequence abolishes the potentiation of transcription due to nucleosome assembly. The nucleosome is present between the first group of EREs and the proximal promoter element, which contains the binding sites for numerous trans-acting factors, including NF-1-like activities. A similar organization is found in all four *Xenopus* vitellogenin genes and in the chicken vitellogenin II gene (*147*). This region has previously been proposed to represent a nucleosome positioning element (*148*). Moreover, the positions of DNase I-hypersensitive sites B1 and B2 in the chicken vitellogenin II promoter documented *in vivo* (at −300 and +1) are consistent with a nucleosome being positioned between them (*143*). Mutational studies of the chicken vitellogenin II promoter using transient assays in which the EREs are progressively brought closer to the proximal promoter element reveal an initial drop in transcription before transcription is potentiated (*140*). Such studies would be consistent with the spacing between the EREs and the proximal promoter element playing an important role in transcription, perhaps due to a requirement for a nucleosomal length of DNA. Deletion of the DNA between the EREs and the proximal promoter element eliminates the potentiation of transcription following nucleosome assembly. Therefore, a nucleosome-mediated static loop potentiates transcriptional stimulation by bringing the estrogen receptor bound to the EREs into juxtaposition with the proximal promoter element and factors bound to it (Fig. 21). Such a model had also been proposed by Elgin and colleagues for activation of the *hsp26* gene in *Drosophila* (*141*). The stimulatory effect of nucleosome assembly was recently demonstrated for a second gene system (*149*).

The vitellogenin B1 gene system provides an example of how the translational positioning of histone–DNA contacts within a nucleosome can allow DNA to be packed, yet leave key regulatory elements accessible to the transcriptional machinery. A second means to accomplish this goal is to make use of the rotational positioning of DNA on the surface of the core histones such that regulatory elements remain accessible from solution. The glucocorticoid receptor and the thyroid hormone receptor provide examples of transcription factors that bind to their regulatory elements in spite of their incorporation into a nucleosome.

The glucocorticoid receptor has a DNA-binding domain with two zinc-binding domains. The nuclear magnetic resonance (NMR) solution structure of the DNA-binding domain reveals the zinc fingers to be arranged very differently from those of TFIIIA (*150*). Each DNA-binding domain uses an α-

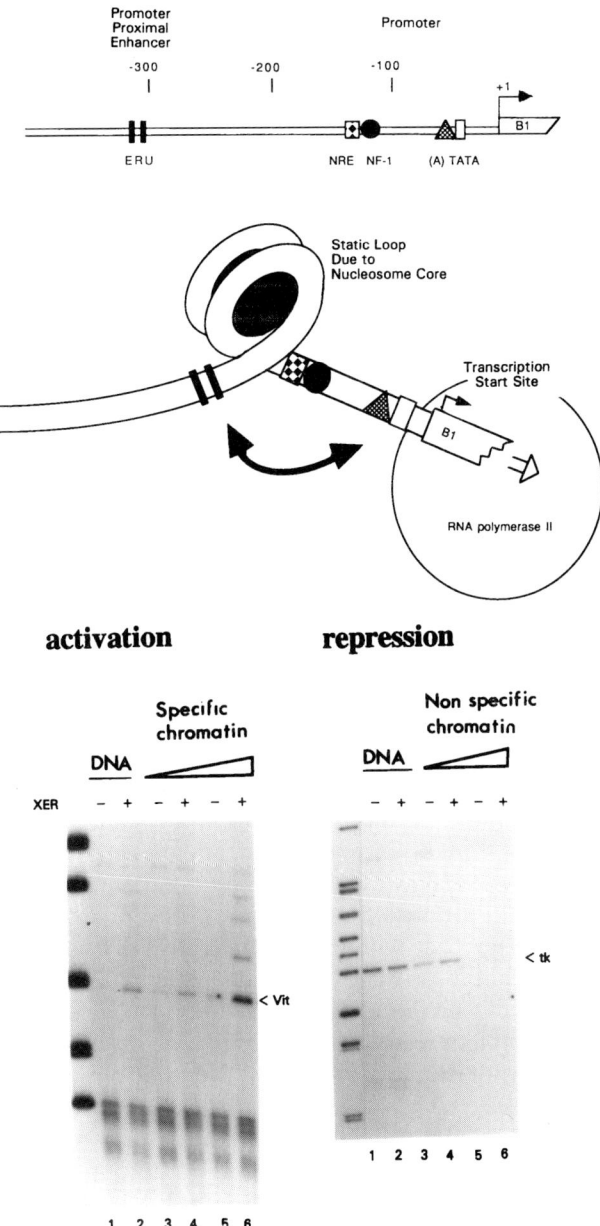

FIG. 21. Role of a static loop created by a nucleosome including part of the *Xenopus* vitellogenin B1 promoter in potentiating transcription. Transcription from a specific chromatin template that positions nucleosomes increases with increasing number of nucleosomes. Transcription from a nonspecific chromatin template in which nucleosomes are not positioned decreases with increasing numbers of nucleosomes. (Reproduced, with permission, from Ref. 3.)

helix in one of the two fingers to interact with a short 6-bp region in the major groove, the other finger is involved in protein–protein interactions. The glucocorticoid receptor associates with DNA as a dimer. The second molecule of the receptor has similar interactions on the same side of the DNA helix, one helical turn away (or 3.4 nm). One well-studied model system in which the glucocorticoid receptor activates transcription from a chromatin template is the regulated transcription of the mouse mammary tumor virus long terminal repeat (MMTV LTR). Hager and colleagues established that the MMTV LTR is incorporated into six positioned nucleosomes, both in episomes and within a mouse chromosome (*151*). Induction of transcription by glucocorticoids requires binding of the glucocorticoid receptor (GR) to the LTR, disruption of the local chromatin structure, and the assembly of a transcription complex over the TATA box (*152*).

Remarkably the GR binds to the nucleosomal MMTV LTR with only a slight reduction in affinity relative to naked DNA (*152–154*). This interaction appears independent of the precise translational position of the nucleosome, GR binding occurs when the nucleosome is at -188 to -45 (*154*) and at -219 to -76 (*153*) or -221 to -78 (*152*) relative to the start of transcription $(+1)$. In both instances the rotational orientation of the GR binding sites will be similar because the separation of nucleosome boundaries is by almost exactly three helical turns of DNA. The latter *in vitro* nucleosome positions compare favorably with those determined *in vivo* (*151*).

Within the nucleosome, the MMTV LTR has several recognition elements for the GR receptor occurring at -175, -119, -98, and -83. The two elements at -119 and -98 face toward the core and remain unbound. The two elements at -175 and -83, which are oriented away from the histone core, are footprinted by the GR *in vitro* (*152–154*). The GR, in fact, seems well suited to interact specifically with nucleosomal DNA. The short recognition elements (6 bp) within the GRE are separated by one helical turn of DNA, causing both to be aligned on the same face of the DNA helix. Consequently, the GR/GRE interaction occurs on only one side of the DNA, much like a prokaryotic repressor, thus possibly circumventing steric interference by the presence of the histone core. Furthermore, the GR homodimer can bind specifically to DNA containing only one GRE half-site, presumably by making both specific and nonspecific contacts to DNA in each half of the dimer, respectively (*155*). Thus highly bent nucleosomal DNA might still provide enough precisely aligned contacts for at least one specific half-site interaction which could then be supplemented by nonspecific contacts. It is also possible that the wrapping of DNA around the histone core brings the two accessible recognition elements close enough for DNA binding of the GR to be stabilized by protein-protein interactions. In any event, the GR represents one of the few examples in which a transcription factor can interact with exposed DNA recognition elements in spite of the large structural distortion of DNA in the nucleosome.

Association of the GR with the nucleosome containing its binding sites appears to have no effect on the integrity of the nucleosome *in vitro*, unlike the apparent disruption of the nucleosome that occurs *in vivo* (152). The binding of other promoter-specific transcription factors (e.g., NF1), which is facilitated by the GR *in vivo*, does not occur *in vitro*. Like TFIIIA, NF1 does not recognize DNA actually in contact with histones (156). The discrepancy in the events believed to occur *in vivo* and *in vitro* might be explained by the absence of molecular machines such as SWI/SNF (157) or coactivator complexes possessing histone acetyltransferase activity, which presumably facilitate chromatin structural changes from the *in vitro* system.

A role for chromatin has also been established in the control of transcription by the thyroid hormone receptor (158, 159). These studies provide a useful example of how the histones can contribute to gene regulation. We had made use of the assembly of minichromosomes within the *Xenopus* oocyte nucleus to examine the role of chromatin in both transcriptional silencing and activation of the *Xenopus* TRβA promoter. Transcription from this promoter is under the control of thyroid hormone and the thyroid hormone receptor (160), which exists as a heterodimer of TR and the 9-*cis*-retinoic acid receptor, RXR. Microinjection of either single-stranded or double-stranded DNA templates into the *Xenopus* oocyte nucleus offers the opportunity for examination of the influence on gene regulation of chromatin assembly pathways that are either coupled or uncoupled to DNA synthesis (161). The staged injection of mRNA encoding transcriptional regulatory proteins and of template DNA offers the potential for examining the mechanisms of transcription factor-mediated transcriptional activation of promoters within a chromatin environment. In particular, it is possible to discriminate between preemptive mechanisms in which transcription factors bind during chromatin assembly to activate transcription, and postreplicative mechanisms in which transcription factors gain access to their recognition elements after they have been assembled into mature chromatin structures. TR/RXR heterodimers bind constitutively within the minichromosome, independently of whether the receptor is synthesized before or after chromatin assembly. Rotational positioning of the TRE on the surface of the histone octamer allows the specific association of the TR/RXR heterodimer *in vitro* (Fig. 22) (158). The coupling of chromatin assembly to the replication process augments transcriptional repression by unliganded TR/RXR without influencing the final level of transcriptional activity in the presence of thyroid hormone.

The molecular mechanisms by which the unliganded thyroid hormone receptor makes use of chromatin in order to augment transcriptional repres-

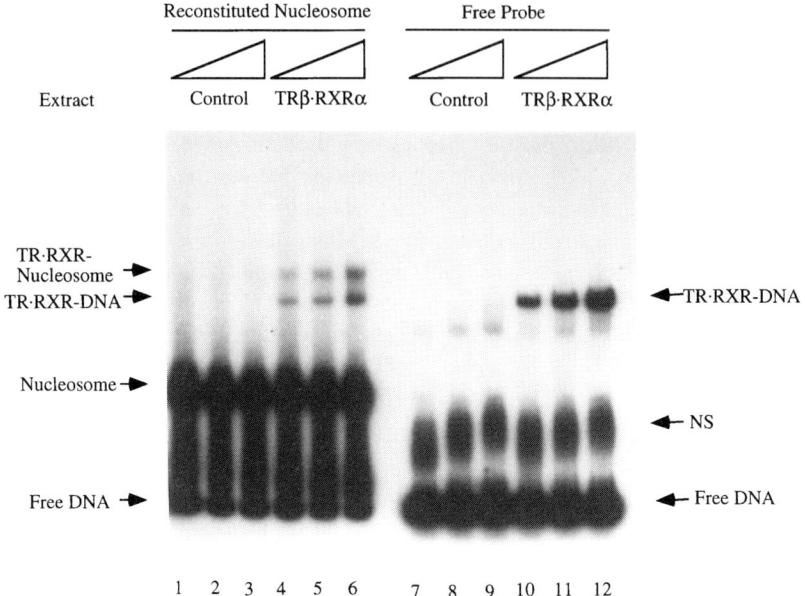

FIG. 22. TR/RXR heterodimer binds to the TRE but not the mutated TRE reconstituted into nucleosome *in vitro*. A 160-bp end-labeled DNA fragment from TRβA promoter (from +163 to +322) containing either the wild-type TRE (TRE) or the mutated TRE (mTRE) was generated by PCR amplification with one of the two primers end-labeled with ^{32}P (position +322), purified, and reconstituted into the nucleosome *in vitro* with histone octamers purified from chicken erythrocytes. The reconstituted nucleosome was then incubated with extract from oocytes with or without (control) overproduction of TRβ/RXRα (lanes 1–6). For a comparison, the binding experiment was also conducted with the end-labeled naked DNA (lanes 7–12). A 20-μl portion of control oocyte extract or oocyte extract with TR/RXR was incubated at room temperature for 20 min and then resolved by a 4% native polyacrylamide gel in 0.5 x TBE. NS, A TR/RXR-independent nonspecific complex.

sion also involve mSin3 and histone deacetylase (*162, 163*). The unliganded thyroid hormone receptor and retinoic acid receptor bind a corepressor NCoR (*164*). NCoR interacts with Sin3 and recruits a histone deacetylase (*162, 163*). All of the transcriptional repression conferred by the unliganded thyroid hormone receptor in *Xenopus* oocytes (*158, 159*) can be alleviated by the inhibition of histone deacetylase using Trichostatin A (*164*), indicative of an essential role for deacetylation in establishing transcriptional repression in a chromatin environment.

The addition of thyroid hormone to the chromatin bound receptor leads to the disruption of chromatin structure (*158, 159*). Chromatin disruption is not restricted to the receptor binding site and involves the reorganization of chromatin structure in which targeted histone acetylation by the PCAF and p300/CBP activators have a contributory role (115, 165). It is possible to separate chromatin disruption from productive recruitment of the basal transcription machinery *in vivo* by deletion of regulatory elements essential for transcription initiation at the start site and by the use of transcriptional inhibitors (158, 159). Therefore chromatin disruption is an independent hormone-regulated function targeted by DNA-bound thyroid hormone receptor. It is remarkable just how effectively the various functions of the thyroid hormone receptor are mediated through the recruitment of enzyme complexes that modify chromatin. These results provide compelling evidence for the productive utilization of structural transitions in chromatin as a regulatory principle in gene control.

E. Nucleosomal Arrays: Higher Order Structure and Transcription Repression

An additional level of constraint to the transcriptional machinery beyond that of the wrapping of DNA in the nucleosome is imposed by the structural dynamics of the chromatin fiber. Several groups have begun to investigate this problem by using biochemically defined DNA templates containing promoter sequences that are reconstituted into defined nucleosomal arrays. One of the advantages of these DNA templates is that they allow the simultaneous determination of the contributions of the core histones to higher order chromatin folding and transcription by RNA polymerases (9, 165). We have determined that salt-dependent compaction of linear and circular nucleosomal arrays can severely repress both transcription initiation and elongation by RNA polymerase III (9). These results suggest that chromatin folding may be a major factor in the inhibition of transcription observed with nucleosomal templates under standard transcription reaction conditions. We have extended these observations to show that removal of histones H2A/H2B from nucleosomal arrays has a major influence both on chromatin compaction and on repression of RNA polymerase III transcription initiation and elongation. In our *in vitro* model system studies, the level of transcriptional repression of octamer reconstitutes in standard (i.e., high Mg^{2+}) transcription buffer was much more pronounced than that observed for the tetramer reconstitutes. Divalent cations induce the folding of nucleosomal arrays (9). This is consistent both with a role for histone octamers in transcriptional repression and with a role for H2A/H2B deficiency in the relief of transcriptional repression. Part of the effect of H2A/H2B deficiency can be traced to increased transcription factor accessibility of the 5S promoter when complexed with a H3/H4 tetramer rather than a histone octamer. Although this H2A/H2B serves to enhance transcription initiation, it is also clear that deficiency in H2A/H2B disrupts the ability of nucleosmal arrays to fold into the higher order structures that are repressive to transcriptional elongation. Thus one of the consequences of depletion of H2A/H2B from active chromatin is to stabilize a more extended chromatin structure that will impose less impediment to RNA polymerase function (10). Because removal of the histone tails or their acetylation also influences chromatin compaction (75, 166), it will be of great interest to examine the consequences of these modifications for the transcription of nucleosomal arrays.

IV. Concluding Remarks

The specific positioning of nucleosomes allows the transcriptional machinery to work effectively in a chromatin environment. Exposure of regulatory elements on the surface of nucleosomes facilitates association of transcription factors with chromatin and the subsequent recruitment of the molecular machines that repress or activate genes through the modification of chromatin structure. Nucleosome positioning also brings specific histones into contact with regulatory elements. This greatly simplifies the problem of disrupting chromatin structures, since entire nucleosomes do not have to be removed before transcription factor access can occur. The disruption of individual histone–DNA interactions within positioned nucleosomes, such as displacement of H1 or H2A–H2B, or acetylation of the amino-terminal tails of the histones, provides examples of how such access can be regulated. Positioned nucleosomes also facilitate the transcription process: the two turns of DNA around a single positioned nucleosome can bring regulatory elements 80 bp apart together on the same surface. Likewise, elements separated by 160 bp can be brought together. This local concentration of transcription factors helps to mediate their activity. It will be interesting to examine whether the positioning of multiple nucleosomes in longer arrays can allow communication over even greater distances. Although we have discussed only a limited number of specific chromatin structures, they provide clear examples of how specific histone–DNA contacts can amplify the range of regulation available to the transcriptional machinery.

ACKNOWLEDGMENT

We thank Ms. T. Vo for manuscript preparation.

REFERENCES

1. R. T. Simpson, *Prog. Nucleic Acids Res. Mol. Biol.* **40**, 143 (1991).
2. D. Pruss, B. Batholomew, J. Persinger, J. Hayes, G. Arents, E. Moudrianakis and A. P. Wolffe, *Science* **274**, 614 (1996).
3. C. Schild, F.-X. Claret, W. Wahli, and A. P. Wolffe, *EMBO J.* **12,** 423 (1993).

4. D. Y. Lee, J. J. Hayes, D. Pruss, and A. P. Wolffe, *Cell* **72**, 73 (1993).
5. K. Ura, H. Kurumizaka, S. Dimitrov, G. Almouzni, and A. P. Wolffe, *EMBO J.* **16**, 2096 (1997).
6. G. Meersseman, S. Pennings, and E. M. Bradbury, *EMBO J.* **11**, 2951 (1992).
7. K. Ura, J. J. Hayes, and A. P. Wolffe, *EMBO J.* **14**, 3752 (1995).
8. K. Ura, K. Nightingale, and A. P. Wolffe, *EMBO J.* **15**, 4959 (1996).
9. J. C. Hansen and A. P. Wolffe, *Biochemistry* **31**, 7977 (1992).
10. J. C. Hansen and A. P. Wolffe, *Proc. Natl. Acad. Sci. U.S.A.* **91**, 2339 (1994).
11. G. Arents, R. W. Burlingame, B. W. Wang, W. E. Love and E. N. Moudrianakis, *Proc. Natl. Acad. Sci. U.S.A.* **88**, 10148 (1991).
12. G. Arents and E. N. Moudrianakis, *Proc. Natl. Acad. Sci. U.S.A.* **90**, 10489 (1993).
13. G. Arents and E. N. Moudrianakis, *Proc. Natl. Acad. Sci. U.S.A.* **92**, 11170 (1995).
14. V. Ramakrishnan, J. T. Finch, V. Graziano, P. L. Lee, and R. M. Sweet, *Nature (London)* **362**, 219 (1993).
15. C. Cerf, G. Lippens, V. Ramakrishnan, S. Muyldermans, A. Segers, L. Wyns, S. J. Waaak, and A. Hallenga, *Biochemistry* **33**, 11079 (1994).
16. T. Richmond, J. T. Finch, B. Rushton, D. Rhodes, and A. Klug, *Nature (London)* **311**, 532 (1984).
17. K. Luger, A. W. Mader, R. K. Richmond, D. F. Sargent, and T. J. Richmond, *Nature* **389**, 251 (1997).
18. T. H. Eickbusch and E. N. Moudrianakis, *Biochemistry* **17**, 4955 (1978).
19. J. J. Hayes and A. P. Wolffe, *Proc. Natl. Acad. Sci. U.S.A.* **89**, 1229 (1992).
20. D. Pruss and A. P. Wolffe, *Biochemistry* **32**, 6810 (1993).
21. A. D. Mirzabekov, V. V. Shick, A. V. Belyavsky, and S. G. Bavykin, *Proc. Natl. Acad. Sci. U.S.A.* **75**, 4184 (1978).
22. A. D. Mirzabekov, S. G. Bavykin, A. V. Belyavsky, V. L. Karpov, O. V. Preobrazhenskaya, V. V. Shick, and K. K. Ebralidse, *Metho. Enzymol.* **170**, 386 (1989).
23. K. K. Ebralidse, S. A. Grachev, and A. D. Mirzabekov, *Nature (London)* **331**, 365 (1988).
24. S. G. Bavykin, S. I. Usachenko, A. O. Zalensky, and A. D. Mirzabekov, *J. Mol. Biol.* **212**, 495 (1990).
25. D. Y. Gushchin, K. K. Ebralidse, and A. D. Mirzabekov, *M. Biol.* **25**, 1400 (1991).
26. S. I. Usachenko, S. G. Bavykin, I. M. Gavin, and E. M. Bradbury, *Proc. Natl. Acad. Sci. U.S.A.* **91**, 6845 (1994).
27. A. Klug and L. C. Lutter, *Nucleic Acids Res.* **9**, 4297 (1981).
28. T. D. Tullius and B. A. Dombroski, *Science* **230**, 679 (1985).
29. J. J. Hayes and A. P. Wolffe, *Trends Biochem. Sci.* **17**, 250 (1992).
30. J. J. Hayes, T. D. Tullius, and A. P. Wolffe, *Proc. Natl. Acad. Sci. U.S.A.* **87**, 7405 (1990).
31. J. J. Hayes, T. D. Tullius, and A. P. Wolffe, *Proc. Natl. Acad. Sci. U.S.A.* **88**, 6829 (1991).
32. J. M. Gale and M. J. Smerdon, *J. Mol. Biol.* **204**, 949 (1988).
33. H. G. Patterton and R. T. Simpson, *Nucleic Acids Res.* **23**, 4170 (1995).
34. H. R. Drew and A. A. Travers, *J. Mol. Biol.* **186**, 773 (1985).
35. A. M. Burkhoff and T. D. Tullius, *Cell* **48**, 935 (1987).
36. A. P. Wolffe and H. R. Drew, *Proc. Natl. Acad. Sci. U.S.A.* **86**, 9817 (1989).
37. R. T. Simpson, F. Thoma, and J. M. Brubaker, *Cell* **42**, 799 (1985).
38. J. H. White and W. R. Bauer, *Cell* **56**, 9 (1989).
39. J. H. White, N. R. Cozzarelli, and W. R. Bauer, *Science* **241**, 323 (1988).
40. D. Rhodes and A. Klug, *Nature (London)* **286**, 573 (1980).
41. J. J. Hayes, J. Bashkin, T. D. Tullius, and A. P. Wolffe, *Biochemistry* **30**, 8434 (1991).
42. J. H. White, R. Gallo, and W. R. Bauer, *J. Mol. Biol.* **207**, 193 (1989).
43. D. Pruss, F. D. Bushman, and A. P. Wolffe, *Proc. Natl. Acad. Sci. U.S.A.* **91**, 5913 (1994).
44. M. E. Hogan, T. F. Rooney, and R. H. Austin, *Nature (London)* **328**, 554 (1987).

45. M. Mahadevan, C. Tsilfidis, L. Sabourin, G. Shutler, C. Amemiya, G. Jansen, C. Neville, M. Narang, J. Barcelo, K. O'Hoy, S. Leblond, J. Earle-MacDonald, P. D. de Jong, B. Wieringa, and R. G. Kornluk, *Science* **255**, 1253 (1992).
46. A. J. M. H. Verkerk, M. Pieretti, J. S. Sutcliffe, Y. H. Fu, D. P. A. Kuhl, A. Pizzuti, O. Reiner, S. Richards, M. F. Victoria, F. Zhang, B. E. Eussen, G. J. B. van Ommen, L. A. J. Blonden, G. J. Riggins, J. L. Chastain, C. B. Kunst, H. Galjaard, C. T. Caskey, D. L. Nelson, B. A. Oostra, and S. T. Warren, *Cell* **65**, 905 (1991).
47. J. S. Godde and A. P. Wolffe, *J. Biol. Chem.* **271**, 15222 (1996).
48. J. S. Godde, S. U. Kass, M. C. Hirst, and A. P. Wolffe, *J. Biol. Chem.* **271**, 24325 (1996).
49. A. J. Varshavsky, V. V. Bakayev, and G. P. Georgiev, *Nucleic Acids Res.* **3**, 477 (1976).
50. K. Hayashi, T. Hofstaetter and Y. Naoko, *Biochemistry* **17**, 1880 (1978).
51. R. T. Simpson, *Biochemistry* **17**, 5524 (1978).
52. J. Allan, P. G. Hartman, C. Crane-Robinson, and F. X. Aviles, *Nature (London)* **288**, 675 (1980).
53. D. Z. Stoynov and C. Crane-Robinson, *EMBO J.* **7**, 3685 (1988).
54. S. C. Albright, J. M. Wiseman, R. A. Lange, and W. T. Garrard, *J. Biol. Chem.* **255**, 3673 (1980).
55. K. Nightingale, S. Dimitrov, R. Reeves, and A. P. Wolffe, *EMBO J.* **15**, 548 (1996).
56. K. L. Clark, E. D. Haley, E. Lai, and S. K. Burley, *Nature (London)* **364**, 412 (1993).
57. S. K. Burley, X. Xie, K. L. Clark, and F. Shu, *Curr. Opin. St. Biol.* **7**, 94 (1997).
58. A. D. Mirzabekov, D. V. Pruss, and K. K. Elbradise, *J. Mol. Biol.* **211**, 479 (1990).
59. R. Losa, F. Thoma, and T. Koller, *J. Mol. Biol.* **175**, 529 (1984).
60. F. A. Goytisolo, S. E. Gerchman, X. Yu, C. Rees, V. Graziano, V. Ramakrishnan, and J. O. Thomas, *EMBO J.* **15**, 3421 (1996).
61. C. Crane-Robinson and O. B. Ptitsyn, *Prot. Eng.* **2**, 577 (1989).
62. J. O. Thomas, C. Rees, and J. T. Finch, *Nucleic Acids Res.* **20**, 187 (1992).
63. J. J. Hayes, R. Kaplan, K. Ura, D. Pruss, and A. P. Wolffe, *J. Biol. Chem.* **271**, 25817 (1996).
64. S. Lambert, S. Muyldermans, J. Baldwin, J. Kilner, K. Lbel, and L. Wijns, *Biochem. Biophys. Res. Commun.* **179**, 810 (1991).
65. T. Boulikas, J. M. Wiseman, and W. T. Garrard, *Proc. Natl. Acad. Sci. U.S.A.* **77**, 121 (1980).
66. J. J. Hayes and A. P. Wolffe, *Proc. Natl. Acad. Sci. U.S.A.* **90**, 6415 (1993).
67. J. J. Hayes, D. Pruss, and A. P. Wolffe, *Proc. Natl. Acad. Sci. U.S.A.* **91**, 7817 (1994).
68. K. Nightingale, D. Pruss, and A. P. Wolffe, *J. Biol. Chem.* **271**, 7090 (1996).
69. J. J. Hayes, *Biochemistry* **35**, 11931 (1996).
70. J. Wong, Q. Li, B. Z. Levi, Y.-B. Shi, and A. P. Wolffe, *EMBO J.* **16**, 7130 (1997).
71. T. J. Richmond, T. Rechsteiner, and K. Luger, *Cold Spring Harbor Symp. Quant. Biol.* **58**, 265 (1993).
72. J. Yao, P. T. Lowary, and J. Widom, *Proc. Natl. Acad. Sci. U.S.A.* **87**, 7603 (1990).
73. J. Yao, P. T. Lowary, and J. Widom, *Biochemistry* **30** 8408 (1991).
74. J. D. McGhee, J. M. Nickol, G. Felsenfeld, and D. C. Rau, *Cell* **33**, 831 (1983).
75. M. Garcia-Ramirez, F. Dong, and J. Ausio, *J. Biol. Chem.* **267**, 19587 (1992).
76. C. S. Hill and J. O. Thomas, *Eur. J. Biol.* **187**, 145 (1990).
77. G. G. Lindsey, S. Orgeig, P. Thompson, N. Davies, and D. L. Meader, *J. Mol. Biol.* **218**, 805 (1991).
78. D. J. Clark and T. Kimura, *J. Mol. Biol.* **211**, 883 (1990).
79. A. C. Lennard and J. O. Thomas, *EMBO J.* **4**, 3455 (1985).
80. V. Graziano, S. E. Gerchman, D. K. Schneider, and V. Ramadrishnan, *Nature (London)* **368**, 351 (1994).
81. F. Thoma, T. Koller, and A. Klug, *J. Chem. Biochem.* **83**, 407 (1979).
82. J. S. Godde and A. P. Wolffe, *J. Biol. Chem.* **270**, 27399 (1995).

83. K. Marushige, V. Ling, and G. H. Dixon, *J. Biol. Chem.* **244,** 5953 (1969).
84. J. Singh and M. R. S. Rao, *J. Biol. Chem.* **262,** 734 (1987).
85. B. P. Cormack and K. Struhl, *Cell* **69,** 685 (1992).
86. T. K. Kim and R. G. Roeder, *J. Biol. Chem.* **269,** 4891 (1994).
87. M. Han, U. J. Kim, P. Kayne, and M. Grunstein, *EMBO J.* **7,** 2221 (1988).
88. H. G. Patterton and R. T. Simpson, *Mol. Cell. Biol.* **14,** 4002 (1994).
89. M. P. Kladde and R. T. Simpson, *Proc. Natl. Acad. Sci. U.S.A.* **91** 1361 (1994).
90. J. L. Kim, D. B. Nikolov, and S. K. Burley, *Nature (London)* **365,** 520 (1993).
91. Y. Kim, J. H. Geiger, S. Hahn, and P. B. Sigler, *Nature (London)* **365,** 512 (1993).
92. J. A. Knezetic and D. S. Luse, *Cell* **45,** 95 (1986).
93. Y. Lorch, J. W. LaPointe, and R. D. Kornberg, *Cell* **49,** 203 (1987).
94. J. L. Workman and R. G. Roeder, *Cell* **51,** 613 (1987).
95. T. Matsui, *Mol. Cell. Biol.* **7,** 1401 (1987).
96. M. Meisternst, M. Horikoshi, and R. G. Roeder, *Proc. Natl. Acad. Sci. U.S.A.* **87,** 9153 (1990).
97. A. M. Imbalzano, H. Kwon, M. R. Green, and R. E. Kingston, *Nature (London)* **370,** 381 (1994).
98. J. Ausio, F. Dong, and K. E. van Holde, *J. Mol. Biol.* **206,** 451 (1989).
99. J. Ausio and K. E. van Holde, *Biochemistry* **25,** 1421 (1986).
100. W. Kruger, C. L. Peterson, A. Sil, C. Coburn, G. Arents, E. N. Moudrianakis, and I. Herskowitz, *Genes Dev.* **9,** 2770 (1995).
101. M. A. Wechser, M. P. Kladde, J. A. Alfieri, and C. L. Peterson, *EMBO J.* **16,** 2086 (1997).
102. H. Kurumizaka and A. P. Wolffe, *Mol. Cell. Biol.* **17,** 6953 (1997).
103. L. Fairall, D. Rhodes, and A. Klug, *J. Mol. Biol.* **192,** 577 (1986).
104. N. P. Pavletch and C. O. Pabo, *Science* **252,** 809 (1991).
105. J. J. Hayes and T. D. Tullius, *J. Mol. Biol.* **227,** 407 (1992).
106. J. M. Berg, *Annu. Rev. Biophys. Chem.* **19,** 405 (1990).
107. D. J. Clark and A. P. Wolffe, *EMBO J.* **10,** 3419 (1991).
108. G. Almouzni, M Mechali, and A. P. Wolffe, *Mol. Cell. Biol.* **11,** 655 (1991).
109. A. Shimamura, D. Tremethick, and A. Worcel, *Mol. Cell. Biol.* **8,** 4257 (1988).
110. D. Tremethick, D. Zucker, and A. Worcel, *J. Biol. Chem.* **265,** 5014 (1990).
111. J. M. Gottesfeld, *Mol. Cell. Biol.* **7,** 1612 (1987).
112. W. R. Bauer, J. J. Hayes, J. H. White, and A. P. Wolffe, *J. Mol. Biol.* **236,** 685 (1994).
113. C. L. Peterson, *Nucleus* **1,** 185 (1995).
114. J. E. Brownell, J. Zhou, T. Ranalli, R. Kobayashi, D. G. Edmondson, S. Y. Roth, and C. D. Allis, *Cell* **84,** 843 (1996).
115. V. V. Ogryzko, R. L. Schiltz, V. Russanova, B. H. Howard, and Y. Nakatani, *Cell* **87,** 953 (1996).
116. J. M. Sogo, H. Stahl, T. Koller, and R, Knippers, *J. Mol. Biol.* **189,** 189 (1986).
117. A. P. Wolffe and D. D. Brown, *Cell* **47,** 217 (1986).
118. A. Worcel, S. Han, and M. L. Wong, *Cell* **15,** 969 (1978).
119. A. P. Wolffe and R. H. Morse, *J. Biol. Chem.* **265,** 4592 (1990).
120. A. P. Wolffe, E. Jordan, and D. D. Brown, *Cell* **44,** 381 (1986).
121. R. Losa and D. D. Brown, *Cell* **50,** 501 (1987).
122. V. M. Studitsky, D. J. Clark, and G. Felsenfeld, *Cell* **76,** 371 (1994).
123. T. E. Shrader and D. M. Crothers, *J. Mol. Biol.* **216,** 69 (1990).
124. S. Pennings, G. Meersseman, and E. M. Bradbury, *Proc. Natl. Acad. Sci. U.S.A.* **91,** 10275 (1994).
125. M. S. Schlissel and D. D. Brown, *Cell* **37,** 903 (1984).
126. A. P. Wolffe, *EMBO J.* **8,** 527 (1989).

127. C. C. Chipev and A. P. Wolffe, *Mol. Cell. Biol.* **12,** 45 (1992).
128. P. Bouvet, S. Dimitrov, and A. P. Wolffe, *Genes Dev* **8,** 1147 (1994).
129. H. Kandolf, *Proc. Natl. Acad. Sci. U.S.A.* **91,** 7257 (1994).
130. A. P. Wolffe, S. Khochbin, and S. Dimitrov, *BioEssays* **19,** 249 (1997).
131. P. Varga-Weisz, T. A. Blank, and P. B. Becker, *EMBO J.* **14,** 2209 (1995).
132. S. Dimitrov and A. P. Wolffe, *Biochim. Biophys. Acta* **1260,** 1 (1995).
133. S. Dimitrov, G. Almouzni, M. Dasso, and A. P. Wolffe, *Dev. Biol.* **160,** 214 (1993).
134. S. Dimitrov, M. Dasso, and A. P. Wolffe, *JCB* **126,** 591 (1994).
135. K. Ura, A. P. Wolffe, and J. J. Hayes, *J. Biol. Chem.* **269,** 27171 (1994).
136. W. Wahli, *Trends Genet.* **4,** 227 (1988).
137. B. Corthesy, F. X. Claret, and W. Wahli, *Proc. Natl. Acad. Sci. U.S.A.* **87,** 7878 (1990).
138. B. Corthesy, I. Corthesy-Theulaz, J. R. Cardinaux, and W. Wahli, *Mol. Endocrinol.* **5,** 169 (1991).
139. S. V. Lyer, D. L. Davis, S. N. Seal, and J. B. E. Burch, *Mol. Cell. Biol.* **11,** 4863 (1991).
140. S. M. Seal, D. L. Davis, and J. B. E. Burch, *Mol. Cell. Biol.* **11,** 2704 (1991).
141. B. Corthesy, J. R. Cardinaux, F. X. Claret, and W. Wahli, *Mol. Cell. Biol.* **9,** 5545 (1989).
142. J. N. J. Philipsen, B. C. Hennis, and G. Ab, *Nucleic Acids Res.* **16,** 9663 (1988)
143. J. B. E. Burch and H. Weintraub, *Cell* **33,** 65 (1983).
144. J. B. E. Burch and M. I. Evans, *Mol. Cell. Biol.* **6,** 1886 (1986).
145. J. B. E. Burch and A. H. Fischer, *Nucleic Acids Res.* **18,** 4157 (1990).
146. B. Corthesy, P. Leonnard, and W. Wahli, *Mol. Cell. Biol.* **10,** 3926 (1990).
147. P. Walker, J. E. Germond, M. Brown-Leudi, F. Givel, and W. Wahli, *Nucleic Acids Res.* **12,** 8611 (1984).
148. U. Dobbeling, K. Ross, L. Klein-Hitpass, C. Morley, U. Wagner, and G. U. Ryffel, *EMBO J.* **7,** 2495 (1988).
149. W. Stunkel, I. Kober, and K. H. Seifart, *Mol. Cell. Biol.* **17,** 4397 (1997).
150. T. Hard, E. Kalleubach, R. Boelens, B. A. Maler, K. Dahlman, L. P. Freedmon, J. Carlstedt-Duke, K. Yamamoto, J. A. Gustafsson, and R. Kaptein, *Science* **249,** 157 (1990).
151. H. Richard-Foy and G. L. Hager, *EMBO J.* **6,** 2321 (1987).
152. T. K. Archer, M. G. Cordingley, R. G. Wolford and G. L. Hager, *Mol. Cell. Biol.* **11,** 688 (1991).
153. T. Perlman and O. Wrange, *EMBO J.* **7,** 3073 (1988).
154. B. Pina, U. Bruggemeier, and M. Beato, *Cell* **60,** 719 (1990).
155. B. F. Luisi, W. X. Xu, Z. Otwinowski, L. P. Freedman, K. R. Yamamoto, and P. Sigler, *Nature (London)* **352,** 497 (1991).
156. P. Blomquist and O. Wrange, *J. Biol. Chem.* **271,** 153 (1996).
157. A. K. Ostlund Farrants, P. Blomquist, H. Kwon, and O. Wrange, *Mol. Cell. Biol.* **17,** 895 (1997).
158. J. Wong, Y.-B. Shi, and A. P. Wolffe, *Genes Dev.* **9,** 2696 (1995).
159. J. Wong, Y.-B. Shi, and A. P. Wolffe, *EMBO J.* **16,** 3158 (1997).
160. M. Ranjan, J. Wong and Y.-B. Shi, *J. Biol. Chem.* **269,** 24699 (1994).
161. G. Almouzni and A. P. Wolffe, *Genes Dev.* **7,** 2033 (1993).
162. L. Alland, R. Muhle, H. Hou, Jr., J. Potes, L. Chin, N. Schreiber-Agus, and R. DePinho, *Nature (London)* **387,** 49 (1997).
163. T. Heinzel, R. M. Laviusky, T. M., Mullen, M., Soderstrom, C. D. Laherty, J. T. Torchia, W.-M. Yang, C. Brard, S. G. Ngo, J. R. Davie, E. Seto, R. M. Eisenman, D. W. Rose, C. K. Glass, and M. G. Rosenfeld *Nature (London)* **387,** 43 (1997).
164. J. Wong, D. Patterton, A. Imhof, Y.-B. Shi, and A. P. Wolffe, *EMBO J.* **17,** 520 (1998).
165. T. E. O'Neill, M. Roberge, and E. M. Bradbury, *J. Mol. Biol.* **223,** 67 (1992).
166. M. Garcia-Ramirez, C. Rocchini, and J. Ausio, *J. Biol. Chem.* **270,** 17923 (1995).

Index

A

ALS, *see* Amyotrophic lateral sclerosis
Alzheimer's disease, neurofilament disorders, 2, 10
Ammonia, glutamine synthetase role in homeostasis, 244, 251–252, 257
Amyotrophic lateral sclerosis
  neurofilament disorders, 2, 10, 13–16
  superoxide dismutase mutation
    incidence in disease, 16–17
    role in neurofilament accumulation, 18
    toxicity, 17–18
Arabinan, structure and degradation, 215, 217
Arabinanase, *Pseudomonas fluorescens* subsp. *cellulosa*
  cellulose-binding domain, 227
  overview, 221–222
Arginine methylation, *see* Protein arginine methylation
Axon, *see* Neurofilament

B

Barbie box
  binding proteins in phenobarbital induction of cytochromes P450, 32–34, 58
  mammalian homologs, 34–35
Barbiturate, *see* Phenobarbital
Basic fibroblast growth factor, arginine methylation, 92–93
bFGF, *see* Basic fibroblast growth factor
Biomass, plant contribution, 212
BM1
  barbie box binding proteins, 32–34, 58
  BM3R1 repression, 30, 33
  phenobarbital induction by repressor binding inhibition, 32–34
BM3
  barbie box binding proteins, 32–34, 58
  BM3R1 repression, 30, 33
  phenobarbital induction by repressor binding inhibition, 32–34
BPAG1, cross-linking neurofilament to actin, 9, 16, 20

C

Cellodextrinase, cellulose-binding domain in *Pseudomonas fluorescens* subsp. *cellulosa*, 219
Cellulase, *see* Cellodextrinase; Endoglucanase; β-Glucan glucohydrolase
Cellulose, structure and degradation, 213–214, 216–217
Cellulose-binding domain
  classification, 223, 225
  function in cellulases, 219, 225, 227–229
  sequence alignment, 224, 226
  structures, 223–225
Cell wall, plants
  composition and structure
    arabinan, 215
    arabinoglactan, 215
    cellulose, 213–214
    galactan, 215
    hemicellulose, 214–215
    overview, 213
  hydrolases
    arabinan-degrading enzymes, 217
    catalytic mechanisms, 231, 233
    cellulases, 216
    diversity, 215–216
    galactan-degrading enzymes, 217
    mannan-degrading enzymes, 217
    xylan-degrading enzymes, 216–217
Chromatin, *see* Nucleosome
Clr, repression of meiotic recombination, 372–373
COX, *see* Cytochrome *c* oxidase
Crossover, meiotic recombination in yeast, 347
CYP1A1, proximal promoter element in phenobarbital induction, 40

CYP2B1, distal phenobarbital-responsive enhancer, 48, 57
CYP2B1/2
  distal phenobarbital-responsive enhancer in phenobarbital induction, 43–44
  proximal promoter element in phenobarbital induction, 35–36, 38, 58
CYP2B2, phenobarbital induction
  distal phenobarbital-responsive enhancer
    assay by *in situ* transient transfection, 45–47
    localization by transfection in primary hepatocytes, 44–45
    transgenic analysis, 42–43
  proximal promoter element, 38–41
Cyp2b9, distal phenobarbital-responsive enhancer
  assay by *in situ* transient transfection, 51–53
  assay by transfection in primary hepatocytes, 53–55
Cyp2b10, phenobarbital induction
  distal phenobarbital-responsive enhancer
    assay by *in situ* transient transfection, 51–53
    assay by transfection in primary hepatocytes, 53–55
  proximal promoter element, 40–41
CYP2C1, distal phenobarbital-responsive enhancer, 44, 47, 50–51
CYP2H1, distal phenobarbital-responsive enhancer, 41, 43
CYP102, *see* BM3
CYP106, *see* BM1
Cytochrome *c* oxidase
  coordinate regulation of genes
    coordinate regulation of nuclear genes, 335–337
    intergenic regulation of mitochondrial and nuclear gene expression, 333–335, 338
    overview of mechanisms, 332–333
  deficiency in disease, 313
  developmental regulation of transcription, 312
  functions, 311
  gene structure of nuclear-encoded subunits
    CpG islands, 321
    exon/intron sizes, 321–323
    muscle-specific regulatory motifs, 330–332
    negative enhancers, 330
    transcription initiation sites, 321, 323, 325–329
    upstream transcription activators, 329
  isoforms in heart and liver, 311–312
  sequence properties of nuclear-encoded subunits
    COX IV, 313–314
    COX Va, 314–315
    COX Vb, 315
    COX VIa, 315–316
    COX VIb, 316, 318
    COX VIc, 318
    COX VIIa, 318–319
    COX VIIb, 319
    COX VIIc, 319–320
    COX VIII, 320
  subunit composition, 311
Cytochrome P450
  abundance of types, 27
  biological functions, 26–27
  induction mechanism, 28
  phenobarbital induction
    Bacillus megaterium, *see* BM1; BM3
    barbie box role, 32–35, 58
    distal phenobarbital-responsive enhancer
      CYP2B1, 48, 57
      CYP2B1/2, DNase I hypersensitivity, 43–44, 58
      Cyp2b10, 51–56
      CYP2B2, 42–51, 55–57
      Cyp2b9, 51–57
      CYP2C1, 44, 47, 50–51
      CYP2H1, 41, 43
      NF-1 binding to phenobarbital-responsive unit, 49–51, 53–55, 59–61
    models, 60–61
    overview, 27–30
    proximal promoter elements and binding proteins
      CYP1A1, 40
      CYP2B1/2, 35–36, 38
      cyp2b10, 40–41
      CYP2B2, 38–41
  reactions in detoxification, 27

D

DNA–histone interactions, *see* Nucleosome
Double-strand breaks, initiation of recombination, 357

E

Endoglucanase, *Pseudomonas fluorescens* subsp. *cellulosa*
  cellulose-binding domain, 219
  genes, 218
  structure, 219
Estrogen receptor, nucleosomal DNA transcription analysis, 409, 411–412

F

Fibrillarin, arginine methylation, 89, 91, 105–106

G

Galactan, structure and degradation, 215, 217
Galactanase A, *Pseudomonas fluorescens* subsp. *cellulosa*
  catalytic mechanism, 238–239
  overview, 222
β-Glucan glucohydrolase, *Pseudomonas fluorescens* subsp. *cellulosa*, 218
Glucocorticoid receptor, nucleosomal DNA transcription analysis, 412, 414–415
Glutamine synthetase
  adipose tissue expression
    functions, 261
    regulation of expression, 260–261
    topography, 260
  ammonia homeostasis role, 244, 251–252, 257
  developmental expression
    chicken, 289–290
    patterns, 263–265
    topographical changes, 265–266
    translation efficiency, 268–269
  gastrointestinal tract expression
    functions, 259
    regulation of expression, 259
    topography, 259
  gene
    avian gene
      cell specificity of expression, elements in determination, 292
      developmental regulation, 289–290
      hormonal inducibility, development, 290–293
      structure, 275
    evolution, 273
    mammalian gene
      downstream regulatory regions, 285, 288
      first intron regulatory regions, 284–285, 287–288
      structure, 275–276
      upstream regulatory regions and functional delineation, 276, 278–281, 283–284, 287
  glutamine homeostasis role, 244–245, 251–252, 293–295
  isoforms, 273
  kidney expression
    functions, 256–257
    regulation of expression, 256
    topography, 256
  kinetic properties, 274
  liver expression
    functions, 251–253
    regulation of expression, 250–251
    topography, 246
  lung expression
    functions, 260
    regulation of expression, 260
    topography, 259–260
  lymphoid tissue expression
    functions, 262
    regulation of expression, 262
    topography, 262
  muscle expression
    functions, 258
    regulation of expression, 257–258
    topography, 257
  nervous system expression
    functions, 254–255
    regulation of expression, 253–254
    topography, 253
  posttranslational regulation
    acetaminophen inactivation, 268

Glutamine synthetase (*cont.*)
    half-lives, *in vitro* versus *in vivo*, 266–267
    oxidation and degradation, 267–268
    reproductive organ expression and functions
        epidydymus, 261
        ovary, 261
        testis, 261
    subcellular localization, 273–274
    tissue distribution in rats and mice, 246–249
    transcriptional regulation
        messenger RNA stability, 271–272
        rate of transcription, 272
        signal transduction pathways, 272–273
    types, 273
Glycerol-3-phosphate acyltransferase, mutants in yeast, 152–153
GS, *see* Glutamine synthetase

H

HCP1/IR1B4, *see* Protein arginine methyltransferase type I
Heat shock protein 70, arginine methylation, 97–98, 103, 123
Hemicellulose, structure and degradation, 214–215, 217
Heterogeneous nuclear ribonuclear protein A1, arginine methylation, 76–77, 89, 91–92, 100, 106
High-moblity group proteins, arginine methylation, 97–98, 103
Histone, *see* Nucleosome
HMG proteins, *see* High-moblity group proteins
hnRNP A1, *see* Heterogeneous nuclear ribonuclear protein A1
HSP70, *see* Heat shock protein 70

I

ICP27, arginine methylation, 96–97
IMPDH, *see* Inositol-5′-monophosphate dehydrogenase
Inositol-5′-monophosphate dehydrogenase
    cell proliferation role, 182–184, 191
    downregulation by p53, 200–201
    expression in malignant transformation, 193–194
    inhibition
        clinical applications, 202–203
        effects
            cell replication, 183, 201
            immune system, 201–202
    isoforms
        gene loci, 196
        gene structures, 196–200
        similarities, 190–191
        tissue distribution, 191–192, 194–195
        transcriptional regulation, 194–196
        transcripts, 194, 199, 201
    kinetic mechanism and parameters, 190
    purification, 190–191
    purine biosynthesis, *de novo*, 184, 186–187
    purine salvage enzyme relationships, 187–189, 194
    substrate binding site, 191
Inositol-1-phosphate synthase
    gene derepression in response to elevated phosphatidylcholine turnover, 163–165, 170–171
    gene regulation with phosphatidylcholine synthesis, 160–163, 167, 169–170
    inositol-sensitive upstream activating sequence regulation model, 165, 167, 169–172
    mutants in *Saccharomyces cerevisiae*
        expression defects, 144–145
        inositol starvation studies, 153–154
        pleiotropic phenotypes of Ino$^-$ and Opi$^-$ mutants, 146
        repression defects, 145–146
        role in phosphatidylinositol synthesis, 142–143
    structure, 143
    transcriptional regulation, 156–158
    yeast genes affecting expression, 143–144
Inositol-sensitive upstream activating sequence
    binding proteins, 158
    gene distribution, 157–158
    gene regulation in nutrient deprivation, 159–160, 169

INDEX 427

modeling of gene regulation, 165, 167, 169–172
mutational analysis, 159

J

Jimpy mice, myelin basic protein, arginine methylation, 119–120

L

Linear element, *Schizosaccharomyces pombe*, 353–354

M

M26
  ade6 transcription, requirement for hotspot recombination, 359
  chromatin structure in recombination, 360–361
  genetic properties of mutation, 355–356
  heptamer sequence
    associated binding proteins, 357–328
    chromosomal context, 359–360
  recombination hotspot, 347, 355
Mannanase A, *Pseudomonas fluorescens* subsp. *cellulosa*
  overview, 221
  structure and catalytic mechanism, 237–238
mat2–mat3, recombination coldspot, 347, 372–373
MBP, *see* Myelin basic protein
Meiosis, *Schizosaccharomyces pombe, see also* M26; *rec* genes
  control of entry, 348–352
  nuclear cytology, 252–355
  overview, 346–348, 373–374
  recombination-deficient mutants and genes, 361–369
  recombination hotspots, 355–361
  region-specific control of recombination, 369–373
Mizoribine, inositol-5′-monophosphate dehydrogenase inhibition and clinical applications, 202–203

MMF, *see* Mycophenolate mofetil
Mycophenolate mofetil, inositol-5′-monophosphate dehydrogenase inhibition and clinical applications, 202
Myelin basic protein, symmetric dimethylation, 67, 70, 73–74, 99–100
  conensus sequence for methylation, 109–110
  functions
    defects in disease, 122–123
    development, 119
    jimpy mice, 119–120
    myelination role, 120–121
  ratio to monomethylation, 118–119
Myosin, arginine methylation, 98–99

N

Neurofilament
  accumulation in disease, 2, 10–11
  axon functions
    caliber modulation, 7
    regeneration role, 9
    transport modulation by NF-H, 7–9, 14–16
  BPAG1 cross-linking to actin, 9, 16, 20
  codon deletion in amyotrophic lateral sclerosis, 13–14
  intermediate filament proteins, 2–4
  phosphorylation
    alteration in disease, 5–6, 19–20
    kinases, 4–7
    phosphatases, 4, 6
    physiological functions, 4–5
    sites, 4–5, 19
  structure, 2–4
  superoxide dismutase mutation, role in accumulation, 18
  transgenic mouse models of disease, 11–13
NF-1, binding to phenobarbital-responsive unit, 49–51, 53–55, 59–61
NodB domain
  function in xylanases, 229, 231
  sequence alignment, 231–232
  xylanase E of *Pseudomonas fluorescens* subsp. *cellulosa*, 219, 221, 229
Npl3, arginine methylation, 97, 105, 124
Nucleolin, arginine methylation, 91, 106–107

428
INDEX

Nucleosome
 arginine methylation of histones, 93–96, 101, 103
 core histones, 381, 383
 dinucleosomal templates in transcription analysis, 406–409
 DNA interactions with histones
  acetylation of histones and transcription efficiency, 409–410
  cross-linking studies, 383–384
  DNA conformation, 389–390
  DNA curvature in positioning histone octamer, 386–388
  hydroxyl radical cleavage studies, 384, 386, 388–389
  linker histone structure, 390, 392–400
  linking-number paradox, 388–389
  surface twist calculations, 388–389
  trinucleotide repeats and human disease, 390
 functional studies using receptors
  estrogen receptor, 409, 411–412
  glucocorticoid receptor, 412, 414–415
  thyroid hormone receptor, 415–417
 micrococcal nuclease cleavage sites, 390, 392, 399
 nucleosomal arrays and transcription repression, 417–418
 overview of transcriptional regulation, 379–380
 transcription factor access to DNA
  TATA-binding protein, 401–402
  TFIIIA, 402–408
Nucleus, cytology during meiosis in yeast, 352–355

O

Odp1, arginine methyltransferase, 84

P

p53, downregulation of inositol-5′-monophosphate dehydrogenase, 200–201
Parkinson's disease, neurofilament disorders, 2, 10

Pat1 kinase, control of entry into meiosis, 348–351
Phenobarbital, induction of cytochromes P450
 Bacillus megaterium, see BM1; BM3
 barbie box role, 32–35, 58
 distal phenobarbital-responsive enhancer
  CYP2B1, 48, 57
  CYP2B1/2, DNase I hypersensitivity, 43–44, 58
  CYP2B2, 42–51, 55–57
  Cyp2b9, 51–57
  Cyp2b10, 51–56
  CYP2C1, 44, 47, 50–51
  CYP2H1, 41, 43
  NF-1 binding to phenobarbital-responsive unit, 49–51, 53–55, 59–61
 models, 60–61
 overview, 27–30
 proximal promoter elements and binding proteins
  CYP1A1, 40
  CYP2B1/2, 35–36, 38
  CYP2B2, 38–41
  Cyp2b10, 40–41
Phosphatidic acid
 derepression of inositol-1-phosphate synthase, 165, 167, 172
 yeast mutants in metabolism, 152–153
Phosphatidylcholine
 choline auxotrophs of *Schizosaccharomyces pombe*, 149–151
 choline-requiring mutants of *Saccharomyces cerevisiae*, 146–147
 synthesis in yeast, 137, 167
Phosphatidylethanolamine, methylation defective mutants in *Saccharomyces cerevisiae*, 147–149, 155, 160
Phosphatidylinositol, *see also* Inositol-1-phosphate synthase
 kinases, 136–137, 139–140
 synthesis in yeast, 135–136
 turnover, 139–140
Phosphatidylserine decarboxylase, mutants in yeast, 151–152, 155
Phosphatidylserine, synthesis in yeast, 141
Phospholipase C, yeast gene, 139
Phospholipase D
 cellular functions and yeast models, 171

INDEX

phosphatidylcholine turnover, 170–171
yeast gene, 137, 139
PRMT1, *see* Protein arginine methyltransferase type I
Protein arginine methylation
　analysis of modified arginines, 68–69
　antibody probes, 69–70
　discovery, 68
　energetic cost, 67–68
　enzymes, *see also* Protein arginine methyltransferase type I; Protein arginine methyltransferase type II
　　type III enzyme, 77, 110
　　type IV enzyme, 77, 110–111
　functions
　　asymmetric methylation
　　　development, 116
　　　nuclear localization, 113–114
　　　nucleic acid binding, 114–115
　　　passive and active methylation modes in modeling, 116–118
　　　signal transduction, 111–113
　　　stabilization of proteins, 116
　　symmetric dimethylation of myelin basic protein
　　　defects in disease, 122–123
　　　development, 119
　　　jimpy mice, 119–120
　　　myelination role, 120–121
　　　ratio to monomethylation, 118–119
　species distribution, 66, 77–78
Protein arginine methyltransferase type I
　discovery, 73–74
　human enzymes
　　expressed sequence tag searches, 81, 83
　　HCP1/IR1B4 identification, 80–81
　　substrates, 81
　physiological functions
　　development, 116
　　nuclear localization, 113–114
　　nucleic acid binding, 114–115
　　passive and active methylation modes in modeling, 116–118
　　signal transduction, 111–113
　　stabilization of proteins, 116
　purification, 70–75, 86–88
　rat enzymes
　　PRMT3, 83
　　3G identification, 78

reaction specificity, 66, 70
sequence homology between species, 79, 83–84
size determination and protein complexes, 86–87
substrates, 66–67, 76, 80, 90
　assays, 88–89
　basic fibroblast growth factor, 92–93
　consensus sequences, 100–101
　fibrillarin, 89, 91, 105–106
　heterogeneous nuclear ribonuclear protein A1, 76–77, 89, 91–92, 100, 106
　high-moblity group proteins, 97–98, 103
　histones, 93–96, 101, 103
　HSP70, 97–98, 103, 123
　ICP27, 96–97
　myosin, 98–99
　Npl3, 97, 105, 124
　nucleolin, 91, 106–107
　potential substrates, 103–109
subunit organization, 88
yeast enzymes
　Odp1, 84
　potential substrates, 103–107
　Rmt1, 84–85, 104, 124
Protein arginine methyltransferase type II
　discovery, 73–74
　purification, 70–75, 86–88
　reaction specificity, 66
　size determination and protein complexes, 86–87
　subunit organization, 88
　symmetric dimethylation of myelin basic protein, 67, 70, 73–74, 99–100
　conensus sequence for methylation, 109–110
　defects in disease, 122–123
　development role, 119
　jimpy mice, 119–120
　myelination role, 120–121
　ratio to monomethylation, 118–119
　tissue distribution, 100
Pseudomonas fluorescens subsp. *cellulosa*
　cell wall hydrolases, *see specific hydrolases*
　taxonomy, 217–218
Purine biosynthesis, *see* Inositol-5′-monophosphate dehydrogenase

Purine salvage
  defects in disease, 189
  inositol-5′-monophosphate dehydrogenase relationships to enzymes, 187–189, 194

R

rad32, recombination-deficient mutants of Schizosaccharomyces pombe, 368
rec genes
  activation of meiotic recombination
    Rec8, 369–371
    Rec10, 369–371
    Rec11, 369–371
  cloning, 363
  recombination-deficient mutants of Schizosaccharomyces pombe
    general features, 364–365
    isolation, 361–363
    rec7, 365–366
    rec8, 366
    rec10, 366
    rec11, 366
    rec12, 366
    rec14, 367
    rec16, 367–368
Recombination, see Meiosis
Rik1, repression of meiotic recombination, 372–373
Rmt1, arginine methyltransferase, 84–85, 104, 124

S

Saccharomyces cerevisiae
  general advantages as model system, 134–135, 171
  phospholipid synthesis, overview, 135–137
  phospholipid turnover, 137, 139–140
SC, see Synaptonemal complex
Schizosaccharomyces pombe
  general advantages as model system, 134–135, 171–172
  phospholipid synthesis, overview, 135–137
  meiosis, see also M26; rec genes
    control of entry, 348–352
    nuclear cytology, 252–355
    overview, 346–348, 373–374
    recombination-deficient mutants and genes, 361–369
    recombination hotspots, 355–361
    region-specific control of recombination, 369–373
Selenazofurin, inositol-5′-monophosphate dehydrogenase inhibition and clinical applications, 203
SOD1, see Superoxide dismutase
SPB, see Spindle pole body
Spindle pole body, reorganization in meiosis, 352
Ste11, control of entry into meiosis, 349–350
Superoxide dismutase
  incidence in amyotrophic lateral sclerosis, 16–17
  role in neurofilament accumulation, 18
  toxicity of mutations, 17–18
swi5, recombination-deficient mutants of Schizosaccharomyces pombe, 368–369
Swi6, repression of meiotic recombination, 372–373
Synaptonemal complex
  linear elements in Schizosaccharomyces pombe, 353–354
  meiosis role, 353–355
  structure, 353

T

TATA-binding protein, access to nucleosomal DNA, 401–402
TBP, see TATA-binding protein
Telomere, chromosome pairing in meiosis, 353
TFIIIA, access to nucleosomal DNA, 402–406
3G, arginine methyltransferase, 78
Thyroid hormone receptor, nucleosomal DNA transcription analysis, 415–417
Tiazofurin, inositol-5′-monophosphate dehydrogenase inhibition and clinical applications, 203

U

UAS_{INO}, see Inositol-sensitive upstream activating sequence
ura4–aim, recombination hotspot, 361

INDEX

X

Xylanase, *Pseudomonas fluorescens* subsp. *cellulosa*
cellulose-binding domain, 227–229
genes, 219, 221
NodB domain of xylanase E, 219, 221, 229, 231

xylanase A
calcium-binding site, 236–237
catalytic domain, structure, 233–234, 236
catalytic mechanism, 233–234, 236

ISBN 0-12-540061-6